ECOSYSTEM DYNAMICS OF THE BOREAL FOREST

The Kluane Project

Edited by

Charles J. Krebs
Stan Boutin
Rudy Boonstra

UNIVERSITY PRESS

2001

OXFORD

UNIVERSITY PRESS

Oxford   New York

Athens   Auckland   Bangkok   Bogotá   Buenos Aires   Calcutta
Cape Town   Chennai   Dar es Salaam   Delhi   Florence   Hong Kong   Istanbul
Karachi   Kuala Lumpur   Madrid   Melbourne   Mexico City   Mumbai
Nairobi   Paris   São Paulo   Shanghai   Singapore   Taipei   Tokyo   Toronto   Warsaw

and associated companies in
Berlin   Ibadan

Copyright © 2001 by Oxford University Press, Inc.

Published by Oxford University Press, Inc.
198 Madison Avenue, New York, New York 10016

Oxford is a registered trademark of Oxford University Press.

Library of Congress Cataloging-in-Publication Data
Ecosystem dynamics of the boreal forest: the Kluane project / edited by Charles J. Krebs,
Stan Boutin, Rudy Boonstra.
    p. cm.
ISBN 0-19-513393-5 ✓
1. Forest ecology—Yukon Territory—Kluane National Park.   2. Kluane National Park
(Yukon)   I. Krebs, Charles J.   II. Boutin, Stan A., 1955–   III. Boonstra, Rudy.
QH106.2.Y84 E36   2000
577.3'09719'1—dc21   00-027425

9 8 7 6 5 4 3 2 1

Printed in the United States of America
on acid-free paper

Dedicated to all the Kluane field workers—students, technicians, and volunteers—whose energy and stamina carried this project to successful completion.

Preface

Every research project is rich in choices, personalities, and defining moments. These events can motivate scientists as much as their thirst for discovery. Our contemporary penchant for brief, obtuse, and dispassionate reports of our scientific research frequently omits the scientific paths taken and often masks the underlying stories. These stories are seldom told, except perhaps while sitting under a tree waiting for an animal to go into a trap, or over a beer at a scientific conference. In our book we wish to impart some of the human aspects of our science endeavors as well as the scientific achievements.

The Kluane Project was a bold attempt to understand the dynamics of a forest ecosystem. A group of us had been studying the vertebrate component of the boreal forest since the mid-1970s. We are largely population ecologists by training and were preoccupied with the intriguing population cycles of snowshoe hares. Our early work pointed to the need to adopt a wider perspective that included all trophic levels if we were to unravel the mechanisms underlying the system. Our approach was one of hypothesis testing by means of experimental manipulation. We were impressed by the elegant community and ecosystem experiments conducted in lakes and marine intertidal systems. Our dream was to try a similar approach in the terrestrial environment at spatial scales far larger than those conducted by other ecologists.

As with any new endeavor, we made many mistakes and learned as we went. In this book we present our scientific findings, but we also attempt to provide some of the thought behind important decisions, some of the hard choices that were made, and some of our mistakes. We feel strongly that our experimental approach is an important way to advance our science and that others must try similar experiments in other ecosystems. The fact that we had a diverse team led to each member seeing problems from a different perspective, and the satisfaction of this type of collaboration had tremendous resonance for us all. If you are studying predators it is not easy to have to give up part of your budget for grouse radios or more vegetation plots, but it pays off in spades when you are repaid with scientific insights into another trophic level.

The book has a scope that is deliberately wide, and we hope to translate some of the excitement we had in our work. The CD-ROM that accompanies the book brings the Kluane region to life, and we hope to capture undergraduate minds such that they will take up the challenges of ecological research. For the graduate student, the book lays out the fascinating complexity of this relatively simple ecosystem but leaves many unanswered questions that are ripe for an assault by rigorous experiments. For seasoned academics, we hope the book will serve as one of the examples that can be used to understand the structure and function of one of the world's most important ecosystems.

They say every picture tells a story. Every scientific project has a story, and this is the Kluane Project story.

Vancouver, British Columbia   C. J. K.
Edmonton, Alberta   S. B.
Scarborough, Ontario   R. B.
September 2000

Acknowledgments

The Kluane Project succeeded because of the hard work and dedication of many people from many parts of Canada as well as colleagues from Australia, Sweden, Switzerland, and Argentina. We thank them for their dedication to this project. Long hours of field work impacts on family life, and we wish to thank all of the partners and offspring of project personnel who helped to support these efforts.

The project was blessed with a number of dedicated, highly competent, and motivated senior field technicians: Frank Doyle, Cathy Doyle, Elizabeth Hofer, Joceyln McDowell, Vilis Nams, Mark O'Donoghue, and Sabine Schweiger. They had a passion for the project and put in enormous effort with little reward apart from that which comes from being able to live at Kluane.

Irene Wingate maintained the southern base in Vancouver and kept people and materials flowing north with diligence and concern for northern conditions. Alice Kenney dreamed of and constructed the CD-ROM that accompanies this book with help from all the scientists whose artistic talents varied from brilliant to zero. Carol Stefan supplied the line drawings that begin each chapter. Ainsley Sykes coordinated the final proofing of the book manuscript and gave up many weekends to patiently work with the authors to correct clerical errors and inconsistencies in various drafts.

We thank the Natural Sciences and Research Council of Canada (NSERC) for both moral and financial support. There were times when we thought that we were employing most of the Yukon, and we thank the Federal Employment Programs in the Yukon for their assistance. Andy and Carole Williams ran the Arctic Institute of North America Kluane Lake Base, without which this research could not have been conducted. The Champaign-Aishihik and Kluane First Nations graciously allowed us to carry out our research on their traditional lands.

Finally, it is unorthodox to acknowledge one of the book editors, but all members of the Kluane Project wish to express their heartfelt thanks to Charles J. Krebs. Charley was the vital force behind this research collaboration. In no small measure, the successes of the study were related to his vision, his leadership, and his ability to keep us all focused on the task for over a decade of field research and writing.

Contents

Contributors   xxi

PART I   INTRODUCTION

1   General Introduction
CHARLES J. KREBS   3

2   The Kluane Region
CHARLES J. KREBS & RUDY BOONSTRA   9

   2.1   Geography and Geology, 10
   2.2   Weather and Climate, 12
      2.2.1   Temperature, 12
      2.2.2   Precipitation, 15
   2.3   Study Area and Vegetation, 15
   2.4   Vertebrate Food Web, 18
   2.5   History of Hare Cycles, 21
   2.6   Large Mammals, 22
   2.7   Human Impacts, 23

3   Trophic Interactions, Community Organization, and the Kluane Ecosystem
A. R. E. SINCLAIR & CHARLES J. KREBS   25

   3.1   Trophic-level Theory, 27
      3.1.1   Community Structure, 27
      3.1.2   Control of Trophic-level Biomass, 27
      3.1.3   Case Studies in Mammalian Communities, 29
   3.2   Interactions of Trophic Levels, 30
      3.2.1   Bottom-up Models, 36

3.2.2   *Top-down Models*, 37

3.2.3   *Herbivore-dominated Models*, 37

3.2.4   *Vegetation-dominated Models*, 37

3.2.5   *Dilution Models*, 37

3.2.6   *Reciprocal Models*, 38

3.3   *Experimental Perturbations in the Boreal Forest*, 38

3.3.1   *Experiment 1: Application of Fertilizer*, 38

3.3.2   *Experiment 2: Addition of Rabbit Chow*, 38

3.3.3   *Experiment 3: Exclusion of Carnivores*, 41

3.3.4   *Experiment 4: Exclusion of Carnivores and the Addition of Rabbit Chow*, 41

3.3.5   *Experiment 5: Exclusion of Hares*, 41

3.3.6   *Experiment 6: Exclusion of Hares and the Addition of Fertilizer*, 41

3.3.7   *Experiment 7: Removal of Vegetation*, 41

3.4   *Discussion*, 42

3.4.1   *Limitations of the Models*, 42

3.4.2   *Models of Trophic-level Interactions in Different Ecosystems*, 42

3.5   *Summary*, 43

4   Experimental Design and Practical Problems of Implementation
STAN BOUTIN, CHARLES J. KREBS, VILIS O. NAMS, A. R. E. SINCLAIR, RUDY BOONSTRA, MARK O'DONOGHUE, & CATHY DOYLE   49

4.1   *Experimental Design*, 50

4.1.1   *Choice of Treatments*, 50

4.1.2   *Size and Spatial Location of Treatments*, 52

4.1.3   *To Replicate or Not to Replicate*, 54

4.2   *Practical Problems of Implementation*, 55

4.2.1   *The Evolution of a Predator-proof Fence*, 55

4.2.2   *Food Supplementation*, 58

4.2.3   *Fertilizer*, 60

4.2.4   *Hare Exclosures*, 61

4.3   *Organization, Personnel, and Equipment*, 61

4.3.1   *Organizational Structure*, 62

4.3.2   *The Problem of Training and Protocols*, 63

4.3.3   *Seen Sheets*, 63

4.3.4   *Winter Work at Kluane*, 64

4.3.5   *Equipment*, 64

4.4   *Summary*, 65

PART II   PLANT DYNAMICS

5   Herbs and Grasses
ROY TURKINGTON, ELIZABETH JOHN, & MARK R. T. DALE   69

5.1   *Hypotheses Regarding Vegetation*, 70

5.2   *Study Area*, 71

5.3   *Tests of Predictions from Three Models*, 72

5.3.1   *Donor Control Model*, 72

    5.3.2   *Herbivore Control Model,* 77

    5.3.3   *Combined Top-down Bottom-up Model,* 78

  5.4   *Which Model Best Describes Boreal Forest Herbs and Grasses?* 84

    5.4.1   *Transient Dynamics and Stability,* 86

    5.4.2   *Impacts of Climatic Warming,* 87

  5.5   *Summary,* 88

6   Shrubs

CHARLES J. KREBS, MARK R. T. DALE, VILIS O. NAMS, A. R. E. SINCLAIR, & MARK O'DONOGHUE   92

  6.1   *The Shrub Community at Kluane,* 93

    6.1.1   *Species Composition,* 93

    6.1.2   *Pattern Changes and Succession,* 93

    6.1.3   *Secondary Chemicals in Kluane Shrubs,* 96

  6.2   *Biomass Dynamics,* 97

    6.2.1   *Methods of Estimation,* 97

    6.2.2   *Impacts of Treatments,* 98

  6.3   *Growth Rates of Shrubs,* 102

    6.3.1   *Methods of Estimation,* 102

    6.3.2   *Impacts of Treatments,* 104

  6.4   *Losses of Twigs to Browsing and Natural Mortality,* 106

    6.4.1   *Methods of Estimation,* 106

    6.4.2   *Impacts of Treatments,* 107

  6.5   *What Limits Primary Production of Shrubs?* 110

    6.5.1   *Succession in Boreal Forest Shrubs,* 110

    6.5.2   *Impact of Hare and Moose Browsing,* 112

    6.5.3   *Role of Secondary Chemicals,* 113

  6.6   *Summary,* 113

7   Trees

MARK R. T. DALE, SHAWN FRANCIS, CHARLES J. KREBS, & VILIS O. NAMS   116

  7.1   *Tree Community at Kluane,* 117

    7.1.1   *Tree Abundance on the Study Area,* 118

    7.1.2   *Tree Growth,* 118

    7.1.3   *White Spruce Cone and Seed Production,* 120

  7.2   *Vegetation Mapping,* 122

    7.2.1   *Fire History of the Study Area,* 123

    7.2.2   *Other Forms of Disturbance,* 128

  7.3   *Succession in the Boreal Forest,* 130

  7.4   *Fertilizer Effects on Trees,* 132

  7.5   *Summary,* 135

PART III HERBIVORES

8   Snowshoe Hare Demography

KAREN E. HODGES, CHARLES J. KREBS, DAVID S. HIK, CAROL I. STEFAN, ELIZABETH A. GILLIS, & CATHY E. DOYLE   141

  8.1   *The Snowshoe Hare Cycle,* 142

8.2   *Methods,* 143
8.3   *Demographic Parameters,* 145
    8.3.1   *Density and Rates of Change,* 145
    8.3.2   *Sex and Age Structure,* 146
    8.3.3   *Reproduction,* 147
    8.3.4   *Survival Rates,* 153
    8.3.5   *Causes of Death,* 162
    8.3.6   *Immigration and Emigration,* 169
8.4   *Impacts of Experimental Treatments,* 169
    8.4.1   *Impacts on Hare Demography,* 171
    8.4.2   *Causation of the Hare Cycle,* 173
    8.4.3   *Efficacy of the Experimental Treatments,* 173
    8.4.4   *Experimental Scale and Methodological Concerns,* 174
8.5   *Interactions with Other Species,* 175
8.6   *Conclusions,* 175

9   The Role of Red Squirrels and Arctic Ground Squirrels
RUDY BOONSTRA, STAN BOUTIN, ANDREA BYROM, TIM KARELS, ANNE HUBBS,
KARI STUART-SMITH, MIKE BLOWER, & SUSAN ANTPOEHLER   179

9.1   *Natural History,* 180
    9.1.1   *Red Squirrels,* 180
    9.1.2   *Arctic Ground Squirrels,* 181
9.2   *Community Interactions and Factors Affecting Population Dynamics,* 181
9.3   *Methods,* 182
    9.3.1   *Red Squirrels,* 182
    9.3.2   *Arctic Ground Squirrels,* 184
9.4   *Results,* 186
    9.4.1   *Red Squirrels,* 186
    9.4.2   *Arctic Ground Squirrels,* 193
9.5   *Discussion,* 201
    9.5.1   *Role of Stochastic Events,* 203
    9.5.2   *Treatment Effects,* 204
    9.5.3   *Role of Squirrels in the Boreal Forest Community,* 209
9.6   *Conclusions,* 210

10   Voles and Mice
RUDY BOONSTRA, CHARLES J. KREBS, SCOTT GILBERT, & SABINE SCHWEIGER   215

10.1   *Natural History and Food Web Links,* 216
10.2   *Community Interactions and Factors Affecting Population Dynamics,* 217
10.3   *Methods,* 218
    10.3.1   *Small Mammal Trapping,* 218
    10.3.2   *Data Analysis,* 219
    10.3.3   *Robustness of Data,* 219
10.4   *Impacts of the Manipulations,* 219
    10.4.1   *Northern Red-Backed Vole,* 220
    10.4.2   Microtus *Voles,* 224

10.5   *What Limits Mice and Vole Populations at Kluane?* 226

    10.5.1   *Northern Red-Backed Vole,* 227

    10.5.2   Microtus *Species,* 234

10.6   *Summary,* 235

11   Forest Grouse and Ptarmigan

KATHY MARTIN, CATHY DOYLE, SUSAN HANNON, & FRITZ MUELLER   240

    11.1   *The Ecological Role of Forest and Alpine Grouse,* 241

        11.1.1   *Trophic Position,* 241

        11.1.2   *Life History,* 241

        11.1.3   *Habitats,* 242

        11.1.4   *Current Understanding of Population Dynamics,* 243

        11.1.5   *Predicted Responses to Experimental Treatments,* 243

    11.2   *Methods,* 244

        11.2.1   *Duration of Grouse Studies,* 244

        11.2.2   *Shakwak Valley Population Trends,* 244

        11.2.3   *Numerical and Reproductive Parameters on Treatment Grids,* 244

        11.2.4   *Data Robustness and Limitations,* 245

    11.3   *Demography of Grouse,* 246

        11.3.1   *Population Trends in the Shakwak Valley,* 246

        11.3.2   *Reproductive Parameters,* 247

        11.3.3   *Population Trends on Control Plots,* 248

        11.3.4   *Relationship of Changes to the Snowshoe Hare Cycle,* 249

    11.4   *Response of Forest Grouse to Experimental Treatments,* 250

        11.4.1   *Fertilizer Addition,* 250

        11.4.2   *Food Addition,* 251

        11.4.3   *Predator Exclosure,* 252

        11.4.4   *Predator Exclosure + Food,* 252

    11.5   *Discussion,* 253

        11.5.1   *Hypotheses Related to Grouse Population Trends,* 253

        11.5.2   *Did the Experiments Change the Dynamics for Grouse?* 253

        11.5.3   *Linkages on the Same Trophic Level,* 253

        11.5.4   *Linkages to Other Trophic Levels,* 254

    11.6   *Conclusions,* 255

        11.6.1   *Comparison with Other Studies,* 255

        11.6.2   *Unexpected Results,* 256

        11.6.3   *Unanswered Questions,* 257

12   Other Herbivores and Small Predators: Arthropods, Birds, and Mammals

JAMES N. M. SMITH & NICHOLAS F. G. FOLKARD   261

    12.1   *Methods,* 263

        12.1.1   *Arthropods,* 263

        12.1.2   *Songbirds and Woodpeckers,* 263

        12.1.3   *Other Herbivores,* 263

    12.2   *Predicted Responses to Experimental Treatments,* 264

    12.3   *Responses by Species Groups,* 264

        12.3.1   *Arthropods,* 264

12.3.2   *Birds,* 265

12.3.3   *Other Herbivores,* 268

12.4   *Discussion,* 269

12.4.1   *Links to the 10-Year Cycle,* 269

12.4.2   *Effects of Fertilization,* 269

12.4.3   *Other Patterns in the Food Web,* 270

12.5   *Summary,* 270

PART IV   MAMMALIAN PREDATORS

13   Coyotes and Lynx

MARK O'DONOGHUE, STAN BOUTIN, DENNIS L. MURRAY, CHARLES J. KREBS, ELIZABETH J. HOFER, URS BREITENMOSER, CHRISTINE BREITENMOSER-WÜERSTEN, GUSTAVO ZULETA, CATHY DOYLE, & VILIS O. NAMS   275

13.1   *Methods,* 278

13.1.1   *Population Monitoring,* 278

13.1.2   *Foraging Behavior,* 279

13.1.3   *Functional Responses,* 280

13.2   *Numerical Responses,* 282

13.2.1   *Density,* 282

13.2.2   *Adult Survival,* 283

13.2.3   *Emigration,* 286

13.2.4   *Recruitment,* 286

13.3   *Social Organization,* 287

13.3.1   *Home Ranges,* 288

13.3.2   *Social Groups,* 289

13.4   *Foraging Behavior,* 289

13.4.1   *Diets,* 289

13.4.2   *Scavenging Behavior,* 293

13.4.3   *Hunting Tactics,* 295

13.4.4   *Group Hunting,* 297

13.4.5   *Use of Trails,* 297

13.4.6   *Habitat Use,* 297

13.5   *Functional Responses,* 300

13.5.1   *Components of Functional Responses,* 301

13.5.2   *Prey Switching,* 304

13.6   *Synthesis and Conclusions,* 307

13.6.1   *Numerical Responses,* 308

13.6.2   *Functional Responses,* 310

13.6.3   *Total Impact of Predation by Coyotes and Lynx on Hares,* 313

13.6.4   *Impact of Predation by Coyotes and Lynx on Alternative Prey,* 313

13.6.5   *Coexistence of Coyotes and Lynx,* 314

13.6.6   *The Specialist–Generalist Contrast,* 316

14   Other Mammalian Predators

MARK O'DONOGHUE, STAN BOUTIN, ELIZABETH J. HOFER, & RUDY BOONSTRA   324

14.1   *Methods,* 325

14.2   Red Fox, 325

14.3   Wolf, 327

14.4   Weasel, 329

14.5   Wolverine, 331

14.6   Marten, 332

14.7   Mink and Otter, 332

14.8   Synthesis and Conclusions, 332

PART V   AVIAN PREDATORS

15   Great Horned Owls
CHRISTOPH ROHNER, FRANK I. DOYLE, & JAMES N. M. SMITH   339

15.1   Methods, 341

      15.1.1   Population Census, 341

      15.1.2   Monitoring Diets, 342

      15.1.3   Radio Telemetry, 342

      15.1.4   Analysis of Telemetry Data, 343

      15.1.5   Statistical Analyses, 344

15.2   Demography, 344

      15.2.1   Reproduction and Population Productivity, 344

      15.2.2   Survival and Emigration, 348

      15.2.3   Estimating Numerical Responses, 351

15.3   Foraging Behavior, 351

      15.3.1   Diet, 351

      15.3.2   Prey Preferences, 354

      15.3.3   Estimating Functional Responses and Predation Impact, 355

15.4   Social Organization, 360

      15.4.1   Social Status and Vocal Activity, 360

      15.4.2   Stability and Size of Home Ranges, 360

      15.4.3   Effect of Territoriality on Spacing of Owls, 361

      15.4.4   Territorial Behavior and Limitation of Population Increase, 361

      15.4.5   Social Behavior and the Time Lag in the Numerical Response, 364

15.5   Responses to Large-Scale Experiments, 366

      15.5.1   Space Use in Territories with Experimental Hot Spots of Prey, 366

      15.5.2   Predator Movements from Poor Patches to Rich Patches, 368

15.6   Discussion, 368

      15.6.1   Large Floating Population When Resources Are Abundant, 368

      15.6.2   Factors Affecting Functional Responses and Predation Impact, 369

      15.6.3   Limitation of Population Growth at Peaks of Cyclic Prey, 371

      *15.6.4   Time Lag in the Numerical Response of a Predator
               of Cyclic Prey,* 371

      15.6.5   Factors Limiting Spatial Aggregation of Predators, 372

16   Raptors and Scavengers
FRANK I. DOYLE & JAMES N. M. SMITH   377

16.1   Methods, 378

      16.1.1   General Approach, 378

16.1.2 *Population Surveys: The Intensive Search Area,* 380

16.1.3 *Reproductive Success,* 380

16.1.4 *Diets,* 380

16.1.5 *Data Analysis,* 381

16.2 *Predicted Responses of Raptors and Corvids,* 381

16.3 *Responses by Groups and Individual Species,* 382

16.3.1 *Large Resident Raptors,* 382

16.3.2 *Large Migratory Raptors,* 385

16.3.3 *Small Migratory Raptors,* 388

16.3.4 *Small Owls,* 391

16.3.5 *Corvids,* 394

16.3.6 *Other Raptors,* 396

16.3.7 *Intraguild Predation,* 396

16.4 *Discussion,* 398

16.4.1 *Functional, Numerical, and Reproductive Responses to Hare Densities,* 398

16.4.2 *Resource Partitioning and Diet Width,* 400

16.4.3 *Intraguild Predation,* 400

16.4.4 *Methodology and Limitations,* 401

16.5 *Summary,* 401

PART VI   COMMUNITY AND ECOSYSTEM ORGANIZATION

17   Testing Hypotheses of Community Organization for the Kluane Ecosystem
A. R. E. SINCLAIR, CHARLES J. KREBS, RUDY BOONSTRA, STAN BOUTIN,
& ROY TURKINGTON   407

17.1 *Experimental Perturbations of the Boreal Forest,* 409

17.1.1 *Direct and Indirect Effects,* 409

17.1.2 *The Experiments,* 409

17.2 *Methods,* 411

17.2.1 *Soil Nitrogen,* 411

17.2.2 *Vegetation,* 411

17.2.3 *Plant Secondary Chemicals,* 412

17.2.4 *Herbivore Biomass,* 412

17.2.5 *Predator Activity,* 412

17.3 *Direct Effects of Trophic-level Perturbations,* 412

17.3.1 *Fertilizer Addition,* 412

17.3.2 *Addition of Hare Food,* 413

17.3.3 *Predator Exclosure,* 416

17.3.4 *Predator Exclosure and Food Addition,* 416

17.3.5 *Hare Exclosure,* 419

17.3.6 *Hare Exclosure Plus Fertilizer,* 421

17.3.7 *Vegetation Removal,* 425

17.4 *Indirect Effects of Trophic-level Perturbations,* 426

17.4.1 *Fertilizer Addition,* 426

17.4.2 *Food Addition,* 426

17.4.3 *Predator Exclosure,* 426

       17.4.4   Food Addition and Predator Exclosure, 427

       17.4.5   Hare Exclosure, 427

   17.5   Discussion, 427

       17.5.1   Direct Effects, 427

       17.5.2   Indirect Effects, 428

       17.5.3   Top-down versus Bottom-up, 430

       17.5.4   Productivity and Biomass Responses, 430

       17.5.5   Other Indirect Effects, 431

       17.5.6   The Role of Secondary Chemicals, 432

       17.5.7   The Dominant Pathways in the Vertebrate Community, 432

   17.6   Conclusion, 433

   17.7   Summary, 433

18   Vertebrate Community Structure in the Boreal Forest:
Modeling the Effects of Trophic Interaction
DAVID CHOQUENOT, CHARLES J. KREBS, A. R. E. SINCLAIR, RUDY BOONSTRA,
& STAN BOUTIN   437

   18.1   Trophic Interaction and Species Coexistence, 438

   18.2   Interaction between Vegetation and Herbivores, 439

   18.3   Interaction between Herbivores and Predators, 440

   18.4   Modeling, 440

       18.4.1   Interaction between Vegetation and Herbivores, 441

       *18.4.2   Equilibrium Conditions for Snowshoe Hares
and Ground Squirrels,* 446

       *18.4.3   Adding Predation to Models of Interaction between Herbivores
and Vegetation,* 450

       18.4.4   The Effect of Herbivore Cycles on Predators, 452

   18.5   Discussion, 458

       18.5.1   Unresolved Modeling Issues, 458

       18.5.2   Mechanisms of Coexistence, 460

       18.5.3   Top-down and Bottom-up Influences on Community Structure, 460

19   Trophic Mass Flow Models of the Kluane Boreal Forest Ecosystem
JENNIFER L. RUESINK & KAREN E. HODGES   463

   19.1   Questions about the Ecosystem Dynamics of the Boreal Forest, 466

   19.2   Modeling the Kluane System, 467

   19.3   Methods for Constructing Mass Balance Models, 467

       19.3.1   Parameterization of Mass Balance Models, 469

   19.4   Results from Mass Balance Models, 474

       19.4.1   Consumption of Plant Biomass by Herbivores, 474

       19.4.2   Phase of Cycle Models, 476

       19.4.3   Seasonal Models, 480

       19.4.4   Food Addition and Predator Reduction Models, 480

       19.4.5   Partial Food Web Models, 483

   19.5   Discussion, 485

       *19.5.1   Do Predator–Prey Dynamics Structure the Boreal Forest
Ecosystem?* 485

        19.5.2 Potential for Other Ecopath Models, 486

        19.5.3 Synthesis and Implications, 487

  19.6 Summary, 488

20   Conclusions and Future Directions

   CHARLES J. KREBS, RUDY BOONSTRA, STAN BOUTIN, & A. R. E. SINCLAIR   491

  20.1 Primary Findings, 492

        20.1.1 Predator Trophic Level, 492

        20.1.2 Herbivore Trophic Level, 493

        20.1.3 Plant Trophic Level, 494

  20.2 Secondary Findings, 494

        20.2.1 Predator Trophic Level, 494

        20.2.2 Herbivore Trophic Level, 495

        20.2.3 Plant Trophic Level, 496

  20.3 Opportunities for Further Work, 497

  20.4 Unsolved Problems, 498

  20.5 Future Boreal Forest Research, 499

Appendix 1   How to Use the Enclosed Kluane CD-ROM   503

Appendix 2   Table of Contents for the Kluane CD-ROM   504

  Index   505

Contributors

SUSAN ANTPOEHLER
Box 10
Bella Bella, B.C. V0T 1B0, Canada

MIKE BLOWER
Box 5957
Whitehorse, Yukon Y1A 5L7, Canada

RUDY BOONSTRA
Division of Life Sciences
University of Toronto at Scarborough,
Scarborough
Ontario M1C 1A4, Canada

STAN BOUTIN
Department of Biological Sciences
University of Alberta, Edmonton
Alberta T6G 2E9, Canada

URS BREITENMOSER
Swiss Rabies Centre
Institute of Veterinary-Virology
University of Bern
Laengass—Str. 122 CH-3012
Bern, Switzerland

CHRISTINE BREITENMOSER-WÜERSTEN
KORA, Thunstrasse 31
3074 Muri b.
Bern, Switzerland

ANDREA BYROM
Manaaki Whenua
Landcare Research New Zealand Limited
P.O. Box 69
Lincoln 8129, New Zealand

DAVID CHOQUENOT
Landcare Research New Zealand Limited
P.O. Box 69
Lincoln 8129, New Zealand

MARK R. T. DALE
Department of Biological Sciences
University of Alberta
Edmonton, Alberta T6G 2E9, Canada

CATHY DOYLE
P.O. Box 129
Telkwa, B.C. V0J 2X0, Canada

F. I. DOYLE
Wildlife Dynamics Consulting
Box 129
Telkwa, B.C. V0J 2X0, Canada

NICK FOLKARD
Department of Zoology
University of British Columbia
Vancouver, B.C. V6T 1Z4, Canada

SHAWN FRANCIS
Applied Ecosystem Management Ltd.
100–211 Hawkins Street
Whitehorse, Yukon Y1A 1X3, Canada

SCOTT GILBERT
Box 5413
Haines Junction
Yukon Y0B 1L0, Canada

ELIZABETH A. GILLIS
Department of Zoology
University of British Columbia
Vancouver, B.C. V6T 1Z4, Canada

SUSAN HANNON
Department of Biological Sciences
University of Alberta
Edmonton, Alberta T6G 2E9, Canada

DAVID S. HIK
Department of Biological Sciences
University of Alberta
Edmonton, Alberta T6G 2E9, Canada

KAREN E. HODGES
Department of Zoology
Tillydrone Avenue
University of Aberdeen
Aberdeen AB24 2TZ, UK

ELIZABETH J. HOFER
Mile 1055 Alaska Highway
Silver Creek, Yukon Territory Y0B 1M0,
Canada

ANNE HUBBS
Natural Resource Service, Alberta
Environment
562 Carmichael Lane
Hinton, Alberta T7V 1S8, Canada

ELIZABETH JOHN
School of Biological Sciences
University of Sussex
Falmer, Brighton, East Sussex
BN1 9QG, UK

TIM KARELS
Department of Biological Sciences
University of Alberta
Edmonton, Alberta T6G 2E9, Canada

CHARLES J. KREBS
Department of Zoology
University of British Columbia
6270 University Blvd.
Vancouver, B.C. V6T 1Z4, Canada

KATHY MARTIN
Department of Forest Sciences
2424 Main Mall
University of British Columbia
Vancouver, B.C. V6T 1Z4, Canada
Canadian Wildlife Service,
5421 Robertson Rd, RR 1 Delta, B.C.
V4K 3N2, Canada

FRITZ P. MUELLER
Wildlife Systems Research
P.O. Box 31106
211 Main Street
Whitehorse, Yukon Y1A 5P7, Canada

DENNIS MURRAY
Department of Fish and Wildlife
Resources
College of Forestry, Wildlife and Range
Sciences
University of Idaho
Moscow, Idaho 83844-1136, USA

VILIS O. NAMS
Department of Environmental Sciences
NSAC, Box 550
Truro, Nova Scotia B2N 5E3, Canada

MARK O'DONOGHUE
Fish & Wildlife Branch
Yukon Department of Renewable
Resources, Box 310
Mayo, Yukon Territory Y0B 1M0, Canada

CHRISTOPH ROHNER
Department of Renewable Resources
University of Alberta
Edmonton, Alberta T6G 2H1, Canada

JENNIFER L. RUESINK
Department of Zoology
University of Washington, Box 351800
Seattle, Washington 98195-1800, USA

SABINE SCHWEIGER
Box 5567
Whitehorse, Yukon Y1A 5H4, Canada

A. R. E. SINCLAIR
Centre for Biodiversity Research
University of British Columbia
6270 University Boulevard
Vancouver, B.C. V6T 1Z4, Canada

J. N. M. SMITH
Department of Zoology
University of British Columbia
Vancouver, B.C. V6T 1Z4, Canada

CAROL I. STEFAN
1016 Country Hills Circle NW
Calgary, Alberta T3K 4W8, Canada

KARI STUART-SMITH
Crestbrook Forest Industries
P.O. Box 4600
Cranbrook, B.C. V1C 4J7, Canada

ROY TURKINGTON
Department of Botany
Center for Biodiversity Research
University of British Columbia
Vancouver, B.C. V6T 1Z4, Canada

GUSTAVO ZULETA
Departamento de Biología
Universidad de Buenos Aires
Ciudad Universitaria, Pab. 2—4to piso
Buenos Aires—1428, Argentina

PART I

INTRODUCTION

1

General Introduction

CHARLES J. KREBS

The boreal forest of North America stretches from Alaska to Newfoundland in an unbroken sweep of more than 5 million km^2. The boreal forest is remarkably uniform in overall appearance, dominated by coniferous trees, and an Alaskan feels quite at home in northern Manitoba or in eastern Quebec. But if you turn a botanist loose in the boreal forest, he or she will tell you that it is remarkably diverse in the dominant tree species, the shrubs, and the grasses and herbs of the forest floor. This botanical diversity is reflected in the various classifications of the subdivisions within the boreal forest, and the wide disagreement about what subdivisions to recognize (Rowe 1972, DEMR 1974, Bonnor 1985, Botkin and Simpson 1990). Figure 1.1 gives a general breakdown of the three main sub-

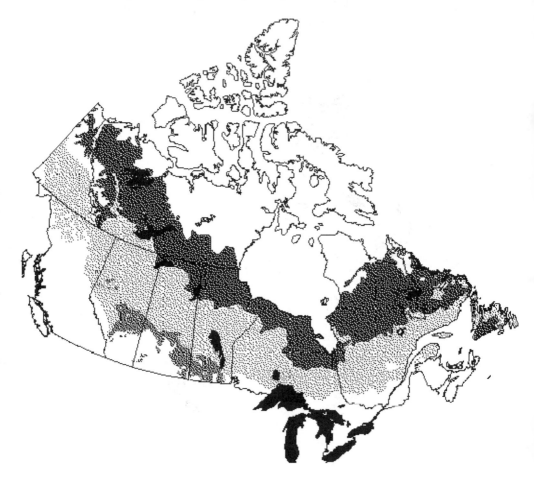

Figure 1.1 Three major subivisions of the North American boreal forest, as defined by the Canadian Forest Service, after Rowe (1972). The middle layer (lightest color) is predominantly boreal forest. The northern edge is a mixture of boreal forest and tundra and the southern edge is a mixture of boreal forest and grasslands. Kluane is part of the Western Mountains stratum of the boreal forest zone. See CD-ROM frame 6 for additional information. Figure modified after Rowe (1972).

divisions within the North American boreal forest and sets the geographical scale for this book.

We focus in this book on one particular part of the boreal forest, the Kluane sector of the Western Mountains region described by Rowe (1972) and Botkin and Simpson (1989). The Kluane region of the boreal forest differs from the surrounding regions in its relatively high elevation and in its location in the climatic rain shadow of the St. Elias Mountains. The white spruce is the only conifer in this area, and typical boreal forest trees such as black spruce, paper birch, lodgepole pine, and larch are absent at Kluane, although they occur within 200 km both northwest and southeast of Kluane. The relatively high elevation of the Kluane area (600–1100 m) is reflected in a colder climate and lower productivity than one finds farther north in central Alaska or in the southeastern Yukon.

In spite of the vegetation differences that occur across the broad expanse of the boreal forest in North America, the vertebrate community is remarkably constant. Among the large mammals, moose and mountain caribou are spread across the entire boreal region, and snowshoe hares and red squirrels predominate among the mid-sized vertebrates. Canada lynx, coyotes, wolves, red foxes, and wolverine are the main mammal predators. Red-backed voles and deer mice are typical small mammals, along with several less common vole and lemming species. Typical boreal forest birds also occur across this 7000-km stretch. Spruce grouse and ruffed grouse occupy the grouse niche, and the three most common songbirds at Kluane—the dark-eyed junco, the yellow-rumped warbler, and Swainson's thrush—are ubiquitous across the boreal forest. Great horned owls and northern goshawks, two common birds of prey at Kluane, are typical raptors across the entire North American boreal forest zone. Figure 1.2 shows the average biomass pyramids for herbivores and carnivores in the Kluane area. Snowshoe hares and the two squirrels are major herbivores, while lynx, coyotes, wolves, and great horned owls are major predators at Kluane. The relative importance of snowshoe hares in this community is shown even more graphically if we convert these biomass values to energy flow through the community. Figure 1.3 illustrates the average energy flow pyramid for the Kluane community.

If the vertebrate community is so similar across this great expanse of boreal forest, there is a good chance that the ecology of this community is governed by a set of general rules common to the Yukon, Quebec, and Alaska. We start with this large assumption. The major objective of the Kluane Project was to study the organization of the vertebrate community in the Kluane region of the Yukon. To do this, we needed to know the abundance of all the major species, their food habits, and how these change over time. The native people of the boreal region have long known about one of the dominant changes in the boreal forest—the 10-year cycle of snowshoe hares and their predators. By focusing our studies on the entire vertebrate assemblage, we hoped to achieve a better understanding of how the cycles of snowshoe hares are controlled and how other vertebrates are affected by the hare cycle. Are red squirrels affected when snowshoe hare numbers decline, or, conversely, are snowshoe hares affected when red squirrels are abundant? How do predators respond to a collapse of their major food source?

In studying community organization, the critical question we need to answer is *what would be the effect of removing a particular species from the community?* If red squirrels were to disappear in the Kluane region because of some disease, for example, what would happen to the other species in this community? If lynx were removed by excessive fur harvests, what would be the effects on other predators and on prey species in this system?

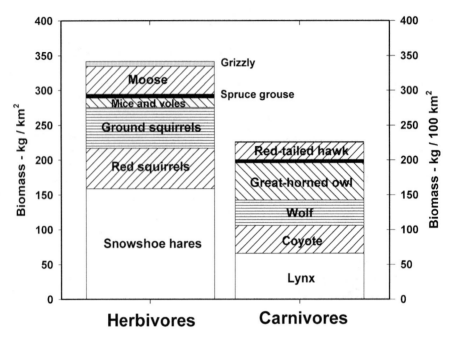

Figure 1.2 Biomass pyramids for herbivore and carnivore species in the Kluane ecosystem. Average values are given for 1987–1995. The herbivore scale is to the left and the carnivore scale to the right. The black bar above the great horned owl is the northern goshawk. The ratio of prey to predators in this ecosystem averages about 150:1. The grizzly bear, an omnivore, is in the herbivore pyramid because the bulk of its diet is herbaceous.

These are complex questions, and ecologists are only now beginning to get some tentative answers for particular ecosystems. If humans were having no impacts on natural communities, such questions about community organization would be of academic interest only. But as human disturbances widen through forestry, agriculture, and tourism, these questions become relevant for conservation, fisheries, and wildlife management.

How can we ecologists study community organization? As the first step, we can describe the food web of the community: who eats whom? This is more difficult than most people imagine because the diets of most vertebrates include many different species and change from winter to summer. There are always problems with diet analysis, but we can approximate the major items in the diet, and some minor items are missed. The second step is to estimate the consumption of each species in the food web by combining the diet information with abundance data. How many snowshoe hares does a lynx eat per day, and how many lynx occupy the study area? How much grass does an arctic ground squirrel eat per day, and how much grass is produced each summer? Once we know all these consumption rates, we should be able to put together an arithmetic balance sheet for the whole community. This simple task has rarely been achieved for any terrestrial vertebrate community. As the third and last step, we can experimentally disturb the system and observe its response. The kinds of experiments we could do on a finite budget in this system are limited to four types of manipulations:

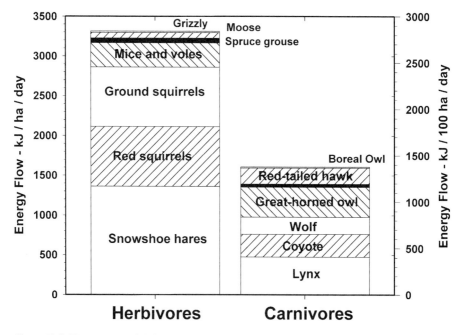

Figure 1.3 Energy pyramids for herbivore and carnivore species in the Kluane ecosystem. Average biomass values were converted to field metabolic energy equivalents with the equations given in Nagy (1987). Average values are given for 1987–1995. The herbivore scale is to the left and the carnivore scale to the right. The black bar above the great horned owl is the northern goshawk.

1. We can add nutrients as fertilizer to change plant production.
2. We can eliminate mammal predators from specific areas by fences and measure the impact of predator elimination on prey populations.
3. We can add supplemental food to the herbivores and measure their responses to enhanced food supplies.
4. We can exclude herbivores from areas and see what happens to vegetation when there is no browsing.

The principle is simple: if you think a community is structured by predation, for example, then you should manipulate predation levels and see what happens to the prey populations. There are two caveats, however, to this simple advice. First, you must do these manipulations on a large enough spatial scale. For most communities we can only guess what scale is large enough. Most ecologists agree that we have not, in general, manipulated large enough areas. In this study we manipulated 1-km^2 blocks of boreal forest, among the largest areas ever manipulated in a terrestrial ecological experiment. Second, the effects of any experiment need time to develop, and ecologists, like all people, are impatient. We have carried out our experiments for 10 years and can address ecological changes on this time scale. For some manipulations, such as logging in the boreal forest, 10 years may not be long enough to see the important impacts. Since most ecological experiments are carried out at the time scale of months to 3 years, our 10-year experiments have extended the range of manipulations in time for this boreal forest community.

In this book we describe the Kluane Boreal Forest Ecosystem Project, which operated from 1986 to 1996 in the southwestern Yukon. We begin by describing the area and its physical setting and then describe the background of the project and the wisdom that had accumulated to 1986 on how this system might operate. The details of the experiments we set up are presented, partly to help the reader appreciate the difficulty of working at $-40°$ and partly to help those contemplating doing similar experiments in the future. Then we examine the three trophic levels of the plants, the herbivores, and the predators in detail to provide some surprises about how the individual species operate within the overall system. Finally, we synthesize these findings in a model of the boreal forest vertebrate community and provide an overview of what we have discovered and what remains to be done. Over the 10 years of this project, nine faculty members from three Canadian universities and 26 graduate students joined with 75 summer assistants and 18 technicians to expend 157 person-years of effort to produce the picture developed here. No one ever thought that ecology was a simple subject like chemistry, but when we began this project we hoped to join forces to make a major advance in our understanding of the boreal forest ecosystem. This book presents our discoveries.

Literature Cited

Bonnor, G. M. 1985. Inventory of forest biomass in Canada. Canadian Forest Service, Petawawa National Forestry Institute, Petawawa, Ontario.

Botkin, D. B., and L. G. Simpson. 1989. The distribution of biomass in the North American boreal forest. *in* Proceedings of the International Conference and Workshop. Global Natural Resource Monitoring and Assessments, Preparing for the 21st century, pages 1036–1045. American Society of Photogrammetry and Remote Sensing, Bethesda, Maryland.

Botkin, D. B., and L. G. Simpson. 1990. Biomass of the North American boreal forest. Biogeochemistry **9:**161–174.

DEMR. 1974. National Atlas of Canada. Department of Energy, Mines and Resources. Ottawa, Ontario.

Nagy, K. A. 1987. Field metabolic rate and food requirement scaling in mammals and birds. Ecological Monographs **57:**111–128.

Rowe, J. S. 1972. Forest regions of Canada. Canadian Forest Service Publication no. 1300, Ottawa, Ontario.

2

The Kluane Region

CHARLES J. KREBS & RUDY BOONSTRA

The Kluane Region of the Yukon has been particularly well studied for two reasons. The Arctic Institute of North America established a research station on the south shore of Kluane Lake in 1959. This is one of few research stations north of 60° in Canada, and it has focused scientific studies in this part of the Yukon for the last 40 years. In 1972 The Canadian government established Kluane National Park just to the west of Kluane Lake, and the designation of the park has brought with it a flurry of studies on the descriptive ecology of the Kluane area. In this chapter we summarize the general setting of the Kluane region and what is known about its physical and biological setting.

2.1   Geography and Geology

The Kluane region of the Yukon is cut by a number of major geological fault systems that usually trend from northwest to southeast. The Denali fault system cuts from central Alaska to northern British Columbia and transects the study area. The Shakwak Fault is part of the Denali system, and it follows the western edge of the Shakwak Trench from the southern end of Kluane Lake toward Haines Junction to the southeast. The eastern slopes of the St. Elias Ranges rise abruptly to 2600 m to form the western edge of the Shakwak Trench, while the Kluane Ranges form the eastern edge, rising to 1600 m. The Shakwak Trench ranges from 8 to 12 km in width just to the south of Kluane Lake. There is no apparent activity along any of these fault blocks at the present time. The major sign of volcanic activity in the Kluane region is the deposit of White River ash from a volcanic eruption near the Klutan Glacier north of Kluane National Park about 1250 years ago in 735 A.D. This ash provides a soil marker about 2–6 cm deep in the valley south of Kluane Lake and has been mapped by Bostock (1952).

Glaciation has been the dominant geological process that has determined the soils and landforms of the Kluane region (Muller 1967). The most recent glaciation event was the Kluane glaciation from about 29,600 years ago to about 12,500 years ago. The entire Shakwak Trench was ice filled at that time. The glacial maximum occurred about 14,000 years ago, and rapid melting followed until the valley was largely ice free by 12,500 years ago (Denton and Stuiver 1967). During deglaciation, large areas of outwash sediments were exposed to winds coming off the retreating glaciers. These sediments were wind blown and deposited as loess on top of the glacial moraines. Loess in the Kluane region varies from 30 to 150 cm deep and is the typical basic soil material of the study area.

The Kluane region lies in the zone of discontinuous permafrost, and scattered throughout the valley are areas underlain by permanently frozen ground. North-facing slopes are more likely to have permafrost, as are areas of peat accumulation and poor drainage. Recent burns typically lose any permafrost they may have had, as do areas cleared for roads or pipelines.

Land use in the Kluane area has been largely centered on placer gold mining and big game hunting. Both of these activities have been reduced or eliminated with the establishment of Kluane National Park in 1972. All of the area south and west of the Alaska Highway between Haines Junction and Burwash Landing is either in the park or in the Kluane Game Sanctuary. The Kluane Game Sanctuary was established in 1942 when the Alaska Highway was built to protect wildlife from overexploitation. Although mining is allowed in the sanctuary, none is currently being done there. Land claims by native groups

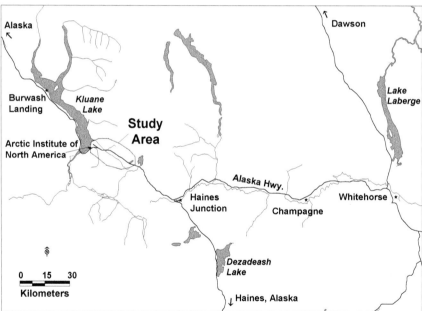

Figure 2.1 Location of the Kluane study area in the southwestern Yukon.

are now completed or nearing completion in this region, and this has changed the status of some of the lands that were formerly protected in the sanctuary. The Alaska Highway bisects our study area (figure 2.1), and thus about one-half of our area is within the protected area of the sanctuary. Hunting of large mammals (moose) still occurs within the sanctuary either by poaching or by native people who are allowed to hunt anywhere for subsistence. The only other major human activity in the Kluane region is fur trapping. The

Yukon maintains a system of exclusive trapping areas under license, and our study area included two trap lines. During the 1980s fur prices were exceptionally low, and in general there was little effort expended in trapping in the Shakwak Trench during our study. Trapping of arctic ground squirrels along the highway occasionally occurred but was minimal, and hunting of snowshoe hares by local people was also minimal. There has been no logging in the Shakwak Trench, and the only tree cutting has been for firewood on a local scale around Silver Creek at the south end of Kluane Lake and on areas to the south of our study zone.

2.2   Weather and Climate

The Kluane region lies in the rain shadow of the St. Elias Mountains with its massive icefields, and this topographic effect, along with the strong seasonality of the high latitude (61°N), dominates the climate (Webber 1974). The southwestern Yukon displays steep environmental gradients owing to its proximity to alpine glaciers and to its high elevation. It lies at the boundary between two major climate systems, that of the cold, dry arctic air masses, and that of the warm Pacific air masses which are modified in transit of the St. Elias Mountains. Because the region is sparsely settled, there are few long-term weather records. There are weather stations at Burwash Airport and at Haines Junction that form the nucleus of the available data. Some weather data were collected at Kluane Research Station (KRS) during the 1960s and 1970s, but there is not a continuous record from this location. Because of mountain topography, no weather station will be typical of all the conditions within our study area. Burwash is typically slightly colder than KRS, and Haines Junction is slightly warmer, and we chose Burwash data as closer to the conditions observed at KRS.

2.2.1   Temperature

The average temperature range for Burwash is shown in figure 2.2 for 1966 to 1996. The cold climate in the Kluane region enforces a short growing season for the vegetation. The average frost-free period for Burwash is only 30 days (range 11–50 days) and for Haines Junction only 21 days (range 0–63 days), so that typically only July can be expected to have continuous freedom from frosts. The cold winter temperatures cause the lakes in the region to freeze early in the autumn. Kluane Lake freezes over completely on average about November 23 (range November 6–December 6) and is free of ice on average about June 7 (range May 25–June 20). The ice thickness on Kluane Lake averages about 130 cm and is almost always more than 1 m thick.

There is considerable variation in temperature from year to year at Kluane, and this variation is superimposed on the general climatic trend toward warmer weather as a result of CO_2 accumulation in the atmosphere. Figure 2.3 shows the average temperature deviations for Burwash Airport for the 10 years of our study at Kluane. The months are grouped into four seasons: winter (November–February), spring (March–April), summer (May–August), and autumn (September–October). Figure 2.3 shows that the summer temperatures in general have been at or above normal since 1986. Winter temperatures, by contrast, were below normal for the three winters from 1988–1989 through the winter of 1990–1991. Most of the temperature deviations during this study were less than 2°C.

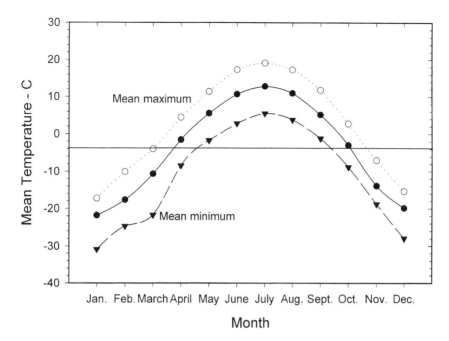

Figure 2.2 Mean monthly temperatures for Burwash Airport for 1966–1996, along with the average maximum and average minimum temperatures for each month. The horizontal line is the annual mean temperature of −3.83°C. (Data from Atmospheric Environment Services, Environment Canada, Whitehorse.)

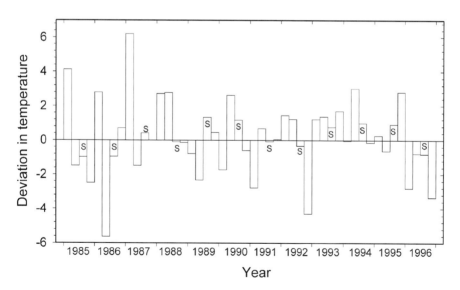

Figure 2.3 Deviations in mean temperature (°C) at Burwash Airport from the long-term temperature average for 1966–1995, for the period of our study (1985–1996). Averages are given for four seasons; S marks the summer season on the histograms. Winter is November–February, and summer is May–August. (Data from Atmospheric Environment Services, Environment Canada, Whitehorse.)

Table 2.1 Precipitation (cm) at the Burwash Airport for 1985–1996 and average precipitation for the period 1966–1996.

Year	Jan	Feb	Mar	Apr	May	June	July	Aug	Sept	Oct	Nov	Dec	Total
Average (1966–95)	1.24	0.75	0.93	1.09	2.52	4.73	6.66	4.20	2.74	1.55	1.32	0.99	28.6
1985	1.12	2.11	0.60	0.32	0.81	3.64	9.44	5.58	4.14	1.26	0.62	1.09	30.7
1986	1.44	1.19	1.11	0.54	3.03	3.06	10.1	2.59	3.24	0.26	0.84	0.58	28.0
1987	0.29	0.86	0.53	1.58	5.14	2.19	3.26	7.54			0.65	0.78	22.8
1988	0.96	0.28	0.20	0.90	6.53	3.90	13.2	5.78	2.28	0.54	1.04	0.32	35.9
1989	1.32	0.38	1.83	0.02	1.03	5.25	0.88	2.98	1.25	2.29	1.28	1.30	19.8
1990	0.88	1.31	1.08	1.94	2.89	7.68	2.79	3.41	3.00	1.64	1.85	1.64	30.1
1991	0.82	0.68	1.03	0.00	0.84	7.15	8.50	3.88	2.82	2.69	2.27	1.38	32.1
1992	0.92	0.32	0.20	1.90	2.81	2.16	11.2	3.74	4.78	0.75	0.49	0.90	30.2
1993	1.45	0.62	0.08	0.17	1.32	2.78	6.53	1.96	2.41	0.74	1.90	0.48	20.4
1994	2.30	0.40	0.94	1.05	0.96	4.18	2.88	3.01	3.78	1.74	1.51	0.26	23.0
1995	0.18	0.94	1.10	0.00	1.98	1.02	9.97	7.49	0.54	0.91	2.68	0.45	27.3
1996	0.56	0.82	0.57	1.07	2.79	0.85	10.2	3.34	1.13	1.73	0.63	1.08	24.8

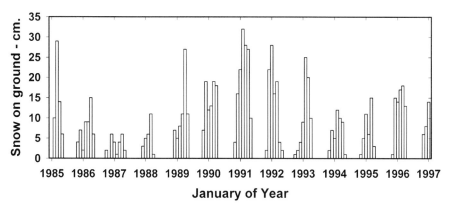

Figure 2.4 Amount of snow (cm) on the ground during the winter months, 1985–1997, at a standard snow course at the Burwash Airport.

2.2.2   Precipitation

The rain shadow of the St. Elias Mountains causes the Kluane area to have less than 30 cm of annual precipitation and makes the region semiarid. About 40–50% of the precipitation falls as snow. The annual precipitation maximum occurs in midsummer, when about half of the days have measurable rainfall. In the winter high-pressure systems tend to predominate, and the very cold weather limits the amount of snowfall. Table 2.1 lists the average annual precipitation for Burwash Airport for 1966–1996, along with the details for 1985–1996. The summers of 1988 and 1991 were particularly wet; the summers of 1989, 1993, and 1994 were particularly dry.

Snowfall is highly variable from year to year, and, in particular, the rate of snow accumulation during the winter changes from year to year. On average, about 1 m of snow falls over the winter, but the exact amount is affected strongly by slight differences in altitude within the study area. Figure 2.4 illustrates the variation in the timing of onset of snow and the variation from year to year in the amount of snow on the ground at the Burwash Airport. Snow depths in the middle of our study area are greater than those normally recorded at Burwash. There is a clear long-term cycle in snow depths, with minimal amounts in the mid-1980s and mid-1990s and maximal amounts around 1991.

2.3   Study Area and Vegetation

The area chosen for the Kluane Boreal Forest Ecosystem Project is a 350 km² section of the Shakwak Trench stretching from the south shore of Kluane Lake down to the Jarvis River (CD-ROM frame 10). This area is bounded to the northeast by the Kluane Hills and to the southwest by the alpine areas of the Kluane Ranges. The study area is thus open to the south for movements of animals within the valley but somewhat more restricted to the north by Kluane Lake, which occupies most of the Shakwak Trench (figure 2.1).

The paleoecological record for our valley has been established by Lacourse and Gajewski (1998), who took lake sediment cores from the deepest part of Sulphur Lake in the southern part of our study area. Their findings are broadly similar to those reported by others for this region (Ritchie 1987, Cwynar 1988). Between 12,000 and 11,250 years B.P., the vegetation was dominated by an open alpine tundra dominated by the presence of *Artemisa* (sage). By 11,250 years B.P., this tundra was replaced by a birch shrub tundra indicative of more continuous vegetative cover. The birch shrub tundra in turn was replaced by a poplar woodland by 10,250 years B.P. It too was replaced, initially by *Juniperus* populations at 9,500 years B.P. and later by spruce at about 8,400 years B.P. The white spruce forest (*Picea glauca*) that occupies our valley was established about 8000 years B.P. Interestingly, black spruce (*P. mariana*) was not recorded extensively in the paleoecological record, though it was found at many other sites in the Yukon, nor is it found in the present day vegetation. Finally, though a significant increase in green alder (*Alnus crispa*) occurred about 6000 years B.P., and alder is still common in the paleoecological record today, we have not recorded it on our study areas. Lacourse and Gajewski (1998) argue that neither long-distance pollen transport, nor its presence in the adjacent vegetative communities, is the explanation.

The vegetation of the study area is typical of the Kluane sector of the boreal forest described by Rowe (1972). White spruce is the dominant tree in the region, and open and closed stands of spruce occupy the majority of the valley floor. The vegetation can be divided into three ecological zones based on elevation (Douglas 1974, 1980): montane valley bottom forests (760–1080 m), subalpine forests (1080–1370 m), and alpine tundra (above 1370 m). The two lower zones are complex mosaics of forests of white spruce, stands of balsam poplar and aspen, and shrub-dominated areas of willow (mostly *Salix glauca*) and dwarf birch. The subalpine vegetation, consisting of open canopy spruce mixed with tall willow shrubs, grade into the low shrub-dwarf plant communities of the alpine.

To measure the vegetation of the valley, we mapped the vegetation and developed a fire history of the area. The fire history is described in chapter 7 (see 7.2.1; CD-ROM frame 49). Based on three bands of LANDSAT image, we created both supervised and unsupervised classification maps of the vegetation. Useful as these images were for a broad classification, they did not provide sufficient detail to serve as a precise map of vegetation types, and their resolution was not appropriate to serve as a base for a fire history study. We decided, therefore, to use air photo interpretation for these purposes.

We flew over the valley for airphotos in July 1992 and the 317 1:10,000 black-and-white photos provide complete coverage of the study area. Using a Bausch and Lomb zoom transfer scope, we produced a 1:32,000 vegetation map from these photos, which used four cover classes for each of three forest types: mature spruce, immature spruce, and aspen. There were also separate categories for water, shrub, wetland, and unvegetated. The resulting map (figure 2.5; Hucal and Dale 1993) has also been imported into the Geographic Information System and is provided on the CD-ROM that accompanies this book (frame 48). Table 2.2 gives the area of the valley occupied by the different habitat types. Closed spruce habitats were judged on the basis of 50% or higher canopy cover of white spruce; areas with less than 50% spruce canopy (but more than 15%) were placed in the open spruce category. The major difference between these two classifications results from the airphoto classification having included a larger section of the subalpine forest zone above 1080 m elevation, virtually all of which is classified as willow-shrub habitat. There is very good agreement between these two independent habitat classifications, which show that the valley vegetation is nearly equally one-third closed spruce, one-third open spruce,

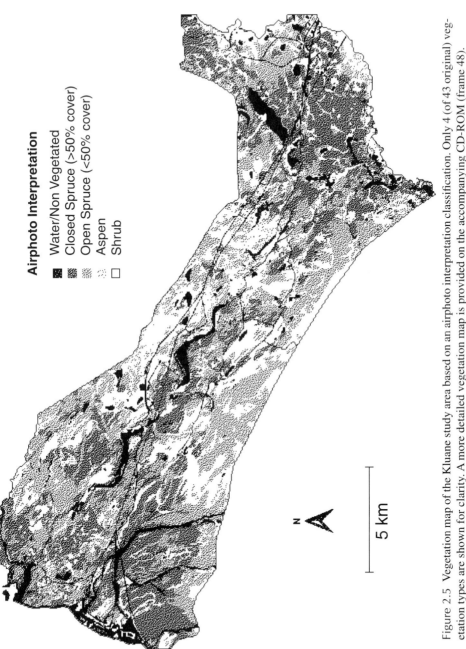

Airphoto Interpretation

- ■ Water/Non Vegetated
- ▨ Closed Spruce (>50% cover)
- ▧ Open Spruce (<50% cover)
- ░ Aspen
- □ Shrub

N

5 km

Figure 2.5 Vegetation map of the Kluane study area based on an airphoto interpretation classification. Only 4 (of 43 original) vegetation types are shown for clarity. A more detailed vegetation map is provided on the accompanying CD-ROM (frame 48).

Table 2.2 Vegetation classes of the Kluane study area.

	LANDSAT Classification		Airphoto Classification	
Vegetation Type	Area (km^2)	Percent	Area (km^2)	Percent
Closed spruce[a]	103.1	32.18	65.27	17.96
Open spruce[a]	96.69	30.18	89.96	24.74
Poplar-aspen	8.78	2.74	16.56	4.55
Willow shrub	82.32	25.69	162.9	44.81
Grass-open	29.50	9.21	28.89	7.94
Totals	320.41	100.00	363.62	100.00

Two estimates are given: one from a LANDSAT supervised classification and one from a more detailed airphoto analysis.

[a]Closed spruce with >50% canopy cover, open spruce with <50% canopy.

and one-third willow shrub. Figure 2.6 illustrates at the level of one study area (predator exclosure) how the two different vegetation classifications compare. Both the airphoto and the LANDSAT classification maps of the entire study area are included on the CD-ROM (frame 48).

In 1993 a spruce bark beetle outbreak began in the Shakwak Trench near Haines Junction and spread north in the next 4 years into our main study area. The spruce bark beetle kills the larger white spruce trees, and, in addition to fire, is a large-scale disturbance factor within the boreal region. Our study was mostly completed before the bark beetle caused severe tree mortality in the Shakwak Trench, and we will leave a discussion of its impact to future studies in the Kluane region.

2.4   Vertebrate Food Web

Boreal forest animal communities are exceptionally similar across the geographical spread of the North American boreal forest from Alaska to Nova Scotia and Newfoundland (figure 2.7). Red squirrels and snowshoe hares are dominant small mammals, along with least chipmunks, red-backed voles, and deer mice. Kluane is unusual in having the arctic ground squirrel as an additional common herbivore. Large mammals include moose, mountain caribou, Dall sheep, wolverines, and grizzly and black bears. The major predators in the system are wolves, coyotes, lynx, and red foxes among the mammals and great horned owls, goshawks, red-tailed hawks, ravens, golden eagles, and hawk owls among the birds.

We have not been able to study all these vertebrate species in the Kluane food web because of scale. Moose, Dall sheep, and caribou have all been excluded from our analysis, as have bears and wolves. All of these large mammals have been studied by others or are the subject of current studies in the Kluane region. None of them is a major player in the system we have analyzed. For example, moose feed in winter on many of the same plants as snowshoe hares, but moose in the Kluane area are rare, probably because of overhunting, and there is no possibility of competition for food between moose and hares at the present time. We have focused our studies on the array of small and medium-sized mammals and birds that may interact over the snowshoe hare cycle.

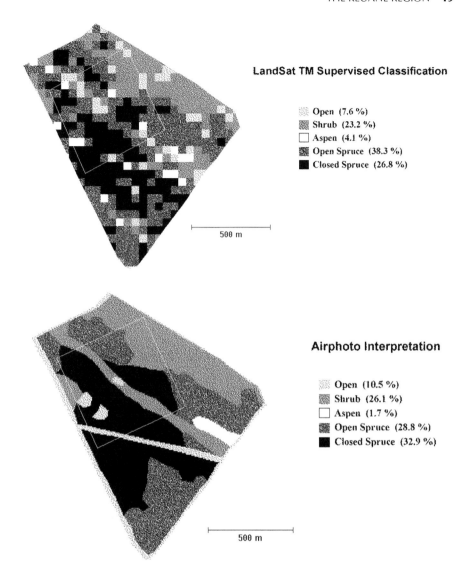

Figure 2.6 Comparison of the LANDSAT supervised classification and the airphoto interpretation of vegetation classes within the predator exclosure area (Beaver Pond fence). The percent area of each class is shown in parentheses. The square outline within the fence area denotes the 36-ha snowshoe hare trapping grid.

Figure 2.8 gives a simplified food web for the vertebrates of the Kluane region of the Yukon. The arrows in this food web indicate the direction of who eats whom. The key purpose of this project is to understand the structure of this food web. To do this, we need to know which linkages between species are critical and which ones are incidental. Critical linkages, if broken, result in a major change in the community, such that common species

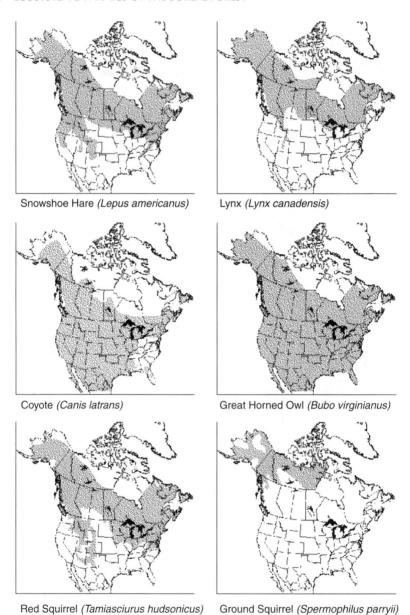

Snowshoe Hare *(Lepus americanus)*

Lynx *(Lynx canadensis)*

Coyote *(Canis latrans)*

Great Horned Owl *(Bubo virginianus)*

Red Squirrel *(Tamiasciurus hudsonicus)*

Ground Squirrel *(Spermophilus parryii)*

Figure 2.7 Range maps of the main herbivores and predators in the Kluane boreal forest community. Note the extensive overlap of these species in the boreal forest zone (see figure 1.1 for comparison, and CD-ROM frame 15). The arctic ground squirrel is unusual in being the only major species in the Kluane community that is not widespread, but restricted to the northwestern sector of the boreal region.

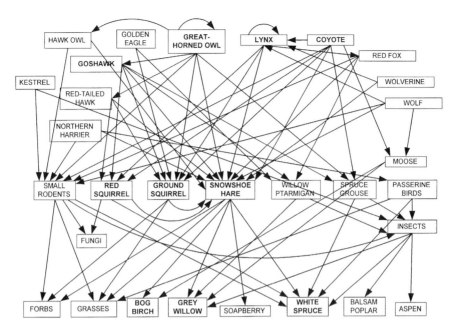

Figure 2.8 A simplified food web for the vertebrates of the Kluane region. Details of the food habits of each species are provided in subsequent chapters.

may become rare or rare species may go extinct. Incidental linkages, if broken, result in negligible or minor changes in the community. The simplified balance of nature model of ecological communities treats this food web as completely critical so that all linkages are essential to its structure. We know that this is incorrect, and in this book we elucidate the resiliency of this vertebrate community and which links are most significant.

2.5   History of Hare Cycles

The central herbivore in the Kluane ecosystem is the snowshoe hare. Hares are well known for their 9- to 10-year population cycles throughout the boreal forest. In this project we used the hare cycle as a perturbation and tried to study how the other species react to this large-scale change. Fortunately, the history of hare cycles in the Yukon is mimicked in the trapping data on lynx, and lynx fur returns have had a long history of use in providing a long-term perspective on hare cycles and in analyzing the structure of these fluctuations (Elton and Nicholson 1942, Keith 1963, Stenseth et al. 1997; CD-ROM frame 14). We make use of this information here to trace the recent history of hare cycles in the Yukon.

Figure 2.9 shows the lynx fur return data for the Yukon Territory from 1920 to 1994, along with similar data for the neighboring provinces of Alberta and British Columbia. Lynx trapping numbers typically peak 1–2 years after the snowshoe hare peak. Recent lynx fur numbers peaked in the Yukon in 1944, 1955, 1964, 1973, 1981, and 1989. Peaks in British Columbia occurred in 1944, 1954, 1962, 1973, 1982, and 1991 and in Alberta

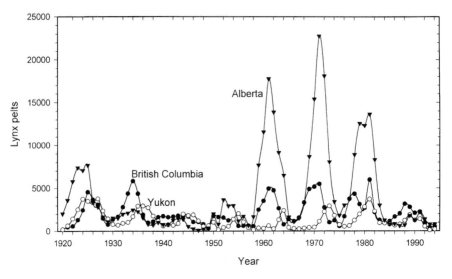

Figure 2.9 Lynx furs traded from the western Canadian provinces from 1920 to 1994. (Data from Statistics Canada, Livestock and Animal Products Statistics, 1995).

in 1943, 1952, 1961, 1971, 1981, and 1991. There is a high degree of synchrony in lynx numbers among these provinces, and we believe that the underlying cyclic patterns are robust to economic and social changes in trapping patterns. We do not think, however, that the amplitude of the fur return cycles should be assumed to reflect biological amplitudes of cycles because of these economic factors. The key point is that periodic fluctuations in lynx and snowshoe hares have been going on in the Yukon and other parts of the boreal forest without major changes for as long as data have been collected. We are thus studying a continuing and repeatable sequence of population changes.

2.6 Large Mammals

We do not attempt to cover the large mammal community in the Kluane region in this book for reasons of scale pointed out earlier. We summarize here a broad overview of these animals and their potential impact on the smaller vertebrates that are the focus of our studies.

Moose are at low numbers in the Kluane region and, in general, moose density is below 1 animal per 3 km^2. The Yukon Wildlife Management Branch has been concerned about the reasons for this low density of moose and has concluded that it is not due to a shortage of food resources. Both bears and wolves prey upon moose, and this predation could be the reason for low moose density. Alternatively, overhunting could be preventing population growth. Whatever the reason, moose are present but not a large element in the Kluane system (Hoefs 1980).

Mountain caribou are not present anywhere on the study area but are found north of Kluane Lake. Like the moose, they exist at very low density and have been the subject of much recent concern in relation to wolf predation and hunting pressure. Seasonal movements of caribou operate on a scale much larger than our study area.

Dall sheep are a particularly spectacular large mammal in the Kluane area and reach high numbers in Kluane National Park. They occupy alpine and subalpine terrain in the Kluane area and have been studied in detail by Hoefs (1980) and Hoefs and Bayer (1983). Because Dall sheep occupy the alpine zone, they have little potential interaction with the boreal forest community that we are studying. Predation on sheep by coyotes, golden eagles, wolves, wolverine, lynx, and grizzlies has been recorded (Hoefs and Cowan 1979), and some of this predation could be an indirect result of the snowshoe hare cycle.

Grizzly bears are common in the Kluane region, and the density of grizzlies in this region is among the highest observed anywhere within their geographic range (Pearson 1975). Density estimates for grizzlies in the Kluane area range from approximately 1 bear per 15 km^2 to 1 bear per 20 km^2 (Environment Canada 1987). Bears hibernate from October to April or early May. Bears are omnivores, and most of their energy comes from eating plants. Soapberry (*Shepherdia canadensis*) and liquorice-root (*Hedysarum* spp.) are prime foods, but bears feed on a wide variety of plants during the summer. Arctic ground squirrels are favorite prey in the subalpine zone, and carrion is widely taken as it becomes available. Bears in the Kluane area are the subject of on-going studies by Parks Canada. From our perspective, bears are a part of the vertebrate community but their low absolute density limits their impact on the scale of the smaller mammals and birds that we have studied here.

Wolves are common throughout the Kluane region but in the mid-1990s were the subject of active control programs by the Yukon government in order to increase moose and caribou populations. Fortunately, these control programs have stopped. Wolves are believed to be dependent primarily on large mammals (moose, caribou, and sheep) for prey and take smaller mammals such as snowshoe hares only incidentally. In terms of impact on smaller mammals, wolves are a minor component of the system we analyzed at Kluane.

2.7 Human Impacts

The major impact of humans on the boreal forest in the Kluane region has been the construction of the Alaska Highway in 1942 and its continuing reconstruction and rerouting since that time. There have been minor fires associated with this construction (see 7.2.1), but the overall impact has not been severe. Traffic on the Alaska Highway is greatly increased in summer but is minimal in winter. Wildlife kills on the highway are typically of ground squirrels along the highway, but some snowshoe hares are killed as well. Predator kills by vehicles are potentially more serious. We recorded three coyote kills from traffic and all but one of five known coyote deaths were associated with the highway. By contrast, only one lynx (of 56 radio collared) was killed along the highway in 1993 by a truck.

There is no agriculture or forestry currently within the study area, and disturbance is limited to a few mining roads that were pushed through into the Kluane Ranges more than 25 years ago. These old roads and the unused sections of the Alaska Highway have been used extensively by this project for both winter and summer travel. Old roads facilitate the movement of predators through the valley, but it is not clear that this affects the biological processes we have studied.

Literature Cited

Bostock, H. S. 1952. Geology of the northwest Shakwak Valley, Yukon Territory. Geological Survey of Canada Memoirs 267. Geological Survey of Canada, Ottawa, Ontario.

Cwynar, L. C. 1988. Late Quaternary vegetation history of Kettlehole Pond, southwestern Yukon. Ecological Monographs **52**:1–24.

Denton, G. H., and M. Stuiver. 1967. Late Pleistocene glacial stratigraphy and chronology, northeastern St. Elias Mountains. Geological Society of America Bulletin **78**:485–510.

Douglas, G. W. 1974. Montane zone vegetation of the Alsek River region, south western Yukon. Canadian Journal of Botany **52**:2505–2535.

Douglas, G. W. 1980. Biophysical inventory studies of Kluane National Park. Unpublished report to Parks Canada, Winnipeg.

Elton, C., and M. Nicholson. 1942. The ten-year cycle in numbers of the lynx in Canada. Journal of Animal Ecology **11**:215–244.

Environment Canada. 1987. Kluane National Park Resource Description and Analysis. Natural Resource Conservation Section, 2 vols. Environment Canada, Parks Prairie and Northern Region, Winnipeg, Manitoba.

Hoefs, M. 1980. Horns and hooves. *in* J. Theberge (ed.). Kluane: pinnacle of the Yukon, page 175. Doubleday, Toronto, Ontario.

Hoefs, M., and M. Bayer. 1983. Demography of a Dall sheep population. Canadian Journal of Zoology **61**:1346–1357.

Hoefs, M., and I. M. Cowan. 1979. Ecological investigation of a population of Dall sheep (*Ovis dalli dalli* Nelson). Syesis **12** (Suppl. 1):1–83.

Hucal, T., and M.R.T. Dale. 1993. Habitat map of the Kluane Boreal Forest Ecosystem Project study area. 1:32000 scale vegetation map (unpublished).

Keith, L. B. 1963. Wildlife's ten-year cycle. University of Wisconsin Press, Madison, Wisconsin.

Lacourse, T., and K. Gajewski. 1998. Post-glacial vegetation history of Sulphur Lake, southwest Yukon. Canadian Association of Geographers, Ottawa, Ontario.

Muller, J. E. 1967. Kluane Lake map area, Yukon Territory. Geological Survey of Canada Memoirs 340. Geological Survey of Canada, Ottawa, Ontario.

Pearson, A. M. 1975. The northern interior grizzly bear, *Ursus arctos* L. Canadian Wildlife Service Report Series No. 34.

Ritchie, J. C. 1987. Postglacial vegetation of Canada. Cambridge University Press, Cambridge.

Rowe, J. S. 1972. Forest regions of Canada. Canadian Forestry Service Publication no. 1300. Canadian Forestry Service, Ottawa, Ontario.

Stenseth, N. C., W. Falck, O. N. Björnstad, and C. J. Krebs. 1997. Population regulation in snowshoe hare and Canadian lynx: asymmetric food web configurations between hare and lynx. Proceedings of the National Academy of Sciences USA **94**:5147–5152.

Webber, B. L. 1974. The climate of Kluane National Park. Project Report no. 16. Atmospheric Environment Service, Environment Canada, Kluane National Park, Haines Junction, Yukon.

3

Trophic Interactions, Community Organization, and the Kluane Ecosystem

A. R. E. SINCLAIR & CHARLES J. KREBS

The conservation of biodiversity depends on functioning natural ecosystems and communities. There are various scientific reasons for conserving natural ecosystems. One is that we need to retain representative portions of the natural world as base line controls for human impacts elsewhere. Another is that we need to preserve species that may later be of benefit to humans. To conserve natural systems, however, we have to understand how disturbances affect natural systems and whether those effects will cause irreversible changes. Alternatively, in human-dominated systems, we need to understand how changes will affect our own livelihood through the stability of the system.

Communities are highly complex, consisting of a large number of species that interact with other species. Studies of communities ask two main questions. First, how are they structured? That is, how many levels, such as plants, herbivores, predators, and top predators, are there? Second, how does the ecosystem work? For example, will the system be affected if we take out one of the species or several of the species? Alternatively, are some species more important to the viability of the community? These questions are relevant to the idea that systems have a large number of species that perform the same function and that there is a large amount of redundancy (Walker 1991). This idea implies that one can afford to lose species without radically altering the integrity and functioning of the community. Redundancy in the function of species in the community is a necessary safety net for that community, and, if we lose that redundancy (i.e., reduce the diversity of species), then we run the risk of a collapse in the system from some small external perturbation.

These questions are also relevant to systems that have already been subjected to changes such as eutrophication of lakes, fragmentation of forests, acid rain pollution of lakes and forests, introduction of exotic species, and creation of protected areas as islands within human ecosystems. Removal of the lower trophic levels, the plants, will always change a community and result in habitat loss. Nearly all human endeavor affects lower levels and, hence, alters habitats. However, when herbivores or predators are removed from a system the effects are less predictable and depend on which species are affected. Does the removal of species at higher trophic levels change the nature of the whole community? We know from studies of some ecosystems such as the marine intertidal and coastal systems of the Pacific Coast of North America that, when top predators (in this case sea otters) are removed from the system, the system changes radically from one type of community to another and stays that way until the top predators return (Estes et al. 1989, Wootton 1994a). Therefore, in this ecosystem the removal of a predator radically alters the community. Not all communities, however, may react in this way when a predator is removed.

Because communities are complex and contain large numbers of species, we cannot study every component and put it together like a jigsaw puzzle. Indeed, it is likely that the sum of the component parts (the dynamics of individual populations) will not tell us how the combination of species will interact with each other. We must therefore study the whole community as a functioning unit and try to find general rules for how communities work. It is unlikely that simple generalities will be appropriate for such complex systems, and we recognize that in reality communities will have complex structure and dynamics. However, we must start our research with a simple idea and then expand on that as we learn more from experimentation. In this chapter we discuss trophic-level theory and lay the theoretical groundwork for the experiments described in the rest of this book.

3.1   Trophic-level Theory

3.1.1   Community Structure

Community structure has been studied by comparing food webs (i.e., the connections between different species in natural history studies of communities) across the world. Over the past decades several hundred descriptive studies of food webs have been exhaustively analyzed (Cohen 1978, Pimm 1982, Briand and Cohen 1987, Paine 1988, Menge and Farrell 1989, Schoener 1989, Hall and Raffaelli 1991, Polis 1991, Schoenly et al. 1991). Despite the incomplete data, a number of general patterns in food webs can be seen (Lawton 1989). Ecologists have suggested rules for how many species one might expect, the types of interactions between species, and the number of trophic levels in different parts of the world.

The functioning of these communities—their dynamics—has also been studied by comparative methods and by perturbation experiments. These studies have focused on how one trophic level affects another or on the nature of the interactions between species in the community. Interactions between species can be either direct or indirect. *Direct interactions* are those in which there is a physical relationship between species, such as a predator eating its prey or direct interference between competitors. *Indirect interactions* along a food chain are those that are one step or more removed. Thus, predation is a direct interaction, but the effect of the prey population on the vegetation as a consequence of that predation one level above is an indirect interaction. Other indirect interactions involve *exploitation competition,* in which one predator affects the food supply of another predator, *apparent competition,* in which two prey species share the same predator, and *indirect mutualism,* in which a predator affects its prey, which in turn affects the competition between another prey and therefore its own predator (Wootton 1994b). Indirect effects can also arise when one species affects the interaction between two other species, sometimes by a change in behavior (Schmitz 1998). Questions asked at this level of generality involve how far perturbations travel as indirect linkages along the food chain. For example, if the main prey for a predator is removed, does the predator switch to a secondary prey, and how does this affect the secondary prey and its own competitors? Alternatively, if a predator is removed, how does this alter the competition between various prey species?

3.1.2   Control of Trophic-level Biomass

Bottom-up hypotheses assume that systems are regulated by nutrient flow from below (White 1978, 1984) because plants are essential to the levels above. Comparative studies have suggested that low soil nutrients and low productivity of plants result in fewer trophic levels, that higher productivity of plants results in more trophic levels, and that a gradient of increasing primary productivity (and more levels) can be traced from high to low latitudes (Oksanen et al. 1981, T. Oksanen 1990, Abrams 1993). In this view, higher trophic levels have neither a regulating effect nor any influence on productivity or biomass on the levels below them (Hawkins 1992, Hunter and Price 1992, Strong 1992). In African terrestrial environments the relationship between nutrients, primary productivity, and secondary productivity has been illustrated by Coe et al. (1976), Botkin et al. (1981), Bell (1982), and McNaughton et al. (1989).

Early ideas of top-down effects can be attributed to Hairston et al. (1960) and Slobodkin et al. (1967), who proposed predator regulation of herbivores. Although their basic premise that green vegetation is available to be eaten was wrong because much of this vegetation is defended, these authors stimulated a number of other ideas. One is the pure top-down hypothesis, which proposes that each trophic level is regulated by the one above with the top predators being self-regulated. This concept was applied to aquatic and marine systems (Menge and Sutherland 1976). If top predators, such as fish in a lake, are removed, then lower predator levels increase, herbivores decrease, and plants increase in biomass. This "cascade hypothesis" (Carpenter et al. 1985, Carpenter and Kitchell 1987, 1988, 1993) recognizes that nutrient availability uniformly raises or lowers these relationships.

Trophic levels could also alternate between top-down and bottom-up regulation (Fretwell 1977, 1987, Oksanen et al. 1981, L. Oksanen 1988, 1990). Thus, predators regulate herbivores which cannot, therefore, regulate their own plant food. Plants are then regulated by nutrients. This idea, called the Fretwell-Oksanen hypothesis, has been applied to the wolf–moose–shrub ecosystem on Isle Royale in Lake Superior (McLaren and Peterson 1994) and to vole communities in Scandinavia (Moen et al. 1993).

The Fretwell-Oksanen hypothesis leads to different predictions in ecosystems with different nutrient levels (Fretwell 1987, Oksanen and Ericson 1987, L. Oksanen 1990, T. Oksanen 1990). Low productivity systems such as on the Arctic tundra support only two trophic levels, plants and herbivores, and in these systems herbivores regulate plants. At higher productivity, such as in temperate terrestrial systems, three trophic levels occur, and plants are nutrient limited. In very productive systems, such as estuaries and some lakes, with four levels, top predators are self-regulating and also regulate lower predators, so that herbivores are regulated from below but also regulate vegetation in turn. Therefore, the plant–herbivore link is a two-way interaction. Two generalities appear from this comparison. First, in systems with even numbers of levels (two, four), herbivores regulate plants and the vegetation is largely herbaceous (grassland, prairie, tundra, etc.). In systems with odd numbers of levels, predators regulate herbivores, and the vegetation, released from severe herbivory, becomes forest or shrub dominated. Second, the number of levels is determined by the productivity of the system through nutrient availability (i.e., ultimately there is a bottom-up influence).

Other models propose that regulatory effects of higher trophic levels depend not so much on productivity as on the level of environmental stress (Menge and Olson 1990), although these two features are quite possibly correlated. Predictions, however, differ depending on which trophic level is the more susceptible to stress. If consumers (herbivores, predators) are more susceptible to harsh climate, for example, than are plants, then under severe environmental stress consumers will be inhibited and plants will be regulated from below. Under benign conditions herbivores should regulate plants (Menge and Sutherland 1976, Connell 1978). The alternative view is that plant defenses are inhibited by stress, such as drought, whereas consumers are less affected. Thus, consumers can regulate plants under stressful conditions (White 1984, Menge and Olsen 1990). The types of stress differ in these two views. The former involves adaptations to harsh environments and is long term, the latter addresses more short-term events. Both concepts are likely to be valid in different circumstances.

In aquatic systems there is evidence that indirect interactions become diluted the farther along the food chain they occur. Thus, a combined top-down, bottom-up hypothesis for aquatic systems suggests that biomass is regulated from below by nutrient availability, but this effect is strongest at the plant level and becomes weaker at progressively higher levels. Equally at the top of the food web, top-down interactions are strong, but these effects weaken with every step down (McQueen et al. 1986, 1989, Pace and Funke 1991). Other mechanisms that reduce the efficiency of predators are interference and territoriality of predators and refuges for prey. These result in attenuation of indirect effects at lower links in the chain, and so bottom-up effects are seen at lower trophic levels (Power 1984, Arditi and Ginsburg 1989, Arditi et al. 1991, Hanski 1991).

These concepts have all been expressed verbally and hence they are imprecise. When we discuss an effect of one trophic level on another as a result of a perturbation, we need to define what type of effect is being considered. Thus, there are immediate or "instantaneous" effects when a change in one trophic level affects another. These may be quite different from long-term consequences if the change persists and the system settles at some new equilibrium. For example, in the intertidal community on the Pacific Coast of North America, gulls feed on goose barnacles (*Pollicepes* spp.) and mussels (*Mytilus*). When gulls were excluded for 2 years from exclosure plots (Wootton 1994a), there was an increase in one prey species (*Pollicepes*) but a decrease in the other (*Mytilus*) as a result of space competition with *Pollicepes*. These are long-term results. The short-term result of gull removal, however, would have been an immediate increase in survival of both prey species. Thus, indirect effects can have complex results farther along the food chain and make them difficult to predict (Yodzis 1988, Wootton 1994b).

3.1.3   Case Studies in Mammalian Communities

Long-term studies have thrown some light on trophic-level dynamics. For example, 40 years of monitoring the large-mammal communities in the Serengeti, Tanzania, revealed the complexities of trophic-level interactions (Sinclair 1995, Mduma et al. 1999). Herbivorous mammals can be divided into migrants and residents. Migrant species, such as the wildebeest (*Connochaetes taurinus*), are regulated through their food supply of grasses (monocots). Although top predators (lions, hyenas) feed on wildebeest, they are not limited by this food source, but rather are limited by the supply of resident herbivores. In turn, top predators limit the density of resident herbivores as well as lower level predators such as cheetah (*Acinonyx jubatus*) and wild dogs (*Lycaon pictus*) (Laurenson 1995). Many of the resident herbivores feed exclusively on herbs (dicots) and determine their composition. Hence, there is a complex series of control pathways. The migrant herbivores are regulated by bottom-up flow from monocots. In contrast, the resident herbivores are regulated by top-down flow from the predators. These conclusions were reached through interpreting perturbations in the ecosystem that acted as seminatural experiments.

Yellowstone National Park, USA, is another system in which the mammalian community has been studied for several decades (e.g., Houston 1982, Keiter and Boyce 1991). This system was previously perturbed by the removal of the top predator, the wolf (*Canis lupus*) a century ago. As in the Serengeti, the dominant migrant herbivore, elk (*Cervus elaphus*), is regulated by food supply. Elk, in turn, determine the vegetation composition, both for the monocots and the dicots, including regenerating young trees such as aspen (Merrill et al. 1994, Singer et al. 1994). The recent reintroduction of wolves to the ecosystem will test whether this bottom-up control persists in the migrant herbivores or whether top-down control takes over in both migrant and resident mammalian herbivores.

The intertidal communities along the Pacific Coast of North America (Estes et al. 1989, Estes and Duggins 1995) show top-down control. Top predators such as sea otters (*Enhydra lutris*) determine the abundance of sea urchins (*Strongylocentrus* spp.), the dominant herbivores, and hence indirectly determine the macroalgae composition and other members of the community.

Conclusions from the intertidal system were obtained from controlled experiments. In contrast, trophic dynamics of terrestrial ecosystems have been difficult to study in controlled experiments because of the large scale and the long time needed to produce results. However, by using the small mammal community in the boreal forest of the Yukon at Kluane, we have been able to generate perturbation experiments to examine the direction of control between trophic levels. These experiments were designed to perturb each trophic level in turn. We approached the problem of predicting the effects of our perturbations in the boreal forest by considering only immediate or instantaneous effects. Then we started from first principles and considered all the possible outcomes from an experimental perturbation as the effects traveled along the food chain.

3.2   Interactions of Trophic Levels

We illustrate in figure 3.1, through a hypothetical example, the way biomass and productivity of trophic levels are related. Each box represents the biomass of a trophic level. The arrows and their numbers indicate the annual flow of biomass, or productivity. Material enters the system from outside through, for example, geological weathering, and exits the system through leaching. The remaining material cycles through the system from the soil to the predators, while each level contributes to the decomposers, which then returns material to the soil nutrients. The input and output rates to the boxes balance out.

Control of the biomass in the boxes can be either from above or below. Thus, an increase in the nutrients of the soil could result in both the flow rates and biomass increasing at each level, indicating a bottom-up control. Equally, one could envisage an increase in the flow rate of material from soil nutrients to plants and an increase in flow rate from plants to herbivores, resulting in a change in herbivore biomass but no change in plant biomass. This would occur by top-down control from herbivores to plants.

There are a number of ways we could measure the responses of one trophic level to changes in another. For example, we could measure growth rate in plants and survivorship in animals. However, to avoid confusion with interpretation, we have used the change in biomass because it can be applied to all trophic levels. We say that one trophic level limits another if changes in one trophic level are associated with changes in the biomass of the adjacent level. There are, of course, a variety of ways in which one could link species

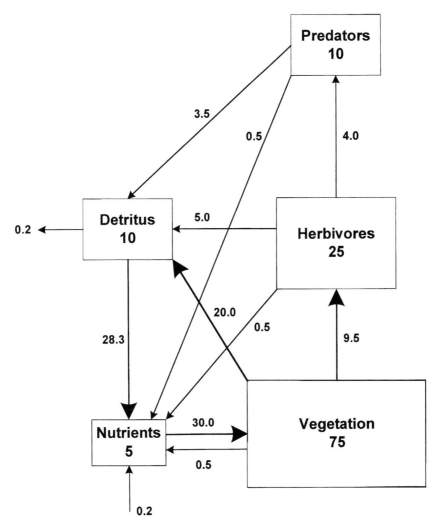

Figure 3.1 Hypothetical flow diagram showing biomass and annual rates of production for nutrient pool (N), vegetation (V), herbivores (H), predators (P), and decomposers (D). The system has annual inputs to the soil by geochemical processes and outputs from decomposers through leaching.

in a food web, but for simplicity, we assume direct linear interactions. We indicate these interactions in the following way:

$$N\updownarrow \leftrightarrow V\updownarrow \leftrightarrow H\updownarrow \leftrightarrow P\updownarrow, \tag{1}$$

where N denotes nutrient concentration, V denotes vegetation biomass (or primary producers), H denotes herbivore biomass (primary consumers), and P denotes carnivore biomass (secondary consumers). The arrows denote trophic-level effects, with a rightward arrow implying that an increase in resource (food) increases the rate of change of biomass of the adjacent consumer level. Similarly, a leftward arrow implies that an increase in consumer biomass decreases the rate of change of the adjacent prey biomass. A vertical arrow implies that a given trophic level has a density-dependent effect on its own rate of growth.

Some of the food web structures one could think of would be biologically implausible. For example, a terminal trophic level that has no negative feedback on its growth would grow infinitely. Accordingly, there must be an intraspecific feedback effect at the end of each plausible chain or else a link between the penultimate and terminal levels. For similar reasons, it would not be biologically plausible to have arrangements such as the following:

$$N \rightarrow V \rightarrow H \rightarrow P \tag{2}$$

because there would be no negative feedback on the rate of growth in any of the trophic levels and so all would grow exponentially over time. Therefore, we assume that there must be some feedback on each level, and possible competition within the trophic levels (i.e., there is always a vertical arrow at each level implied in our models). The algebraic formulations for instantaneous change at each trophic level are illustrated in table 3.1 for three basic types of bottom-up, top-down and reciprocal control (Sinclair et al. 2000). There are, however, 27 biologically plausible models (table 3.2), all of which have negative feedbacks at every link in the web, either due to self-limited growth or an impact from the next highest level. All models comprise various combinations of the equations in table 3.1 (J. M. Fryxell, personal communication). The majority of the models are highly stable, each level equilibrating monotonically or with a few minor fluctuations. Figure 3.2 illustrates the time dynamics of three of the models. Figure 3.2a shows the stable, pure bottom-up model 1. Figure 3.2b shows the pure reciprocal model 27 with some damped fluctuation. Only three models show more marked fluctuation, and they all involve reciprocal interactions between plants and herbivores with predators imposing top-down control (models 7, 20, 22). Figure 3.2c illustrates model 20 involving double dilution effects (see table 3.2).

We have assumed in our equations the simplest (linear) density-dependent relationships. For illustrative purposes, consider interactions between plants (V) and herbivores (H). A left arrow implies that changes in herbivore density affects the rate of plant biomass change, near equilibrium, but not vice versa. Such a situation implies that herbivore growth is limited by something other than food, even though herbivores do consume plants. For example, hares feed on white spruce, and the higher the density of hares, the heavier the browsing impact on the trees. However, hare survival is little affected by spruce biomass. Similarly, a right arrow implies that changes in plant density affect the rate of change of herbivore biomass, near equilibrium, but not vice versa. Biologically this occurs when red squirrels feed on cones, hares eat senescent willow leaves, and carnivores rely on carrion. A double-headed horizontal arrow between plants and herbivores implies that herbivores respond to plant abundance, and plants, in turn, respond to herbivore abundance, forming a reciprocal relationship (Caughley 1976). Finally, a density-dependent form of intraspecific competition implies that each trophic level has its own intrinsic limits.

Table 3.1 Minimal algebraic expressions for bottom-up, top-down, and reciprocal interactions between trophic levels.

Bottom-up Interactions

$$\frac{dN}{dt} = r_0 - a_{00}N$$

$$\frac{dV}{dt} = a_{10}NV - a_{11}V^2$$

$$\frac{dH}{dt} = a_{21}VH - a_{22}H^2$$

$$\frac{dP}{dt} = a_{32}HP - a_{33}P^2$$

Top-down Interactions

$$\frac{dN}{dt} = r_0 - a_{00}N - a_{01}NV$$

$$\frac{dV}{dt} = r_1V - a_{12}VH - a_{11}V^2$$

$$\frac{dH}{dt} = r_2H - a_{23}HP - a_{22}H^2$$

$$\frac{dP}{dt} = r_3P - a_{33}P^2$$

Reciprocal Interactions

$$\frac{dN}{dt} = r_0 - a_{00}N - a_{01}NV$$

$$\frac{dV}{dt} = a_{10}NV - a_{12}VH - a_{11}V^2$$

$$\frac{dH}{dt} = a_{21}VH - a_{23}HP - a_{22}H^2$$

$$\frac{dP}{dt} = a_{32}HP - a_{33}P^2$$

r_i is the per capita rate of increase of trophic level i; a_{ij} is the community matrix coefficient for trophic level j acting on trophic level i. N = soil nutrient pool, V = plant biomass, H = herbivore biomass, P = predator biomass. All models use some combination of these equations.

Table 3.2 Models for instantaneous change at each trophic level.

	Model							
Largely Bottom-up								
Pure bottom-up	1	N	→	V	→	H	→	P
Bottom-up, nutrient reciprocal	2	N	↔	V	→	H	→	P
Bottom-up, vegetation reciprocal	3	N	→	V	↔	H	→	P
Bottom-up, herbivore reciprocal	4	N	→	V	→	H	↔	P
Largely Top-down								
Pure top-down	5	N	←	V	←	H	←	P
Top-down, nutrient reciprocal	6	N	↔	V	←	H	←	P
Top-down, vegetation reciprocal	7	N	←	V	↔	H	←	P
Top-down, herbivore reciprocal	8	N	←	V	←	H	↔	P
Herbivore Dominant								
Herbivore dominant	9	N	←	V	←	H	→	P
Herbivore dominant, vegetation dilution	10	N	→	V	←	H	→	P
Herbivore dominant, nutrient reciprocal	11	N	↔	V	←	H	→	P
Herbivore dominant, predator reciprocal	12	N	→	V	←	H	↔	P
Herbivore–vegetation co-dominant	13	N	←	V	↔	H	→	P
Vegetation Dominant								
Vegetation dominant	14	N	←	V	→	H	→	P
Vegetation dominant, herbivore dilution	15	N	←	V	→	H	←	P
Vegetation dominant, nutrient reciprocal	16	N	↔	V	→	H	←	P
Vegetation dominant, predator reciprocal	17	N	←	V	→	H	↔	P
Largely Dilution								
Herbivore dilution	18	N	→	V	→	H	←	P
Vegetation dilution	19	N	→	V	←	H	←	P
Herbivore–vegetation joint dilution	20	N	→	V	↔	H	←	P
Largely Reciprocal								
Reciprocal, herbivore bottom-up	21	N	↔	V	↔	H	→	P
Reciprocal, predator top-down	22	N	↔	V	↔	H	←	P
Reciprocal, nutrient bottom-up	23	N	→	V	↔	H	↔	P
Reciprocal, vegetation top-down	24	N	←	V	↔	H	↔	P
Reciprocal, vegetation bottom-up	25	N	↔	V	↔	H	↔	P
Reciprocal, herbivore top-down	26	N	↔	V	←	H	↔	P
Pure reciprocal	27	N	↔	V	↔	H	↔	P

We assume self-limitation where there is no left arrow affecting a trophic level. A left arrow indicates that one level is changed by that above, a right arrow indicates that one level is changed by that below, and a two-way arrow indicates that two levels affect each other. "Dilution" means that a level is influenced by both lower and higher levels, whereas "dominant" implies that a level affects both lower and higher levels.

N = nutrient pool; V = vegetation biomass; H = herbivore biomass; P = predator biomass.

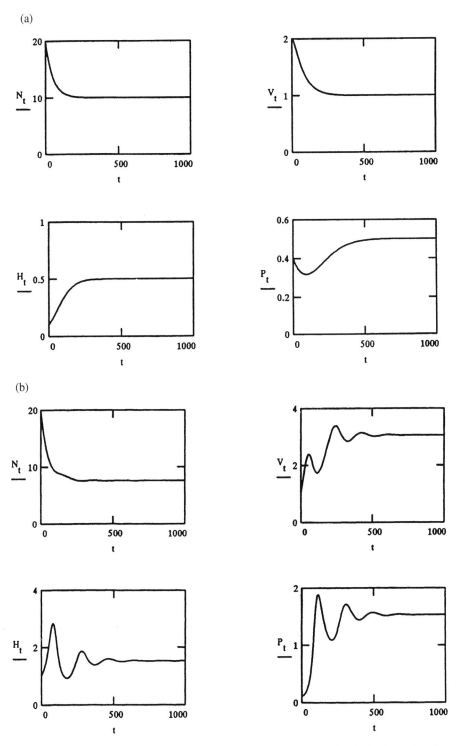

Figure 3.2 The trajectory through time over 1000 intervals of 3 of the algebraic models described in table 3.2. (a) Pure bottom-up model, (b) pure reciprocal model, and (c) double dilution model. The compartments show nutrient pool (N), vegetation (V), herbivores (H), and predators (P).

(c)

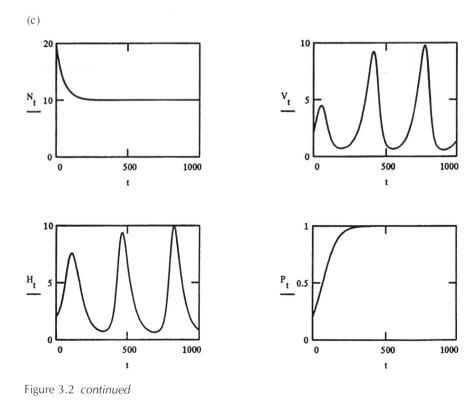

Figure 3.2 *continued*

The 27 possible ways trophic levels can interact directly with each other can be conveniently grouped by the dominant direction of control in the system and by the dominant trophic level. Although other groupings can be produced just as easily, we prefer the six groups described next and shown in table 3.2.

3.2.1 Bottom-up Models

Bottom-up models have an uninterrupted direction of control from lower to higher levels (figure 3.2a). They all predict that nutrient availability determines the biomass in the higher levels and thus assume that supply of nutrients is limiting.

Such models, therefore, would be most relevant in nutrient-poor environments where plants have difficulty obtaining nutrients. In terrestrial systems this would apply to polar tundra (L. Oksanen 1988, 1990), subarctic boreal forests with acidic soils (this study), and nutrient-poor sclerophyll forests in Australia. The soils of much of Australia are so poor that nutrient content of leaves in individual trees and the densities of mammalian foliovores reflect the local nutrient condition of the soil the trees grow in (Braithwaite et al. 1983, Braithwaite 1996). In aquatic systems nutrients are low in high-latitude, cold oligotrophic lakes, in acidic marshes and peat bogs, and in warm open-ocean systems.

3.2.2   Top-down Models

Top-down models have an uninterrupted direction of control from the top level downward. Thus, predators control the herbivore biomass, and herbivores control plant biomass. They all predict that nutrients are not limiting and that community structure is determined by type of predator species and predation.

These models could be applicable to aquatic systems, particularly the more eutrophic lakes with many trophic levels. Larger fish species determine the presence or absence of small fish species, which in turn determine which zooplankton occur (Menge and Sutherland 1976, Estes et al. 1977, Simenstad et al. 1978, Paine 1980, Power 1990, Menge 1992, Carpenter and Kitchell 1993, Rosemund et al. 1993, Wootton 1994a).

3.2.3   Herbivore-dominated Models

In herbivore-dominated models the herbivore level controls both the predator and plant levels. Interactions between plants and nutrients can be either top-down or bottom-up. These would apply to systems where herbivores dominate. Generally, such systems are grasslands. In tropical savannas, such as the Serengeti, large grazing mammals structure the vegetation and determine predator biomass (McNaughton 1985, Semmartin and Oesterheld 1996). Predators do not limit the dominant herbivores (Sinclair 1975, Sinclair and Arcese 1995). The dominant effects of herbivores are also seen in temperate prairies (Huntly 1991) and in arctic and antarctic tundra (Leader-Williams 1988, Albon and Langvatn 1992). In many of these systems, the large mammalian herbivores become numerous and develop seasonal migrations.

3.2.4   Vegetation-dominated Models

Vegetation-dominated models predict that soil nutrients are determined by plant biomass and that herbivores are limited by food supply. Such models would apply to forests where trees dominate the biomass. In particular, tropical forests absorb most of the nutrients in the soil, leaving the soil low in nutrients (Connell 1978). At the same time, herbivores in these forests, largely insects, consume only a small fraction of the vegetation because plants are protected by structural and chemical defenses (Reichle et al. 1973, Sinclair 1975). Similar processes occur in temperate forests (Feeny 1968).

One model (13) has both vegetation and herbivores as codominants. This would predict high vegetation biomass, but its structure would be strongly influenced by herbivores because of its high palatability. Such special conditions are not obviously identified in nature, but may represent subarctic and alpine systems where vegetation and herbivores dominate (Oksanen et al. 1981, Batzli 1983, Keith 1983, Fretwell 1987, L. Oksanen 1990, Moen et al. 1993, Marquis and Whelan 1994, Krebs et al. 1995).

3.2.5   Dilution Models

Dilution models imply that both the lower and higher levels limit the intermediate trophic level. Thus, there are both bottom-up and top-down effects (figure 3.2c). Such a process may be seen in aquatic systems, where both nutrient content of lakes and the predator community affect trophic dynamics (McQueen et al. 1986, 1989). Insect-parasitoid–dominated systems may also be dilution systems, where parasitoids control herbivorous insects and nutrients control vegetation, which in turn affects herbivores (Lawton and Strong 1981, Strong et al. 1984, Price et al. 1990, Gomez and Zamora 1994).

3.2.6 Reciprocal Models

Reciprocal models suggest that there are two-way interactions between most of the trophic levels. The last model (27; figure 3.2b) predicts reciprocal effects at all levels. These models, therefore, could apply to most ecosystems.

3.3 Experimental Perturbations in the Boreal Forest

We tested the models described above using experimental perturbations of the boreal forest food web. The models can be distinguished by unique sets of predictions produced from seven perturbations that systematically reduced or enhanced each trophic level. The predictions were concerned with the subsequent direction of change in the populations or biomass of other levels. The main components of the system are described elsewhere, but in essence the plant level is characterized by herbaceous dicots, woody shrubs such as willow and birch, and white spruce. The herbivore level is dominated by the snowshoe hare, which exhibits a 10-year cycle of abundance, and by ground squirrels, red squirrels, and various vole species. The predator level is composed of carnivores such as lynx and coyote and various raptors, notably the great horned owl.

Predictions from the models on the changes in biomass are indicated in table 3.3 as an increase, decrease, or no change. However, where two perturbations are applied simultaneously, both acting in the same direction, then an additional prediction can be made on the magnitude of change relative to either single perturbation. The seven experiments are explained below. Details of how they were set up are outlined in chapter 4.

3.3.1 Experiment 1: Application of Fertilizer

Fertilizer was applied from the air to two 1-km^2 blocks of forest. This should increase the soil nutrient pool (N). Models that suggest that plants are responsive to the nutrient pool (i.e., those with a double arrow between N and V) predict increases in biomass of higher trophic levels to varying lengths of the food chain. Models with a left arrow predict that plants are not limited by nutrients and should not respond to fertilizer inputs.

3.3.2 Experiment 2: Addition of Rabbit Chow

Commercial rabbit chow was applied ad libitum to two areas. This food is eaten readily by hares and ground squirrels. The treatment has the effect of artificially increasing the food supply for herbivores independently of the natural food. Models that propose herbivores respond to food supply predict an increase in herbivores and a subsequent decrease in natural food. Some of these models also predict that predators should increase. The remaining models assume that herbivores are not limited by food supply and so do not respond to food addition.

Table 3.3 Predictions of the direction of change in biomass at each of the trophic levels from each of the seven experimental treatments according to the 27 models.

Model	Exp. 1				Exp. 2				Exp. 3			Exp. 4			Exp. 5		Exp. 6		Exp. 7
	N	V	H	P	N	V	H	P	N	V	H	N	V	H	N	V	N	V	N
1	+	+	+	+	0	0	+	+	0	0	0	0	0	+	0	0	+	+	0
2	+	+	+	+	0	0	+	+	0	0	0	0	0	+	0	0	?	+	+
3	+	+	+	+	0	−	+	+	0	0	0	0	−	+	0	+	+	++	0
4	+	+	+	+	0	0	+	+	+	0	+	0	0	++	0	0	?	+	0
5	+	0	0	0	0	0	0	0	+	−	+	+	−	+	−	+	?	+	+
6	+	+	0	0	+	−	+	0	+	−	+	++	−	++	−	+	?	++	+
7	+	0	0	0	0	0	0	0	+	−	+	+	−	+	−	+	?	+	+
8	+	0	0	0	0	0	0	0	0	0	0	0	0	0	0	+	?	+	+
9	+	0	0	0	0	0	0	0	0	0	0	0	0	0	−	+	?	+	0
10	+	+	0	0	0	0	0	0	0	−	+	0	−	+	0	+	?	++	+
11	+	+	0	0	0	0	0	0	0	0	0	0	0	+	0	+	?	++	0
12	+	+	0	0	0	−	0	0	0	0	0	0	−	+	−	+	+	++	+
13	+	0	0	0	+	0	+	+	0	0	+	+	−	+	0	0	?	0	+
14	+	0	0	0	0	0	+	0	0	0	+	0	0	++	0	0	?	0	+
15	+	0	0	0	0	0	+	0	0	0	+	0	0	++	0	0	+	+	+
16	+	+	+	0	0	0	+	+	0	0	+	0	0	++	0	0	?	0	+
17	+	0	0	0	0	0	+	0	0	0	0	0	0	+	0	0	+	+	0
18	+	+	+	0	0	0	0	0	0	−	+	0	−	+	0	0	+	+	+
19	+	+	+	0	0	0	+	0	0	−	0	+	−	++	0	+	?	+	+
20	+	+	+	0	0	−	+	0	0	0	0	0	0	+	0	+	+	++	0
21	+	+	+	+	+	−	+	+	0	0	0	+	−	++	−	+	?	+++	0
22	+	+	+	0	+	−	+	0	+	−	+	++	−	++	−	+	?	++	+

(continued)

Table 3.3 (Continued)

Model	Exp. 1				Exp. 2				Exp. 3			Exp. 4			Exp. 5		Exp. 6		Exp. 7
	N	V	H	P	N	V	H	P	N	V	H	N	V	H	N	V	N	V	N
23	+	+	+	+	0	−	+	+	0	−	+	0	−−	++	0	+	+	++	0
24	+	0	0	0	+	−	+	+	+	−	+	++	−−	++	−	+	?	+	+
25	+	+	+	+	0	0	+	+	0	0	+	0	0	++	0	0	?	+	+
26	+	+	0	0	0	0	0	0	+	−	+	+	−	+	−	+	?	++	+
27	+	+	+	+	+	−	+	+	+	−	+	++	−−	++	−	+	?	++	+

N = nutrient pool; V = vegetation biomass; H = herbivore biomass; P = predator biomass.
Experiment 1 = fertilizer addition, Exp. 2 = food addition, Exp. 3 = predator exclosure, Exp. 4 = predator exclosure + food addition, Exp. 5 = hare exclosure, Exp. 6 = hare exclosure + fertilizer, Exp. 7 = vegetation exclosure.
+ = biomass increase, − = biomass decrease, ++ or −− = double effect on biomass of two treatments, 0 = no change, and ? = unpredictable. Because the predator fence eliminates mammalian predators, no predictions are made for that level in the fence treatments. Similarly, the hare exclosures preclude predictions for the herbivore and predator levels, and vegetation removal precludes predictions for all three higher levels.

3.3.3   Experiment 3: Exclusion of Carnivores

Carnivores were excluded from a 1-km^2 area by wire fencing. This fence is permeable to hares and squirrels through small holes in the fence. We also covered a small part of this area (10 ha) with an overhead monofilament screen. This had the effect of partially deterring great horned owls from depredating animals below the screen. For our purposes the fence reduced the predator trophic level and allowed us to see how the effects traveled along the food chain.

3.3.4   Experiment 4: Exclusion of Carnivores
and the Addition of Rabbit Chow

A similar 1-km^2 area was used for experiment 4. Carnivores were excluded and ad libitum rabbit chow provided. The removal of predators is predicted to increase herbivores. In addition, for those models where herbivores are responsive to their food supply, we predict an additional increase in herbivore biomass relative to that change from experiment 3. The remaining models predict a change in herbivores similar to that in experiment 3.

3.3.5   Experiment 5: Exclusion of Hares

Hares are the dominant herbivore biomass in the system, and they were excluded by fencing a 4-ha area. This has the effect of removing the top two trophic levels. The predictions are that vegetation biomass should increase and soil nutrients decrease.

3.3.6   Experiment 6: Exclusion of Hares
and the Addition of Fertilizer

This experiment is similar to experiment 5 but with the addition of fertilizer as in experiment 1. The two perturbations are predicted to have opposite effects on soil nutrients: fertilizer adds nutrients, but removal of hares results in less nutrients. Because the outcome could be any value depending on absolute amounts of inputs and outputs, no qualitative prediction for soil nutrients can be made. However, for vegetation biomass one can predict two alternative responses. In models that assume that plants respond to nutrient levels, there should be a greater increase in plant biomass relative to experiment 5. In contrast, the other models predict no difference in plant biomass changes between experiments 5 and 6.

3.3.7   Experiment 7: Removal of Vegetation

In ten 1-m^2 plots on each of five sites, vegetation was killed by herbicide and left in situ. The perimeter was cut with a spade so as to kill roots from outside the plot. Soil nutrient levels were measured relative to an equal number of immediately adjacent control plots with intact vegetation. Soil nutrients should increase in plots where the vegetation has been killed.

In summary, 21 models make unique sets of predictions for these experiments. There are also three pairs of unique sets (5 and 8, 12 and 19, and 6 and 26). The three pairs of models with the same predictions occur because one perturbation is missing from our boreal forest experiments (for logistical reasons): the direct addition of herbivores. If this experiment had been achieved, then all these models could have been discriminated.

3.4   Discussion

3.4.1   Limitations of the Models

We employed both removal and addition experiments. Removal experiments are more powerful because, to analyze the effect of a factor, it must be removed. Addition experiments are not the reverse of removals (Royama 1977), but they are necessary in this context to determine the presence of right-arrow regulatory effects. The double perturbation experiments where both addition and removal were applied allow us to evaluate the relative strengths of simultaneous top-down and bottom-up effects.

In general, the most informative experiments are those that influence higher rather than lower trophic levels. This is because perturbations at lower levels do not necessarily feed up—only some models predict this result. In contrast, perturbations on higher levels should always feed down to lower levels.

Our assumptions of the effect of food addition may be incorrect in two ways. One possibility is that rabbit chow could act more as an attractant, causing hares and other species to commute to the food from outside the experimental area. Such commuting would have the effect of increasing the herbivore trophic level directly through a behavioral response instead of through a reproductive response to food. A second possibility is that food addition would attract herbivores to eat more of the artificial food and so would decrease the impact of herbivores on the vegetation. In both cases predictions can be altered accordingly, and the number of unique sets does not alter.

We started this analysis of models and their predictions by assuming simple linear, first order interactions. Many other models, including nonlinear interactions, such as saturating functional responses, could also be suggested, and this would be the next step. We consider that these more complex models would lead to basically similar predictions. We started with experimental tests of simple models because they are easier to reject than complex models. We should first test simple models with our results before turning to complex models (Hairston and Hairston 1997). Density manipulation experiments in these different ecosystems is the most effective way of determining the generality of these top-down and bottom-up models (Dwyer 1995).

3.4.2   Models of Trophic-level Interactions in Different Ecosystems

Although we cannot yet assign individual models to particular ecosystems, we can speculate on the classes of models that may be applicable in different ecosystems. Nutrient-poor sclerophyll forests such as those of Australia have strong bottom-up effects (Braithwaite et al. 1983, Braithwaite 1996) and would be largely represented by right-arrow models such as 1–4, 16, 20–23, and 25. Tropical forests, where vegetation dominates (Connell 1978, 1983), could be represented by models 14–17, 21–22, and 24–25.

In subarctic tundra systems, where vegetation and herbivores dominate (Oksanen et al. 1981, Fretwell 1987, L. Oksanen 1990, Moen et al. 1993, Marquis and Whelan 1994, Krebs et al. 1995), models 9–17 could apply. As one moves to lower latitudes, such as in the temperate grasslands of the North American prairies (Huntly 1991) and in the tropical savannas of the Serengeti (Sinclair 1975, McNaughton 1985, Sinclair and Arcese 1995), herbivores dominate, and models with arrows leading from the herbivores could apply (e.g., models 9–13, 21, 23–24, and 26).

Both bottom-up and top-down (dilution) effects at the herbivore level could operate in insect herbivores (models 15–18, and 20) (Hairston et al. 1960, McQueen et al. 1986, Pace and Funke 1991, Harrison and Cappuccino 1995). However, insect-dominated systems may also be controlled by predators or parasitoids (Lawton and Strong 1981, Strong et al. 1984, Price et al. 1990, Gomez and Zamora 1994, Spiller and Schoener 1994, Floyd 1996, Moran and Hurd 1998), and models 5–8, 22, 24, and 26 might apply.

In aquatic ecosystems, models with mainly left arrows (largely top-down effects; e.g., 5–8, 22, 24, and 26) may apply (Menge and Sutherland 1976, Estes et al. 1977, Simenstad et al. 1978, Paine 1980, McQueen et al. 1986, Power 1990, Menge 1992, 1995, Carpenter and Kitchell 1993, Rosemund et al. 1993, Wootton 1994a). Indirect effects are well known in aquatic and marine systems (Schoener 1993, Wootton 1993, 1994b, Menge 1997) but far less evident in terrestrial systems. The most general of the models is 27, the pure reciprocal model, because it could represent many if not all ecosystems.

3.5 Summary

Models of community organization involve variations of the top-down (predator control) or bottom-up (nutrient limitation) hypotheses. Verbal models, however, can be interpreted in different ways, leading to confusion. Therefore, we predict from first principles the range of possible trophic-level interactions and define mathematically the instantaneous effects of experimental perturbations. Some of these interactions are logically and biologically unfeasible. The remaining set of 27 feasible models is based on an initial assumption, for simplicity, of linear interactions between trophic levels. Many more complex and nonlinear models are logically feasible but, for parsimony, simple ones are tested first.

We have described a series of seven experiments designed to test the predictions of the models and distinguish between them. These experiments were conducted on the vertebrate community in the boreal forest at Kluane, Yukon. With these experiments we can distinguish 21 models and 3 pairs of models, giving 24 unique sets of predictions. In chapter 17 we apply these models to the results discussed in detail in the intervening chapters of this book.

Literature Cited

Abrams, P. A. 1993. Effect of increased productivity on the abundances of trophic levels. American Naturalist **141**:351–371.
Albon, S. D., and R. Langvatn. 1992. Plant phenology and the benefits of migration in a temperate ungulate. Oikos **65**:502–513.

Arditi, R., and L. R. Ginsburg. 1989. Coupling in predator-prey dynamics: ratio-dependence. Journal of Theoretical Biology **139**:311–326.

Arditi, R., L. R. Ginsburg, and H. R. Akcakaya. 1991. Variation in plankton densities among lakes: a case for ratio-dependent predation models. American Naturalist **138**:1287–1296.

Batzli, G. O. 1983. Responses of arctic rodent to nutritional factors. Oikos **40**:396–406.

Bell, R.H.V. 1982. The effect of soil nutrient availability on community structure in African ecosystems. *in* B. J. Huntley and B. H. Walker (eds). Ecology of tropical savannas, pages 193–216. Springer-Verlag, New York.

Botkin, D. B., J. M. Mellilo, and L. S-Y. Wu. 1981. How ecosystem processes are linked to large mammal population dynamics. *in* C. W. Fowler and T. D. Smith (eds). Dynamics of large mammal populations, pages 373–387. Smith, John Wiley and Sons, New York.

Braithwaite, L. W. 1996. Conservation of arboreal herbivores: the Australian scene. Australian Journal of Ecology **21**:21–30.

Braithwaite, L. W., M. L. Dudzinski, and J. Turner. 1983. Studies of the arboreal eucalypt forests being harvested for woodpulp at Eden, New South Wales. II. Relationship between the fauna density, richness and diversity and measured variables of habitat. Australian Wildlife Research **10**:231–247.

Briand, F., and J. E. Cohen. 1987. Environmental correlates of food chain length. Science **238**:956–960.

Carpenter, S. R., and J. F. Kitchell. 1987. The temporal scale of variance in limnetic primary production. American Naturalist **129**:417–433.

Carpenter, S. R., and J. F. Kitchell. 1988. Consumer control of lake productivity. Bioscience **38**:764–769.

Carpenter, S. R., and J. F. Kitchell (eds). 1993. The trophic cascade in lakes. Cambridge University Press, Cambridge.

Carpenter, S. R., J. F. Kitchell, and J. R. Hodgson. 1985. Cascading trophic interactions and lake productivity. BioScience **35**:634–639.

Caughley, G. 1976. Plant-herbivore systems. *in* R. M. May (ed). Theological history, pages 94–113. Saunders, Philadelphia.

Coe, M. J., D. H. Cumming, and J. Phillipson. 1976. Biomass and production of large African herbivores in relation to rainfall and primary production. Oecologia **22**:314–354.

Cohen, J. E. 1978. Food webs and niche space. Princeton University Press, Princeton, New Jersey.

Connell, J. H. 1978. Diversity in tropical rainforests and coral reefs. Science **199**:1302–1310.

Connell, J. H. 1983. On the prevalence and relative importance of interspecific competition: evidence from field experiments. American Naturalist **122**:661–696.

Dwyer, G. 1995. Simple models and complex interactions. *in* N. Cappuccino and P. W. Price (eds). Population dynamics: new approaches and synthesis, pages 209–227. Academic Press, New York.

Estes, J. A., and D. O. Duggins. 1995. Sea otters and kelp forests in Alaska: generality and variation in a community ecological paradigm. Ecological Monographs **65**:75–100.

Estes, J. A., D. O. Duggins, and G. B. Rathbun. 1989. The ecology of extinctions in kelp forest communities. Conservation Biology **3**:252–264.

Estes, J. A., N. S. Smith, and J. F. Palmisano. 1977. Sea otter predation and community organization in the Western Aleutian Islands, Alaska. Ecology **59**:822–833.

Feeny, P. P. 1968. Effect of oak leaf tannins on larval growth of the winter moth *Operophtera brumata*. Journal of Insect Physiology **14**:801–817.

Floyd, T. 1996. Top-down impacts on creosotebush herbivores in a spatially and temporally complex environment. Ecology **77**:1544–1555.

Fretwell, S. D. 1977. The regulation of plant communities by food chains exploiting them. Perspectives in Biology and Medicine **20**:169–185.

Fretwell, S. D. 1987. Food chain dynamics: the central theory of ecology? Oikos **50**:291–301.

Gomez, J. M., and R. Zamora. 1994. Top-down effects in a tritrophic system: parasitoids enhance plant fitness. Ecology **75**:1023–1030.

Hairston, N. G., and N. G. Hairston. 1997. Does food web complexity eliminate trophic-level dynamics? American Naturalist **149**:1001–1007.

Hairston, N. G., F. E. Smith, and L. B. Slobodkin. 1960. Community structure, population control and competition. American Naturalist **94**:421–425.

Hall, S. J., and D. Raffaelli. 1991. Food-web patterns: lessons from a species-rich web. Journal of Animal Ecology **60**:823–842.

Hanski, I. 1991. The functional response of predators: worries about scale. Trends in Ecology and Evolution **6**:141–142.

Harrison, S., and N. Cappuccino. 1995. Using density-manipulation experiments to study population regulation. *In* N. Cappuccino and P. W. Price (eds). Population dynamics: new approaches and synthesis, pages 131–148. Academic Press, New York.

Hawkins, B. A. 1992. Parasitoid-host food web and donor control. Oikos **65**:159–162.

Houston, D. B. 1982. The northern Yellowstone elk: ecology and management. MacMillan, London.

Hunter, M. D., and P. W. Price. 1992. Playing chutes and ladders: bottom-up and top-down forces in natural communities. Ecology **73**:724–732.

Huntly, N. 1991. Herbivores and the dynamics of communities and ecosytems. Annual Review of Ecology and Systematics **22**:477–503.

Keiter, R. B., and M. S. Boyce. 1991. The greater Yellowstone ecosystem: redefining America's wilderness heritage. Yale University Press, New Haven, Connecticut.

Keith, L. B. 1983. Role of food in hare population cycles. Oikos **40**:385–395.

Krebs, C. J., S. Boutin, R. Boonstra, A. R. E. Sinclair, J. N. M. Smith, M. R. T. Dale, K. Martin, and R. Turkington. 1995. Impact of food and predation on the snowshoe hare cycle. Science **269**:1112–1115.

Laurenson, M. K. 1995. Implications for high offspring mortality for cheetah population dynamics. *in* A.R.E. Sinclair and P. Arcese (eds). Serengeti II: dynamics, management and conservation of an ecosystem, pages 385–399. University of Chicago Press, Chicago.

Lawton, J. H. 1989. Food webs. *in* J. M. Cherrett (ed). Ecological concepts, pages 43–78. Blackwell Scientific Publications, Oxford.

Lawton, J. H., and D. R. Strong. 1981. Community patterns and competition in folivarous insects. American Naturalist **118**:317–338.

Leader-Williams, N. 1988. Reindeer on South Georgia: the study of an introduced population. Cambridge University Press, Cambridge.

Marquis, R. J., and Whelan, C. J. 1994. Insectivorious birds increase growth of white oak through consumption of leaf-chewing insects. Ecology **75**:2007–2014.

McLaren, B. E., and R. O. Peterson. 1994. Wolves, moose and tree rings on Isle Royale. Science **266**:1555–1557.

McNaughton, S. J. 1985. Ecology of a grazing ecosystem: the Serengeti. Ecological Monographs **55**:259–294.

McNaughton, S. J., M. Osterheld, D. A. Frank, and K. J. Williams. 1989. Ecosystem level patterns of primary productivity and herbivory in terrestrial habitats. Nature **341**:142–144.

McQueen, D. G., M. R. S. Johannes, J. R. Post, T. J. Stewart, and D. R. S. Lean. 1989. Bottom-up and top-down impacts on freshwater pelagic community structure. Ecological Monographs **59**:289–309.

McQueen, D. G., J. R. Post, and E. L. Mills. 1986. Trophic relationships in freshwater pelagic ecosystems. Canadian Journal of Fisheries and Aquatic Sciences **43**:1571–1581.

Mduma, S. A. R., A. R. E. Sinclair, and R. Hilborn. 1999. Food regulates the Serengeti wildebeest: a forty-year record. Journal of Animal Ecology **68**:1101–1122.

Menge, B. A. 1992. Community regulation: under what conditions are bottom-up factors important on rocky shores? Ecology **73**:755–765.

Menge, B. A. 1995. Indirect effects in marine rocky intertidal interaction webs: patterns and importance. Ecological Monographs **65**:21–74.

Menge, B. A. 1997. Detection of direct versus indirect effects: were experiments long enough? American Naturalist **149**:801–823.

Menge, B. A., and T. M. Farrell. 1989. Community structure and interaction webs in shallow marine hard-bottom communities: tests of an environmental stress model. Advances in Ecological Research **19**:189–262.

Menge, B. A., and A. M. Olsen. 1990. Role of scale and environmental factors in regulation of community structure. Trends in Ecology and Evolution **5**:52–57.

Menge, B. A., and J. P. Sutherland. 1976. Species diversity gradients: synthesis of the roles of predation, competition and temporal heterogeneity. American Naturalist **110**:351–369.

Merrill, E. H., N. L. Stanton, and J. C. Hik. 1994. Responses of bluebunch wheatgrass, Idaho fescue, and nematodes to ungulate grazing in Yellowstone National Park. Oikos **69**:231–240.

Moen, J., H. Gardfjell, L. Oksanen, L. Ericson, and P. Ekerholm. 1993. Grazing by food-limited microtine rodents on a productive experimental plant community: does the "green desert" exist? Oikos **68**:401–413.

Moran, M. D., and L. E. Hurd. 1998. A trophic cascade in a diverse arthropod community caused by a generalist arthropod predator. Oecologia **113**:126–132.

Oksanen, L. 1988. Ecosystem organization: mutualism and cybernetics or plain Darwinian struggle for existence? American Naturalist **131**:424–444.

Oksanen, L. 1990. Predation, herbivory, and plant strategies along gradients of primary productivity. *in* D. Tilman and J. Grace (eds). Perspectives on plant consumption, pages 445–474. Academic Press, New York.

Oksanen, L., and L. Ericson. 1987. Concluding remarks: trophic exploitation and community structure. Oikos **50**:417–422.

Oksanen, L., S. D. Fretwell, J. Arruda, and P. Niemala. 1981. Exploitation ecosystems in gradients of primary productivity. American Naturalist **118**:240–261.

Oksanen, T. 1990. Exploitation ecosystems in heterogeneous habitat complexes. Evolutionary Ecology **4**:220–234.

Pace, M. L., and E. Funke. 1991. Regulation of planktonic microbial communities by nutrients and herbivores. Ecology **72**:904–914.

Paine, R. T. 1980. Food webs: linkage, interaction strength and community infrastructure. Journal of Animal Ecology **49**:667–685.

Paine, R. T. 1988. Food webs: road maps of interactions or grist for theoretical development? Ecology **69**:1648–1654.

Pimm, S. L. 1982. Food webs. Chapman and Hall, London.

Polis, G. A. 1991. Complex trophic interactions in deserts: an empirical critique of food web theory. American Naturalist **138**:123–155.

Power, M. E. 1984. Depth distributions of armoured catfish: predator-induced resource avoidance? Ecology **65**:523–528.

Power, M. E. 1990. Effects of fish in river food webs. Science **250**:811–814.

Price, P. W., N. Cobb, T. P. Craig, G. W. Fernandes, J. K. Itami, S. Mopper, and R. W. Preszler. 1990. Insect herbivore population dynamics on trees and shrubs: new approaches relevant to latent and eruptive species and life table development. *in* E. A. Bernays (ed). Insect-plant interactions, volume 2, pages 1–38. CRC Press, Boca Raton, Florida.

Reichle, D. E., R. A. Goldstein, R. I. Van Hook, and G. J. Dodson. 1973. Analysis of insect consumption in a forest canopy. Ecology **54**:1076–1084.

Rosemund, A. D., P. J. Mulholland, and J. W. Elwood. 1993. Top-down and bottom-up control of stream periphyton: effects of nutrients and herbivores. Ecology **74**:1264–1280.

Royama, T. 1977. Population persistence and density dependence. Ecological Monographs **47**:1–35.

Schmitz, O. J. 1998. Direct and indirect effects of predation and predation risk in old-field interaction webs. American Naturalist **151**:327–342.

Schoener, T. 1989. Food webs from the small to the large. Ecology **70**:1559–1589.

Schoener, T. W. 1993. On the relative importance of direct versus indirect effects in ecological communities. *in* H. Kawanabe, J. E. Cohen, and K. Iwasaki (eds). Mutualism and community organization: behavioral, theoretical and food-web approaches, pages 365–411. Oxford University Press, New York.

Schoenly, K., R. A. Beaver, and T. A. Heumier. 1991. On the trophic relations of insects: a food-web approach. American Naturalist **137**:597–638.

Semmartin, M., and M. Oesterheld. 1996. Effect of grazing pattern on primary productivity. Oikos **75**:431–436.

Simenstad, C. A., J. A. Estes, and K. W. Kenyon. 1978. Aleuts, sea otters and alternate stable state communities. Science **200**:403–411.

Sinclair, A. R. E. 1975. The resource limitation of trophic levels in tropical grassland communities. Journal of Animal Ecology **44**:497–520.

Sinclair, A. R. E. 1995. Serengeti past and present. *in* A. R. E. Sinclair and P. Arcese (eds). Serengeti II: Dynamics, management and conservation of an ecosystem, pages 1–30. University of Chicago Press, Chicago.

Sinclair, A. R. E., and P. Arcese (eds). 1995. Serengeti II: dynamics, management and conservation of an ecosystem. University of Chicago Press, Chicago.

Sinclair, A. R. E., C. J. Krebs, J. M. Fryxel, R. Turkington, S. Boutin, R. Boonstra, P. Seccombe-Hett, P. Lundberg, and L. Oksanen. 2000. Testing hypotheses of trophic level interactions: a boreal forest ecosystem. Oikos **89**:313–328.

Singer, F. J., L. C. Mark, and R. C. Cates. 1994. Ungulate herbivory of willows on Yellowstone's northern winter range. Journal of Range Management **47**:435–443.

Slobodkin, L. B., F. E. Smith, and N. G. Hairston. 1967. Regulation in terrestrial ecosystems, and the implied balance of nature. American Naturalist **101**:109–124.

Spiller, D. A., and T. W. Schoener. 1994. Effects of top and intermediate predators in a terrestrial food web. Ecology **75**:182–196.

Strong, D. R. 1992. Are trophic cascades all wet? Differentiation and donor-control in speciose ecosystems. Ecology **73**:747–754.

Strong, D. R., D. Simberloff, L. G. Abele, and A. B. Thistle (eds). 1984. Ecological communities: conceptual issues and the evidence. Princeton University Press, Princeton, New Jersey.

Walker, B. H. 1991. Biological diversity and ecological redundancy. Conservation Biology **6**:18–23.

White, T. C. R. 1978. The importance of a relative shortage of food in animal ecology. Oecologia **3**:71–86.

White, T. C. R. 1984. The abundance of invertebrate herbivores in relation to the availability of nitrogen in stressed food plants. Oecologia **63**:90–105.

Wootton, J. T. 1993. Indirect effects and habitat use in an intertidal community: interaction chains and interaction modifications. American Naturalist **141**:71–89.

Wootton, J. T. 1994a. Predicting direct and indirect effects: an integrated approach using experiments and path analysis. Ecology **75**:151–165.

Wootton, J. T. 1994b. The nature and consequences of indirect effects in ecological communities. Annual Review of Ecology and Systematics **25**:443–466.

Yodzis, P. 1988. The indeterminacy of ecological interactions as perceived through perturbation experiments. Ecology **69**:508–515.

4

Experimental Design and Practical Problems of Implementation

STAN BOUTIN, CHARLES J. KREBS, VILIS O. NAMS, A. R. E. SINCLAIR, RUDY BOONSTRA, MARK O'DONOGHUE, & CATHY DOYLE

In this chapter we summarize the experimental design and explore the reasons for particular choices among competing designs. For any given experimental design, the difficulty is in the details of implementing the design, and here we present the practical issues we faced. Part of this discussion may be useful to anyone contemplating a long-term, multidisciplinary ecological field study, but the thrust of this chapter is to discuss the experiments we did at Kluane Lake from 1986 to 1996, why we did them, and how.

4.1   Experimental Design

The experimental design is the Achilles heel of all field ecological research, and it always involves compromise because many ecological factors can impinge on any population or community and one cannot study all of them. Three choices dictate the experimental design. One must first decide on the treatments to be studied. Given the treatments, decisions about the size of the treatment unit and the spatial interspersion of the treatments are next. And finally, one must decide on the level of replication of the treatments. All of these issues have been discussed extensively in the statistical literature (e.g., Fisher 1966, Sokal and Rohlf 1995), and we apply these general statistical principles to the particular question of understanding how the boreal forest community functions.

4.1.1   Choice of Treatments

To understand how a community is structured and how it functions, a food web must be constructed (see figure 2.8). Two choices then must be made. Over a series of years a detailed description of the linkages in the food web could be constructed in the hope that natural variation will be large enough to infer the structure behind the linkages. In the boreal forest, the 10-year cycle of snowshoe hares is a well-known natural perturbation, and consequently this strategy could be used with some effect. Alternatively, you can perturb the system to see how it responds to the manipulations. The history of ecology is littered with decisions of this critical juncture, and it is abundantly clear that the second strategy maximizes our rate of progress toward understanding how communities work (Paine 1984, Krebs 1991, Menge 1995). We note that it is possible to balance these two strategies because it is critical to have a good description of what happens on control areas as well as on the manipulated sites.

In theory any of a wide variety of community variables can be manipulated, and it is not immediately clear what is best for achieving understanding. Most ecologists gravitate toward food supply and predation as two key resources in community organization, and these are the two main variables we chose to manipulate. But we could have manipulated parasites or diseases, temperature or precipitation, soil chemistry, or fire. The crux of this decision is the empirical test: how much understanding can we achieve with the choice of a few variables? At the end of this book we return to this question, but in principle ecology will progress by natural selection favoring those kinds of manipulations that increase our understanding and rejecting those kinds that shed little light on how a community is organized.

If one chooses to manipulate food supplies and predation levels, there is still the decision whether to direct treatments toward the entire trophic level or to concentrate on the keystone species, which in the boreal forest is the snowshoe hare. In principle for the reasons discussed in chapter 3, it is best to manipulate on a broad scale and direct treatments toward the entire trophic level. It is practical in a vertebrate community to manipulate the plant and the herbivore trophic levels, but there are severe limitations on how much vertebrate predators can be manipulated. In theory trophic levels can be manipulated up or down, but, again, practical constraints intervene to direct manipulations at increasing plants and herbivores and decreasing predators. One could ask how herbivores would respond to cutting their food resources in half, and this may be an interesting experiment but impossible to actually do in the boreal forest. Similarly, one could ask how herbivores would respond to a doubling of their predators, but this would be technically a most difficult experiment to carry out. Because great horned owls occupy territories covering all of the forest every year (see chapter 15), releasing owls into this ecosystem would not add to the predator population. We were constrained by these kinds of considerations to increasing food resources and reducing predator impacts in this ecosystem.

We selected food addition based on our previous experience that this treatment increased snowshoe hare densities two- to three-fold (Sinclair et al. 1988, Smith et al. 1988). Consequently, this manipulation was chosen as the mechanism of increasing the size of the herbivore trophic level. This was the natural complement to the fertilization and predator exclusion experiments, which manipulated the plant and the predator trophic levels. We knew when we began in 1986 that rabbit chow was eaten readily by ground squirrels, gray jays, moose, and grizzly bears but was eaten less by voles, mice, and red squirrels, who respond more to seeds (Gilbert and Krebs 1981, Boutin 1990, Klenner and Krebs 1991, Schweiger and Boutin 1995). Because we were constrained financially from adding both rabbit chow and seeds to areas, we chose to add rabbit chow only, knowing that it may not supplement the food resources of all the herbivores equally. Consequently, the first major manipulation was to add commercial rabbit chow year-round to selected field sites.

We knew less about how to increase the food resources to the plants. Adding fertilizer is the chosen mechanism in all of modern agriculture and intensive forestry, but the consequences of fertilization on all the component species of the boreal forest was not known when we began this study. We searched the literature and found that, although many forest fertilization experiments had been carried out, no one had looked at the entire plant community because trees were the usual target object. We chose a standard agricultural NPK fertilizer with the ratio 35:10:5 and with nitrogen in the form of ammonium nitrate. We chose a level of 17.5 g nitrogen/m^2 as a target level of addition from Alaskan studies on the response of trees to this level of nutrient addition (Haag 1974, Shaver and Chapin 1980, Van Cleve and Oliver 1982). In chapter 5 we explore on a small scale the impact of fertilization over a broad range of levels. Fortunately, our choice of level was nearly optimal for many species of plants in this part of the boreal forest.

Vertebrate predators such as lynx operate on a large spatial scale, and the only feasible manipulation is to reduce their numbers on target areas. There are two ways to do this. Either predators can be removed continuously from the target area or they can be prevented from entering with some type of barrier. We rejected the predator removal option, partly for aesthetic and conservation reasons (we did these studies at the edge of Kluane National Park) and partly because the logistics are more difficult than one might presume (not all animals enter traps, etc.). By creating a sink by predator removal, we would also face the possibility of continuous immigration of surplus predators from surrounding areas, potentially altering control populations. Predator exclosures are a better design. We discovered a new type of electric fencing made by Gallagher Fencing of New Zealand in 1986, and decided to attempt to fence mammalian predators out of target areas. In principle this was simple, and in the end it worked very well. The details of the problems we ran into are described in the next section.

Avian predators were more difficult to exclude, and a mechanical barrier was the only option available. We were unable to construct an effective barrier against birds of prey, for detailed reasons given below, and this was our largest failure to achieve the original experimental design. Consequently, we were unable to reduce avian predation rates on any of the target areas, and the predator exclusion areas must only be considered as mammalian predator exclusion areas.

We carried out a fourth large-scale experiment designed to examine the potential interaction between food and predation. We combined rabbit chow addition and electric fencing on one area. This experiment does not fit directly into the community trophic-level approach to the Kluane project, but it was pivotal to furthering our understanding of the snowshoe hare cycle. We had many hours of discussion about whether we should replicate the predator exclusion experiment or do this interactive experiment with reduced mammalian predation and increased food resources for herbivores.

Finally, we wanted to know what would happen to the vegetation of the boreal forest if the dominant herbivores, snowshoe hares, were excluded from an area. We constructed two hare exclosures, each 4 ha, to provide a small area with reduced browsing and grazing pressure. By eliminating hares from the area, we also had to eliminate moose browsing, but this was not a problem in general because moose were relatively rare in our system (see 2.6).

In addition to these imposed perturbations, we studied a system with large natural perturbations. For example, spruce seed crops vary over several orders of magnitude among years, and we had no control over when these occurred. Berry crops in the boreal forest vary dramatically from year to year. And, of course, there was the snowshoe hare cycle. These natural perturbations would propagate through our experimental perturbations and give us additional understanding of how the boreal forest community operates.

In summary, we had four available treatments designed to kick each trophic level: add supplemental food, eliminate mammalian predation, add nutrients as fertilizer, and exclude hare browsing. In addition to these experimental treatments, natural variations in cone crops, berry crops, and snowshoe hare abundance could impact populations under study.

4.1.2   Size and Spatial Location of Treatments

We have so far discussed the treatments imposed on target sites but have not discussed the size of the areas used and their spatial interspersion. Our choice was to make the experimental unit a 1-km^2 block of forest. We did this for two reasons, one ecological and one logistical. Most ecological manipulations have been done on small spatial scales, typically a few square meters (Tilman 1989). We wanted to increase the scale of manipulation closer to a landscape level. We also wanted to manipulate areas that had a large enough population of the main herbivores that we could detect significant changes in their populations. From previous experience we knew that herbivores were affected by experimen-

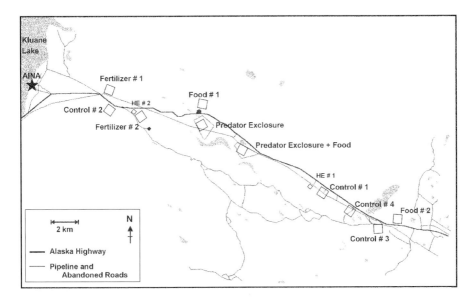

Figure 4.1 Map of the study area for the Kluane Boreal Forest Ecosystem Project. The Arctic Institute of North America Kluane Research Station is at the left edge of the map, on the shores of Kluane Lake.

tal treatments well beyond the boundaries of a trapping grid. We decided for hares that a 35-ha trapping area was a reasonable size for population estimates, and this grid size effectively trapped hares over a 60-ha area. We focused on snowshoe hares, arctic ground squirrels, and red squirrels and calculated that, on 1 km², populations would be sufficiently large that we could detect effects of ±20% change in the herbivore trophic level. This scale is a herbivore scale. Plant experiments can clearly be done at a smaller scale (chapter 5). Direct manipulation of predator numbers would require much larger areas and is currently impossible with a limited budget.

The spatial location of the treatments and their control areas was a compromise of interspersion of treatments and controls (as recommended by Hurlbert 1984) and the problem of access. Figure 4.1 shows the study area in the Shakwak Trench just southeast of Kluane Lake. Road access in this area is limited by the Alaska Highway, which transects the valley and sections of the old Alaska Highway as it was upgraded since World War II. Our decisions on the spatial locations of the treatments followed these four rules:

1. All 1-km² blocks must be predominantly open-spruce or closed spruce forest.
2. Every experimental unit must be at least 1 km from the nearest unit.
3. Fertilization plots must be close to the airstrip at the Arctic Institute Base to facilitate aerial fertilization.
4. Vehicle access must be feasible within at least 500 m of the area.

Most of these decisions are self-explanatory. The 1-km spacing rule arose from previous studies which suggested that snowshoe hares almost never moved as far as 1 km. This assumption has turned out to be wrong (see chapter 8), but it is true that few individuals move as far as 1 km. The greatest restriction on randomization was the necessity to put the fertilizer addition areas close to the only available airstrip (see figure 4.1).

We made other operational decisions that were not random. We separated the three areas that had food addition by at least 4 km so that individual predators would not have more than one of these areas within their home range. We had four control areas scattered throughout the valley. In the final analysis we used nearly all the habitat along the 30-km segment of the Alaska Highway that had suitable forest cover.

4.1.3 To Replicate or Not to Replicate

Given the treatments and the size of the experimental units, our final decision was how much replication to include in the project. We faced the classical statistical dilemma of whether to study many small areas or a few large units. Having decided to use large land units, we were forced by logistics to minimize replication. We began by replicating the control areas four times because we wanted a good description of what was occurring throughout the valley. We replicated the nutrient addition and the food supplementation treatments twice because this was the maximum we could afford to do on our budget. We could not replicate the predator exclusion areas (electric fences) for logistic reasons detailed below. We could not afford to build or keep up more than one electric fence per treatment. This is the Achilles heel of our study: we have only one replicate of the mammalian predator exclusion treatment and one replicate of the predator exclusion + supplemental food treatment (figure 4.1). Table 4.1 summarizes the treatments applied and the number of replicates of each.

Our defense for low replication is simply that we could not do more than this. This is a common problem for all large-scale ecological experiments (Carpenter et al. 1995). The reason for doing large-scale experiments is often to see how large of an effect is produced by a treatment. Interesting effects can then be searched more thoroughly in follow-up experiments. There is a series of statistical procedures that can be used to test for changes in time-series data. Randomized intervention analysis is one of the key methods that can be used (Carpenter et al. 1989). One key to interpreting the results of unreplicated experi-

Table 4.1 Experimental treatments used in the Kluane Boreal Forest Ecosystem Project, 1986–1996.

Treatment	No. of Replicates	Details
Food supplementation	2	Commercial rabbit chow, year-round ad lib, spread over 35-ha area
Mammalian predator exclosure	1	Electric fence, unable to effectively exclude birds of prey
Predator exclosure + food	1	Electric fence around 1 km^2 and rabbit chow ad lib spread over central 35-ha area
Nutrient addition	2	NPK fertilizer added in spring
Hare exclosure	1	4-ha exclosure, moose also excluded
Hare exclosure + nutrient addition	1	4-ha exclosure in fertilizer 2 area

All experimental areas are 1 km^2 except as noted.

ments is the amount of variation observed among replicated control populations, but a more important factor is the level of understanding of the natural history of the system under study. Terrestrial vertebrate communities are among the best known in the world for the linkages among the elements in the food web. We think this is more important for the understanding of our experiments than statistical p-values (Yoccoz 1990). For statistical purposes it is possible to generate p-values by assuming that the variance among controls would also apply to the variance among replicates of the fencing treatments, if we had been able to replicate these.

We were also encouraged by our earlier studies of this boreal forest community (Krebs et al. 1986, Sinclair et al. 1988, Smith et al. 1988), which showed great uniformity among snowshoe hare populations on study areas spaced >100 km apart in the Kluane region. This reflects the general synchrony that occurs in the hare cycle throughout western Canada and Alaska.

4.2   Practical Problems of Implementation

We discuss here some of the practical difficulties we had to overcome to get our treatments to work effectively in the Yukon boreal forest. These details illustrate the problems of trying to do large-scale field manipulations.

4.2.1   The Evolution of a Predator-proof Fence

High-tensile electric fencing was developed in New Zealand for deer farming, and when we began this study we assumed it would be highly effective against mammalian predators. The fence was energized by batteries with 8600 volts pulsed at 4 msec at 1-sec intervals. We began with a 1.8-m tall design with alternating positive and negative wires spaced at 25 cm. The first fence was built in the summer of 1987 at a site called Beaver Pond that had been trapped for hares during the 1976–1985 study. The construction was spearheaded by Tony Sinclair and a large crew of undergraduates and technicians and was completed by late August 1987. Fortunately, this was just after the low of the hare cycle, so predator numbers were minimal. During the winter of 1987–1988, we could snow track predators as they approached the electric fence. We found to our dismay that two lynx, one red fox, and one wolverine walked right through the fence while it was fully operating. Two problems had clearly arisen that we had not anticipated. First, the winter fur of these mammals seemed to insulate them from electric shock. Second, snow is a poor conductor of electricity, so when they did make contact with a live wire there was no ground to pass the current. Fortunately, the predators that moved into the fenced area left almost immediately.

We had a second problem in the winter of 1987–1988 because heavy snowfalls and drifting, along with some icing, shorted out the lower wires of the fence. We found we had to dig out the largest snowdrifts and developed a method of turning off the lower wires of the fence as they became snow covered. We also found we needed to add height to the fence to prevent animals from jumping over it when snow depth was greatest.

In the summer of 1988 we built a second electric fence to use for the predator exclosure + food addition treatment at a site called Hungry Lake. One side of this fence ran along an unused pipeline clearing, which facilitated construction. This fence was built to a new standard of 2.3 m height, and we added more wires to reduce the chances of animals squeezing between them.

Nevertheless, we found in autumn 1988, when the snow returned, that lynx were still walking through the electric fences. By this time hare numbers were increasing rapidly, along with predator numbers, so we faced an emergency situation. Stan Boutin developed a new design of fence by adding 5-cm chicken wire to the inside of the electric fence to serve both as a barrier but more importantly as a ground wire. We first put chicken wire up to 1 m height but then added it to 2 m height to cover deep snow years (figure 4.2). To prevent lynx from climbing over and coyotes from digging under the fence, we placed live wires 20 cm from the chicken wire at ground level, at 1 m, and at 2 m. Once the lower wires were covered by snow in winter, they were turned off. The chicken wire posed one further problem. The holes were not large enough to allow anything except voles to move freely through it. We did not want to confound our experiment with the "fence effect" (Krebs et al. 1969, Ostfeld 1994), so we cut holes in the wire large enough for hares to pass through (CD-ROM frames 62–64). The chicken-wire solution worked well and the problem was largely solved because once lynx received an electric shock, they learned immediately to avoid the fence and the predator-exclusion experiments were effective. The electric fences were completely effective by December 1988. Although the electric fences protected hares inside the fence from mammalian predators, hares could move through the fence, and this caused some problems with lynx in particular hunting hares along the fence and catching hares that moved outside of the electric fence.

We had originally intended to eliminate avian predation on a small subset (10 ha) of the predator exclosure. We began in summer 1987 to set out a physical barrier by stringing old gill nets in the trees of the target area (CD-ROM frame 65). Because so much gill netting is damaged each year in the Pacific salmon fishery, there was an unlimited supply of netting that could be had for nearly no cost. The work of stringing it in the spruce forest was formidable, and many students labored mightily to cover a few hectares with netting. To our dismay, it all collapsed under the weight of snow and hoarfrost in the first winter, 1987–1988. Our next attempt in 1988 was to replace the netting with monofilament fishing line strung out in parallel lines about 45 cm apart. We used both steel-reinforced and standard monofilament line and again found that over the winter hoarfrost built up on the lines and caused them to sag or break. We covered about 10 ha with fishing line on one site (predator exclosure treatment), but in spite of all the effort, we judged these mechanical barriers to be relatively ineffective in preventing avian predation on herbivores. Even where the netting or monofilament remained intact, we found that goshawks could walk under the net to kill hares and squirrels. Some reduction in predation occurred under the monofilament for ground squirrels (see chapter 9). We do not know of any simple way to keep avian predators out of large areas of forest and, in spite of much effort, we failed to manipulate this guild of predators in the way we had planned.

Moose were one problem that continually plagued the electric fence. Periodically, about every month or two, a moose would walk through the electric fence. We checked the fences every day to make sure they were operating correctly, so we would immediately find the breakage and repair it. Clearly the moose would receive a shock in doing this, because once inside, they would often not go near the electric fence to attempt to leave. The evolutionary history of moose had not preadapted them to understanding electric fences, and this problem was an intermittent headache of maintenance. The timing of

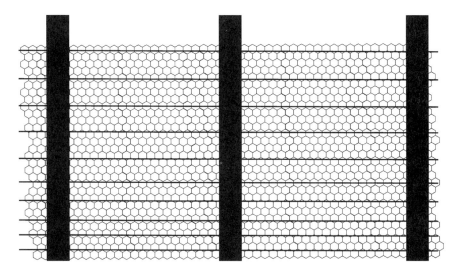

Figure 4.2   Schematic illustration of the electric fence for the mammalian predator exclosure experiments. The 2.2-m fence was lined with chicken wire, which acted as a ground for predators challenging the fence. Holes in the chicken wire would permit hares to come and go from the fenced area but were too small for predators to use. Two such fences were built around 1-km² forest blocks.

the moose entering the fence seemed impeccable. The Kluane Project was funded by the Natural Sciences and Engineering Research Council of Canada, and, as part of the review process, a committee visited our field sites in the summer of 1989. There was considerable skepticism among the committee that we could carry out these large-scale experiments, so of course we were concerned that everything was in working order. Frank Doyle did one last check of the fences on the morning of the review, and all was in order. The review committee was led to the fence, and Stan Boutin suggested that they might wish to test the fence themselves. The normal way to do this was with a voltmeter, but a simple test was to hold a piece of grass and slide it along the wire until one could feel the tingle of the electrical pulse. No one took Stan up on this offer, which was fortunate because the fence was completely dead. Not more than 100 m from where the group stood, a moose had just walked through the fence and shorted the wires. Fortunately, Jamie Smith saved the day by leading the group off to look at an interesting bird he had sighted. After the committee recommended that the project be continued, word of this episode leaked back to Jack Millar, the committee chairman, who claimed that the committee knew about the calamity all along! We owe a lot to the support of the members of the original review committee—Jack Millar, Lloyd Keith, and Dan Keppie—who showed foresight in giving us the chance to do this work.

A second problem with the electric fence was trees falling across it during storms. This occurred several times a year during the first 2 years, and again we would recognize the problem because the fence would short out.

Overall, the electric fences worked remarkably well once we overcame the initial problems, with only a few predators gaining temporary access. We had not, however, anticipated the amount of work involved in fence maintenance, which was a nearly full-time job in winter. Batteries for the fences had to be changed every 3 days in winter. Solar-powered electrical chargers are not an option in the Yukon winter. These maintenance costs, in addition to the high costs of the fencing itself, prevented us from building replicate fences for these two treatments in 1989.

4.2.2 Food Supplementation

Providing commercial rabbit chow ad libitum to three treatment areas also turned out not to be a simple job (CD-ROM frames 58–60). The problem arose because grizzly bears and moose also love to eat rabbit chow, and, while we did not mind feeding a moose or two, we did not want to conduct a grizzly feeding experiment. We carried out a feeding experiment from 1976–1984 (Smith et al. 1988) using commercial pheasant feeders and fed only in the winter when bears were hibernating. These simple feeders were attacked and destroyed on occasion by moose and bears, so that when we began this study we decided we needed a better method of feeding. One way to keep moose away is to build a corral or fence, and this works well when the fence is strong, but it will not deter bears. We decided to use old 45-gallon oil drums for feeders and to cable these to spruce trees so they could not be moved. Grizzly bears simply destroyed these drums and ripped them off the trees to get at the chow.

Rudy Boonstra suggested the next prototype, the use of road culverts. We used heavy steel road culverts 30 cm diameter and 4–5 m long and placed the rabbit chow in the middle of the culvert. Hares and squirrels could enter and leave easily to get access to the chow, but bears and moose could not. Again, we underestimated grizzly bears. They simply destroyed these culverts or turned them on end to gain access to the chow. We could find no way of rigidly fixing them to the ground so that bears could not rip them up, so we had to abandon culverts in the summer of 1987.

Finally, Stan Boutin, in what is probably the best practical use of optimal foraging theory, recognized that the solution was to broadcast feed. We began in 1988 to spread rabbit chow on four trails transecting the feeding grids (figure 4.3). Chow was spread in summer by a mechanical fertilizer spreader pulled on the back of an all-terrain vehicle (ATV), and in winter by snowmobile. The mechanical spreader sprayed chow over a strip of 4–5 m wide with an average of 2 pellets/10 cm^2 (or 25 g/m^2). Because the chow was spread out and not concentrated, bears and moose were more reluctant to take the time to feed on such small and dispersed items. We still had visits from both bears and moose on the two open feeding grids, but the threat of attack was considerably reduced. About once a summer a bear would tree one of our workers on the food areas, and the most difficult time occurred in 1995, when a particularly aggressive grizzly kept workers off food 1 grid for 4 weeks. On the Predator exclosure + food addition area, there were no problems feeding because bears and moose were typically kept out by the electric fence. From 1990 onward we also spread chow by hand under trees to prevent avian predators from attacking hares feeding along the snowmobile trails in the open in winter.

The frequency of feeding was partly determined by hare numbers and the rate of disappearance of chow, with some constraints from weather. In general we fed every 3–4 days when hare numbers were high and once a week when numbers were low. The amount

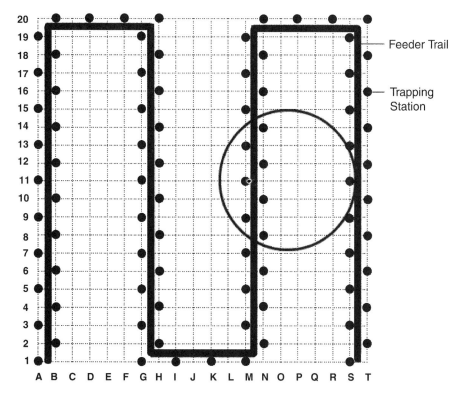

Figure 4.3 Schematic illustration of the method of adding supplemental food to treatment areas. A snowshoe hare trapping grid (20 × 20 stakes) was provided with supplemental rabbit chow in four strips as illustrated. A fertilizer spreader in summer spread the chow over a 4-m strip up and down these rows, and a snowmobile was used during winter. Food was added as the supplies on the ground became exhausted, and consequently frequency of feeding was affected both by hare density levels and by the weather. Three such areas were provided with rabbit chow from 1987 to 1995. The large circle shows a typical snowshoe hare home range of 5 ha to illustrate that all hares would have access to a feeder row.

of feed was adjusted largely by the frequency of feeding rather than by the amount spread per unit area. Table 4.2 gives the amount of rabbit chow put out during this study.

The feeding treatments were highly successful with only minor problems once we determined how to spread the feed. None of the animals using the feed relied entirely on chow for their food supplies, so this food addition must be thought of as a supplementation. The food experiments, as we discuss in later chapters, were specifically designed to supplement the herbivore trophic level by increasing it independently of the plant trophic level. Snowshoe hares in particular feed extensively on woody plants in winter and herbs in summer in addition to the rabbit chow we added, so that natural food plants are impacted by the treatment of supplementing with chow.

Table 4.2 Amount of rabbit chow supplied to the food addition grids throughout the study.

Time Period	Food 1 Grid	Food 2 Grid	Fence + Food Grid
Winter 1988–89	3160	2320	2240
Summer 1989	2240	1960	4600
Winter 1989–90	3800	4400	4080
Summer 1990	3380	3620	5120
Winter 1990–91	3220	4260	4920
Summer 1991	4500	5280	5620
Winter 1991–92	2640	4100	5760
Summer 1992	1680	2080	4880
Winter 1992–93	1160	780	3120
Summer 1993	1140	1160	3620
Winter 1993–94	1360	1240	1900
Summer 1994	2400	2380	5840
Winter 1994–95	1280	1520	2240
Summer 1995	3600	3720	8040
Winter 1995–96	3080	3580	1440

Units are kilograms delivered for each 6-month period.

4.2.3 Fertilizer

In 1985 we conducted a pilot fertilizer experiment. With the help of Scott Gilbert, we fertilized a 5-ha area at the center of fertilizer 2 grid by hauling fertilizer by hand and spreading it by hand from buckets. This experience taught us that we needed to mechanize fertilization and that we needed to cover a large area. Susan Hannon planned the fertilizer treatments by canvassing the leading boreal forestry specialists and reviewing the literature. She quickly found that the boreal forest was believed to be nitrogen limited, but recognized that we did not know if phosphorous and potassium might also be limiting. We elected to put out commercial NPK fertilizer to cover the possibility of phosphorous and potassiium being in short supply.

In 1987 we began to add nutrients to the soil and thus to the vegetation of two 1-km^2 blocks of boreal forest by means of aerial fertilization. Given that we wished to put out NPK fertilizer (35:10:5) at 17.5g N/m^2 on 100 ha, we needed to distribute 55 metric tons of fertilizer to each of the two areas at the start of the growing season. Fertilizer was brought by truck to the airstrip at the Kluane Research Station, and at the end of May each year a commercial agricultural aircraft from Alberta came north to spread the fertilizer (CD-ROM frames 53 and 54). In 1988 we put out a series of test plots to measure the uniformity of fertilizer fall on the ground and were convinced that the pilots were doing a very good delivery job. The growth stimulus to the plants, particularly the grasses (see chapter 5) was so strong that one could visually inspect the areas fertilized to see the uniformity of the vegetation response.

Aside from the amount of work involved, the aerial fertilization was carried out without any significant problems. Because of budget restrictions, we had to reduce the fertilizer amount by half in 1989. In 1987, we used only nitrogen fertilizer at 25 g/m^2 ammonium nitrate, but in all the other years from 1988 through 1994 we used the full amount of NPK fertilizer. We stopped aerial fertilization after the 1994 treatment.

The only significant problem we had with the fertilizer treatments was with horses on fertilizer 1. A local Kluane outfitter let his horses free-range for 10 months of the year from 1987 to 1990, and they concentrated their grazing on the enhanced plant growth of the fertilizer 1 area. About 15 horses were involved and the result is that net productivity of grasses and herbs was reduced on this area in those years. In 1990 new restrictions were placed on free-range horses, and they were moved to another location. Fertilizer 2 area was never affected by this problem.

4.2.4   Hare Exclosures

In 1988 Rudy Boonstra recognized the need for areas without hare browsing, and consequently we built two hare exclosures, one within fertilizer 2 area and one in open spruce forest just to the north of control 1 (figure 4.1; CD-ROM frame 56). We constructed these exclosures of 5-cm plastic-mesh construction-type fencing (L-77 Vexar, Dupont Canada), 1.5 m high, which was convenient for building because it was light and long lasting. We used two heights of this mesh, so that the total fence reached 3 m height. The high-visibility orange mesh would, we hoped, prevent moose from knocking down the fence and walking through it. This turned out to be only partly correct, and we had a moose inside the fertilizer 2 hare exclosure twice, in 1988 with only minor impact and in 1994 (after a tree fell on the fence) with significant feeding impact. We also found that snowshoe hares would chew through the plastic mesh. The resulting immigrant hares were immediately removed, since we kept hare traps in these exclosures at all times, so the experimental design was only slightly affected. To solve this problem in the winter of 1989–1990, we were forced to put 4-cm mesh chicken wire 1.5 m high around these exclosures to reinforce the bottom strand of plastic mesh. Hares did not attempt to chew through the metal chicken wire. Ground squirrels and red squirrels could move freely over or under the fence, so these species were not excluded, and voles and mice could move through the mesh freely. The only other problem with these hare exclosures was trees that occasionally fell across the fence during windstorms, but fortunately this happened only a few times.

4.3   Organization, Personnel, and Equipment

Our original plan was to have three full-time technicians in charge of the Kluane Project year-round because faculty members could not be away from their teaching duties during the academic year. It quickly became clear that we required more people to achieve our goals. Meeting these needs was relatively simple in the summer because we could hire summer students. In most years we used three summer interns hired by the University of Calgary for the Arctic Institute of North America and hired an additional two to five summer students from the University of British Columbia and one or two students from the University of Alberta and the University of Toronto. In any given year there were also two to five graduate students working on thesis projects related to the Kluane Project as well. In winter we quickly found that the snow tracking of mammalian predators could produce a gold mine of information (see chapter 13), and we needed more winter workers. We were able to use the Yukon government's winter unemployment programs from 1988 through 1996 to employ, on average, three or four additional workers through each winter season. Winter snow tracking is a specialized craft, and it required people with a range of talents for working in difficult winter conditions. We were extremely fortunate to find excellent winter snow trackers in the Yukon throughout this study, and this permitted us to achieve the goals we had set out for winter predator studies.

Over the 10 years of this study we have tallied a total of 157 person-years of effort to complete the work we summarize here. It is to these many individuals who contributed over the years that this project owes its success.

4.3.1   Organizational Structure

As the project began in 1986 we grappled with the problem of designing an organizational structure that would work effectively to accomplish our research goals with a minimum of bureaucracy. We decided on a four-tier structure. The project leader, Charles Krebs, was the top tier. His job was to coordinate the entire project to achieve a balanced experiment. His authority, we all agreed, had to be absolute in making decisions about conflicts within the project and about allocating resources. The second tier included all the faculty cooperating in the project, and these were assigned to particular sectors. Charles Krebs was in charge of snowshoe hares. Stan Boutin was in charge of predator research and red squirrels. Rudy Boonstra was in charge of mice, voles, and ground squirrels. Jamie Smith was in charge of predatory birds and passerines. Tony Sinclair took charge of building the electric fences and organized much of the shrub studies each spring. Susan Hannon and Kathy Martin covered the grouse and ptarmigan portion of the project, and Susan Hannon did all the original work organizing the fertilizer treatments. Mark Dale was in charge of mapping the vegetation, the fertilizer experiments after Susan Hannon got them started, and the tree work. Roy Turkington was in charge of the herbaceous studies and that component of the fertilizer trials. We designed redundancies among the faculty to cover for sabbatical leaves and other absences due to university demands on our time.

Field technicians were organized around a head technician living at the Kluane Research Station, who was in charge of the fieldwork. Vilis Nams was head technician from 1987 to 1990, Sabine Schweiger from 1990 to 1992, Mark O'Donoghue from 1992 to 1995, and Jocylyn McDowell from 1995 to 1996. The head technician reported directly to the project leader Charles Krebs and coordinated the other field workers. Field technicians were assigned particular aspects of the project. Frank Doyle was the key person for avian predator studies, and he maintained the electric fences and the feeding experiments. Cathy Doyle was the key technician for the snowshoe hare studies and the radio telemetry work. Elizabeth Hofer was in charge of winter snow tracking and winter work on mammalian predators. Irene Wingate, based in Vancouver, served as logistics technician, and she kept the field crew supplied with everything from pencils to electric fencing wire. All of these technicians interfaced with the graduate students who worked on particular aspects of the overall project. Any conflicts that arose were referred to the head technician and if necessary to the project leader for a final decision.

Each autumn we had a 2- to 3-day annual meeting in Vancouver or Edmonton of all the faculty participating, the head technician, and as many of the other technicians as we could afford to bring south. These meetings were critical in setting priorities because we reviewed progress each year, pinpointed problems, and agreed on solutions to be implemented. At this meeting we allocated the budget to the different subsections of the project as a group. We were fortunate in having a congenial group that always kept the overall goals in view, and in which conflicts over funding or technical help were infrequent. Most important, when there was a conflict within the group, we all agreed that the decision of the project leader was final.

4.3.2   The Problem of Training and Protocols

One problem that arose immediately in this study was the repeatability of methods. With changes of summer students from year to year and slow turnover of technicians and graduate students, we were concerned that our standard methods did not drift. One way to achieve consistency of field methods is to have one generation of students train the next, and in general this worked quite well for procedures such as hare live trapping or vegetation transects. But we wanted to make these procedures more concrete, and we began in 1989 to publish each year *The Kluane Handbook.* The handbook contained a detailed description of the field procedures for every type of data we were collecting from shrub biomass sampling to winter snow tracking, and it evolved from a 47-page handbook in 1989 to a 117-page handbook in 1995. We included in the handbook a general description of the Kluane Project and a detailed statement of the procedures as well as the means of data entry into computers after the field work was completed. We gave each worker a personal copy of the handbook each year and asked each of the faculty and technicians to revise their sections at the end of each summer field season to make the procedures clearer to newcomers. This was a useful means of standardizing our procedures.

In spite of this, we still encountered some data that were impossible to accept. Because there were periods of intensive field work, some data would not be analyzed until several months or even a year after collection, and anomalies would then turn up that were impossible to sort out. Although these problems were relatively minimal in this project, we believe that this is the primary problem in large field projects and that data management and data archiving are worth as much effort as possible to make sure that methods do not drift from year to year.

4.3.3   Seen Sheets

Vilis Nams and Frank Doyle developed one of the innovations we used in this project as a way of getting information on uncommon species in the vertebrate community. "Seen sheets" provided a way of monitoring populations by tallying random encounters with animals seen while doing other jobs in the forest. We used 160 field workers from 1986 to 1996 to record 25 species of the less intensively studied vertebrates in the Kluane Region. Each worker would record, for every day in the field, the number of hours in the field, their mode of transport (foot, truck, or snowmobile) and the numbers of each species seen. We trained summer students in the identification of raptors and other species of birds with which they were unfamiliar. We obtained, on average, 7700 h of observations each year through the study, and we could index species abundances annually by the average number seen per hour. Confidence limits were placed on these indices by bootstrapping techniques. We recognized that observers varied greatly in their sighting abilities, but because we had some observers that spanned much of the study, we had a control group of high-quality observers to use as a baseline for calibration (Hochachka et al. 2000). These data form a valuable addition to our understanding of community dynamics in this system and are used extensively in chapters 11, 12, 14, and 16.

4.3.4 Winter Work at Kluane

Wind and cold were problems during periods of extreme weather in midwinter, and at times long stretches would go by when certain work could not be done. These severe periods were fortunately not common enough to prevent us from doing research in winter. The winter is a special time in the north, and our technicians and students adapted well to the lack of southern amenities such as running water and regular electricity.

One special problem we encountered with this study was winter isolation. This is a common, well-described problem in the north in winter, called "cabin-fever," becoming "bushed," and a variety of colorful euphemisms. In a camp with 8–10 workers who see each other day after day with little outside visitation, personal conflicts could escalate beyond reason. This was a problem in about half of the winters of this study, and we found it difficult to do much about it. One way of alleviating winter problems was for the faculty to visit as much as possible during the winter, but this was constrained by teaching commitments. Because we had e-mail working throughout the project, we could communicate rapidly with the field crews from our southern university posts, and this was a valuable asset. But we never could overcome the overall problem of winter isolation. There were times when we wished that instead of an ecologist we had a resident psychiatrist on the project staff. We think the key to isolated winter work is the personalities of the people involved, but we are not sure that we can determine from knowledge of how people operate in an open summer environment how they will respond to winter isolation.

4.3.5 Equipment

The final practical problem is that of equipment. Field ecological projects are more typically personnel limited than equipment limited, but some pieces of equipment were essential. The major items were all transportation equipment—trucks, snowmobiles, and ATVs. Trucks were always in short supply, and their use on gravel roads and during the winter caused great wear and tear. Research grants for trucks are rare, and we typically made do with older vehicles that were kept running by Donjek Upton, our resident mechanic. Snowmobiles were continually wearing out, partly because most of them were not designed for continuous use in the winter, and we often put up to 4000 km on one snowmobile during the winter season. ATVs were used in summer to carry feed and spread it on the food grids and were useful for electric fence maintenance in summer as well.

Electrical power in winter was essential for running computers and providing some lights for laboratory work, and this was yet one more problem we had to overcome. We found that conventional gasoline generators would not run 8 h a day without excessive wear and rapid failure. We obtained a small diesel generator that provided 4 kilowatts of power, but had trouble keeping it operational. Electrical generating equipment is oriented toward large units generating more than 30 kilowatts, and these diesel units operate well in the summer at Kluane. But too much fuel is needed to run this large a generator during the winter for small power demands, and the net result was that we were forced to rely on smaller generators that were less reliable.

4.4   Summary

To unravel the structure underpinning vertebrate community dynamics in the boreal forests of Kluane, we carried out four kinds of manipulations to analyze both top-down (predator) and bottom-up (soil nutrient) forces. The two major manipulations were to provide additional food as rabbit chow, to fence out mammalian predators, and to combine both these two treatments on one area. The third type of manipulation was to add nutrients in the form of NPK fertilizer to two areas of forest, and the final manipulation was to fence snowshoe hares out of blocks of habitat to provide areas free of hare browsing. All of the major manipulations were done on experimental units of 1 km^2 because we wished to work at a large spatial scale. We were unable to replicate the predator fence experiment and the predator fence + food addition experiment because of costs and logistics.

There were a whole series of practical problems that arose while we were attempting these manipulations, and we outline them briefly. The predator-proof electric fence proved illusory until we lined it with chicken wire as a ground to the 8600-volt live wires, and then it worked excellently. Food supplementation had to overcome the attractions of grizzly bears and moose, and by spreading the feed we increased the cost of foraging for these large herbivores and minimized their impact. Fertilizer was spread aerially with no serious problems and good, uniform coverage.

Problems of personnel largely centered on maintaining protocols with changing student assistants, and we wrote a methods handbook to facilitate the transfer of methods from one year to the next. The work carried on winter and summer, and winter work was particularly stressing on personnel because of isolation and difficult weather conditions. Nevertheless, we overcame these many problems to carry out the experimental manipulations with the success outlined in the remainder of this book.

Literature Cited

Boutin, S. 1990. Food supplementation experiments with terrestrial vertebrates: patterns, problems, and the future. Canadian Journal of Zoology **68**:203–220.

Carpenter, S. R., S. W. Chisholm, C. J. Krebs, D. W. Schindler, and R. F. Wright. 1995. Ecosystem experiments. Science **269**:324–327.

Carpenter, S. R., T. M. Frost, D. Heisey, and T. K. Kratz. 1989. Randomized intervention analysis and the interpretation of whole-ecosystem experiments. Ecology **70**:1142–1152.

Fisher, R. A. 1966. The design of experiments, 8th ed. Hafner Publishing, New York.

Gilbert, B. S., and C. J. Krebs. 1981. Effects of extra food of *Peromyscus* and *Clethrionomys* populations in the southern Yukon. Oecologia **51**:326–331.

Haag, R. W. 1974. Nutrient limitations to plant production in two tundra communities. Canadian Journal of Botany **52**:103–116.

Hochachka, W. M., K. Martin, F. Doyle, and C. J. Krebs. 2000. Monitoring vertebrate populations using observational data. Canadian Journal of Zoology **78**:521–529.

Hurlbert, S. H. 1984. Pseudoreplication and the design of ecological field experiments. Ecological Monographs **54**:187–211.

Klenner, W., and C. J. Krebs. 1991. Red squirrel population dynamics. I. The effect of supplemental food on demography. Journal of Animal Ecology **60**:961–978.

Krebs, C. J. 1991. The experimental paradigm and long-term population studies. Ibis **133**(suppl.):1–6.

Krebs, C. J., B. S. Gilbert, S. Boutin, A. R. E. Sinclair, and J. N. M. Smith. 1986. Population biology of snowshoe hares. I. Demography of food-supplemented populations in the southern Yukon, 1976–84. Journal of Animal Ecology **55**:963–982.

Krebs, C. J., B. L. Keller, and R. H. Tamarin. 1969. *Microtus* population biology: demographic changes in fluctuating populations of *M. ochrogaster* and *M. pennsylvanicus* in southern Indiana. Ecology **50**:587–607.

Menge, B. A. 1995. Indirect effects in marine rocky intertidal interaction webs: patterns and importance. Ecological Monographs **65**:21–74.

Ostfeld, R. S. 1994. The fence effect reconsidered. Oikos **70**:340–348.

Paine, R. T. 1984. Ecological determinism in the competition for space. Ecology **65**:1339–1348.

Schweiger, S., and S. Boutin. 1995. The effects of winter food addition on the population dynamics of *Clethrionomys rutilus*. Canadian Journal of Zoology **73**:419–426.

Shaver, G. R., and F. S. I. Chapin. 1980. Response to fertilization by various plant growth forms in an Alaskan tundra: nutrient accumulation and growth. Ecology **61**:662–675.

Sinclair, A. R. E., C. J. Krebs, J. N. M. Smith, and S. Boutin. 1988. Population biology of snowshoe hares III. Nutrition, plant secondary compounds and food limitation. Journal of Animal Ecology **57**:787–806.

Smith, J. N. M., C. J. Krebs, A. R. E. Sinclair, and R. Boonstra. 1988. Population biology of snowshoe hares II. Interactions with winter food plants. Journal of Animal Ecology **57**:269–286.

Sokal, R. R., and F. J. Rohlf. 1995. Biometry, 3rd ed. W. H. Freeman, New York.

Tilman, D. 1989. Ecological experimentation: strengths and conceptual problems. *in* G. E. Likens (ed). Long term studies in ecology, pages 136–157. Springer Verlag, New York.

Van Cleve, K., and L. K. Oliver. 1982. Growth response of postfire quaking aspen (*Populus tremuloides,* Michx.) to N, P, and K fertilization. Canadian Journal of Forest Research **12**:160–165.

Yoccoz, N. G. 1990. Use, overuse, and misuse of significance tests in evolutionary biology and ecology. Bulletin of the Ecological Society of America **72**:106–111.

PART II

PLANT DYNAMICS

5

Herbs and Grasses

ROY TURKINGTON, ELIZABETH JOHN, & MARK R. T. DALE

P lants in the boreal forest are an important component of the ecosystem for two main reasons. First, the plants as vegetation form the physical surroundings for both herbivores and carnivores and are the basis of the physical structure of the community. Second, as primary producers, they provide the energy and nutrients to the herbivores on which higher trophic levels depend. Therefore, understanding the factors that limit the quantity and the quality of plants is fundamental. Our studies focus on the herbaceous vegetation but primarily on two grasses, *Festuca altaica* and *Calamagrostis lapponica,* which dominate small meadows scattered throughout the white spruce forest, and on the four herbs, *Lupinus arcticus, Anemone parviflora, Mertensia paniculata,* and *Achillea millefolium* var. *borealis,* which are relatively abundant in the forest understory (see the appendix following this chapter). These plants provide a source of high-quality food to the herbivores.

Soil nutrients, especially nitrogen, often limit the productivity of boreal forest vegetation (Bonan and Shugart 1989) and may control vegetation standing crop. Plants differ in their abilities to respond to increased nutrient levels, and community composition usually changes after fertilization as more competitive species begin to dominate. The nutrient availability level also influences a species' ability to produce defensive chemical compounds against herbivory (Coley et al. 1985) and the ability to regrow after herbivory (Hilbert et al. 1981, Maschinski and Whitham 1989). Conversely, herbivory may have a direct effect on vegetation quantity and quality. Herbivory has long been known to influence species composition in some plant communities (Huntly 1991) due to differential plant palatability and differences in plants' abilities to tolerate herbivory. Many plants produce defensive chemical compounds in response to herbivory (Palo and Robbins 1991, Tallamy and Raupp 1991). The snowshoe hare has distinct preferences among the summer forage species available in the boreal forest (Bryant et al. 1991).

5.1 Hypotheses Regarding Vegetation

For the purposes of the Kluane study, we considered the soil to be a trophic level. The herbaceous vegetation, along with the soil nutrient pool, forms two of the four trophic levels recognized in this system. To understand some of the inter-trophic linkages between components of the system, three hypotheses regarding the vegetation were tested: that vegetation was controlled by (1) nutrient availability alone (bottom-up, or donor control), (2) herbivores alone (top-down control), and (3) both nutrient availability and herbivores. This involved three major experimental treatments: fertilization, herbivore exclusion, and fertilization plus herbivore exclusion. These treatments allowed us to make specific predictions about changes in plant biomass, or standing crop, under the three different hypotheses (table 5.1). In some cases, however, response was also assessed using changes in rates of plant turnover, plant nutrient content, or secondary compounds. In addition to the direct predictions in table 5.1, some subsequent predictions were also made; these predictions are formalized below.

Hypothesis 1: Vegetation is controlled by nutrient availability alone.
This bottom-up, donor control, hypothesis makes five predictions:

1. With fertilization there will be an overall increase in the total biomass or standing crop of herbaceous vegetation.

Table 5.1 Predictions of the direction of change in plant biomass, or standing crop, from each of the three experimental treatments according to the three models of community organization.

Model	Fertilizer Added	Herbivores Excluded	Fertilizer Added, Herbivores Excluded
Bottom-up, donor control	+	0	+
Top-down, herbivore control	0	+	+
Interactive control	+	+	+

+ = biomass increase, 0 = no change.

2. Alternatively, there will a greater nutrient content of vegetation in fertilized areas.
3. There will be no increase in vegetation standing crop when herbivores are excluded.
4. Vegetation composition will change as more competitive species replace stress tolerators in fertilized areas.
5. If vegetation is removed from a plot, there will be an increase in soil nutrient levels.

Hypothesis 2: Vegetation is controlled by herbivores alone.
This top-down hypothesis makes two predictions:

1. There will be no increase in vegetation standing crop with fertilizer, especially during the snowshoe hare peak.
2. Assuming snowshoe hares are the major herbivore, then in hare exclosures, vegetation standing crop will increase and secondary compound content will decrease.

Hypothesis 3: Vegetation is controlled by both nutrient availability and herbivory.
This interactive hypothesis makes four predictions:

1. There will be an increase in plant productivity in fertilizer plots, but biomass will remain unchanged.
2. On fertilizer plots, grazing intensity will increase due to increased quality of forage.
3. In hare exclosures, vegetation standing crop will increase and secondary compound content will decrease.
4. In hare exclosures, soil nutrient levels will be lower than in control plots immediately outside of the exclosures because the larger, protected and ungrazed plants will extract more nutrients from the soil.

5.2 Study Area

The vegetation in our area is dominated by white spruce (*Picea glauca*), with a shrub understory dominated by gray willow (*Salix glauca*) and dwarf birches (*Betula nana* and *B. glandulosa*). The herb layer (including dwarf shrubs) includes *Lupinus arcticus, Festuca altaica, Calamagrostis lapponica, Mertensia paniculata, Anemone parviflora, Achillea millefolium* var. *borealis, Linnaea borealis, Arctostaphylos uva-ursi, Epilobium angustifolium,* and *Solidago multiradiata* (CD-ROM frame 44). Some of the studies used the already established grids, especially the two 4-ha hare exclosure grids described in chapter 4; one of these was fertilized. Most of the other studies described were done at two sites in areas of moderately open white spruce forest with a well-developed herbaceous understory. The first site was at Boutillier summit (km 1690.4, Alaska Highway), about 50 m into the forest to the north of the highway, and the second site was about 3 km to the south near a Microwave tower.

5.3 Tests of Predictions from Three Models

The experimental treatments for most of the Kluane study were either the 4-ha herbivore exclusion grids or the larger grids of approximately 1 km². However, most of the herbaceous work was done at a smaller scale either as a subsample within a major grid (e.g., predictions 1.1, 1.3, 2.2, 3.3, 3.4), or on 5 m × 5 m or smaller plots. We adopted this approach to minimize the strong heterogeneity that was so visually apparent in the distribution and abundance of the herbaceous vegetation; to have included this heterogeneity by using larger sampling areas would have undoubtedly obscured real patterns that occur within the herbaceous vegetation. We typically chose sites that had most of the major herbaceous species present and in which each of the study species were typically at least 10% of the vegetation cover. This reduced effects introduced by sites that had unusual species composition or had a lot of bare ground.

5.3.1 Donor Control Model

The five predictions that we tested from the overall model that vegetation is controlled by nutrient availability had a variety of results.

Prediction 1.1: Fertilization Will Result in an Overall Increase in the Amount of Herbaceous Vegetation   This prediction was verified for most species, but rejected for *Anemone* and *Lupinus.*

The density and biomass of selected species were estimated on both of the fertilized grids and immediately adjacent to the grids. Sampling was done in both closed spruce and open spruce forest at all sites. Fertilization began in 1987 but detailed sampling did not begin until 1993 (Turkington et al. 1998). Sampling was done during July when herb biomass is at a peak. The probability of *Mertensia* being present and the number of flowering stems on each *Mertensia* increased when plots were fertilized (figure 5.1). For the other species, *Achillea millefolium* and *Solidago,* fertilization had little effect. The dry mass of most species, whether flowering or nonflowering (figure 5.1), increased in closed sites when fertilized. Only *Epilobium* increased (nonflowering) in an open site. In addition fertilizer had a significant effect on the growth of *Festuca altaica* and *Calamagrostis lapponica* after only one season (D. Hik, unpublished data).

Arii (1996) investigated the effects of fertilization on *Achillea, Anemone, Festuca, Lupinus,* and *Mertensia. Achillea* and *Festuca* increased biomass and leaf number with fertilization, but *Mertensia* and *Lupinus* remained unchanged (figures 5.2, 5.3). At the highest level of fertilization, there was high mortality (almost 80%) in the *Anemone* population (figure 5.3). The individualistic response of these species means that prediction 1.1 cannot be applied to all herbaceous species.

Prediction 1.2: Vegetation in Fertilized Areas Will Have Greater Nutrient Content but No Increase in Standing Crop   This prediction was verified for *Anemone, Epilobium, Lupinus, Mertensia,* and *Achillea.*

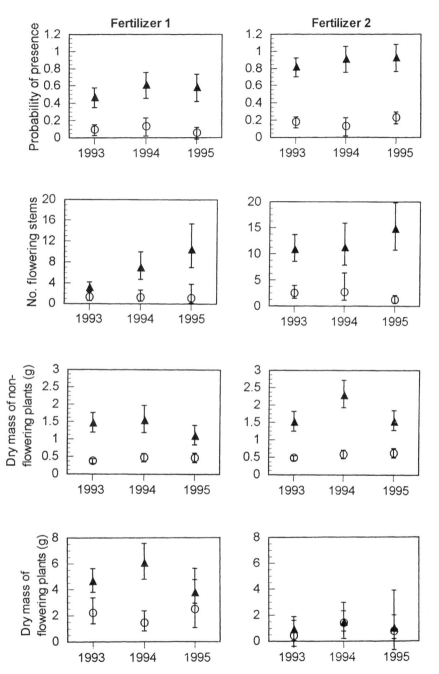

Figure 5.1 The abundance and growth of *Mertensia paniculata* in closed white spruce forest on the fertilizer 1 and fertilizer 2 grids (▲) and adjacent to the grids in unfertilized (○) plots. Error bars are 95% confidence limits.

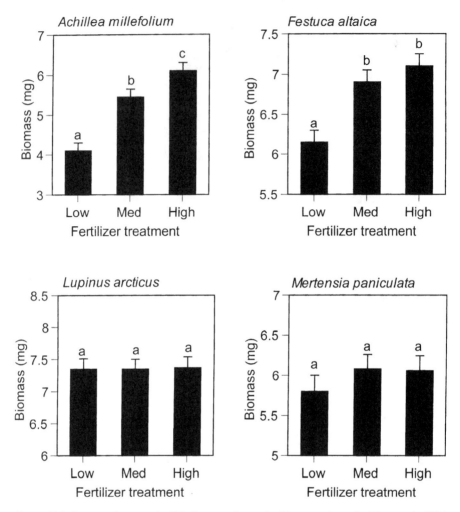

Figure 5.2 Average dry mass ($\pm$ SE) (log transformed) of four species at final harvest in 1994, growing at three levels of fertilization (low = not fertilized; medium = 200 g/m^2; and high = 400 g/m^2 of water-soluble 6:8:6 NPK). Within a panel, bars that share a common letter are not significantly different ($p > .05$; least significant difference).

We analyzed the effect of fertilizer treatment on the levels of available nitrogen and free amino acids in *Anemone, Epilobium, Lupinus, Mertensia,* and *Achillea* during the 1991 growing season. Fertilized and unfertilized plants were sampled at three dates through the growing season, and on each date, young, old, and medium-aged leaves were removed for analysis. Fertilized plants of all species had increased nitrogen content. The percentage of nitrogen decreased as the season progressed in both control and treatment plots. The amino acids serine, asparagine, γ-aminobutyric acid, cystine, arginine, and valine all showed substantial increases after fertilization; other amino acids did not increase relative to controls. Asparagine and arginine, which are used to transport and store nitro-

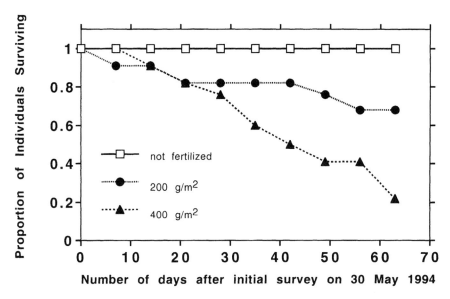

Figure 5.3 Differential survival of three natural populations of *Anemone parviflora* ($n = 12$ for each population) in 1994, growing at three levels of fertilization. Fertilizer was added as a solution of 6:8:6 NPK.

gen, were the two most abundant amino acids in these plants and, with fertilizer addition, increased more (8.5× versus a maximum of 3×) than the other amino acids. Changes in the levels of specific amino acids depended on the species and on the date of sampling.

Prediction 1.3: Standing Crop Will Not Increase When Herbivores Are Excluded
Extensive vegetation sampling of both hare exclosure grids was conducted in the summers of 1991–1995, typically in late June. At each site we established eight 40-m transects, each of which was bisected by the exclosure fence. Leaf area index was estimated in 1 m × 50 cm quadrats placed at 4-m intervals along each transect. No consistent significant differences were found between the herbaceous vegetation inside and outside of the fences since 1992, when a number of species showed weak responses to the fences; these surveys were done immediately after a peak in snowshoe hare densities.

Prediction 1.4: Vegetation Composition Will Change as More Competitive Species Replace Stress Tolerators Sixteen 5 m × 5 m plots were chosen at both the Boutillier and Microwave sites for a 2 × 2 factorial experiment (with and without fertilizer, with and without grazing enclosures) (John and Turkington 1995, Turkington et al. 1998). We applied a high nitrogen fertilizer once per year from 1990 to 1995, at the same rate as on the main treatment grids. This was done just after snow melt, early in the growing season in late May and early June. Twice each summer a survey of the percent cover of all plants less than 1 m tall was made in each plot. To minimize trampling effects, each plot was divided into five untrodden experimental and sampling zones, 5 m long, with 15-cm wide

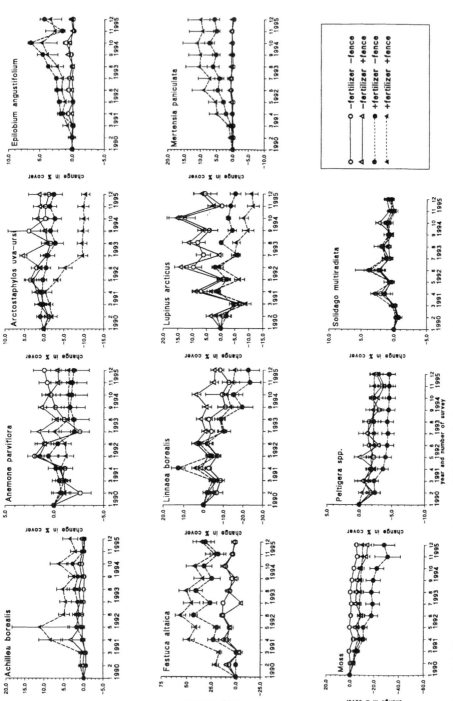

Figure 5.4 Change in percent cover (mean ± SE) for 11 herbaceous and dwarf shrub species in a 2 × 2 factorial experiment in plots at the Microwave and Boutillier Summit sites from 1990 to 1995. Treatments (fertilizer and fences) were first imposed in May 1990, and fertilizer was added each year in late May.

walkways between each zone. Fences were 1 m high and made of galvanized chicken wire with 2.5-cm mesh, supported by 2-m high T-bars, and were firmly stapled to the ground to prevent animals from getting under the fence. Unfenced plots also had T-bars with string between them to prevent casual trampling by data collectors. No fence damage was recorded, and there was no evidence that large mammals, such as moose, entered the exclosures during the winter.

We investigated the effects of exclosure, fertilizer addition, and time for 11 of the most frequently occurring species. The eleven species analyzed showed a range of responses to the treatments (figure 5.4; CD-ROM frame 45). Ten showed a significant fertilizer effect, with four increasing in response to fertilizer (*Mertensia, Festuca, Achillea,* and *Epilobium*) and six decreasing over the 6-year period (*Anemone,* various species of moss, *Peltigera, Linnaea, Arctostaphylos,* and *Lupinus*). Not all fertilizer effects were immediately apparent, with some species showing an initial positive or neutral response to fertilizer, which later reversed (e.g., *Linnaea*). Fertilization altered the species composition of small meadows in favor of *Calamagrostis* over *Festuca* (D. Hik, unpublished data). Fertilizer effects on herbs in the boreal forest are highly individualistic and do not follow a single prediction like hypothesis 1. However, in general, it is evident that long-term fertilization causes a shift from a herb-dominated community to one dominated by grasses.

In general, the untreated (control) community is rather static from year to year (figure 5.4). But even in the control plots, some species such as *Anemone* and *Lupinus* fluctuate from year to year. *Achillea* and *Arctostaphylos* responded fairly quickly to fertilizer addition, but this effect diminished with time, and no significant differences were detectable in surveys 11 and 12 in 1995. The response of *Linnaea* to fertilizer did not become apparent until later surveys.

Prediction 1.5: If Vegetation Is Removed from a Plot, Soil Nutrient Levels Will Increase   In these northern spruce forests, soils develop in a cool and moderately humid climate. The decomposition of plant remains is slow, and the surface of the soil is covered by a mat (mostly spruce needles) of partly decomposed, acidic, plant material, which is quite sharply separated from the underlying soil. Beneath this mor litter layer, the upper soil is leached and typically grayish. The leached layer is acidic and low in nitrates and most other plant nutrients. In 1996 P. Seccombe-Hett (unpublished data) identified three hundred 1-m^2 plots beside the hare exclosure grid. In 150 of them vegetation was killed by applying Roundup (Monsanto Corp.); the other 150 were left untreated. Soils were sampled for nitrate nitrogen in early July, 4 weeks after vegetation removal. Nitrate nitrogen levels were significantly higher ($p < .1$) in those plots in which the vegetation had been removed.

5.3.2   Herbivore Control Model

We tested two predictions that follow from the idea that vegetation is controlled solely in a top-down manner by the amount of herbivory.

Prediction 2.1: Fertilizer Will Not Increase Vegetation Standing Crop   This prediction has largely been rejected in our previous discussion of prediction 1.1. However, after six growing seasons (using the same 5 m $\times$ 5 m plots as in prediction 1.4), four species,

Achillea, Festuca, Arctostaphylos, and mosses, responded significantly to the exclusion of herbivores (figure 5.4). Responses varied, but notably, *Achillea, Festuca,* and *Mertensia* increased when fertilized, but they had an even greater increase when fenced as well. This indicates that some portion of the additional productivity due to fertilization was being consumed in the unfenced plots. Therefore, although hares affect the standing crop of these three species, they do not affect the standing crop for the other five species monitored. These data are more consistent with prediction 3.1. In general, where herbivore density is at natural levels, the impact of herbivores on vegetation biomass is slight; however, even fairly moderate applications of fertilizer will induce noticeable changes in vegetation. Fertilizing causes a major decline in moss in those plots not protected from grazers and a decline in *Arctostaphylos* in those plots protected from grazers (figure 5.4). *Lupinus* (a legume), *Linnaea,* and *Arctostaphylos* decrease, while *Mertensia* and *Festuca* increase, when fertilized and fenced. One species, *Solidago multiradiata,* was remarkable in its lack of response to treatments, showing no tendency at all to respond to the experimental manipulations even after 6 years.

Prediction 2.2: Vegetation Standing Crop Will Increase in Hare Exclosures   Tests of prediction 1.3 demonstrated that there were no changes in vegetation biomass or species composition inside the hare exclosures. An additional study in the summer of 1991 (John and Turkington 1995) was conducted using four paired plots at each of five of the hare trapping grids that provided a range of snowshoe hare densities. Each pair consisted of a fenced and adjacent unfenced plot, each 1 m $\times$ 1 m. After 8 weeks, above-ground biomass was harvested from all plots and sorted to species. At all sites, there was less *Festuca* and *Achillea* in the unfenced plots, and at four of the five sites there was less total biomass outside than inside. Analyses revealed that even though there were apparently some effects by herbivores these responses occurred only when snowshoe hare densities were artificially high.

5.3.3   Combined Top-down Bottom-up Model

Many community studies suggest that neither the top-down nor the bottom-up model of community organization is adequate and that one needs to consider a combination of these two models (see chapter 3). Here we were able to test four predictions that flow from a combined model.

Prediction 3.1: Plant Productivity Will Increase in Fertilizer Plots but Biomass Will Remain Unchanged   This prediction was verified for *Mertensia* but rejected for *Anemone* and *Lupinus.*

We must discriminate between *plant standing crop* or *biomass* (i.e., how much vegetation present at any particular time [see prediction 1.1]), and *plant productivity* (i.e., how much is produced). The productivity of a system can be increased by fertilization, but the standing crop remains unchanged if the additional productivity is consumed by herbivores. Addressing this prediction required detailed and frequent monitoring of populations of individual species. Two studies were done. The first used the 5 m $\times$ 5 m grids described earlier (prediction 1.3). Within each of the thirty-two 5 m $\times$ 5 m plots, we selected three populations each of *Anemone* and *Mertensia.* These are common herbaceous species in the forest understory and are eaten by snowshoe hares. The location of each population was permanently marked, and all plants were individually labeled. Initially we began monitoring 700 *Anemone* and 854 *Mertensia*. Populations were surveyed three times per season. For each stem we recorded the number and size of leaves and the number of leaf buds and flowers. The total leaf length per stem was calculated from the sum of the leaf lengths (John and Turkington, 1997).

Both species responded more strongly to fertilizer addition than to the exclusion of herbivores. *Mertensia* produced more stems, more flowering stems (figure 5.5a), more leaves per stem (figure 5.5b), and increased total leaf length per stem (figure 5.5c) for nonflowering stems in the fertilized plots. Analysis of variance detected a significant fertilizer and exclosure effect in most years of the study. *Anemone* showed contrasting responses at the individual and population levels; although individual stems produced slightly (and significantly) more leaves in fertilized plots (figure 5.5e), the population density declined (figure 5.5d). There were no strong effects on either leaf size or flowering. There is evidence of higher leaf turnover in fertilized plants (a proportionally higher production of buds, more rapid decline in leaf number at the end of the growing season, and greater fluctuation in leaf number throughout the season), but there must ultimately be a higher death than birth rate, both at the leaf and the stem level, leading to the observed population decline.

In a second study, Graham (1994) and Graham and Turkington (2000) investigated the effects of various treatments on the population dynamics of *Lupinus* using a $2 \times 2 \times 2$ factorial cross of $\pm$ fertilizer, $\pm$ simulated herbivory, and $\pm$ neighbor removal applied to 1-m^2 quadrats. These quadrats were fenced to exclude hares and other large mammalian herbivores. They were surveyed every second week during the 1991 and 1992 growing seasons, and 11 different demographic and growth variables were measured. The *Lupinus* populations were remarkably unresponsive to treatments, even by the end of the second season (figure 5.6; CD-ROM frame 45). In addition, Graham tagged 320 clumps of *Lupinus* and at regular intervals recorded the condition of the leaves. This measure was used to calculate an index (live summer leaf-days) of plant availability to herbivores by counting the mean number of leaves that are available as food and how many days those leaves had been alive and available for consumption (Graham 1994, Graham and Turkington 2000). By the end of the second growing season the only significant effect was a decrease due to fertilization, where the mean live summer leaf-days for the fertilized treatment was 14070 and in controls 25561 (figure 5.7).

Prediction 3.2: Grazing Intensity Will Increase on Fertilized Plots Due to Increased Quality of Forage   To assess this prediction requires a careful examination of figure 5.4. If grazing intensity increases in fertilized plots, then two conditions must be met. First, the amount of change between the treatments no fertilizer, no fence and no fertilizer, with fence should be less than the difference between fertilizer added, no fence and fertilizer added, with fence. This describes a situation where the addition of fertilizer alone, without herbivores, will change the vegetation biomass (either an increase or decrease), but the change will be moderated in fertilized plots that have the herbivores present because they remove some of the presumed higher quality vegetation produced. The degree of moderation will be greatest where the intensity of grazing is greatest. Second, the percent cover of species in fertilized plots must be higher in those plots that are fenced compared

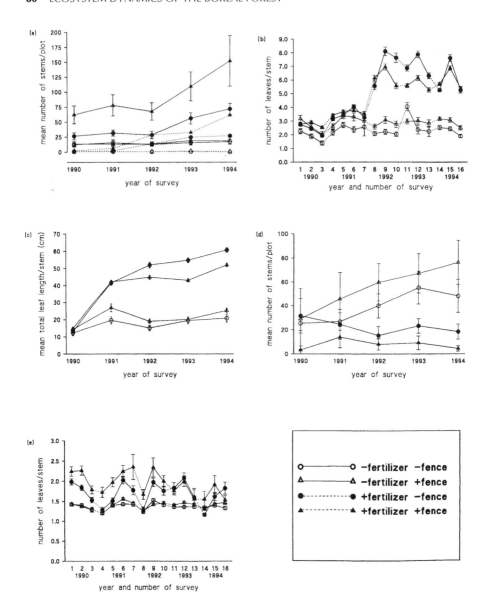

Figure 5.5 The effects of fertilization and small mammal exclosures on the growth of *Merten-sia* (a–c) and *Anemone* (d, e). (a) Mean (± SE) number of stems per plot, by treatment in the mid-season survey. The contribution of flowering plants to the total is shown by the dotted lines. (b) Mean (± SE) number of leaves per stem for the whole population through all surveys. Solid lines join data from the same year, dotted lines connect data from different years. (c) Mean (± SE) total leaf length per stem (cm) for the mid-season surveys of each year. (d) Changes in mean number of stems per plot, expressed as a proportion of the original number of stems in each treatment. (e) Mean (± SE) number of leaves per stem. Solid lines join data from the same year; dotted lines connect data from different years.

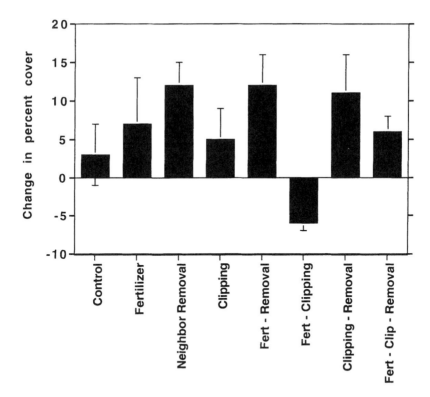

Figure 5.6 Mean (± SE) change in percent cover of *Lupinus arcticus* in response to fertilization, clipping, and removal of neighbors, from June 1991 to August 1992.

to unfenced plots (i.e., fertilized, no fence should have lower percent cover values than fertilized, fenced). In figure 5.4, five species meet these criteria—mosses, *Festuca, Achillea, Mertensia,* and *Peltigera*—and the prediction can be accepted for these species.

In a second study, Dlott (1996) did a transplant experiment to investigate the combined effects of fertilizer and herbivory on plant growth (CD-ROM frame 45). Eight species were transplanted into an experimental design of ± herbivory (fences) crossed with no added nitrogen, 12 g N/m^2 or 35 g N/m^2, on each of three grids (fertilizer 1, food 1, and predator exclosure + food) representing different densities of grazing animals. Survival (figure 5.8), number of leaves (figure 5.9), and plant height were surveyed weekly in 1992 and monthly in 1993. Percent cover surveys were taken in 1993 to investigate treatment effects on existing vegetation. Results indicate that there is no top-down effect at natural nutrient levels, and differences inside and outside exclosures were only significant when grazing pressure and nutrients were experimentally high (CD-ROM frame 45). This result is consistent with short-term exclosure experiments described in John and Turkington (1995).

The effects of fertilizer addition and clipping intensity on the potential of plants to regrow after grazing were examined (Hicks and Turkington 2000). Individuals of *Festuca, Achillea,* and *Mertensia* were transplanted to plots and subjected to three levels of clip-

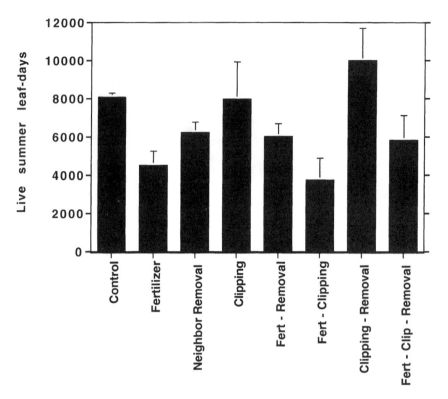

Figure 5.7 The effects of fertilization, clipping, and removal of neighbors on the availability of leaves of *Lupinus arcticus,* measured as live summer leaf-days (± SE), to herbivores. This index is the number of days each leaf is alive during the summer of 1992 summed over all leaves.

ping (none, 50% leaf loss, and 100% leaf loss) and two levels of fertilizing (none and 15 g N/m^2). Plant growth was assessed using biweekly measurements of total leaf area per plant, and individuals were harvested at the end of the growing season to obtain final biomass estimates. After 100% leaf loss, unfertilized *Festuca* grew as much new leaf area as the unclipped plants and ended up approximately half the size of the unclipped plants (figure 5.10). Unfertilized *Mertensia* regrowth was significantly increased when clipped because it could fully compensate for tissue loss by the end of one growing season. When fertilizer was added, clipped individuals of each species regrew significantly less than the unclipped controls

Prediction 3.3: Vegetation Standing Crop in Hare Exclosures Will Increase and Secondary Compound Content Will Decrease   This prediction is the same as prediction 2.2, discussed above.

Prediction 3.4: Soil Nutrient Levels in Hare Exclosures Will Be Lower Than in Control Plots Immediately Outside of the Exclosures   This prediction was tested by

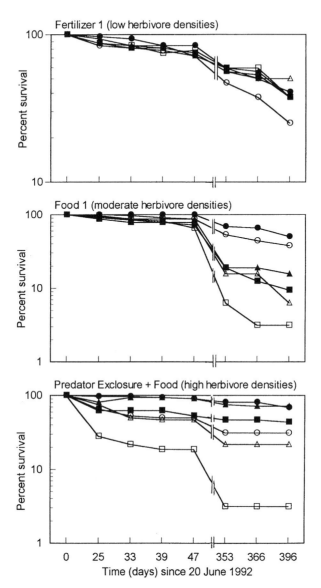

Figure 5.8 Differential survival of transplanted seedlings at the fertilizer 1 (lowest hare density), food 1, and predator exclosure + food (highest hare density) grids. On each grid, plots were either fenced (filled symbols) or unfenced (open symbols) and treated with one of three levels of fertilizer application: none ($\bigcirc$,$\bullet$), 11.6 g N/m^2 ($\triangle$,$\blacktriangle$), and 35 g N/m^2 ($\square$,$\blacksquare$).

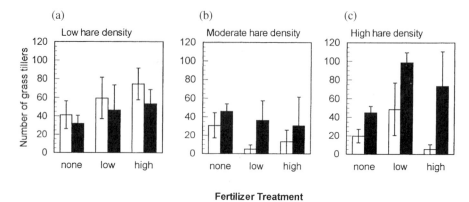

Figure 5.9 Number (± SE) of grass tillers in plots on three grids providing different densities of snowshoe hares in July 1993. (a) Fertilizer 1 (lowest hare density), (b) food 1 (moderate hare density), and (c) predator exclosure + food (highest hare density). Solid bars represent fenced plots; open bars represent no fence. Low fertilizer is 11.6 g N/m² and high is 35 g N/m².

Seccombe-Hett (see prediction 1.5). She showed that soil nitrate nitrogen levels were higher ($p < .05$) inside the hare exclosure than outside.

5.4   Which Model Best Describes Boreal Forest Herbs and Grasses?

The majority of the data presented supports the hypothesis that herbaceous vegetation standing crop is controlled by nutrient availability alone, the bottom-up model of community organization (table 5.2; Turkington et al. 1998). With fertilization there was an overall increase, up to threefold after 3 years, in the amount of herbaceous vegetation, especially among the grasses *Festuca* and *Calamagrostis*. In addition, there was a greater nutrient content of vegetation in fertilized areas. The one study designed specifically to

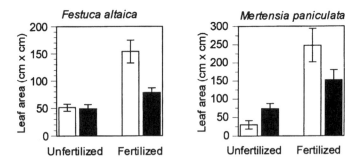

Figure 5.10 Leaf area (± SE) regrowth (cm²) of *Festuca altaica* and *Mertensia paniculata* after a clipping (filled bars) treatment on fertilized and unfertilized plots in 1995, compared with new growth on unclipped (clear bars) plants.

Table 5.2 Predicted, and overall direction of change in plant biomass (or nutrient content), or standing crop, from each of the three experimental treatments according to the three hypotheses.

Model	Fertilizer Added		Herbivores Excluded		Fertilizer Added, Herbivores Excluded	
	Predicted	Observed	Predicted	Observed	Predicted	Observed
Bottom-up, donor control	+	Mostly verified (1.1, 1.2)	0	Verified (1.3)	+	Rejected (1.3)
Top-down control	0	Mostly rejected (2.1)	+	Rejected (2.2, 3.3)	+	Rejected (1.3)
Interactive control	+	Some support (3.1, 3.2)	+	Rejected (2.2, 3.3)	+	Rejected (1.3)

Numbers in parentheses refer to predictions in text. + = biomass increase, 0 = no change. There was variability in species' response to treatments in most cases; this table summarizes the overall direction of change.

investigate the rate of turnover of standing crop in response to fertilization did not support the hypothesis of higher turnover. This study used *Lupinus,* which, in hindsight, was an inappropriate choice of test species. Of the many common species in the area, a few (e.g. mosses, *Peltigera, Linnaea,* and *Arctostaphylos*) had variable responses to the addition of fertilizer, but only *Anemone* and *Lupinus* consistently failed to increase in abundance when fertilized. It was initially unexpected that soils inside the herbivore exclosure should have more nitrate nitrogen than soils in off-grid plots. However, results from some of the other studies showed that when vegetation is grazed, some of the species increase their relative growth rates. This additional growth due to grazing outside the exclosures may reduce nutrient levels to below those in the ungrazed exclosures.

The remainder of the studies indicate that the impact of mammalian herbivory, primarily by hares, on the boreal forest understory abundance and composition is minimal compared to that of fertilizer addition. Only *Achillea, Festuca, Arctostaphylos,* and mosses responded significantly to the exclusion of herbivores. We have been able to show some herbivory effects, but only when herbivore numbers were artificially elevated or when the vegetation was fertilized and the plant was significantly more attractive to hares. Herbivory effects are evident in the 5 m × 5 m plots where the results clearly show that the intensity of grazing increases when plots are fertilized. In addition, Dlott's (1996) study demonstrated herbivory effects, but only when herbivore numbers were experimentally elevated to artificially high levels.

Overall there is little evidence that natural levels of mammalian herbivory limit herbaceous vegetation of the Kluane region or affect its composition. In this area, herbivory has a major impact on the woody vegetation (see chapter 6), so the lack of response seen in the herbaceous vegetation could be due to a number of factors: (1) although snowshoe hares eat herbaceous vegetation, they do not eat enough to have an effect on plant population processes, relying on the woody vegetation for the bulk of their diet, even in summer. (2) If hare populations are more limited by a shortage of winter than summer food, then their populations may be maintained at a level below that at which an effect on the summer vegetation is seen. (3) Although apparently available, a large portion of the herbaceous vegetation may well be unavailable due to chemical defense. (4) These experiments were set up during a period of high snowshoe hare densities (1990) and in the 2 subsequent years the density of hares declined dramatically. It is quite possible that in the decline phase of the snowshoe hare cycle the hare's impact on its food supply is less than at other times. Ideally, this work would have begun 2–3 years earlier when snowshoe hare densities were increasing. This would have allowed us to assess the impact of herbivory during the 3 or 4 years of highest herbivore densities rather than the 2 years we were able to monitor.

Another of the primary predictions made was that, in response to fertilization, vegetation composition would change as more competitive species replace stress tolerators. Boreal plant communities are generally nutrient limited, particularly by nitrogen (Chapin and Shaver 1985, Bonan and Shugart 1989). When fertilizer is added to such low-nutrient communities, it will differentially promote the growth of some species, leading to changing light and moisture conditions for their neighbors. The effects range from causing a dramatic increase in abundance to causing a decline. These differential changes in biomass will in turn affect competitive relationships and lead to shifts in relative abundance and species diversity. *Achillea, Festuca,* and *Mertensia* increased in percent cover when fertilized, while *Anemone* declined. *Festuca* shows the most dramatic increases in abundance, from 20% to greater than 50% cover in fertilized plots. It is likely that the declines after fertilizer addition for some species are due to competition from the more rapidly growing species, especially *Festuca.* However, Arii's (1996) study demonstrated that *Anemone* suffered increased mortality with increasing levels of fertilizer, with or without neighbors, so mortality can be due to competition or due to the direct effects of fertilizer alone. In general, these results support current theory about the nature and process of plant competition for nutrient resources (Grime 1977, Tilman 1988), in which, as the ratio or total abundance of resources changes, species composition also changes. The results reported here are largely consistent with those of Nams et al. (1993), which showed an increase in *Festuca, Achillea,* and *Epilobium* with fertilization.

Increased nutrient uptake may be diverted into structures other than vegetative biomass. For example, reproduction may increase with nutrient levels. The total number of flowering stems of *Mertensia* increased only after 2 years, suggesting that there is a time lag between the acquisition of resources and the production of flowering stems. A high investment in flowering may divert resources from vegetative structures. However, as flowering stems in *Mertensia* are taller and have more leaves than nonflowering stems, their production may be an advantage in competing in a more productive community. *Mertensia* seedlings were not observed within our study plots, despite the large number of flowers produced. In recent long-term experiments on graminoids, Shaver and Chapin (1995) similarly found an increased production of ramets after 1 year of fertilization, followed by increased flowering in the second year.

5.4.1 Transient Dynamics and Stability

As the plant community changes in response to fertilization, we predicted that the initial responses will be transient, and more permanent shifts in vegetation composition will not be evident until later. The surveys in the 5 m × 5 m plots were completed for seven growing seasons, and they permit only an assessment of this prediction. The results of our study verify this prediction. For example, the increase in leaf production in *Mertensia* was apparent only in the third year of the study. *Anemone* initially responded positively to nutrient addition, and it was only after several seasons that its decline became apparent. In part, this may have been due to the long-term nature of the changes being induced by the treatments, such as increased biomass of other species and the gradual accumulation of nutrients in fertilized plots. The fact that initial responses to treatments may not reflect the more permanent long-term responses raises an important issue for experimental field ecology. Grubb (1982) in the United Kingdom and Inouye and Tilman (1995) and Tilman (1988) in Cedar Creek, Minnesota, have demonstrated such immediate "transient dynamics." In each of these cases, early surveys yielded useful information on the potential for short-term interference, but more permanent, stable responses to treatments did not materialize until later. Indeed, at Cedar Creek, changes in species composition were still occurring 10 years after the manipulations had taken place (Inouye and Tilman 1995). Such studies are particularly critical in systems where the species are long-lived and have limited rates of dispersal and establishment. In systems such as the boreal forest understory, it should be expected that new equilibrium conditions would be reached very slowly.

5.4.2   Impacts of Climatic Warming

Several authors have suggested that the application of nutrients to northern communities may produce some of the same effects in the plant community that might be caused by global environmental change (Aerts and Berendse 1988, Berendse and Jonasson 1992, Jonasson 1992, Berendse 1994). Global changes such as increasing CO_2 concentrations, increasing deposition of nitrogen and sulfur pollutants, and rising temperatures will have crucial impacts on nutrient cycles, leading to changes in primary production and species composition. Berendse and Jonasson (1992) argue that climate change will increase the supply of nutrients by stimulating decomposition processes and increase the rate of soil carbon accumulation. These changes will, of course, be modified by the interactions between plants and their environment. Reported responses from other systems such as heathland (Aerts and Berendse 1988, McGraw and Chapin 1989) and tundra (Chapin and Shaver 1985, Henry et al. 1986) show that some species will respond strongly to an increased nutrient supply at the expense of others, leading to an increase in biomass but a loss of diversity. In our system we might initially expect that bryophytes, lichens, prostrate growth forms, and low nutrient-requiring species will be suppressed or eliminated by faster growing, more upright clonal species such as the graminoids, *Mertensia paniculata* and *Achillea millefolium* var. *borealis.* Clearly, species and vegetation types with low nutrient-uptake demands will be the most sensitive to the predicted changes. At this stage it is also clear that the shrubs, chiefly *Salix glauca* and *Betula glandulosa,* would increase as well, at least initially, and one could only speculate as to the consequences of this additional food supply on the snowshoe hare cycle.

The majority of the data presented supports the bottom-up hypothesis that herbaceous vegetation standing crop is controlled by nutrient availability alone. With fertilization there was an overall increase in the amount of herbaceous vegetation, especially among the grasses, and a greater nutrient content of vegetation. *Festuca* showed the most dramatic increases in abundance, while *Anemone* and *Lupinus* consistently failed to increase in abundance when fertilized. It is likely that the declines after fertilizer addition for some species are due to competition from the more rapidly growing species. The studies also indicate that the vegetation is not under top-down control and that the impact of mammalian herbivory, primarily by hares, on the vegetation is very low. An effect can be induced when herbivore numbers have been artificially elevated or when the vegetation has been fertilized and the plants are significantly more attractive to hares.

5.5 Summary

Plants in the boreal forest form the physical structure of the community, and they provide the energy and nutrients to the herbivores on which higher trophic levels depend. Our studies focus on the herbaceous vegetation, primarily two grasses, *Festuca altaica* and *Calamagrostis lapponica* and four herbs *Lupinus arcticus, Anemone parviflora, Mertensia paniculata,* and *Achillea millefolium* var. *borealis.* These plants provide a source of relatively high-quality food to the herbivores. Understanding the factors that limit the quantity and the quality of plants is fundamental. Soil nutrients, especially nitrogen, often limit the productivity of boreal forest vegetation and may control vegetation standing crop. Conversely, herbivory may have a direct effect on vegetation quantity and quality. To understand some of the inter-trophic linkages between components of the system, three hypotheses regarding the vegetation were tested: that vegetation was controlled by (1) nutrient availability alone (bottom-up, or donor control), (2) by herbivores alone (top-down control), and (3) by both nutrient availability and herbivores. This involved three major experimental treatments: fertilization, herbivore exclusion, and fertilization plus herbivore exclusion. These treatments allowed us to make specific predictions about changes in plant biomass, or standing crop, under the three different hypotheses. The results support the bottom-up or donor control model for the herbaceous vegetation in the Kluane region of the boreal forest.

Appendix

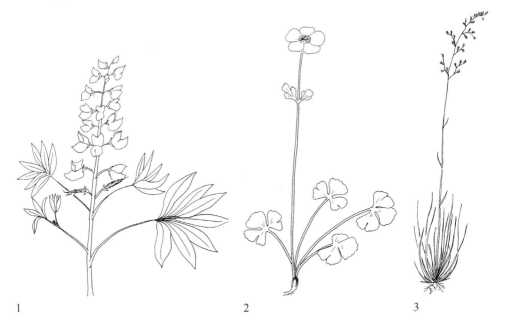

1   2   3

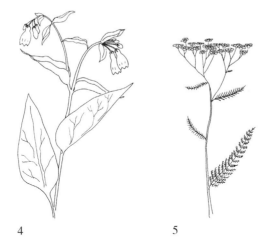

4   5

1. Arctic lupine (*Lupinus arcticus*)
2. Northern anemone (*Anemone parviflora*)
3. Northern rough fescue (*Festuca altaica*)
4. Bluebell (*Mertensia paniculata*)
5. Yarrow (*Achillea millefolium*)

Drawings by Shona Ellis.

Literature Cited

Aerts, R., and F. Berendse. 1988. The effects of increased nutrient availability on vegetation dynamics in wet heathlands. Vegetatio **76**:63–69.

Arii, K. 1996. Factors restricting plant growth in a boreal forest understory: a field test of the relative importance of abiotic and biotic factors. MSc thesis. University of British Columbia, Vancouver.

Berendse, F. 1994. Competition between plant populations at low and high nutrient supplies. Oikos **71**:253–260.

Berendse, F., and S. Jonasson. 1992. Nutrient use and nutrient cycling in northern ecosystems. *in* F. S. Chapin III, R. L. Jefferies, J. F. Reynolds, G. R. Shaver, J. Svoboda, and E. W. Chu (eds). Arctic ecosystems in a changing climate; an ecophysiological perspective, pages 337–356. Academic Press, San Diego, California.

Bonan, G. B., and H. H. Shugart. 1989. Environmental factors and ecological processes in boreal forests. Annual Review of Ecology and Systematics **20**:1–28.

Bryant, J. P., P. J. Kuropat, P. B. Reichardt, and T. P. Clausen. 1991. Controls over the allocation of resources by woody plants to chemical antiherbivore defense. *in* R. T. Palo and C. T. Robbins (eds). Plant defenses against mammalian herbivory, pages 83–102. CRC Press, Boston.

Chapin III, F. S., and G. R. Shaver. 1985. Individualistic growth response of tundra plant species to environmental manipulations in the field. Ecology **66**:564–576.

Coley, P. D., J. P. Bryant, and F. S. Chapin III. 1985. Resource availability and plant antiherbivore defense. Science **230**:895–899.

Dlott, F. K. 1996. Components of regulation of boreal forest understory vegetation: a test of fertilizer and herbivory. MSc thesis, University of British Columbia, Vancouver.

Graham, S. A. 1994. The relative effect of clipping, neighbours, and fertilization on the population dynamics of *Lupinus arcticus* (Family Fabaceae). MSc thesis, University of British Columbia, Vancouver.

Graham, S. A., and R. Turkington. 2000. Population dynamics response of *Lupinus articus* to fertilization, neighbour removal and clipping in the understory of the boreal forest. Canadian Journal of Botany **78**:753–758.

Grime, J. P. 1977. Evidence for the existence of three primary strategies in plants and its relevance to ecological and evolutionary theory. American Naturalist **111**:1169–1194.

Grubb, P. J. 1982. Control of relative abundance in roadside *Arrhenatheretim:* results of a long-term garden experiment. Journal of Ecology **70**:845–861.

Henry, G. H. R., B. Freedman, and J. Svoboda. 1986. Effects of fertilization on three tundra plant communities of a polar desert oasis. Canadian Journal of Botany **64**:2502–2507.

Hicks, S., and R. Turkington. 2000. Compensatory growth of three herbaceous perennial species: the effects of clipping and nutrient availability. Canadian Journal of Botany **78**:759–767.

Hilbert, D. W., D. M. Swift, J. K. Detling, and M. I. Dyer. 1981. Relative growth rates and the grazing optimization hypothesis. Oecologia **51**:14–18.

Huntly, N. 1991. Herbivores and the dynamics of communities and ecosystems. Annual Review of Ecology and Systematics **22**:477–503.

Inouye, R. S., and D. Tilman. 1995. Convergence and divergence of old-field vegetation after 11 yr. of nitrogen addition. Ecology **76**:1872–1887.

John, E., and R. Turkington. 1995. Herbaceous vegetation in the understorey of the boreal forest: does nutrient supply or snowshoe hare herbivory regulate species composition and abundance? Journal of Ecology **83**:581–590.

John, E., and R. Turkington. 1997. A 5-year study of the effects of nutrient availability and herbivory on two boreal forest herbs. Journal of Ecology **85**:419–430.

Jonasson, S. 1992. Plant responses to fertilization and species removal in tundra related to community structure and clonality. Oikos **63**:420–429.

Maschinski, J., and T. G. Whitham. 1989. The continuum of plant responses to herbivory: the influence of plant association, nutrient availability and timing. American Naturalist **134**: 1–9.

McGraw, J. B., and F. S. Chapin III. 1989. Competitive ability and adaptation to fertile and infertile soils in two *Eriophorum* species. Ecology **70**:736–749.

Nams, V. O., N. F. G. Folkard, and J. N. M. Smith. 1993. Effects of nitrogen fertilization on several woody and non woody boreal forest species. Canadian Journal of Botany **71**:93–97.

Palo, R. T., and C. T. Robbins (eds). 1991. Plant defenses against herbivory. CRC Press, Boston.

Shaver, G. R., and F. S. Chapin III. 1995. Long term responses to factorial, NPK fertilizer treatment by Alaskan wet and moist tundra sedge species. Ecography **18**:259–275.

Tallamy, D. W., and M. J. Raupp (eds). 1991. Phytochemical induction by herbivores. John Wiley & Sons, New York.

Tilman, D. 1988. Plant strategies and the dynamics and structure of plant communities. Monographs in population biology, no. 26. Princeton University Press, Princeton, New Jersey.

Turkington, R., E. John, C. J. Krebs, M. R. T. Dale, V. O. Nams, R. Boonstra, S. Boutin, K. Martin, A. R. E. Sinclair, and J. N. M. Smith. 1998. The effects of NPK fertilization for nine years on the vegetation of the boreal forest in northwestern Canada. Journal of Vegetation Science **9**:333–346.

6

Shrubs

CHARLES J. KREBS, MARK R. T. DALE, VILIS O. NAMS, A. R. E. SINCLAIR, & MARK O'DONOGHUE

Shrubs are an important component of the vegetation of the boreal forest because they provide complex structure where trees are absent or added dimensions in the forest between the herb layer and the tree layer. Shrubs are the winter food of the key species of herbivores in the boreal zone. Snowshoe hares and moose rely on browse from shrubs to get them through the winter period. One of the objectives of the Kluane Project was to obtain a good description of the changes in biomass and utilization of shrubs during the hare cycle, and in this chapter we present a summary of what we discovered.

6.1 The Shrub Community at Kluane

We include here the woody component of the plant community that grows between about 10 cm and 3–4 m in height in the Kluane boreal forest. We exclude from this discussion small trees (discussed in chapter 7) and the dwarf woody plants such as *Arctostaphylos uva-ursi,* which can be a dominant form of ground cover. In this section, we describe first the species that occur at Kluane and their relative abundances, the successional sequence in the shrub community, and the chemical defenses shrubs use against herbivores.

6.1.1 Species Composition

The shrub community in the Kluane region is dominated by gray willow (*Salix glauca*). For the 1700 shrub clip plots that we measured on control areas from 1987 to 1996, gray willow is 98.1% of the above-ground shrub biomass, bog birch (*Betula glandulosa*) is 1.25%, *Potentilla fruticosa* is 0.33%, and soapberry (*Shepherdia canadensis*) is 0.14%, on average. There are two other species of shrub willows in the Kluane area, but they are restricted in distribution (*S. alaxensis, S. scouleriana*).

Different experimental areas within the study region have highly variable shrub communities. Table 6.1 (Beals 1960) summarizes the prominence values for shrubs from the different treatment areas. Gray willow is common and is the dominant shrub on all the areas. A few differences stand out. Bog birch is prevalent on the two fertilizer treatments, food 2, and on the fence grid but nearly absent on control 1 and hare exclosure 1. Both soapberry and *Potentilla* are patchy in the study area.

The patchy nature of the shrub vegetation is difficult to portray with conventional measures and techniques, and we had to develop new methods to describe site heterogeneity. In this part of the boreal forest, it would be possible for a snowshoe hare to live in a 5-ha home range dominated by bog birch with soapberry very common. In other areas of the valley, no birch or soapberry would occur at all in the same size of home range, and the most general statement one can make is that every hare would have abundant gray willow within its home range anywhere in the valley.

6.1.2 Pattern Changes and Succession

Vegetation pattern analysis describes the spatial heterogeneity of the vegetation as well as the way this heterogeneity changes with time (Dale and Zbigniewicz 1997). We examined the effects of the experimental manipulations on the spatial pattern of the two major

Table 6.1 Prominence values of the major tall shrub species in the treatment areas in 1987–1988.

Grid	Salix glauca	Betula glandulosa	Potentilla fruticosa	Shepherdia canadensis
Control 1	92	4	14	0
Control 2	183	84	68	0
Fertilizer 1	57	164	33	0
Fertilizer 2	131	123	0	1
Food 1	106	15	0	0
Food 2	98	169	5	3
Fence	104	234	1	0
Fence + food	123	10	0	0
Hare exclosure 1	102	0	4	4
Hare exclosure 2	101	58	1	2

Prominence is measured by the relative cover and the relative frequency of the species in quadrats (Beals 1960). Because these values are means of only two 100-m transects, they give only a general view of the variation among sites. (Data from M. Zbigniewicz, personal communication.)

shrub species, *Salix glauca* and *Betula glandulosa,* before and after the 1989 population peak of the snowshoe hare. In this context, *spatial pattern* refers to the predictability of the locations of plants. A simple pattern is a regular alternation of high-density patches and low-density gaps. The intensity of such a pattern is the difference in density between the two phases. The scale of the pattern is the average of the patch and gap sizes (Dale and MacIsaac 1989). The scale of pattern of the vegetation may be an important habitat characteristic for herbivores, because for the same average density of plants, larger scales of pattern mean greater distances between patches of food plants or of cover.

Vegetation may have more than one scale of pattern, as when the patches occur in clusters. The effect of herbivores may be to break up patches into smaller units, causing a new small scale of pattern to develop in the vegetation. We predicted that treatments that increased snowshoe hare density would decrease the intensity of shrub pattern and cause the appearance of smaller scales of pattern. Treatments that decrease browsing or enhance the plants' ability to grow should increase intensity and cause the loss of small-scale pattern. We also predicted that moderate herbivory would decrease shrub patch size. Therefore, we investigated both pattern scale and intensity and patch size and used the data collected before and after the hare population peak to test our predictions.

We selected level areas occupied by shrub vegetation 0.5–2 m in height. Within each area we established two or more transects of 1001 contiguous quadrats, each 10 cm × 10 cm, and these were sampled in 1988 before the hare peak and in 1993. We recorded ocular estimates of the cover of all species in each quadrat.

We compared the 2 years by looking at the number of nonempty quadrats in each year and the density in them. We used two-sample *t* tests to compare the average densities of quadrats that were not empty in both years (Dale and Zbigniewicz 1997). The *t*-tests on the quadrat densities showed an overall positive effect on shrub cover attributable to fertilizer addition, even in the presence of herbivores, and to herbivore exclosure. The high and prolonged hare peak, caused by food addition and predator exclosure, reduced shrub cover, especially of *Betula,* due to a smaller proportion of the quadrats being occupied. The increase at fertilized sites could not be attributed to an increase in nonzero quadrats. The increase in *Betula* in the herbivore exclosure with fertilizer was due to increases both in the number of occupied quadrats and in the density. In the food-only grids and the untreated grids, the proportion of quadrats occupied decreased, while the average density in those occupied increased.

To investigate spatial scale, the data were analyzed using Hill's (1973) three-term local quadrat variance (3TLQV) because it is the best method to detect the scale of the pattern (Lepš 1990). The method calculates variance as a function of *block size,* the number of quadrats that are combined into larger units. Peaks and shoulders in the plot of variance as a function of block size reflect scales of pattern in the data (Dale and Blundon 1990). We concentrated on the smallest and most obvious scales of pattern revealed by the plots of variance. We compared years by looking at the intensity of individual peaks in the variance plot and at the total variance over the range of block sizes examined. The positions of peaks in the variance graphs were also compared between years to see whether the scales of pattern had shifted or whether scales had been gained or lost. Where there was a good match between the positions of variance peaks, we compared the intensity of pattern at that scale (for the calculation of intensity, see Dale and MacIsaac 1989).

Most of the sites showed some increases in total variance attributable to the proportional change in total cover. There were few dramatic changes in the 3TLQV graphs: most of the peak shifts are small, as are changes in intensity. For both species, the average scale of pattern was between 3 and 4 m. There was no consistent evidence of the appearance or disappearance of small scales of pattern.

Whereas Hill's 3TLQV analysis is used to detect the scale of pattern, Galiano's (1982) new local variance detects patch size by producing peaks in its variance plot at block sizes equal to the sizes of the patches or the gaps, whichever is smaller. In our data, the patches were almost always the smaller phase, and we looked at the smallest block sizes that produced clear peaks in the plot of variance.

There are some clear trends in patch size, such as an increase in *Betula* patch size at the three fertilized sites. At the control and food addition sites, patch sizes decreased or the variances associated with smaller sizes increased, showing that the smaller patches became more common (Dale and Zbigniewicz 1997).

The conclusion is that our early predictions were not supported by the data. The peak density of the herbivore between the years sampled seems to have had little effect on the pattern of the food plants. The intensity of pattern increased slightly at most sites as the cover in occupied quadrats increased. This applied particularly to sites that experienced normal or near normal peak densities. In spite of high rates of twig browsing during the peak, at most sites the basic characteristics of the spatial pattern recovered quickly. Only where food addition and predator exclosure enhanced and prolonged the hare density peak was there a sharp decline in the intensity of spatial pattern of the preferred winter food plant *Betula.* The addition of fertilizer produced favorable conditions for the plants' regrowth, whereas the combination of food addition and predator exclosure produced a clear effect at Hungry Lake, strongly reducing pattern intensity and patch size for *Betula.* The spatial pattern of these shrubs is resilient to normal changes in herbivory and therefore may persist for decades through several hare population cycles.

6.1.3 Secondary Chemicals in Kluane Shrubs

Plant defense theory argues that shrubs that are browsed by herbivores should attempt to defend themselves chemically to reduce herbivore damage (Bryant et al. 1994, Coley et al. 1985). Earlier studies (Sinclair and Smith 1984, Sinclair et al. 1988) have shown that phenolic compounds change over the hare cycle and are the most sensitive compounds to browsing. Phenolic compounds have been identified in other birch species (Reichardt et al. 1984), but they appear to be at low levels in willow species. A crude index of phenolic compounds can be obtained from methanol extraction. A 20-g fresh weight sample was taken from the twigs of gray willow and bog birch collected in the autumn for growth measurements. One 2-g sample was ground in a blender and then soaked in methanol for 2 days. The solvent was decanted and replaced with fresh methanol twice more. The combined solvent was then evaporated and the remaining extract weighed, and the results were expressed as a percentage (gram extract per gram wet weight of twig). We were unable to do replicate samples for many of the treatments, and our evaluation of significant changes in these indices of secondary chemical levels must rely on the replicates done on two control and two fertilizer grids. For birch and willow, differences of 4% or more among years or among treatments are approximately statistically significant.

For bog birch, this crude methanol extract showed a pronounced cycle coinciding with the hare cycle (table 6.2). All treatments except fence + food showed an initial low index in 1986 and 1987, followed by an increase in 1988, a peak in 1989, and still high but declining values in 1990. The index then fell to low values in 1991 and remained there until 1994, the last year of records. Birch values on the fence + food treatment remained low throughout the hare peak, but then dropped to even lower levels after 1991. Fertilizer treatments showed the same cycle as other treatments. These changes in secondary chemicals in birch are large (table 6.2).

In contrast, for gray willow there was much less apparent change from 1987 to 1994 (table 6.3). Values increased sharply for controls and food and fence treatments from 1987

Table 6.2 Crude methanol extract of secondary chemicals from bog birch (*Betula glandulosa*) current annual growth taken as a pooled sample from winter twigs in May of each year.

Year	Controls	Fertilized	Food	Predator Exclosure	Predatore Exclosure + Food	Hare Exclosure + Fertilizer
1986	26.2					
1987	26.7	22.3	25.1	22.3		
1988	39.1	27.1	30.1	38.3	27.9	20.6
1989	40.1	36.4	43.2	39.7	25.9	40.1
1990	?	30.4	31.4	35.1	28.5	30.8
1991	?	18.8	26.5	22.5	29.2	26.3
1992	23.8	20.5	20.5	18.9	18.9	19.3
1993	19.7	19.4	21.5	25.3	23.4	21.1
1994	26.3	19.8	22.5	20.0	23.6	19.5

Data are expressed as a percentage of the wet twig weight. The average standard deviation for replicate samples was 1.75, but on most areas only a single sample was analyzed. Unfortunately, there was no birch on the hare exclosure that was not fertilized.

Table 6.3 Crude methanol extract of secondary chemicals from gray willow (*Salix glauca*) current annual growth taken as a pooled sample from winter twigs in May of each year.

Year	Controls	Fertilized	Food	Exclosure Predator	Predator Exclosure + Food	Hare Exclosure	Hare Exclosure + Fertilizer
1986							
1987	14.4	13.0	13.3	17.8		15.8	
1988	19.1	17.1	21.8	20.5	13.9	19.7	13.3
1989		18.2	22.4	19.0			15.2
1990	16.3	15.9	17.4	18.0	14.9	15.8	12.1
1991	18.2	18.7	18.7	17.4	17.6	19.7	14.1
1992	17.2	15.6	17.1	16.6	17.1	17.5	13.6
1993	17.0	16.9	17.8	16.3	17.9	17.1	14.5
1994	15.5	13.2	15.5	15.8		16.8	13.1

Data are expressed as a percentage of the wet twig weight. The average standard deviation for replicate samples of controls and fertilized plots was 1.78, but on most areas only a single sample was analyzed.

to 1988 and then declined gradually to 1994. The index values for fence + food were lower than those for the controls or the fertilized grids. There was no clear cycle in the willow methanol extracts for fertilizer treatments, in contrast to the results for birch. Where hares were excluded, application of fertilizer appeared to result in a lower value of extract compared with that for the control area, but the differences were small.

There was thus a major difference in the secondary chemical responses of the two main shrubs at Kluane to the snowshoe hare cycle. Bog birch increased the level of chemical defense as hare numbers increased but did not maintain these high levels during the years of snowshoe hare decline from 1991 to 1994. The gray willow results, in contrast, suggest little change in secondary chemical levels through the hare cycle and only minor changes associated with the treatments.

6.2 Biomass Dynamics

Because of the intensity of browsing on shrubs associated with the snowshoe hare cycle, we put considerable effort into measuring the biomass dynamics of bog birch and gray willow through the 1986–1996 period. Because earlier studies by Keith et al. (1984) and Smith et al. (1988) indicated that snowshoe hares rarely browsed large twigs, we divided biomass for both birch and willow into small twigs (<5 mm diameter) and large twigs (>5 mm diameter). In this section we discuss the methods we used to measure biomass and the effects of the treatments on biomass of the two major shrubs in this part of the boreal region.

6.2.1 Methods of Estimation

In our previous studies we used nondestructive sampling to measure browsing responses over a hare cycle (Smith et al. 1988). We decided to change in this study to destructive sampling ("clip plots") because the repeatability of nondestructive methods would be low with so many individual observers involved. We used two general methods of destructive sampling to estimate standing biomass at the end of winter.

Quadrat Sampling We began in 1987 by setting out random quadrats of 1 × 2 m on two control areas, two fertilizer areas, and one hare exclosure. In 1990 we began sampling the predator exclosure + food area as well. The variance of 1 × 2 quadrats was so large that we looked for a better quadrat shape in 1988. The basic problem was that many quadrats contained no shrubs at all. We found in a trial analysis that long, thin quadrats, 10 m × 20 cm, could reduce sampling variation (CD-ROM frame 46). And so from 1989 onward we used these long, thin quadrats.

In spite of a statistical power analysis that indicated a sample size of 50 quadrats would give us precision of ±20% of the mean, we continued to find high variability from year to year in the estimates of standing crop. This variability resulted from habitat hetero-geneity on a fine spatial scale. We classified the habitat around each sampling point into six categories, but we were unable to improve precision by this stratification. Because we were unable to sample more than 50 quadrats in each sampling area, we had to be content with the data obtained.

Transect Sampling In 1993 we adopted a second approach to biomass determination. This approach was based on six transects of 6 × 600 m within each sampling area. In each transect the size of each willow or birch bush was measured (basal diameter, height, num-ber of main stems). By destructive sampling of a series of bushes of variable sizes, we de-veloped multiple regressions for each area to predict standing crops of these shrubs from these measurements. In all cases regressions with high levels of predictability were ob-tained. These transects had to be sampled only once if we assumed that the standing crop of large stems did not change much from year to year. We could then compute the stand-ing crop in any given year by correcting the biomass estimates by the spring ratios of small twigs to total biomass (these ratios were obtained from the clip plots). Because these spring ratios could be estimated precisely, this gave us biomass estimates of higher preci-sion than we obtained from simple clip-quadrat sampling. The limiting assumption of no change in large twig biomass would be true over a few years but would not hold over the time scale of succession (15+ years).

In all our analyses of bog birch and gray willow, we separated two size classes of twigs. Small twigs are <5 mm in diameter and represent the growth point of the shrubs. These form the main winter food for snowshoe hares, and thus we are particularly interested in the growth dynamics of this size class for these shrubs. Large twigs are >5 mm diame-ter and are typically not browsed by snowshoe hares or by moose. Some large twigs are girdled each year and otherwise die from natural causes.

6.2.2 Impacts of Treatments

The different treatment areas differed considerably in their average standing crop of gray willow and bog birch, as indicated in table 6.1, and these differences were present before any of the treatments were applied to experimental areas. From previous studies (Smith et al. 1988), we had expected the pattern of change in shrub biomass shown in fig-

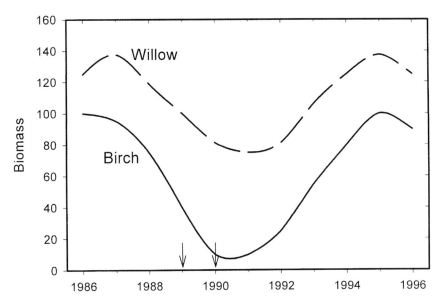

Figure 6.1 The expected pattern of biomass changes in small twigs (<5 mm diameter) of gray willow and bog birch in the Kluane area. Arrows indicate the peak density of snowshoe hares. Snowshoe hares can depress standing crops each winter by browsing, and, because they prefer bog birch over gray willow, we expected the impact to be much greater in bog birch. We also expected a time lag of 1–2 years in the recovery of the vegetation after the hare peak passed.

ure 6.1. Both willow and birch biomass should be depressed by snowshoe hare browsing, and this depression should be more severe in the preferred winter food plant, bog birch (Sinclair and Smith 1984).

Because there was so much variation in average standing crop of shrubs on the different treatment areas, we standardized the shrub biomass data. We used the midpoint of our data (1990) as the standard and calculated all shrub biomass relative to 1990. Figure 6.2 shows the relative standing crop of all shrubs on the two control sites and illustrates a completely unexpected pattern of biomass change. Shrub biomass increased as hare numbers increased, and did not decrease as predicted in figure 6.1. Shrub biomass peaked 2–3 years after hares peaked in abundance, and it appears that snowshoe hares, in fact, stimulated shrub growth as a side effect of their browsing. We had not anticipated that browsing would enhance productivity, as shown in figure 6.2. Note that figure 6.2 includes total shrub above-ground biomass.

Shrub biomass as well as species composition varied greatly among different sites. Figure 6.3 illustrates this for the two major shrub species, gray willow and bog birch, on each of the study sites and emphasizes three important points. First, bog birch was much less abundant overall than gray willow. Only 6.6% of the shrub biomass on all the areas studied was bog birch, and even on sites with the greatest birch abundance, birch reached only 15% of total shrub biomass. Second, the control areas in particular had very little birch

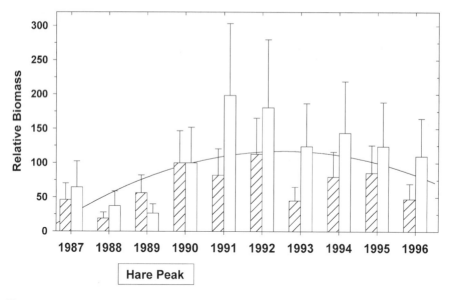

Figure 6.2 Relative biomass of all shrubs combined on two control areas at Kluane, with a second-degree polynomial regression and 95% confidence limits. Biomass peaked in the springs of 1992 and 1993, 3 years after the snowshoe hare peak. Biomass was standardized to spring 1990 = 100% on each area in order to compare them. For control 1, average biomass dry weight per square meter was 173 g, and for control 2 it was 913 g in 1990.

(1.2% of total shrub biomass). On some areas birch was virtually absent and thus cannot be a required food for hares. Third, small twigs made up a small fraction (9.7% on average) of the total standing crop of these shrubs.

Changes in standing crop of small twigs of willow and birch from year to year were highly variable because they were the result of two conflicting pressures: browsing offtake by hares and growth stimulation by hares (figure 6.2). The expected patterns are thus not easy to see in these data. One way to investigate the changes in standing crop is to determine the rate of change of standing crop from one year to the next. We define the rate of change as:

$$\lambda_t = \frac{\text{biomass in May of year } t + 1}{\text{biomass in May of year } t}.$$

Table 6.4 gives the average values of these rates of change for large and small twigs of willow and birch, and figure 6.4 plots the yearly changes for both species of shrubs.

Two important points emerge from these data. Table 6.4 shows that virtually all of these rates of change were positive, so that both large and small branches of both species were increasing in biomass each year, on average, by about 20–25%. This is a reflection of the pattern shown in figure 6.2 of an increase in biomass over most of the study period. We can decompose this trend for large and small (<5 mm diameter) branches of birch and willow. Figure 6.4 shows that for bog birch the rates of change of small twigs became neg-

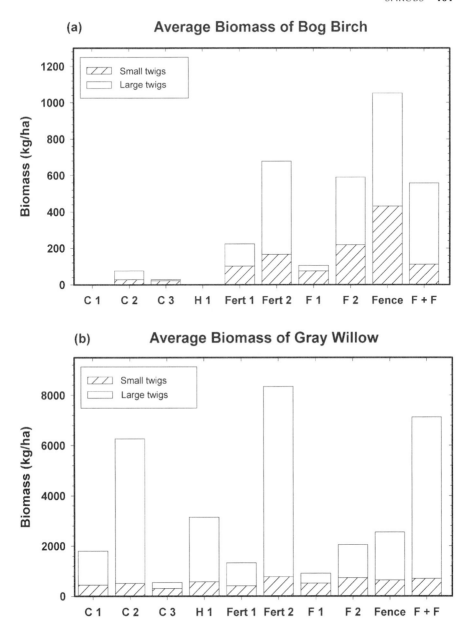

Figure 6.3 Average standing crop at the end of winter for the two major shrub species at Kluane Lake. Small twigs are <5 mm diameter; large twigs are all other above-ground stems. (a) Bog birch; (b) gray willow. C1 = control 1, H1 = hare exclosure 1, fert 1 = fertilizer 1, F1 = food 1, F2 = food 2, F+F = fence + food treatment. Data are averaged over all years.

Table 6.4 Average values of the finite rate of change of biomass per year for bog birch and gray willow for the period 1987–1996.

Treatment	Bog Birch		Gray Willow	
	Small Twigs	Large Branches	Small Twigs	Large Branches
Control 1	0.75[a]	1.63	1.03	1.41
Control 2	1.48	1.39	1.19	1.32
Hare exclosure 1	—	—	1.10	1.36
Fertilizer 1	1.20	1.48	1.16	1.19
Fertilizer 2	1.07	1.13	1.18	1.32
Fence + food	1.15	1.07	1.37	1.11
Grand mean	1.23	1.27	1.17	1.28

A rate of change of 1.0 indicates no change in biomass from year to year.
[a]Very small samples for birch due to restricted amounts present.

ative on control areas from 1988 through 1993, following the predictions shown in figure 6.1. For willow there is no apparent pattern and no relation to the snowshoe hare peak in 1989–1990. Willow apparently compensated for the average hare browsing pressure, in contrast to the prediction shown in figure 6.1, while bog birch did not.

Two processes combine to produce these effects on shrub biomass. Growth over summer adds biomass to both large and small twigs, and browsing as well as natural deaths cause losses to standing crops in both winter and summer. From the above analysis, we can see that, on average, the growth process seemed to outweigh the loss processes. We now turn to the estimation of these two components.

6.3   Growth Rates of Shrubs

In a previous study we developed a new nondestructive method for measuring the growth of individual tagged twigs by photographic means (Krebs et al. 1986). In this more extensive study, the photographic method became too laborious, and we developed a new method of destructive sampling to obtain an index of small twig growth for bog birch and gray willow.

6.3.1   Methods of Estimation

Each autumn, after the leaves had fallen, we collected from each of the study areas a sample of 200 live twigs of both birch and willow. These were frozen until the following May when we had time to measure them. For each twig we clipped off the terminal shoot at a diameter of 5 mm and discarded the larger pieces. We inspected each 5-mm twig for new growth from the previous summer and clipped off all this new growth. New growth was easy to distinguish on the basis of color of bark, the presence of resin glands in birch, and hairs in willow. The index of growth measured for each 5-mm twig was defined as:

$$\text{Growth index} = \frac{\text{dry weight of current annual growth on the 5-mm twig}}{\text{dry weight of the complete 5-mm twig}}$$

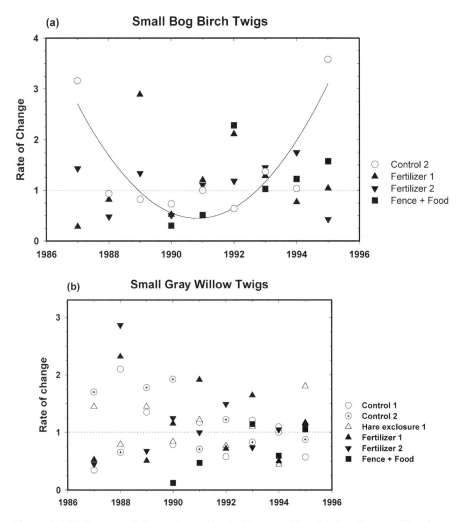

Figure 6.4 Finite rates of change in small-twig biomass of bog birch and gray willow from the various treatment areas. (a) Bog birch. Rates are <1 when hares are abundant. Curve is a second-degree polynomial fitted to control 2 data. (b) Gray willow. No trend is apparent.

and expressed as a percentage. This is not, strictly speaking, a growth rate because the twig itself also increased in diameter during the summer, and we measured only the extension growth component. Nevertheless, the true growth rate of the twig must be equal to or greater than this index of growth. The experimental unit was a single twig, and we did not take more than one twig from a single bush when we collected them in the autumn. We could have collected these twigs in spring instead of autumn, but we wanted to sample them before the snowshoe hares had removed their winter browse. All growth estimates were made on the basis of dry weights. We did not record any direct measure of large branch growth rates for shrubs.

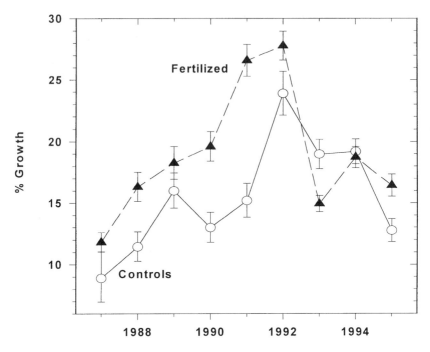

Figure 6.5 Growth rates of terminal branches of 5-mm twigs of bog birch on control and fertilized areas from 1987 growth year to 1995 growth year. Error bars indicate 95% confidence limits for each estimate. Snowshoe hares reached a peak in 1989 and 1990.

6.3.2 Impacts of Treatments

Bog Birch   Areas with small amounts of bog birch became impossible to sample once hares became abundant because they ate almost all the available birch. Consequently, we do not have samples of birch from all treatments in all years. There was a strong cycle in bog birch growth rates, with peak growth occurring 1 or 2 years after the hare peak had passed (figure 6.5). This cycle in growth was evident on both the fertilized areas and on the control areas. On average, over the entire study, fertilized birch twigs showed a 26% higher growth index (20.5%) than unfertilized twigs (16.2%). This difference masks 2 years (1993, 1994) in which fertilized growth rates were at or below control growth rates during the low of the hare cycle. Growth rates of birch on the fenced grid were no different from those on the controls, but the other treatments affected growth rates in unexpected ways. The fence + food treatment had the highest growth of 5-mm birch twigs (25.2% per year), a rate 55% above the controls. In contrast, the food 1 grid showed reduced birch growth (12.1% per year), only 74% that of the controls. The hare exclosure + fertilizer treatment showed birch growth equal to the fertilized plots (20.4%), so that there was no evidence that excluding hares from this plot either increased or decreased birch growth over that expected on fertilization alone. These results are summarized in figure 6.6.

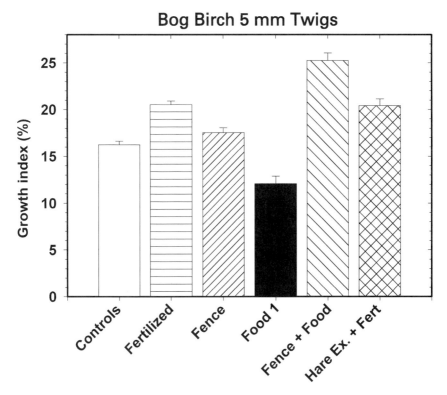

Figure 6.6 Average growth indices for 5-mm bog birch twigs for the various treatments, with 95% confidence limits. Averages were taken over 1987–1995 growth years.

Gray Willow Gray willow is the most common shrub in the Kluane region, so there was never any difficulty obtaining samples of 5-mm twigs for estimating summer growth rates. There was a strong cycle in willow growth rates on the fertilized grids, with peak growth occurring 1 or 2 years after the hare peak had passed (figure 6.7). This cycle in growth was not evident on the control areas, which showed a nearly linear trend toward lower growth rates with time. On average, over the entire study, fertilized willow twigs showed a 30% higher growth index (20.0%) than unfertilized willow twigs (15.4%). Growth rates of willow on the fenced grid were no different from those on the controls, but the other treatments affected growth rates in unexpected ways. The fence + food treatment had the highest growth of 5-mm willow twigs (23.2% per year), a rate 51% above the controls. The food 1 grid also showed increased willow growth (22.7% per year), 48% above that of the controls. The hare exclosure + fertilizer treatment showed willow growth equal to the fertilized plots (20.2%), so that there was no evidence that excluding hares from this plot either increased or decreased willow growth over that expected from fertilization alone. These results are summarized in figure 6.8. The patterns shown by birch and willow are identical except for the food 1 grid, which had decreased birch growth but increased willow growth.

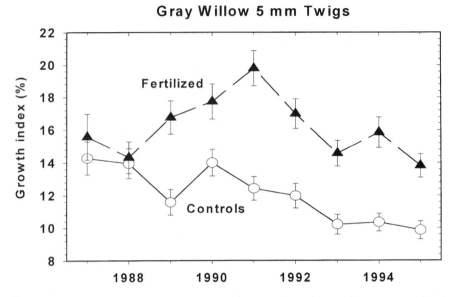

Figure 6.7 Growth rates of terminal branches of 5-mm twigs of gray willow on control and fertilized areas from 1987 to 1995 growth years. Error bars indicate 95% confidence limits for each estimate.

We interpret these effects as fertilization effects. Adding rabbit chow to the food grids also adds nutrients, either directly by the breakdown of uneaten chow or indirectly by the urine and feces of hares at high density. The amount of growth increase achieved by fertilization seems to be nearly the maximum that can be obtained for this ecosystem, and adding more fertilizer would have achieved little gain.

6.4   Losses of Twigs to Browsing and Natural Mortality

The demography of 5-mm twigs from willow and birch shrubs is affected by two principal sources of loss: browsing by snowshoe hares and natural mortality. Browsing by hares leaves a characteristic angular cut from the chisel teeth, but other forms of loss are more vague, and consequently natural mortality in our terminology includes all forms of death not caused by browsing. Moose browsing could be identified, but moose were so rare on our study areas that moose browsing was never more than a trace source of loss. In this section we discuss how the sources of loss changed over the years of the study and how they were affected by the snowshoe hare cycle.

6.4.1   Methods of Estimation

We determined the fate of 5-mm twigs by tagging 400 twigs from different bushes on each of 9 study areas. We studied both birch and willow on all areas except for food 2 and

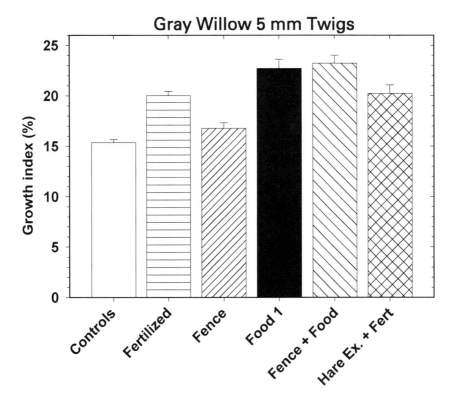

Figure 6.8 Average growth indices for 5-mm gray willow twigs from the various treatments, with 95% confidence limits. Averages were taken over the 1987–1995 growth years.

control 3. Each May we inspected each twig and classified it as intact, completely browsed by hare or moose, partially browsed, natural mortality, or accidental mortality (caused by our tagging or bending) (CD-ROM frame 47). Partial browsing removed part but not all of the live buds on the 5-mm twig, and we did not attempt to estimate the fraction removed in this study (as we did in previous work; see Smith et al. 1988). As twigs grew, we moved the aluminum numbered tags to keep each twig approximately 5 mm in diameter. Virtually all the browsing of willow and birch twigs occurs in the winter months between September and May, and we assumed all losses of twigs to be winter losses. On some areas birch was much less abundant than willow, and our sample sizes deviated from 200 of each species.

6.4.2 Impacts of Treatments

Bog Birch   Hares prefer to eat bog birch in winter, and consequently the browsing pressure on this shrub is intense in the Kluane region. Table 6.5 gives the percentages of birch twigs that were completely browsed by hares on the different areas during this cycle. Loss

Table 6.5 Percentage of 5-mm terminal twigs of bog birch (*Betula glandulosa*) browsed by snowshoe hares each winter.

Grid		Percentage Completely Browsed								
	1986–87	1987–88	1988–89	1989–90	1990–91	1991–92	1992–93	1993–94	1994–95	1995–96
Control 1	0	6	77	67	71	3	0	0	0	21
Control 2	0	6	41	79	91	43	1	0	3	4
Fertilizer 1	0	5	41	74	67	0	0	0	0	2
Fertilizer 2	0	2	31	46	58	9	1	1	4	1
Food 1	0	9	24	59	57	4	0	1	1	2
Fence	0	2	20	28	52	10	0	2	0	2
Fence + food	0	?	?	83	88	25	5	0	1	3
Hare exclosure 2[a]	0	0	34	5	0	0	0	0	0	1
Total controls	0.0	6.0	55.6	76.9	85.7	33.0	0.9	0.1	2.1	8.5
Total fertilized	0.0	3.5	37.0	62.9	61.7	5.4	0.7	0.6	2.6	1.1

[a]This grid should have no browsing by hares if the exclosure is perfectly operational.

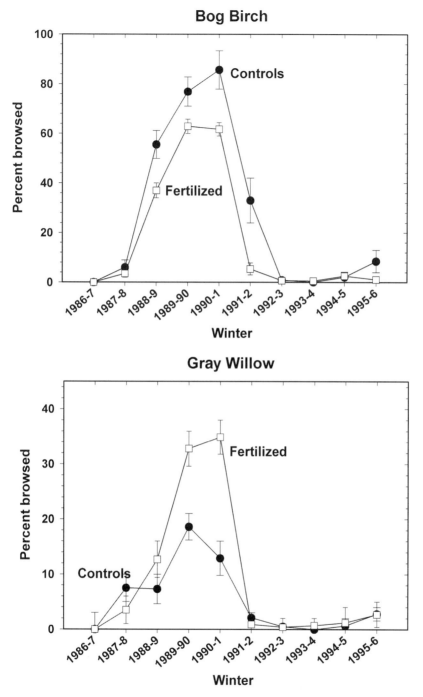

Figure 6.9 Percentage of 5-mm terminal twigs of bog birch and gray willow that were completely browsed by snowshoe hares over winter on control and fertilized areas. Two control areas and two fertilized areas are combined. Sample sizes were approximately 400 for each point. Error bars indicate 95% confidence limits.

rates due to browsing increase to 80–90% in the peak winters. Birch is not completely eliminated in this community, at least partly because of protection from snow. Once birch is buried by snow, hares do not have access to it until spring. Birch also grows well in large, open areas where hares do not often venture. Figure 6.9 shows that fertilized birch twigs were browsed at a slightly lower rate than control birch twigs, but these differences were not statistically significant. The browse rate for birch was high on the fence + food grid, which had many more hares than the controls.

In addition to complete browsing of tagged twigs, we recorded partial browsing in which the twig retained some growth buds. Partial browsing varied in tandem with complete browsing, and in the years of maximum browsing an additional 17–19% of birch twigs were partially browsed. In the 2 peak years of the hare cycle, this means that virtually every live bog birch twig had some browsing damage.

Natural mortality of small twigs also occurred, but it was always low in comparison to browsing offtake. On average, 3.3% of 5-mm birch twigs died from natural causes each winter. There was considerable variation from year to year (range 0–20%), but this variation was not associated with any particular treatment or year. In comparison with losses due to snowshoe hare browsing, natural twig deaths were relatively rare events, only about one-tenth as frequent as browsing losses.

Gray Willow   Gray willow was browsed at much lower rates than bog birch (table 6.6). Only on the fence + food grid with very high hare densities in 1990–1991 did the rate of complete browsing exceed 50% of the marked twigs removed in one winter. On average, about 20–30% browsing of willow twigs occurred in the peak phase. Willow utilization seemed particularly low on the control areas, with less than 20% removal at the maximum.

In addition to complete browsing of tagged twigs, we recorded partial browsing in which the twig retained some growth buds. Partial browsing varied in proportion to complete browsing, and in the years of maximum browsing, an additional 8–9% of willow twigs were partially browsed.

Natural mortality of small willow twigs also occurred. On average, 6.3% of 5-mm willow twigs died from natural causes each winter. There was considerable variation from year to year (range 0–20%), but this variation was not associated with any particular treatment or year. For gray willow the losses due to snowshoe hare browsing are, on average, almost the same as natural twig deaths. Averaged over 10 years, the probability of loss per year for a 5-mm willow twig is about 5–6% for browsing and 5–6% for natural mortality.

6.5   What Limits Primary Production of Shrubs?

In the broad sense, primary production in the boreal forest is limited by temperature, soil nutrients, and browsing. In this section we discuss how nutrients and browsing interact, within the confines set by temperature, to alter primary production of shrubs, particularly gray willow and bog birch in the Kluane region.

6.5.1   Succession in Boreal Forest Shrubs

Bog birch and gray willow are present from the earliest successional stages after fire in the Kluane region of the boreal forest. They presumably reach their peak biomass in

Table 6.6 Percentage of 5-mm terminal twigs of gray willow (*Salix glauca*) browsed by snowshoe hares each winter.

Grid	Percentage Completely Browsed									
	1986–87	1987–88	1988–89	1989–90	1990–91	1991–92	1992–93	1993–94	1994–95	1995–96
Control 1	0	14	9	18	17	1	1	0	1	4
Control 2	0	1	4	15	8	3	0	0	1	1
Fertilizer 1	0	3	14	41	47	1	1	0	2	2
Fertilizer 2	0	5	10	19	20	1	0	1	1	4
Food 1	0	5	22	41	47	6	0	0	6	10
Fence	0	7	4	22	11	7	1	0	0	2
Fence + food	0	?	?	47	63	16	3	0	2	3
Hare exclosure 1[a]	0	1	1	0	0	0	0	0	0	0
Total controls	0.0	7.5	7.3	18.6	12.9	2.1	0.5	0.0	0.6	2.8
Total fertilized	0.0	3.5	12.7	32.8	34.9	0.9	0.4	0.7	1.2	2.7

[a]This grid should have no browsing by hares if the exclosure is perfectly operational.

the early tree stage of succession, and once the forest begins to close canopy they begin to lose out, possibly to root competition by white spruce. In late successional stages mosses may dominate the forest floor and, in these stands, willow and birch are much reduced in abundance. Because the time frame of succession is so long, no one has made any direct observations on these trends. Competition for water would appear to be of minor importance in the Kluane boreal forest area, since there is typically sufficient summer rain as well as snow melt to replenish soil water. Competition for light would also seem to be minimal in most stages of succession except for the latest ones, and presumably the relative abundance of shrubs reflects more the balance between competitive ability for soil nutrients and browsing pressure by herbivores.

The addition of fertilizer increased the growth rate of birch and willow 25–30% each year. There was no indication of any differential effect on these two species, at least in biomass growth. We do not know the long-term consequences of fertilization, but it was clear that in 10 years only small changes occurred in the shrub community, in contrast to the herb community (see chapter 5). Browsing seemed to produce many more dramatic impacts than nutrient addition in this slow system.

6.5.2   Impact of Hare and Moose Browsing

By far the strongest pressure on the shrubs at Kluane is browsing, and almost all the browsing is done by snowshoe hares; moose are relatively rare. Snowshoe hares prefer bog birch to gray willow and gray willow to all the other shrubs. This preference for birch is, we think, the reason for the low relative abundance of birch in the forests of Kluane. Birch in the entire valley is only about one-tenth as abundant as willow. If hares could be excluded from an area, we think birch would be much more common. We did, however, see no sign of a birch resurgence on the hare exclosure plots, and we think this reflects the slow rate of change in the boreal forest ecosystem. (We had only one hare exclosure + fertilization treatment [4 ha] that had a very dense stand of birch. Unfortunately, we did not have a measurement of birch before setting up the treatment, nor were we doing clip plots to estimate the biomass of shrubs inside this plot during our study.)

Moose browsing on Isle Royale, Michigan, has been shown to have a strong impact on both tree and shrub communities. These impacts could be measured by comparing fenced and unfenced plots 40 years after they were set up (Pastor et al. 1988, McInnes et al. 1992). Snowshoe hares were able to enter the fenced plots in Michigan, but they were never the dominant browser in this system. Shrub biomass was lower in these fenced plots, presumably because of tree competition, and trees increased about 50% in biomass inside the moose exclosures after 40 years. Shrub biomass in our Kluane plots averaged 6250 kg/ha dry weight, about twice the average biomass of the Michigan plots, which had a completely different suite of species. We would guess from these results in the southern boreal zone that exclosures for Kluane hares would have to operate for at least 50 years to measure similar kinds of effects, if they would occur.

Although hare browsing exerts a dominant effect on bog birch, we were unable to see strong impacts on gray willow. There is considerable browsing on willow at the hare peak, but the large biomass of willow (95% of the shrub biomass is gray willow) reduces the impact of the hare browsing. Willow shrubs also seem to suffer more natural losses of branches, and in this sense may be preadapted to an approximately equivalent amount of loss from browsing.

One of the most striking results of our studies on the shrub community at Kluane has been the finding that hare browsing seems to stimulate shrub production. We presume that this occurs either through nutrient recycling with a time lag of 2–3 years after the hares peak in abundance or as a physiological response of the shrubs to browsing itself. This stimulation effect is shown clearly in the fence + food treatment, which had the highest observed growth rates for both willow and birch twigs in spite of having no direct nutrient addition as fertilizer. The high densities of hares on this grid (see chapter 8) explain the growth stimulation. These results resemble the findings of McNaughton (1985), who showed that grazing in the Serengeti increased primary production.

6.5.3   Role of Secondary Chemicals

Secondary compounds appear to be responding directly to the influence of browsing by hares—the heavier the browsing, the higher the values of crude methanol extract. However, where browsing was extremely high in the fence + food treatment, secondary compounds were inhibited, much as the shrub growth was depressed. Thus, regrowth appears to be a compensatory response to browsing, and the secondary compounds appear to be a possible deterrent to further browsing. There is experimental evidence that such extracts do inhibit both feeding behavior and digestive abilities (Sinclair et al. 1982, 1988; Rodgers and Sinclair 1997). This effect is most apparent in bog birch and less so in gray willow. Because bog birch is the preferred species of winter food for hares, chemical defense is perhaps of higher value to this plant.

Fertilizer had the effect of reducing the secondary compounds in both species, although the effect was not large. The result is consistent with hypotheses proposing that secondary compounds may function to protect nutrients that are hard for plants to obtain (Coley et al. 1985), where nutrients are provided, there is less stimulus for the plants to produce secondary compounds to defend the nutrients.

6.6   Summary

Because snowshoe hare browsing can be severe at the peak of the hare cycle, we had expected shrub biomass to decline as hares increased. In contrast, we found that total shrub biomass increased with increased browsing, so that over the 10 years of study there was a net increase in shrub biomass. Browsing by hares seemed to stimulate primary production of shrubs in this system. The spatial pattern of shrub-dominated areas also recovered quickly after the snowshoe hare peak.

Hares prefer to eat bog birch in winter, and browsing rates reached 80–90% in the peak winters of 1989–1990 and 1990–1991. Biomass of small birch twigs decreased as hares increased, but the same pattern was not seen in small willow twigs. Browsing on gray willow twigs was always much less than browsing on birch and reached peaks of 20–40% on most areas. Bog birch would not exist in forested sites at Kluane if it was not protected by snow cover for much of the winter, and we suggest that hare browsing is responsible for the relatively low abundance of birch in the Kluane region.

Fertilization increased the growth rates of all the shrubs by about 25–30% over control values. Fertilized willow twigs were eaten at a higher rate than control twigs, but the opposite tendency was shown by bog birch.

Excluding hares from areas had little impact on any of our measures of biomass or growth in willow or birch, and we think that processes in the Kluane ecosystem are too slow to show impacts in less than 50 years of hare exclusion.

Literature Cited

Beals, E. 1960. Forest bird communities in the Apostle Islands of Wisconsin. Wilson Bulletin **72**:156–181.

Bryant, J. P., R. K. Swihart, P. B. Reichardt, and L. Newton. 1994. Biogeography of woody plant chemical defense against snowshoe hare browsing: comparison of Alaska and eastern North America. Oikos **70**:385–395.

Coley, P. D., J. P. Bryant, and F. S. Chapin III. 1985. Resource availability and plant antiherbivore defense. Science **230**:895–899.

Dale, M. R. T., and D. J. Blundon. 1990. Quadrat variance analysis and pattern development during primary succession. Journal of Vegetation Science **1**:64–153.

Dale, M. R. T., and D. A. MacIsaac. 1989. New methods for the analysis of spatial pattern in vegetation. Journal of Ecology **77**:78–91.

Dale, M. R. T., and M. W. Zbigniewicz. 1997. Spatial pattern in boreal shrub communities: effects of a peak in herbivore densities. Canadian Journal of Botany **75**:1342–1348.

Galiano, E. F. 1982. Détection et mesure de l'hétérogénéité spatiale des espèces dans les pâturages. Acta Oecologia Plantarum **3**:269–278.

Hill, M. O. 1973. The intensity of spatial pattern in plant communities. Journal of Ecology **61**:225–235.

Keith, L. B., J. R. Cary, O. J. Rongstad, and M. C. Brittingham. 1984. Demography and ecology of a declining snowshoe hare population. Wildlife Monographs **90**:1–43.

Krebs, C. J., A. R. E. Sinclair, R. Boonstra, and J. N. M. Smith. 1986. A photographic technique for estimating browse growth and utilization. Wildlife Bulletin **14**:286–288.

Lepš, J. 1990. Comparison of transect methods for the analysis of spatial pattern. *in* F. Krahulec, A.D.Q. Agnew, S. Agnew, and J. H. Willems (eds). Spatial processes in plant communities, pages 71–82. Academia Press, Prague.

McInnes, P. F., R. J. Naiman, J. Pastor, and Y. Cohen. 1992. Effects of moose browsing on vegetation and litter of the boreal forest, Isle Royale, Michigan, USA. Ecology **73**:2059–2075.

McNaughton, S. J. 1985. Ecology of a grazing ecosystem: the Serengeti. Ecological Monographs **55**:259–294.

Pastor, J., R. J. Naiman, B. Dewey, and P. McInnes. 1988. Moose, microbes, and the boreal forest. BioScience **38**:770–777.

Reichardt, P. B., J. P. Bryant, T. P. Clausen, and G. D. Wieland. 1984. Defense of winter-dormant Alaska paper birch against snowshoe hares. Oecologia **65**:58–69.

Rodgers, A. R., and A. R. E. Sinclair, 1997. Diet choice and nutrition of captive snowshoe hares (Lepus americanus): interactions of energy, protein, and plant secondary compounds. Ecoscience **4**:163–169.

Sinclair, A. R. E., and J. N. M. Smith. 1984. Do plant secondary compounds determine feeding preferences of snowshoe hares? Oecologia **61**:403–410.

Sinclair, A. R. E., C. J. Krebs, and J. N. M. Smith. 1982. Diet quality and food limitation in herbivores: the case of the snowshoe hare. Canadian Journal of Zoology **60**:889–897.

Sinclair, A. R. E., C. J. Krebs, J. N. M. Smith, and S. Boutin. 1988. Population biology of snow-shoe hares III. Nutrition, plant secondary compounds and food limitation. Journal of Animal Ecology **57**:787–806.

Smith, J. N. M., C. J. Krebs, A. R. E. Sinclair, and R. Boonstra. 1988. Population biology of snowshoe hares II. Interactions with winter food plants. Journal of Animal Ecology **57**:269–286.

7

Trees

MARK R. T. DALE, SHAWN FRANCIS, CHARLES J. KREBS, & VILIS O. NAMS

7.1 Tree Community at Kluane

The tree flora of the Kluane Valley is remarkably depauperate, even for the boreal forest, which is not known for high diversity of tree species. Three species of trees are present. The only species of conifer is the white spruce, *Picea glauca* (Moench) Voss. The absence of its congener, the black spruce, *Picea mariana* (Mill.) BSP, which is widespread and common in other parts of the boreal forest, is surprising, particularly since black spruce occur to the north and to the east of Kluane. The other two tree species in the valley are angiosperms of the genus *Populus:* trembling aspen, *Populus tremuloides* Michx., and balsam poplar, *Populus balsamifera* L. In spite of their ability to spead clonally, the two species of *Populus* are comparatively rare in the valley, making spruce the dominant tree in area, cover, and biomass. This chapter is therefore primarily about white spruce. Although some herbivores feed on parts of the spruce trees (e.g., red squirrels eat the seeds), these trees are more important to the other organisms in the ecosystem because of the physical structure they create. The second way in which spruce are important is as primary producers because they are responsible for a large portion of the net primary productivity in the valley. Although this production does not benefit all the herbivores directly, it affects them indirectly through the accumulation of litter, which provides cover and in turn affects their food plants through its physical and chemical properties.

An obvious feature of the tree vegetation of the study area is its heterogeneity. Much of the valley is covered by spruce forest, but it varies from open stands to closed canopy and has a range of ages due to the disturbance regime, of which fire is the dominant feature. Fire initiates forest succession and maintains the heterogeneous character of the boreal forest vegetation mosaic. In general, the North American boreal forest fire regime is characterized by large, high-intensity, stand-replacing wildfires with short return intervals (Wein and MacLean 1983, Johnson 1992). Few other natural disturbance mechanisms control vegetation dynamics on such large scales in this region. To assess the vegetation of the valley and its spatial heterogeneity, we undertook several studies, including a fire history of the valley.

Some historical climate information is available for the region in the form of a dendroclimatic study by Allen (1982). Like most areas in northwestern North America, our region has experienced a warming trend since the end of the Little Ice Age in about 1820. The current climate is, however, still cold and dry because of the rain shadow of the adjacent St. Elias Mountains. It has a mean annual temperature of $-3°C$ and mean annual precipitation of about 230 mm, most falling as rain during the summer.

The predominant wind direction during summer is from the southwest, but local wind patterns are influenced by glacial effects, topography, and valley orientation. The large side valleys, which enter perpendicular to the main trench (the Alsek, Slims, and Jarvis rivers), can funnel katabatic and glacial winds into the trench, creating brief wind storms and spectacular dust clouds along loess-filled stream beds. These wind characteristics may be important during forest fires.

The frequency of lightning storms can strongly influence the frequency and spatial pattern of fires on a landscape (Johnson 1992). Lightning, however, is rare in the study area compared with neighboring systems because the Kluane region lies outside the major "lightning belt" of central Yukon Territory (Hawkes 1983). This low frequency of lightning may also be an orographic effect of the St. Elias Mountains.

117

Table 7.1 Tree abundance on the study grids estimated from T-square sampling.

Grid	White Spruce		Trembling Aspen		Balsam Poplar	
	Small Trees	Large Trees	Small Trees	Large Trees	Small Trees	Large Trees
Control 1		421 ± 80	0			0
Control 2						
Control 3	907 ± 104	544 ± 92	0		0	0
Food 1	768 ± 122	414 ± 76	545 ± 256	0	137 ± 131	48 ± 123
Food 2						
Fertilizer 1	985 ± 142	109 ± 38				
Fertilizer 2						
Fence	685 ± 204	164 ± 28	611 ± 282	44 ± 15	0	0
Fence + food						

Small trees are >10 cm height and <10 cm dbh. Large trees are >10 cm dbh. Estimates are stems per hectare (± 1 SE).

7.1.1   Tree Abundance on the Study Area

Tree abundance was estimated on each of the control and treatment areas by T-square sampling (Krebs 1999). Stems were estimated separately for small and large trees, with 10 cm diameter at breast height (dbh) as the point of separation. Small trees shorter than 10 cm were not measured in these samples. Because the vegetation of the valley is so patchy (see CD-ROM frame 12), we attempted to cover the entire area of each grid to provide an average value for tree density.

Table 7.1 gives the estimates of tree abundance for five of the three control areas and the six experimental areas, and table 7.2 gives the overall habitat classification for each of these nine areas from the airphoto (see 2.3, table 2.2). The tree component of the vegetation differs substantially among these nine areas. The most dense closed spruce forest occurs on control 2, control 3, food 2, fertilizer 2, and fence; the most open forest types occur on fertilizer 1, food + fence, food 1, and control 1.

7.1.2   Tree Growth

We estimated tree growth only for white spruce. Spruce twigs grow from the distal end, and each year of growth is marked by a growth ring that circles the stem. The most recent year of growth may have lateral buds, but these buds do not grow into branches until the next year. The length of each year's growth provides a measure of the tree's vigor. We can measure growth in branch length for the last 3–4 years and thus quantify year-to-year variation in growth rates as well as fertilizer effects. We used relative growth within each tree to control somewhat for variation among individual trees. We adopted 1986 as the base year and expressed all growth as relative to growth in that year; for example,

$$\text{Relative growth for 1990} = \frac{\text{Length of branch growth in 1990}}{\text{Length of same branch growth for 1986}}$$

In 1989 two observers measured 120 branches in duplicate to calculate repeatability of the measurements on growth. Repeatability was 0.98, so there seemed to be no need to mea-

Table 7.2 Percentage of vegetation classes of the nine major study sites in the Kluane study area.

Vegetation Type	Control 1	Control 2	Control 3	Food 1	Food 2	Fertilizer 1	Fertilizer 2	Fence	Fence + Food
Closed spruce (50–100% cover)	18.87	61.45	55.28	7.26	48.03	11.73	46.34	32.91	0.57
Open spruce (25–50% cover)	34.37	18.43	15.66	54.91	28.07	44.26	20.59	28.81	30.99
Poplar–aspen	2.16	3.09	0.00	7.97	0.29	5.77	0.25	1.74	5.88
Willow shrub	21.00	11.05	18.28	27.27	11.66	32.68	27.56	26.08	51.09
Grass-open	22.96	5.98	10.01	2.59	11.89	5.56	5.26	10.45	11.47
Water	0.64	0.00	0.77	0.00	0.05	0.00	0.00	0.00	0.00
Totals	100	100	100	100	100	100	100	100	100

Estimates were obtained from airphoto analysis as described in Chapter 2. Each study site is a 1-km^2 block centered on the snowshoe hare study grid. Percentages are given for the five main habitat types and water.

Table 7.3 White spruce branch extension growth on two control areas
at Kluane, 1986–1995.

Year	Control 1 Mean Growth	SE	Control 2 Mean Growth	SE
1987	0.936	0.030	0.882	0.030
1988	0.949	0.036	0.823	0.027
1989	1.069	0.048	0.881	0.031
1990	1.080	0.063	0.900	0.042
1991	1.030	0.056	0.881	0.045
1992	0.973	0.058	0.876	0.045
1993	0.903	0.037	1.190	0.128
1994	0.870	0.043	1.285	0.143
1995	0.658	0.045	1.058	0.090

Growth rates are expressed as ratios relative to 1986 = 1.00, the year the study began. We measured 75 trees on each area in each year.

sure everything twice. We also sampled three branches on each tree to measure variation within one tree as well as variation between different trees. These two sources of variation were nearly equal, and the optimal sampling design was thus to measure only one branch per tree.

Table 7.3 gives the estimated white spruce growth ratios for two control areas for 1987–1995. Spruce growth varied markedly from year to year; 1987 and 1988 were years of relatively poor growth, and 1990 and 1991 were years of good growth. During the last 2 years of the study, some growth rates deteriorated because of spruce bark beetle attacks (see 12.3).

7.1.3   White Spruce Cone and Seed Production

Spruce trees tend to produce some cones each year, but there are certain years, called *mast years,* in which all of the trees produce huge numbers of seeds. This occurs at irregular intervals but over large geographic areas. Spruce seeds are an important food source for red squirrels, chipmunks, mice, voles, and passerine birds. We sampled seed production by spruce in two ways: by counting cones and by collecting seeds in buckets. We used both methods because red squirrels tend to harvest many cones before they open each year. As a consequence, counting cones on the trees in August tells us how much is produced by the trees, and counting seeds falling to the ground over the next 8 months tells us how much is available to other small mammals and birds after the red squirrels have finished harvesting.

We collected seeds in buckets 28 cm in diameter, placed at 86 hare-trapping stations on control 1, food 1, fertilizer 1, fence, and fence + food grids. The buckets were set out in late August and collected the following May. The buckets were covered with wire to prevent animal access over winter. Only intact spruce seeds were counted.

We estimated cone production in the first week of August of each year on all the seed grids listed above as well as on control 2 and fertilizer 2 grids. On each grid, 86 spruce

Table 7.4 White spruce cone crop index and seed production on control areas, Kluane Lake, 1986–1995.

Year	Cone Index per Tree			Number of Seeds per m^2		
	Mean Count	Lower 95% CL	Upper 95% CL	Mean Count	Lower 95% CL	Upper 95% CL
1986	77	70	85	194[a]	80[a]	376[a]
1987	39	36	43	68[a]	33[a]	145[a]
1988	57	50	63	31	16	68
1989	0	0	0	0	0	0
1990	35	28	42	34	5	150
1991	2	1	3	2	0	3
1992	72	62	83	129	67	223
1993	164	146	183	675	416	1090
1994	2	1	3	20	7	36
1995	102	90	115	460	256	696

Control 1 and Control 2 were counted for cones, but only Control 1 was used for seed production estimates. Year is the year of seed production; $N = 86$ for all counts.

[a] Estimated from regression of seed counts on cone counts from 1988 to 1995.

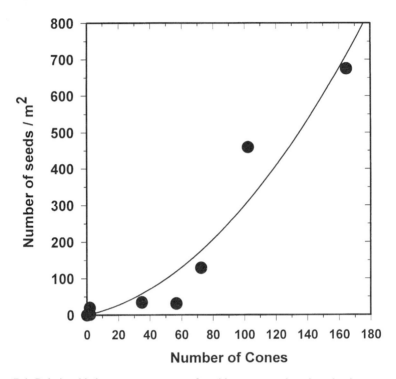

Figure 7.1 Relationship between cone counts for white spruce and seed production per square meter for 8 years, 1988–1995, on control 1 grid. Cones were counted in August while still maturing, and seeds were counted as an accumulation falling to the ground over the following winter months. Squirrels harvest cones before they release seeds, and for low cone crop years the seeds that fall to the ground are less than one would predict from cone production counts. Regression line is a second-degree polynomial, $y = 0.9210\,x + 0.02080\,x^2$ with $R^2 = .93$.

trees that were 5 cm dbh or larger were counted, and the same trees were recounted each year. The number of new cones in the top 3 m of the tree was counted using binoculars. If the total number of cones exceeded 100, we took a photograph of the top of the tree using a telephoto lens and, using a magnifying glass, counted cones on the photographs.

Table 7.4 gives the white spruce cone counts and the seed production for control areas from 1986 to 1995. High cone counts occurred in 1986, 1993, and 1995. Complete cone failure occurred in 1989, 1991, and 1994. Figure 7.1 shows the relationship between cone counts and seed production for white spruce on the control areas. Seed production from small cone crop years is less than one would predict from the regression. This could result from red squirrels harvesting a high fraction of the cones in low cone crop years or could result from measurement error.

7.2 Vegetation Mapping

The vegetation can be divided into three ecological zones based on elevation (Douglas 1980): montane valley bottom forests (760–1080 m), subalpine forests (1080–1370 m), and alpine tundra (above 1370 m). The two lower zones are complex mosaics of forests of white spruce, stands of aspen and balsam poplar, and shrub-dominated areas of willow (*Salix* spp.) and dwarf birch (*Betula glandulosa* Michx.). The subalpine vegetation, consisting of open-canopy spruce mixed with tall willow shrubs, grades into the low shrub–dwarf plant communities of the alpine.

To assess the processes leading to the heterogeneous vegetation of the valley, we mapped the vegetation and developed a fire history of the area. Based on three bands of LANDSAT image, we created both supervised and unsupervised classification maps of the vegetation (see Dale 1990). Based on this classification, closed and open spruce account for 32% and 30% of the area, with shrubland covering another 26% and *Populus*-dominated stands only 3%. A revised version of a supervised classification (Kenney and Krebs unpublished data) was imported into the SPANS Geographic Information System, (GIS) which was then used to make comparisons of individual animal's home ranges and other areal characteristics. A digital elevation model of the valley was also produced for use in the GIS environment. Useful as these images were, they did not provide sufficient detail to serve as a precise map of vegetation types, and their resolution was not appropriate to serve as a base for a fire history study. We decided, therefore, to use airphoto interpretation for these purposes. The valley was flown for airphotos in July 1992 and the 317 1:10,000 black-and-white photos provide complete coverage of the study area (see 2.3, table 2.2).

7.2.1   Fire History of the Study Area

In addition to providing the information for a 1:32,000 vegetation map (Hucal and Dale 1993), the airphotos were used as the basis for the valley's fire history. To begin the fire history analysis, a 1:50,000 universal transverse mercator (UTM) grid was set out on acetate airphoto overlays using a transfer scope. Based on interpretation of canopy-height differences on these airphotos, distinct stand margins were identified and sample sites chosen.

The field sampling for the fire history took place during the summers of 1994 and 1995. Along distinct fire margins, fire-scarred trees were found and their locations recorded on the airphotos. At these locations, we collected the following: (1) disks from fire-scarred trees, (b) two or more increment cores (at 30 cm height) from large canopy trees in the post-fire regenerating stand and in the adjacent unburned stand, (3) height and diameter at breast height from all trees cored or cut, (4) physical site description, and (5) general stand information. The cores from unburned trees were collected to provide an estimate of the regeneration lag following fire. Obtaining complete tree cores in stands greater than 200 years old was difficult because of heart rot.

We dated all fire-scarred tree disks (362) and tree increment cores (more than 1500) using the techniques of Yamaguchi (1991) and McBride (1983). Relative ring-width patterns were noted, and a tree marker-year chronology was developed to cross-date samples using the method of Yamaguchi (1991). This method allows the detection of missing and false rings and creates an accurate tree chronology.

We reconstructed individual fires by transferring fire scar and tree increment core dates to acetate overlays using the 1:32,000 scale vegetation map as a base layer. Individual fires were reconstructed back to 1800; the reconstruction of earlier fires is inaccurate due to a shortage of fire scar evidence and spatial resolution. In the absence of fire scar evidence for stands originating before 1800, the time since fire was estimated as 25 years more than the age of the oldest tree found. The increment of 25 years was based on the average post-fire regeneration lag we determined from regenerating stands, and it agrees well with the value reported by Hawkes (1983) for the neighboring Kluane National Park.

Acetate fire boundary overlays were digitized in the "v.digit" module of the GRASS 4.1 (U.S. Army Corps of Engineers 1993) geographic information system. A separate data layer was constructed for each fire year back to 1800, and the annual area burned was calculated with the GIS. We then converted these vector maps to raster maps and combined with the "r.patch" command to produce a time-since-fire map for the period 1800–present. In this map, the most recent fires have definite boundaries and overlie earlier events. Areas of overlap can then be calculated and displayed to produce a fire frequency map for this time period. For areas originating before 1800, with no fire scar evidence and unknown fire boundaries, a single stand origin data layer was digitized and patched with the area burned from 1800–present. This combination of stand origin dating methods provided a complete time-since-fire map for all areas below treeline.

Fire history statistics for 1800–1994 are summarized in table 7.5 (CD-ROM frame 50). There has not been a wildfire in the study area since 1956. The last fire event to burn more than 200 ha occurred in 1929 near Kloo Lake, and this was the only fire with written documentation. Glover (1929), on a routine Royal Canadian Mounted Police patrol, noted "A forest fire was burning in the vicinity of Kloo Lake and considerable timber had fallen across the government road." Figure 7.2 displays the location of all areas burned since 1800.

Table 7.5 gives the area burned annually in the valley from 1800 to 1994. Some of the larger fire years were synchronous with large fire events (late 1840s and 1880s) across western North America, as indicated by fire history studies performed at Jasper National Park (Tande 1979), the Boundary Waters Canoe Area (Heinselman 1973), and the Bitterroot National Forest (Arno 1976). These fire years may be characterized by particular weather conditions (Johnson and Wowchuck 1993). Some of the large fire years in the Shakwak Trench, however, are not synchronous with the rest of the continent, indicating that small-scale, localized weather systems may also be important. Similarly, large fire years in the Kluane National Park are not all synchronous with those in the Shakwak Trench (Hawkes 1983). The extreme topography of the Kluane region may be responsible for highly variable conditions over small distances.

Fire sizes are highly variable, with a few large fires being responsible for most of the area burned. This fire regime is characteristic of conditions throughout the boreal forest of North America (Johnson 1992). The largest individual fire event during the past 200 years affected 12.46% of the forested area within the study site. Large areas of very old, contiguous subalpine and upper-montane forest along the south study boundary may have arisen from a single fire that affected more than 6000 ha. Generally, individual fires appear to be smaller than in other areas of the boreal forest with more subdued topography. Individual fire events larger than 10000 ha are relatively common in other parts of the boreal forest.

Individual burn patterns are complex and variable. Some recent burns with easily detectable margins display classic elliptical shapes (Anderson 1983, Alexander 1985). Other burns have irregular, complex margins and do not display patterns associated with strong

Table 7.5 Fire history of the 350-km^2 main study area at Kluane from 1800 to 1994 (fires are listed in reverse chronological order).

Year of Fire	Area Burned (km^2)	Percent of Total Study Area	Interval between Fires	No. of Spatially Discrete Fires
1956	<0.01	<0.01	39	1
1953	0.01	<0.01	3	1
1951	0.01	<0.01	2	1
1943	0.04	0.01	8	3
1939	0.11	0.03	4	1
1937	0.16	0.05	2	1
1936	1.22	0.35	1	3
1934	1.10	0.31	2	3
1932	0.51	0.14	2	2
1930	0.04	0.01	2	1
1929	7.38	2.09	1	2
1928	1.47	0.42	1	2
1925	0.07	0.02	3	1
1924	2.60	0.74	1	5
1923	0.77	0.22	1	2
1921	4.21	1.19	2	2
1920	9.03	2.56	1	7
1919	3.72	1.05	1	1
1918	0.23	0.07	1	1
1916	0.19	0.05	2	2
1915	1.18	0.33	1	2
1914	0.16	0.05	2	1
1913	1.56	0.44	1	2
1912	0.10	0.03	1	1
1908	2.66	0.75	4	1
1907	0.57	0.16	1	2
1906	8.46	2.40	1	2
1904	0.10	0.03	2	1
1903	1.80	0.51	1	2
1902	0.01	<0.01	1	1
1898	0.15	0.04	4	2
1893	0.61	0.17	5	3
1892	8.64	2.45	1	2
1891	0.43	0.12	1	1
1888	10.74	3.04	3	3
1885	2.75	0.78	3	3
1883	1.06	0.30	2	1
1881	0.51	0.14	2	2
1878	32.13	9.10	3	2
1877	14.18	4.02	1	3
1875	3.17	0.90	2	2
1872	9.92	2.81	3	4
1871	2.36	0.67	1	1
1865	21.09	5.97	6	1
1861	0.03	0.01	4	1
1858	2.57	0.73	3	1
1855	0.09	0.02	3	1
1853	0.04	0.01	2	1

(continued)

Table 7.5 (*Continued*)

Year of Fire	Area Burned (km^2)	Percent of Total Study Area	Interval between Fires	No. of Spatially Discrete Fires
1849	3.62	1.03	4	1
1848	0.05	0.02	1	1
1847	15.98	4.53	2	2
1845	43.97	12.46	1	1
1844	6.22	1.76	1	1
1836	23.10	6.55	8	2
1822	3.76	1.07	14	1
1820	2.97	0.84	2	1
1818	0.27	0.08	2	1
1815	0.55	0.16	4	1
1814	0.48	0.14	1	1
1806	14.07	3.99	8	1
Totals	274.99	77.92	—	105

wind-driven fires. Irregularities in burn patterns appear to be partly associated with terrain complexity. The Jenny Lake area has a network of eskers and associated glacial features that may have dramatically influenced burn patterns in this area. In the Jarvis River area, the orientation of some fires is perpendicular to the main valley orientation and may be associated with strong winds funneling out of the Jarvis River valley.

Fire behavior in our valley cannot be fully reconstructed due to the long time periods involved, but the following observations can be made. Fires are generally stand replacing, with the exception of meadow areas, where individual trees were found to have survived several low-intensity events with little direct evidence of the fire on the ground. One other interesting feature was the existence of *permanent residuals* or unburned patches within the fire margins. Many of these residuals were found in the Sulphur Lake-Sulphur Creek area, which is quite wet. Many residuals have survived several fires as indicated by differently aged fire scars along their margins. This evidence suggests that many small fires have followed the same boundaries and were contained by the same topographic features.

There has been extensive overlap of fires since 1800. Based on forest age-class information from the Shakwak Trench, the total area burned since 1800 is 16874 ha, or 47.80% of the forested area. Based on individual fire reconstruction the total area burned is much greater—27,499 ha or 77.90% of the forested area—because this calculation takes into account the extensive overlap. This difference is due primarily to the spatial distribution of fires on the landscape: some areas burn frequently, while others burn rarely. More recent fires thus mask regeneration from older events. Large fires occurred in the 1840s, 1860s, and 1870s but are responsible for little of the current forest age-class composition because most of these areas have reburned since.

The fire regime appears to have changed over the past 200 years from large, infrequent events during the 1800s to small, relatively frequent fires during the early 1900s (CD-ROM frame 49). It is possible that smaller fires occurred during the 1800s but that more recent fires have erased the evidence of these events. Fires of the magnitude experienced

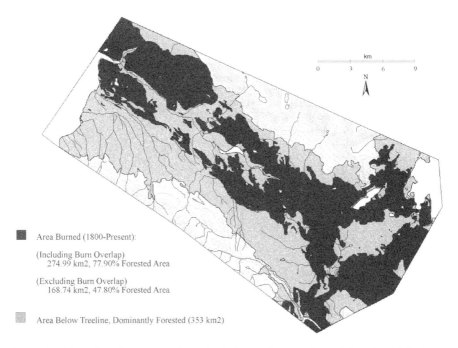

Area Burned (1800-Present):

(Including Burn Overlap)
274.99 km2, 77.90% Forested Area

(Excluding Burn Overlap)
168.74 km2, 47.80% Forested Area

Area Below Treeline, Dominantly Forested (353 km2)

Figure 7.2 Location of all areas in the main study area that were burned since 1800 (after Francis 1996).

in the 1800s did not occur in the 1900s. This change in fire frequency and size around 1900 cannot be directly attributed to human intervention: active fire suppression did not begin until the 1950s.

The building of the Alaska Highway in 1942–1943 appears to have had little effect on the fire regime. Small fires adjacent to the Old Alaska Highway (1–4 ha in size) were dated to 1943 and 1953, suggesting some human association. These fire events and a small fire (1–2 ha in size) on the shore of Sulphur Lake in 1956 were the most recent fires detected. Our evidence does not suggest that human land-use patterns directly caused the shift in fire regime between the 1800s and 1900s. Instead, dendroclimatic data, glacial geomorphological evidence, and older burn patterns suggest a natural change in the fire regime.

Unlike the recent fire history, the reconstruction of older individual fires is nearly impossible due to the loss of evidence over time. Only seven fire scars were found to date pre-1800 events accurately, and, of these, five date a large 1767 fire that probably burned about 4000 ha in the valley bottom. All forest stands sampled displayed some evidence of fire and are therefore assumed to be of fire origin.

Figure 7.3 displays the current forest age-class distribution of the entire study site, grouped into 20-year age classes. Two age classes, the 1870s and 1650s, dominate the distribution of forest ages and give a bimodal appearance to the graph. Nearly 40% of the forested area is made up of these two age classes, possibly indicating that large areas

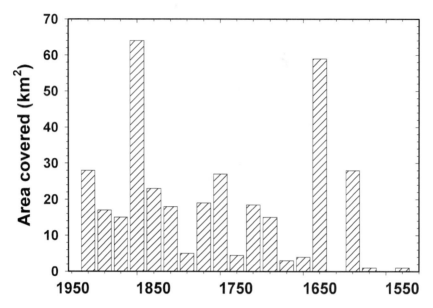

Figure 7.3 The current forest age-class distribution of the entire study site, grouped into 20-year age-classes (after Francis 1996).

burned during these periods. Twenty-seven percent (9508 ha) of the forested area has been fire-free for more than 300 years.

The map in figure 7.4 (CD-ROM frame 49) shows the geographic distribution of 100-year forest age-class elements throughout the study area. The entire south side of the valley is dominated by forests more than 300 years old, while the valley bottom and north boundary are dominated by younger age classes and a more heterogeneous pattern of stand ages. The spatial difference in age-class distribution between the north and south sides of the valley gives rise to the bimodal appearance of figure 7.3, where the south boundary is dominated by old forests and the valley bottom is dominated by younger forests.

7.2.2   Other Forms of Disturbance

In addition to wildfire, flooding, landslides, wind throw, soil movements, people, and insects are all disturbance agents in this forested landscape. Each operates at different scales and has different effects on the forest. During the last 200 years, these agents appear to have played relatively minor roles in shaping the current vegetation mosaic within the study area.

Flooding and landslide hazards near Kluane Lake have been reviewed by Clague (1979, 1981). High gradient streams flowing from the mountains across alluvial fans and aprons at the edge of the Shakwak Trench can cause localized flooding. The only area in the valley that has been largely affected by these events is near the southeast shore of Kluane Lake. Silver Creek and associated streams have caused washouts on the Alaska Highway many times in the past, which have also disturbed the surrounding forests. Also, the

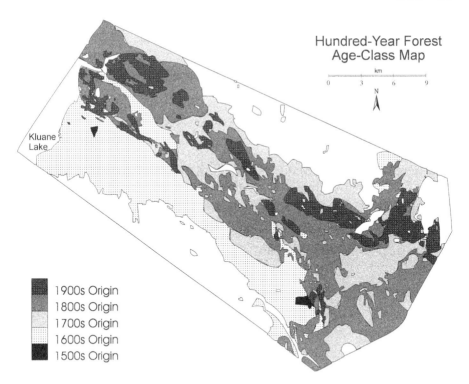

Figure 7.4 The geographic distribution of 100-year forest age-class elements throughout the study area (after Francis 1996).

entire shoreline area on which the KRS is situated is underlain by alluvial deposits and dissected by numerous abandoned river channels. This area has been stabilized for a long time; the stand origin for this alluvial fan was dated to approximately 1650.

Large patches of wind throw damage were not found in the study area. Only localized patches along stand margins and exposed ridges displayed effects of this kind of disturbance, which in some forests can be extensive, if infrequent (Hemstrom and Franklin 1982, Baker and Veblen 1990). Associated with wind throw are low-gradient soil movements. These were detected in many locations throughout the valley but primarily along slopes of the Kluane hills and the Kluane range. It appears that through some mechanism of permafrost melt or simple colluvial processes, soil cohesion is lost and downhill creep begins. Waterlogging may also be a factor. The net effect is that shallow root systems of spruce become displaced and trees become more disposed to wind throw.

Human impacts on the valley have historically been small. Besides the few transportation corridors which run through the valley bottom and some minor fuel wood harvesting and lumbering around Jenny and Kloo Lakes, the study area remains little altered by direct human impact.

From 1994 to 1997 much of the valley and southern portions of neighboring Kluane National Park Reserve experienced a major spruce beetle (*Dendroctonus rufipennis*

Kirby) outbreak, affecting an area orders of magnitude larger than any single fire event in recent history. Although an event similar to this may have occurred in the past, little evidence for a major historical forest insect outbreak of this magnitude was detected. Large numbers of dead, unburned boles are not present as would remain after such major canopy mortality. Hawkes (1983) provides a picture of a spruce stand thinned by spruce beetles in the 1940s when only some of the trees in the stand were killed.

Although spruce beetle outbreaks and forest fires may operate on similar temporal and spatial scales, they can have very different ecological consequences. As a forest disturbance agent, spruce beetles differ from wildfire in four major aspects: (1) spruce beetles select individual trees, whereas wildfire is more nonselective, (2) spruce beetles create a more heterogeneous stand age-structure through host tree selection, (3) spruce beetle mortality results in a greater abundance of large-diameter, coarse woody material carry-over than fire, which creates structurally complex forest stands, and (4) spruce beetles can cause massive canopy mortality without removing forest floor duff, an important factor in spruce seedling germination.

The history of interaction between spruce beetles and fire can only be speculated, but large, infrequent fires or large, infrequent spruce beetle outbreaks will both result in rapid shifts from old to young forests. Spruce beetles and fires may therefore operate at similar spatial and temporal scales in this system. Recently burned areas less than 100 years old are generally not susceptible to spruce beetle damage, because the trees are still young and vigorous (Baker and Veblen 1990, Veblen et al. 1991). In this sense, spruce beetle damage and recently burned areas display a spatially non-overlapping distribution. This is visible in the study area, with stands of damaged and healthy green conifers forming a patchwork on the landscape. We do not know whether this spatially non-overlapping pattern of disturbance perpetuates itself through time. There is no direct evidence that beetle damage in spruce forests leads to large conflagrations.

7.3   Succession in the Boreal Forest

The Kluane region is a harsh growing environment, and recovery from disturbance events is a lengthy process. Post-fire regeneration lag time, particularly for spruce, can be very long and highly variable. Drier sites are able to regenerate relatively quickly and densely in less than 25 years, but low-lying, hydric sites can take 50–75 years to recover even sparsely from fire and thus remain shrub dominated for most of this time. Possible reasons for this include decreased transpiration rates due to vegetation removal, which results in an elevated water table and melting permafrost caused by forest duff consumption, again resulting in periodically saturated soils. Depending on the intensity of the fire, *Populus* and *Salix* may resprout from surviving below-ground parts immediately after fire (Hawkes 1983, Rowe 1983). Sites dominated by these deciduous species may be undergoing a slow succession toward spruce forest.

White spruce establishes only from seed and not by vegetative means. Seed production begins when the trees are 40 years or more in age (Rowe 1955), but cone and seed production is highly variable between years and locations. There may be 2–6 years between years of good seed production or even 10–12 years in some locations (Zasada et al. 1992). The timing of fires relative to good seed years is important for reestablishment,

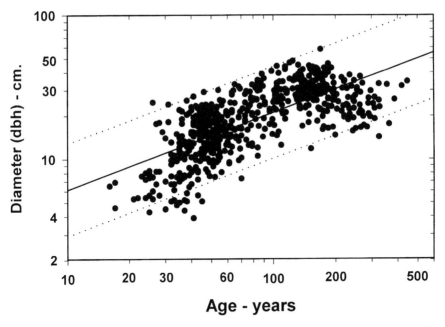

Figure 7.5 Relationship between age (years) and diameter at breast height (dbh, cm) for white spruce from the Kluane study area (N = 624). The log-log regression is log (dbh) = 0.5382 log (age) + 0.2480 (r^2 = .49). The 95% confidence limits for an individual predicted value of diameter at breast height are given by the dotted lines.

particularly because white spruce seeds have limited longevity and dormancy (Putman and Zasada 1986). Seed availability may be an important limiting factor that contributes to the long lags observed. It is often suggested that exposed mineral soil is necessary for seedling establishment, but there is evidence that downed woody material such as rotting logs will also support seedlings (Rowe 1955, Putman and Zasada 1986).

One factor that contributes to the variability of tree size as a function of age and to the generally slow establishment of spruce is the impact of snowshoe hares. Hares normally avoid eating small spruce trees, preferring willow and birch, probably because of the higher concentrations of the antifeedant chemical camphor in the juvenile trees compared to mature white spruce (Sinclair et al. 1988). At high hare densities, however, the apical shoot of the spruce is frequently eaten (Sinclair et al. 1993). In one sample area, of the 63 small spruce trees in a 20 m × 30 m plot, 55 trees had been clipped at least once and 10 twice or more. It is not clear what the overall effect of this clipping is on the trees' ultimate size and rate of growth, but it may be several hare cycles (decades) before a tree is tall enough to escape this periodic herbivory.

To determine the variability in white spruce growth rates, we compared age and diameter at breast height of 543 spruce trees measured throughout the study area. Because of differences among sites and genetic differences among trees, the linear relationship between the two is highly variable. Figure 7.5 shows a log-log plot of age and diameter at

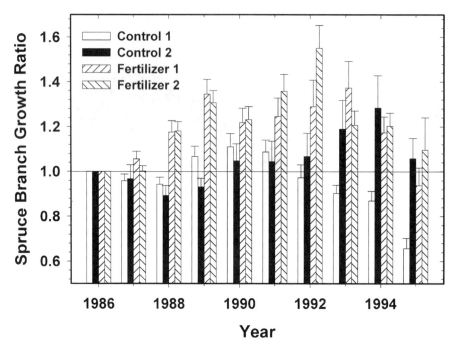

Figure 7.6 White spruce lateral branch extension growth on fertilized and control areas, 1986–1995. Extension growth is expressed as a ratio of the year's growth to the standard year of 1986 (+ 1 SE). Two control grids and the two fertilized grids were measured ($n = 50$ trees in each area).

breast height for white spruce in the Kluane area, which improves the linear regression fit over that of the untransformed data. This regression gives an average 19.5 cm dbh at 100 years, which is similar to the values found by Jozsa et al. (1984) for white spruce at Swan Hills (55° N) and Fort Vermilion (58° N) in Alberta. This indicates that, although the biological processes do occur slowly in our valley, they are comparable to those at other sites in the northern boreal forest.

7.4   Fertilizer Effects on Trees

Because one of the experimental treatments was the addition of fertilizer, we wanted to determine its effects on the growth and reproduction of white spruce. Seventy-five trees were studied in open spruce vegetation on control and fertilizer grids by measuring the annual growth of a single branch on each tree each year. The growth was compared, in each case, to the growth of the same branch in 1986 before the fertilizer was added. We also looked for fertilizer effects on reproductive effort by counting new cones each year.

The addition of fertilizer had a clear effect on the vegetative growth of the spruce trees. After the fertilizer treatment began in 1987, the branches grew 15–50% more than in the

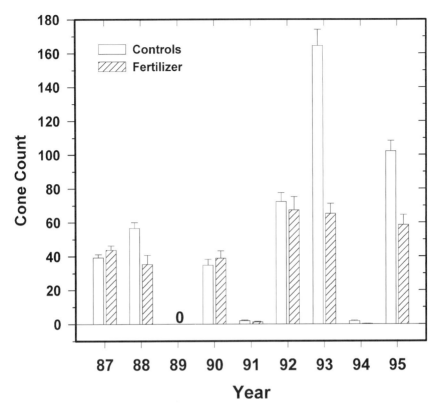

Figure 7.7 Average white spruce cone counts for 172 trees from control areas and 172 trees from fertilized areas (+ 1 SE). Counts were done in August of each year. Cone counts are an index of cone production and not an absolute estimate. There were no cones produced in 1989.

prefertilizer year. Figure 7.6 illustrates the strength of this response. On average, from 1987 to 1995 fertilized trees grew 22% per year more than control trees.

Although fertilizer increased vegetative growth, there was no significant effect on cone production (figure 7.7). Indeed, if there was any trend, it was for control trees to have more cones than fertilized trees, although these differences were not statistically significant. Spruce trees normally exhibit mast years, with little or no seed production between mast years. The addition of fertilizer did not seem to change the timing of masting, since fertilized trees produced many cones in the same years as control trees and no cones in the low years of 1989, 1991, and 1994. In spite of the similar cone production of spruce on control and fertilized areas, seed production was significantly higher on the fertilized area (figure 7.8). Over all 8 years, seed production was 44% higher on the fertilized areas. If the cone failure years of 1989, 1991, and 1994 are excluded from these data, the increase in seed production due to fertilization is 61%. The increased seed production on fertilized areas could have two explanations. First, red squirrels were less abundant on fertilized ar-

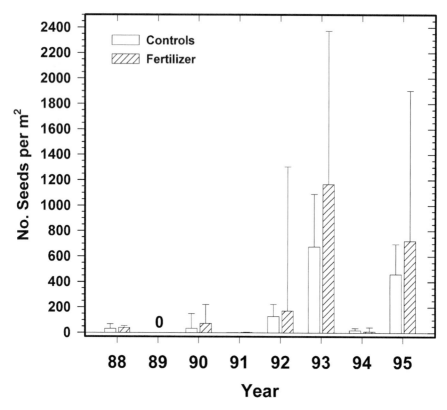

Figure 7.8 Seed production from white spruce, measured by 86 seed buckets on control 1 and fertilizer 1 grids, 1988–1995. Seeds falling to the ground were collected in the buckets over the winter and were counted the next spring. Year is the year of seed production. Error bars show 95% upper confidence limits.

eas, and thus fewer cones were harvested before they could shed seeds in the autumn. Second, cones on fertilized trees contained more seeds, either because of a higher fraction of fertile seeds in each cone or, alternatively, because they produced larger cones with more seeds. Unfortunately, we have no measurements of the viability of white spruce seeds or the number of viable seeds per cone for either controls or fertilized trees.

Fertilization could also affect the palatability of trees to herbivores. Small white spruce trees contain camphor, which acts as an antifeedant against snowshoe hares (Sinclair et al. 1988). The addition of fertilizer and the exclusion of herbivores both decreased the amount of camphor in the distal branches of small spruce (figure 7.9). The dynamics and importance of these effects are yet to be worked out. Aspen and balsam poplar are also protected to varying degrees by antifeedant chemicals that are produced in response to browsing by hares (Bryant 1981, Jogia et al. 1989). The chemistry of these plants may also be affected by the addition of fertilizer, but because of their rarity in the Kluane forests, we did not examine either *Populus* species for fertilizer effects.

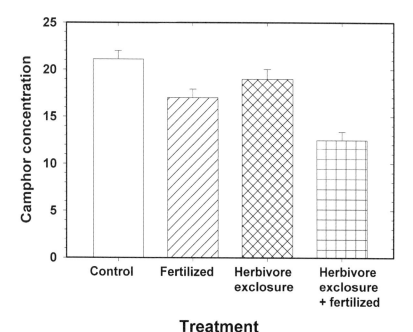

Figure 7.9 Camphor concentration in the distal branches of small white spruce trees in four treatment areas at Kluane (+1 SE). Camphor deters snowshoe hares from feeding on small spruce trees. Herbivore exclosures excluded snowshoe hares and moose after 1988. Camphor concentration is micrograms per gram dry mass. Measurements were made in spring 1995 (after G. Sharam, unpublished).

7.5   Summary

The boreal forest of the Shakwak Trench is dominated by white spruce, with considerably fewer trembling aspen and balsam poplar. The density of trees varies greatly in the nine main study areas. Much of the spatial variability of the tree vegetation is due to fire. Fire size is highly variable, with most burned area resulting from a few large fires in the 1840s, 1870s, and 1880s. Many of these large fires are synchronous with large fire events across western North America, indicating the importance of large-scale, characteristic weather systems coordinating extreme fire weather. The fire regime has changed over the past 200 years from large, infrequent fires to small, relatively frequent fires. There has not been a major fire in the Shakwak Trench since 1929. Human impact on this changing fire regime appears to be small.

Other disturbances have had little impact on the forests of the Kluane region. A spruce bark beetle outbreak began in the region in 1994 near the end of this study. Little evidence was found for a spruce beetle outbreak of the current magnitude having occurred in the last 100 years.

Vegetation recovery from fire and disturbance events is a lengthy and highly variable process. Some of the trees (*Populus*) can establish by surviving below-ground parts, but the dominant tree, white spruce, establishes only from seed. Reestablishment of white spruce may take 75 years or more in some habitats. Although several factors ensure that the spruce trees recruit and grow slowly, their average size at 100 years (19 cm dbh) is comparable with trees at other northern boreal sites.

White spruce trees showed a strong vegetative growth response to fertilizer, growing on average 22% more each year. In contrast, cone production was identical in fertilized and control spruce trees, and mast years occurred at the same time on all areas. Seed production, however, was higher on fertilized plots, with 44–61% more seeds falling to the ground on the fertilized site. This increase in seed production could be a result of fewer red squirrels on the fertilized grids or could be due to cones of fertilized trees containing more seeds. Fertilizer alters white spruce needle chemistry, reducing the concentration of the antifeedant chemical camphor.

Literature Cited

Alexander, M. E. 1985. Estimating the length-to-breadth ratio of elliptical forest fire patterns. *In* Proceedings of the Eighth Conference on Fire and Forest Meteorology, 29 April–2 May, Detroit, Michigan, pages 287–304. Society of American Foresters, Bethesda, Maryland.

Allen, H. D. 1982. Dendrochronological studies in the Slims River valley, Yukon Territory. MSc Thesis, University of Calgary.

Anderson, H. E. 1983. Predicting wind-driven wildland fire size and shape. Research Paper INT-305. USDA Forest Service, Washington, DC.

Arno, S. F. 1976. The historical role of fire on the Bitterroot National Forest. Research Paper INT-87. USDA Forest Service, Washington, DC.

Baker, W. L., and T. T. Veblen. 1990. Spruce beetles and fires in the nineteenth-century subalpine forests of western Colorado, U.S.A. Arctic and Alpine Research **22**:65–80.

Bryant, J. P. 1981. Phytochemical deterrence of snowshoe hare browsing by adventitious shoots of four Alaskan trees. Science **213**:889–890.

Clague, J. J. 1979. An assessment of some possible flood hazards in Shakwak Valley, Yukon Territory. Paper 79-1B, pages 63–70. Geological Survey of Canada, Ottawa.

Clague, J. J. 1981. Landslides at the south end of Kluane Lake, Yukon Territory. Canadian Journal of Earth Science **18**:959–971.

Dale, M. R. T. 1990. Two-dimensional analysis of spatial pattern in vegetation for site comparison. Canadian Journal of Botany **68**:149–158.

Douglas, G. W. 1980. Biophysical inventory studies of Kluane National Park. Unpublished report to Parks Canada, Prairie Region, Winnipeg.

Francis, S. R. 1996. Linking landscape pattern and forest disturbance: fire history of the Shakwak Trench, southwest Yukon Territory. MSc thesis. University of Alberta.

Glover, J. 1929. Champagne-Kluane District RCMP Patrol Report, August 20, 1929. RG18, volume 3662, file G567–20. *in* Support material for the Kluane Wildlife Management Database by G. Lotenberg. Unpublished report prepared for Parks Canada, Historic Sites, 1995.

Hawkes, B. C. 1983. Fire history and management study of Kluane National Park. Report prepared for Parks Canada, Prairie Region, Winnipeg.

Heinselman, M. L. 1973. Fire in the virgin forests of the Boundary Waters Canoe Area, Minnesota. Quaternary Research **3**:329–382.

Hemstrom, M. A., and J. F. Franklin. 1982. Fire and other disturbances of forests in Mount Rainier National Park. Quaternary Research **18**:32–51.

Hucal, T., and M. R. T. Dale. 1993. Habitat map of the Kluane Boreal Forest Ecosystem Project study area. 1:32 000 scale vegetation map. Unpublished.

Jogia, M. K., A. R. E. Sinclair, and R. J. Anderson. 1989. An antifeedant in balsam poplar inhibits browsing by snowshoe hares. Oecologia **79**:189–192.

Johnson, E. A. 1992. Fire and vegetation dynamics: studies from the North American boreal forest. Cambridge University Press, Cambridge, Massachusetts.

Johnson, E. A., and D. R. Wowchuck. 1993. Wildfires in the southern Canadian Rocky Mountains and their relationship to mid-tropospheric anomalies. Canadian Journal of Forest Research **23**:1213–1222.

Jozsa, L. A., M. L. Parker, P. A. Bramhall, and S. G. Johnson. 1984. How climate affects tree growth in the boreal forest. Northern Forestry Information Report NOR-X-225. Canadian Forest Service, Edmonton, Alberta.

Krebs, C. J. 1999. Ecological methodology. Benjamin/Cummings, Menlo Park, California.

McBride, J. R. 1983. Analysis of tree rings and fire scars to establish fire history. Tree-Ring Bulletin **43**:51–67.

Putman, W., and J. C. Zasada. 1986. Direct seeding techniques to regenerate white spruce in interior Alaska. Canadian Journal of Forest Research **16**:660–664.

Rowe, J. S. 1955. Factors influencing white spruce reproduction in Manitoba and Saskatchewan. Technical Note 3. Department of Northern Affairs and National Resources, Canada, Forestry Branch, Ottawa.

Rowe, J. S. 1983. Concepts of fire effects on plant individuals and species. in R. W. Wein and D. A. MacLean (eds). The role of fire in northern circumpolar ecosystems, pages 135–154. John Wiley and Sons, Toronto.

Sinclair, A. R. E., M. K. Jogia, and R. J. Anderson. 1988. Camphor from juvenile white spruce as an antifeedant for snowshoe hares. Journal of Chemical Ecology **14**:1505–1514.

Sinclair, A. R. E., J. M. Gosline, G. Holdsworth, C. J. Krebs, S. Boutin, J. N. M Smith, R. Boonstra, and M. R. T. Dale. 1993. Can the solar cycle and climate synchronize the snowshoe hare cycle in Canada? Evidence from tree rings and ice cores. American Naturalist **141**:173–198.

Tande, G. F. 1979. Fire history and vegetation pattern of coniferous forests in Jasper National Park, Alberta. Canadian Journal of Botany **57**:1912–1931.

U.S. Army Corps of Engineers. 1993. Geographic resources analysis support system, version 4.1 (GRASS 4.1). U.S. Army Corps of Engineers, Construction Engineering Research Laboratory, Champaign, Illinois.

Veblen, T. T., K. S. Hadley, M. S. Reid, and A. J. Rebertus, 1991. The response of subalpine forests to spruce beetle outbreak in Colorado. Ecology **72**:213–231.

Wein, R. W, and D. A. MacLean (eds). 1983. The role of fire in northern circumpolar ecosystems. John Wiley and Sons, Toronto.

Yamaguchi, D. K. 1991. A simple method for cross-dating increment cores from living trees. Canadian Journal of Forest Research **21**:414–416.

Zasada, J. C., T. L. Sharik, and M. Nygren. 1992. The reproductive process in boreal forest trees. in H. H. Huggat, R. Leemans, and G. B. Bonan (eds). A systems analysis of the global boreal forest, pages 85–125. Cambridge University Press, Cambridge.

PART III

HERBIVORES

Snowshoe Hare Demography

KAREN E. HODGES, CHARLES J. KREBS, DAVID S. HIK,
CAROL I. STEFAN, ELIZABETH A. GILLIS, & CATHY E. DOYLE

8.1   The Snowshoe Hare Cycle

Snowshoe hares, *Lepus americanus,* exhibit continent-wide cyclic fluctuations in abundance. Peak densities occur every 8–11 years, and densities fluctuate 5 to 25-fold during a cycle. Snowshoe hares are typically the dominant herbivore in boreal forests, and their cyclic fluctuations have widespread ramifications for the shrubs and trees that they eat, for the transient and resident predators that eat them, and for the other forest herbivores that may compete with hares for food or that may serve as alternative prey for predators (Elton 1924, Finerty 1980, Keith 1990, Krebs et al. 1992, Royama 1992).

Both mortality rates and reproductive rates of snowshoe hares also show regular cycles. Mortality rates are highest during the decline phase and lowest during the late low and early increase phases (Krebs et al. 1986b, Trostel et al. 1987, Keith 1990). The number of litters per breeding season, the proportion of females pregnant with each litter, and the number of leverets per litter all show cyclic changes, with the highest annual reproductive output occurring during the late low and early increase phases (Ernest 1974, Cary and Keith 1979, Stefan 1998).

Dispersal rates are not as well known for snowshoe hares. Dispersal appears to vary through the cycle, but the highest dispersal rates have variously been found to occur in the peak and early decline phases (Windberg and Keith 1976, Boutin et al. 1985) or in the increase phase (Keith and Windberg 1978, Wolff 1980). This discrepancy may be due to the different techniques that have been used for assessing dispersal (e.g., loss rates, removal grids, radio telemetry locations). Juvenile hares apparently disperse more than adults do (Dolbeer and Clark 1975, Windberg and Keith 1976, Keith et al. 1984, Boutin et al. 1985, Gillis and Krebs 1999).

Various explanations have been proposed to explain the demographic changes that lead to the numeric hare cycle. One set of proposals implicates the variation in food supply through the cycle, arguing that food shortage leads to reduced reproduction and increased starvation as hares spend more time searching for browse and are in poorer physical condition (Pease et al. 1979, Vaughan and Keith 1981, Keith 1983). Snowshoe hares eat woody shrubs and trees in the winter and forbs, grasses, and leaves of shrubs in the summer. Because of the heavy browsing that occurs during peak hare densities, winter forage availability is lowest during the early decline phase (Pease et al. 1979, Sinclair et al. 1988, Smith et al. 1988). Additionally, secondary compounds increase with plant regrowth and may lead to relative food limitation as plants regrow after heavy browsing (Bryant 1981, Fox and Bryant 1984). Food limitation, either absolute or relative, has therefore been suggested as a potential initiator of the cyclic declines.

An alternative hypothesis argues that patterns of predation explain the cycle. Most hares die of predation and most hares die as juveniles; survival is lowest during the decline phase (Boutin et al. 1986, Keith 1990, Krebs et al. 1995, Stefan 1998, Gillis and Krebs 1999). Hares are the predominant prey species for lynx, coyotes, goshawks, and great horned owls (Keith et al. 1977, O'Donoghue et al. 1997). These predators show both numeric and functional responses to the hare cycle, with the numeric responses typically lagging 2–4 years behind the population changes of the snowshoe hares (Keith 1990, Boutin et al. 1995, O'Donoghue et al. 1998). Leverets are often killed by red squirrels and ground squirrels (O'Donoghue 1994, Stefan 1998). Some leverets and adults die of exposure or starvation during the decline phase of the cycle, but these causes of death seldom occur at other phases (Boutin et al. 1986, Keith 1990, Stefan 1998). The predation hypothesis suggests that these regular shifts in mortality are capable of driving the cycle.

Three lines of evidence suggest that food explanations are incomplete. Food shortages are not reliably detected during cyclic peaks and declines (Sinclair et al. 1988, Smith et al. 1988), food addition experiments have failed to stop cyclic declines (Krebs et al. 1986a, 1986b), and patterns of plant chemical defenses lag behind the numeric changes of hares, which suggests that the changes in plant chemistry cannot cause the cyclic decline (Sinclair et al. 1988). Additionally, it is unclear whether food limitation can have a sufficiently large impact on hare survival rates; the predation hypothesis has a similar problem, in that changes in mortality patterns cannot explain the regular cyclic changes in reproduction.

There are several hypotheses that link the effects of food and predation on snowshoe hare population dynamics. In one scenario, cyclic declines are thought to be initiated by the scarcity of food, which increases starvation rates and also makes hares more accessible to predators; predation is then thought to lengthen the decline and sustain the low phase (Keith 1974, 1981, 1990). Alternatively, food and predation may interact throughout the cycle in their impacts on hare dynamics, rather than influencing hares sequentially (Krebs et al. 1992). A third hypothesis suggests that predators affect hare foraging behavior and physiology, with high predation pressure causing hares to have poorer diets, increased stress, and reduced fecundity (Hik 1994, 1995, Boonstra et al. 1998a). Additionally, the impacts of food and predation on hare demography may be affected by the levels of parasitic infestation of hares (Sovell and Holmes 1996, Murray et al. 1997, 1998).

The Kluane experiments were designed to evaluate the relative impacts of food and predation on hare demography. Because neither factor alone seems able to explain all of the demographic changes, they were manipulated in a factorial fashion to examine how each factor affects hare demography and how they interact in their effects on snowshoe hares.

8.2   Methods

Our primary objective was to determine the effects of food, predation, and their interaction on the demography of snowshoe hares through a population cycle. The fertilization and food addition treatments increased food availability by increasing plant growth and providing artificial food, respectively, and the two predator exclosure treatments reduced the risk of predation. We focused on twice-yearly trapping of hares for population estimates, coupled with radio telemetry for detailed study of survival and causes of death; in some years, hare reproduction was studied using maternity cages. Numerous additional questions were also addressed by individual researchers, so the precise questions asked in each year at each study site varied (table 8.1).

To obtain population estimates and to establish rates of population increase, we conducted trapping sessions on each experimental grid in March-April and October. Each session lasted 3–7 trap nights, but early in the project it became clear that animals caught night after night were losing weight from the trapping, so grids were trapped a maximum of 2 consecutive nights, then allowed to rest for at least 2 nights before additional trapping. Each grid had 86 Tomahawk traps (Tomahawk Live Trap Co., Tomahawk, Wisconsin) located along four traplines (CD-ROM frame 24). Traps were baited with alfalfa and

Table 8.1 Snowshoe hare demographic parameters measured during a population cycle near Kluane Lake, Yukon.

	Type of Information			
Treatment Grid	Population Censuses, Increase Rates[a] (Trapping)	Survival, Mortality Causes[b] (Radio Telemetry)	Reproduction, Leveret Survival[c] (Maternity Cages + Telemetry)	Juvenile Survival, Dispersal[d] (Radio Telemetry)
Control 1 (Sulphur)	1987–1998	1988–1996	1988–1989, 1992, 1994–1996	1995–1996
Control 2 (Silver)	1987–1998	1991–1996	1990–1992	—
Control 3 (Chitty)	1988–1990, 1993–1996	1988, 1993–1996	1989–1990, 1994–1996	1989–1990, 1995–1996
Control 4 (Lloyd)	—	1992, 1994–1995	1992, 1994, 1996	1989–1990
Control areas off-grid	—	1992–1996	1992, 1994–1995	1995–1996
Fertilizer 1 (Flint)	1988–1996	1988–1996	—	—
Fertilizer 2 (Grizzly)	1988–1996	1993–1996	—	—
Food addition 1 (Gravel Pit)	1988–1996	1988–1996	1989–1990, 1995	1989–1990, 1995–1996
Food addition 2 (Agnes)	1988–1996	1992–1996	1989–1990, 1995	1989–1990, 1995–1996
Predator exclosure (Beaver Pond)	1988–1996	1988–1996	1988	—
Predator exclosure + food (Hungry Lake)	1988–1996	1988–1996	1991–1992	—

The lettered footnotes indicate the primary literature summarizing these studies. The peak occurred in 1989–1990, and the lowest densities were in 1993. Leveret survival was monitored from birth until weaning (30 days), and juvenile survival was postweaning until March of the following year.

[a]Krebs et al. (1992, 1995, 1996), Boutin et al. (1995), Hodges et al. (1999a).
[b]Krebs et al. (1992, 1995, 1996), Hodges et al. (1999a).
[c]O'Donoghue and Krebs (1992), O'Donoghue (1994), Krebs et al. (1995), Stefan (1998).
[d]O'Donoghue and Bergman (1992), Gillis (1999), Gillis and Krebs (1999).

snow or apple for moisture. At peak hare densities, some grids had additional traps to reduce trap saturation. Traps were 30–60 m apart, and lines of traps were 150 m apart, for an effective grid size of about 60 ha. Trapping sometimes occurred at other times to replace radio collars or to obtain reproductive information. Initially, traps were left open 24 h and checked at dawn, but because this schedule was stressful to hares and because many squirrels entered traps in the day, we quickly shifted to setting traps at dusk and checking them at dawn. For each hare caught, we recorded ear tag (Monel #3, National Band and Tag Co., Newport, Kentucky), weight, right hind foot length, reproductive condition (males, scrotal or abdominal testes; females, lactating, not lactating) (CD-ROM frame 25).

To monitor adult survival, some hares >1000 g were radio collared with 40 g radio collars equipped with mortality sensors (Lotek, Newmarket, Ontario). Radio-collared hares were monitored every 1–2 days to provide survival estimates, and all animals that died were located to determine the cause of death. The predators leave distinct signs, such as scat, tracks, feathers, or pellets, and also have distinct methods of eating hares. In about 50% of cases we were able to assign cause of death to a particular species of predator (CD-ROM frame 37) and in another ~10% of cases we could distinguish avian from mammalian predation but could not identify the particular predator species.

To estimate natality and survival of leverets, we trapped pregnant females 1–14 days before parturition and held them in 60 × 60 × 120 cm chicken wire and wood cages until they gave birth (CD-ROM frame 26). Leverets were counted, ear tagged, weighed, and sexed. This method provided information on timing of reproduction and litter sizes. We calculated stillborn rates from the number of leverets born dead; necropsies were used to confirm that dead leverets died before birth (e.g., lungs not inflated, no internal trauma). At least half of the leverets in each litter were radio tagged by gluing 2–2.5 g transmitters to their backs (Biotrack, Wareham, England). The radio-tagged leverets were located daily until weaning (~4 weeks) for estimates of survival and identification of cause of death (O'Donoghue 1994, Stefan 1998). In this chapter, we present natality results following the analysis of Stefan (1998); these estimates differ from those in Krebs et al. (1995) and Boonstra et al. (1998a) because of a reappraisal of the timing of the first litter of the season. In 1995 some juveniles were fitted with 25 g radio collars (Biotrack) to obtain estimates of postweaning juvenile survival and dispersal (Gillis 1999).

8.3   Demographic Parameters

8.3.1   Density and Rates of Change

The main peak in hare densities occurred in 1989–1990, but on control 3 peak spring densities were in 1988 (figure 8.1). Peak spring densities on control sites were 1.6–2.0 hares/ha, and densities dropped to 0.01–0.1 hares/ha by spring 1993. Peak autumn densities ranged higher, with a maximum control density of 3.1 hares/ha on control 3. Averaged across all control sites, the cyclic amplitude was 18-fold. Through the rest of this chapter, we refer to the years 1989 and 1990 as peak, 1991 and 1992 as decline, 1993 and 1994 as low, and 1995 and 1996 as increase.

All experimental treatments resulted in higher hare densities, especially during the decline and low phases (figures 8.2, 8.3; CD-ROM frame 52). Averaged across the entire cycle, fertilization increased hare densities 1.3-fold, food addition 3.1-fold, removing mammalian predators 2.0-fold, and the predator exclosure + food manipulation 9.7-fold (CD-ROM frame 52). Both fertilized sites had their peak snowshoe hare densities in spring 1990, with densities of 1.8 and 2.2 hares/ha; they reached low densities of 0.3 and 0.7 hares/ha in 1993. The food addition grids had peak hare densities a year later, in 1991; the densities of 5.1 and 6.6 hares/ha were 3–5 times higher than peak control densities. The low densities on food grids were 0.2 and 0.5, again higher than the lowest densities on the control areas. The hares on the predator exclosure reached a peak density of 1.8 hares/ha in 1990, declined to 0.2 hares/ha in 1993, increased in 1994 and 1995, then decreased again in 1996. Hares on the predator exclosure + food treatment reached a high density of 6.1 hares/ha in 1990 and dropped to a low density of 1.0 hares/ha in 1993.

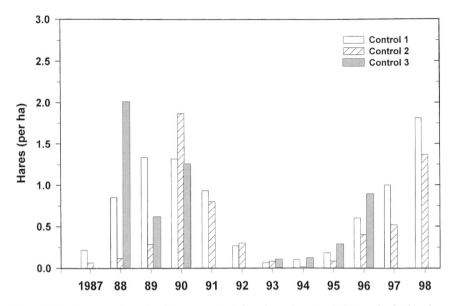

Figure 8.1. Snowshoe hare densities on control sites through a population cycle. Spring densities were calculated using the average of Jolly-Seber and jackknife estimators for trapping sessions conducted in March and April and assuming an effective grid size of 60 ha. Control 3 was not trapped in 1987, 1991, 1992, 1997, or 1998.

Population growth rates were also affected by the experimental manipulations. On control sites, hare populations declined for 4–5 years, from 1990 through 1994 (figure 8.4). The decline was most rapid from 1991 through 1993. Hares then showed some increase in 1994–1995, and a much greater rate of increase in 1995–1996. Hares on fertilizer treatments showed a population decline similar to that of control hares, but, unlike the control treatments, they showed high rates of increase during 1993–1994, then continued to increase at lower rates during 1994–1996. Hare populations on food addition treatments declined only from 1991 through 1993, but these decline rates and the subsequent rates of increase were similar to those of the control populations. The hare population on the predator exclosure had a decline similar to the control hares, then increased for 2 years before declining again in 1995–1996. The predator exclosure + food hare population declined slightly from 1990 to 1992, but the major crash in hare numbers occurred during the winter of 1992–1993.

8.3.2 Sex and Age Structure

Juvenile hares composed more of control hare populations during the low and early increase phases (~85–90%; table 8.2) than during the decline phase (~55–65%). The oldest hares (4 years old) were trapped during the decline. Neither fertilization nor food addition affected population age structures. The populations inside the predator exclosure fences had lower proportions of juveniles, and a higher proportion of hares reached ages

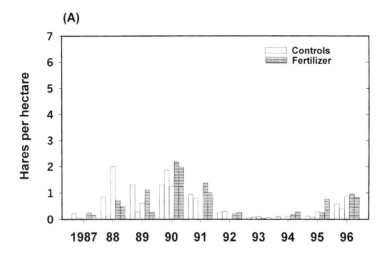

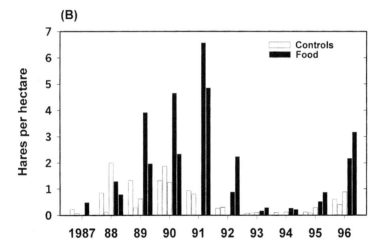

Figure 8.2 Snowshoe hare densities on fertilizer (A), food addition (B), predator exclosure (C), and predator exclosure + food (D) sites. Spring densities were calculated using the average of Jolly-Seber and jackknife estimators for trapping sessions conducted in March and April and assuming an effective grid size of 60 ha.

older than 2 years; the oldest hares reached ages of 5 and 6 years. Sex ratios in the breeding population remained around 1:1 on all treatments.

8.3.3 Reproduction

On control areas, the pregnancy rate varied among years and litter groups (table 8.3). Two litter groups were produced in the decline, three litter groups in late increase and peak

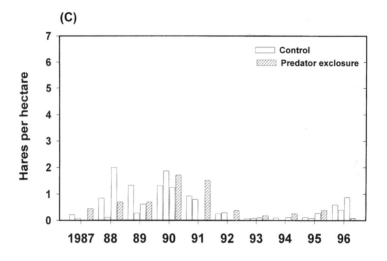

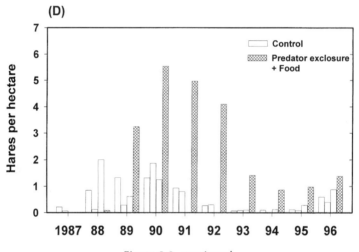

Figure 8.2 *continued*

years, and four litter groups in the late low and early increase. We did not use maternity cages for the fourth litter, but we could tell if a fourth litter was produced in a year from late summer trapping of pregnant females or autumn trapping of small juveniles. All females trapped in 1994 through 1996 were pregnant for the first three litter groups. In other years, 77–100% of females were pregnant with each litter group, and the lowest pregnancy rates usually occurred for the last litter group of the breeding season. Pregnancy rates were slightly higher on food grids relative to control areas in 1989 and 1990, but all females on both treatments were pregnant with all three litters in 1995. Unlike hares on the control areas, hares on the predator exclosure + food grid had three litters in both 1991 and 1992. On control areas, stillborn rates varied among years, with the highest rate

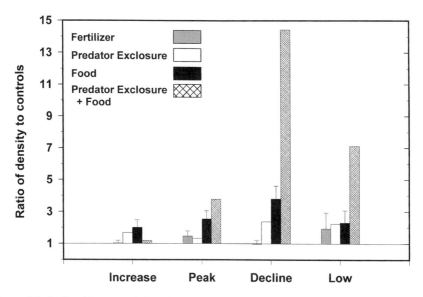

Figure 8.3 Ratio of hare population sizes on manipulated sites to mean control densities. Any value other than one indicates a treatment effect. The standard error bars are given for the replicated fertilization and food treatments.

(30.4%) occurring for litter 2 of 1991. Few leverets were stillborn in the first litter of any year (<8%), but stillborn rates increased in later litters in all years. Stillborn rates on the food grids were consistently double the rates on control areas, reaching maxima of 44.6% (1989) and 23.5% (1990) for litter 3. We do not know whether the maternity cage methodology influenced stillborn rates.

Mean parturition dates for the first litter generally were in the fourth week of May (table 8.3). On control areas, the mean parturition date was about 7 days later in 1991 and 1992 than in 1994 through 1996; 1994 first litter parturition dates were 10–19 days earlier than in all other years. Relative to hares on control sites, hares on food addition sites had earlier mean parturition dates by 4–5 days during the peak (1989 and 1990) and by 12 days in the first year of the increase (1995). During the decline, hares on the predator exclosure + food treatment site gave birth 5–10 days earlier than did hares on control areas.

Mean litter size differed among years and litter groups on control areas (table 8.3). Except for the first litter, hares had smaller litters during the decline than during the increase phase of the cycle. The average size of the first litter (3.6) did not differ among years; litter 1 was consistently smaller than the average litter sizes for both litter 2 (5.8) and litter 3 (5.3). Litter sizes did not differ significantly between control and food grids at peak hare densities, but during the decline phase litter sizes remained high on the predator exclosure + food grid while decreasing by 25% for litter 2 on control areas. Sex ratios did not differ from 1:1 for any litter group in any treatment.

On control areas, total reproductive output was lowest during the decline (6.9 leverets/female per summer) and highest during the late low and early increase phases (18.9 leverets/female per summer) of the snowshoe hare cycle (figure 8.5a). Reproductive output

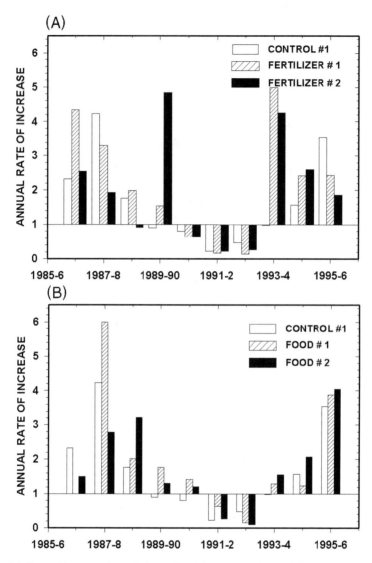

Figure 8.4 Rate of increase through the cycle. Values were calculated from spring densities as $(t + 1)/t$. A = fertilizer 1 and 2; B = food 1 and 2; C = predator exclosure; D = predator exclosure + food.

did not differ between food addition and control areas during the peak or early increase phases. However, although reproductive output decreased dramatically on the control areas during the population decline (from 12.8 in 1990 to 7.0 in 1992), it remained high on the predator exclosure + food treatment (15.5 in 1991 and 16.3 in 1992). This difference was due to hares having smaller and fewer litters on the control areas than on the predator exclosure + food grid. We do not know if reproductive output remained high on the food addition sites during the decline phase.

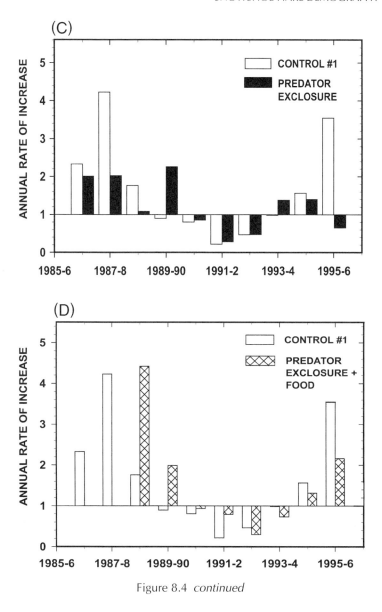

Figure 8.4 *continued*

The number of young weaned per female showed a similar pattern (figure 8.5b). On control sites, the most young were weaned during the early increase phase (9.7 weaned young/female in 1995), whereas the number of weaned young declined in each year of the population decline. Females on the food addition sites weaned more young than control females in 1990 (peak densities), but weaned fewer young in 1989 and 1995. Hares on the predator exclosure + food treatment weaned 6.2 young in 1992, while the control females we observed failed to wean any young.

Table 8.2 The age structure and sex ratio of snowshoe hare populations in spring.

Grid/Year	No. of Hares Caught	Mean Age (Years ± SE)	% Young	Age of Oldest Hare(s)	% Females
Control					
1989	72	1.24 ± 0.05	77.8	3	41.7
1990	187	1.57 ± 0.05	55.6	3	34.2
1991	87	1.47 ± 0.08	65.5	4	39.1
1992	30	1.60 ± 0.17	63.3	4	40.0
1993	12	1.17 ± 0.11	83.3	2	50.0
1994	14	1.21 ± 0.16	85.7	3	42.9
1995	28	1.11 ± 0.06	89.3	2	37.0
1996	106	1.11 ± 0.03	89.6	3	51.0
Fertilizer					
1989	60	1.08 ± 0.04	91.7	2	46.7
1990	102	1.22 ± 0.05	80.4	3	52.5
1991	61	1.39 ± 0.07	62.3	3	51.7
1992	26	1.65 ± 0.17	57.7	3	50.0
1993	5	2.40 ± 0.60	40.0	4	40.0
1994	27	1.00 ± 0.00	100.0	1	33.3
1995	53	1.08 ± 0.04	92.5	2	47.2
1996	90	1.18 ± 0.05	84.4	3	51.7
Food					
1989	116	1.08 ± 0.03	92.2	2	45.6
1990	154	1.21 ± 0.04	81.2	3	54.9
1991	355	1.45 ± 0.04	70.7	4	53.5
1992	72	1.56 ± 0.12	69.4	5	54.2
1993	20	1.55 ± 0.21	70.0	4	60.0
1994	26	1.35 ± 0.19	84.6	5	40.0
1995	53	1.08 ± 0.05	94.3	3	53.8
1996	213	1.15 ± 0.03	87.3	4	49.8
Predator Exclosure					
1989	40	1.30 ± 0.08	72.5	3	40.0
1990	181	1.34 ± 0.05	75.1	4	45.0
1991	80	1.58 ± 0.09	60.0	4	58.8
1992	23	1.48 ± 0.15	60.9	4	52.2
1993	10	1.70 ± 0.30	60.0	3	60.0
1994	15	1.53 ± 0.27	73.3	4	46.7
1995	24	1.33 ± 0.18	79.2	5	50.0
1996	5	1.60 ± 0.25	40.0	2	40.0
Predator Exclosure + Food					
1989	92	1.05 ± 0.02	94.6	2	54.3
1990	219	1.17 ± 0.03	83.1	3	43.6
1991	185	1.56 ± 0.05	55.1	3	43.2

Table 8.2 (*Continued*)

Grid/Year	No. of Hares Caught	Mean Age (Years ± SE)	% Young	Age of Oldest Hare(s)	% Females
1992	135	1.50 ± 0.07	68.9	4	54.5
1993	52	1.87 ± 0.12	38.5	4	65.4
1994	39	1.72 ± 0.17	64.1	4	46.2
1995	36	1.36 ± 0.13	75.0	4	55.6
1996	71	1.42 ± 0.11	74.7	6	56.5

These values are derived from trapping censuses conducted in March–April of each year. Hare ages were assigned on the basis of the size, weight, reproductive characteristics, and time of year when each hare was first caught. On January 1, we added a year to each hare's age. We do not report 1988 data because we would not have been able to age older hares accurately because of the lack of several years of prior trapping. Percent young is the percentage of hares in March that were born in the previous summer.

8.3.4   Survival Rates

Estimates of survival rates of adult hares were lowest during the decline (in 1991–1992, 30-day survival was 0.64, for an annual survival of 0.5%; figure 8.6) and highest during the increase phase (in 1988–1989 and 1995–1996, 30-day survival was 0.91, for an annual survival of 32%). Hares on the fertilizer treatments had survival patterns similar to hares on control sites. Hares on food addition sites had particularly low 30-day survival (0.70, 0.62) during the decline, but had slightly higher survival than control hares did during the other years. Hares on both the predator exclosure and predator exclosure + food treatments had higher survival (corrected for hares leaving the fences) than hares not protected from predators. Each of the fenced treatments had only one year when 30-day hare survival was lower than 0.90. On the predator exclosure the lowest survival was 0.83 in 1991–1992, and on predator exclosure + food the lowest survival was 0.89 in 1992–1993. In both cases, these were the years with the greatest decline rate. Hares on control sites had 30-day survival rates higher than 0.90 in only 2 years (1988–1989, 1995–1996), both of which were years of population increase.

Preweaning survival rates of leverets until 30 days varied among litter groups, among years, and among treatments (table 8.4). In most years on control areas, leverets born in the second litter of the year had lower survival than leverets born in other litters. Survival was low and variable at peak hare densities, lowest during the decline, and high during the increase phase. Survival rates did not differ between control and food addition areas, except in 1995 when survival was 12–44% lower on the food addition grids. In 1992, a decline year, none of the radio-tagged leverets on control areas survived. In contrast, leverets on the predator exclosure + food grid had survival rates of 0.21–0.43 in this year, which are still low compared with the survival rates observed during the increase phase on control sites (0.37–0.71). Throughout the cycle, of the leverets that were killed, 70% died in their first week of life and 23% in the second week.

Postweaning juvenile survival was measured in one year (1995–1996) of the increase phase (Gillis 1999). Juveniles born in the first two litters had survival rates similar to adults, but juveniles born in the third and fourth litters had poorer survival (table 8.5). During this year, food addition had no statistically noticeable effect on juvenile survival.

Table 8.3 Parturition data for snowshoe hares.

					Control			
	1988	1989	1990	1991	1992	1994	1995	1996
Mean Birth Date								
Litter 1	24 May ± 0.6 (6)	25 May ± 0.7 (8)	24 May ± 0.6 (14)	29 May ± 1.1 (7)	31 May ± 3.0 (6)	12 May ± 0.8 (11)	22 May ± 2.4 (6)	22 May ± 0.7 (12)
Litter 2	30 June (1)	2 Jul ± 0.9 (13)	26 Jun ± 0.7 (27)	6 Jul ± 0.7 (11)	6 Jul ± 4.5 (2)	18 Jun ± 1.0 (9)	20 Jun ± 1.3 (9)	27 Jun ± 0.9 (14)
Litter 3	30 Aug ± 2.0 (3)	7 Aug ± 1.3 (7)	31 Jul ± 0.4 (19)	—	—	25 Jul ± 1.2 (10)	28 Jul ± 1.4 (8)	2 Aug ± 0.6 (13)
Range of Birth Date								
Litter 1	23 May–27 May	23 May–29 May	20 May–29 May	27 May–3 Jun	29 May–11 Jun	9 May–17 May	16 May–2 Jun	19 May–28 May
Litter 2	30 June	26 Jun–7 Jul	21 Jun–5 Jul	2 Jul–9 Jul	26 Jun–11 Jul	14 Jun–24 Jun	13 Jun–27 Jun	22 Jun–4 Jul
Litter 3	30 Jul–5 Aug	2 Aug–13 Aug	28 Jul–5 Aug	—	—	20 Jul–2 Aug	28 Jul–2 Aug	30 Jul–5 Aug
Pregnancy Rate %								
Litter 1	—	93.8 (32)	89.4 (67)	100 (7)	77 (9)	100 (14)	100 (17)	100 (29)
Litter 2	—	96.8 (31)	96.2 (52)	85 (13)	100 (9)	100 (8)	100 (17)	100 (21)
Litter 3	—	82.4 (34)	86.4 (50)	0 (13)	0[a]	100 (8)	100 (12)	100 (27)

Stillborn Rate

Litter 1	—	0 (29)	7.3 (55)	0 (20)	0 (13)	3.1 (32)	0 (17)	0 (40)
Litter 2	—	7.7 (78)	4.5 (156)	30.4 (46)	0 (6)	0 (48)	1.6 (62)	6.3 (80)
Litter 3	—	6.3 (32)	13.6 (81)	—	—	4.7 (64)	8.9 (45)	1.3 (79)

Litter Size

Litter 1	4.3 ± 0.21 (6)	3.6 ± 0.3 (8)	3.9 ± 0.2 (14)	3.3 ± 0.4 (7)	3.2 ± 0.2 (6)	3.2 ± 0.3 (11)	3.0 ± 0.5 (6)	3.6 ± 0.3 (12)
Litter 2	5 (1)	6.0 ± 0.4 (13)	5.8 ± 0.3 (27)	4.2 ± 0.6 (11)	4.5 ± 1.5 (2)	5.9 ± 0.4 (10)	6.9 ± 0.5 (9)	5.8 ± 0.3 (14)
Litter 3	6.0 ± 0 (3)	4.4 ± 0.4 (7)	4.3 ± 0.3 (19)	—	—	6.4 ± 0.3 (11)	5.6 ± 0.6 (8)	6.1 ± 0.5 (13)

% Females

Litter 1	46.2 (26)	42.1 (19)	53.2 (47)	68.8 (16)	50.0 (12)	42.4 (33)	55.6 (18)	51.2 (41)
Litter 2	40.0 (5)	54.2 (72)	56.9 (144)	63.0 (46)	50.0 (6)	44.2 (52)	51.6 (62)	44.4 (81)
Litter 3	33.0 (18)	53.3 (30)	45.7 (70)	—	—	37.5 (64)	48.9 (45)	50.6 (79)

(continued)

Table 8.3 (*Continued*) Parturition data for hares on food and predator exclosure + food treatments.

	Food			Predator Exclosure + Food	
	1989	1990	1995	1991	1992
Mean Birth Date					
Litter 1	20 May ± 0.6 (8)	20 May ± 0.5 (26)	10 May ± 1.5 (6)	—	21 May ± 0.5 (15)
Litter 2	23 Jun ± 0.6 (17)	24 Jun ± 0.8 (28)	10 Jun ± 1.7 (5)	28 Jun ± 1.0 (9)	27 Jun ± 0.4 (18)
Litter 3	31 Jul ± 0.8 (15)	26 Jul ± 0.7 (22)	20 Jul ± 0.6 (5)	5 Aug ± 0.4 (6)	2 Aug ± 0.4 (19)
Range of Birth Date					
Litter 1	19 May–23 May	17 May–25 May	6 May–15 May	—	16 May–24 May
Litter 2	20 Jun–28 Jun	19 Jun–2 Jul	5 Jun–15 Jun	24 Jun–3 Jul	25 Jun–30 Jun
Litter 3	21 Jul–2 Aug	21 Jul–2 Aug	18 Jul–28 Jul	4 Aug–6 Aug	31 Jul–6 Aug
Pregnancy Rate %					
Litter 1	91.8 (73)	99.2 (119)	100 (27)	92.8[b]	92.8[b]
Litter 2	100 (67)	97.0 (67)	100 (19)	97.5[b]	97.5[b]
Litter 3	96.6 (58)	87.0 (77)	100 (15)	88.1[b]	88.1[b]

Stillborn Rate					
Litter 1	3.6 (28)	13.6 (86)	0 (25)	—	0 (55)
Litter 2	6.2 (114)	9.6 (135)	14.6 (41)	5.7 (70)	9.1 (132)
Litter 3	44.6 (72)	23.5 (119)	11.8 (34)	0 (28)	5.4 (111)
Litter size					
Litter 1	3.8 ± 0.4 (9)	3.7 ± 0.1 (26)	4.2 ± 0.4 (6)	—	4.1 ± 0.3 (15)
Litter 2	6.3 ± 0.5 (18)	6.1 ± 0.3 (27)	7.2 ± 0.4 (6)	7.8 ± 0.4 (9)	7.5 ± 0.3 (20)
Litter 3	4.8 ± 0.4 (16)	5.6 ± 0.2 (23)	5.7 ± 0.9 (6)	4.7 ± 0.6 (6)	5.9 ± 0.4 (20)
% Females					
Litter 1	47.6 (21)	56.8 (74)	56.0 (25)	—	37.0 (54)
Litter 2	57.4 (94)	43.8 (121)	50.0 (40)	50.7 (69)	50.4 (129)
Litter 3	62.5 (40)	49.5 (93)	46.9 (32)	66.7 (27)	51.4 (109)

Sample sizes are in parentheses. For stillborn rate and percent females, the sample sizes are numbers of leverets. For all others, sample sizes are numbers of adult females. Means are ± 1 SE.

[a]Although trapping data were not available for this litter, no third litter juveniles were captured later in the season.

[b]Pregnancy rates were not measured directly; these values were estimated from values for hares on food and control sites in 1989 and 1990 and used to calculate reproductive output in these years. Pregnancy rates were similarly estimated for control sites in 1988.

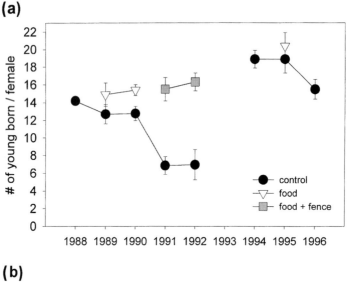

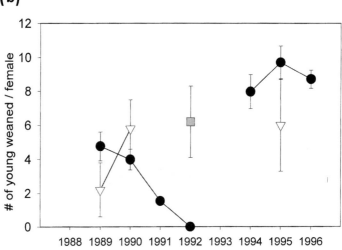

Figure 8.5 Snowshoe hare reproductive output and recruitment through the cycle. (a) Values are total young per female per summer, calculated from pregnancy rates and mean litter size per litter group. Litter 4 pregnancy rates and litter sizes for 1994 and 1995 were estimated based on values for last litters in other years, because the fourth litter was not measured directly. Pregnancy rates for control hares in 1988 and predator exclosure + food hares in 1991 and 1992 were based on averages for hares in 1989 and 1990 (see table 8.3). Standard errors were calculated from litter sizes. (b) Recruitment to 30 days. Values are calculated from reproductive output and survival to 30 days. For the food-addition grids in 1989, there was no survival estimate for litter 1, so we used the conservative value of survival of the second litter (0.15). This value is conservative because in most years hares in the second litters had lower survival than hares in first litters.

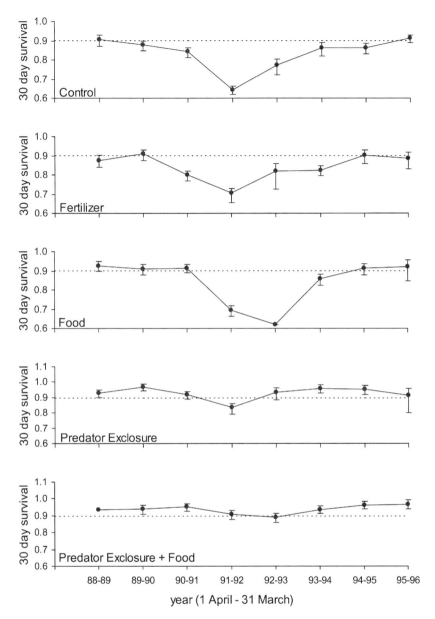

Figure 8.6 Adult snowshoe hare survival. Values are 30-day survival and 95% CLs (based on Pollock et al. 1989, equation 3); the line at 0.90 is given for comparison between treatments. A 30-day survival of 0.90 corresponds to an annual survival rate of 27.8%. Survival estimates were calculated from 1 April through 31 March. The point for fertilizer grids in 1995–1996 is based on the period 1 April to 29 February, since many radio collars were removed in the spring trapping census, making March survival estimates suspect. The 1995–1996 estimate for predator exclosure + food is based on the period 1 April–8 November, since a coyote inside the fence inflicted heavy mortality thereafter. Within each year, sample sizes were 42–168 hares (control), 26–131 (fertilizer), 44–114 (food), 25–113 (predator exclosure), and 35–94 (predator exclosure).

Table 8.4 Preweaning survival of snowshoe hares.

	1989	1990	1991	1992	1994	1995	1996
Control							
Litter 1	0.73 ± 0.13 (12)	0.27 ± 0.13 (11)	0.47 ± 0.12 (17)	0 (9)	0.50 ± 0.22 (27)	0.71 ± 0.11 (19)	0.61 ± 0.08 (39)
Litter 2	0.22 ± 0.09 (23)	0.13 ± 0.05 (41)	0 (21)	0 (4)	0.37 ± 0.08 (46)	0.61 ± 0.07 (49)	0.54 ± 0.06 (66)
Litter 3	0.18 ± 0.12 (11)	0.51 ± 0.11 (24)	—	—	0.66 ± 0.07 (50)	0.60 ± 0.09 (31)	0.56 ± 0.07 (61)
Food							
Litter 1	—	0.45 ± 0.10 (31)				0.49 ± 0.15 (14)	
Litter 2	0.15 ± 0.07 (28)	0.15 ± 0.07 (33)				0.17 ± 0.08 (22)	
Litter 3	0.15 ± 0.13 (15)	0.57 ± 0.11 (27)				0.48 ± 0.10 (28)	
Predator Exclosure + Food							
Litter 1				0.21 ± 0.11 (26)			
Litter 2				0.37 ± 0.09 (39)			
Litter 3				0.43 ± 0.09 (36)			

Kaplan-Meier techniques were used to estimate survival (± SE) of radio-tagged leverets from birth until 30 days old; hares were typically weaned between 4 and 5 weeks of age. The numbers of hares monitored are given in parentheses.

Table 8.5 Survival rates of postweaning juvenile hares.

		Control			*Food*	
	n	30-Day Survival	95% CL[a]	*n*	30-Day Survival	95% CL
Litter 1	16	0.93	0.86–0.97	8	0.85	0.00–0.94
Litter 2	13	0.95	0.88–0.99	9	0.94	0.84–0.99
Litter 3	18	0.78	0.00–0.86	10	0.89	0.00–0.98
Litter 4	4	0.41	0.00–0.89	8	0.75	0.00–0.87

These data were collected from radio-tagged animals in 1995–1996, during the population increase. Survival is estimated by the Kaplan-Meier estimator.

[a]We calculated 95% confidence limits using Greenwood's standard error (Pollock et al. 1989).

Simple Leslie matrix models of the demography of the cycle indicate that 30-day survival rates of postweaning hares must be above ~0.90 for the population to increase (figure 8.7). Above that survival rate, preweaning juvenile recruitment (number of leverets
surviving to 30 days per female per year) has a strong impact on the rate of increase, but
it seems that survival of hares >28 days old is the critical parameter for determining

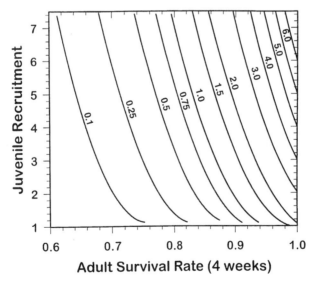

Figure 8.7 The impact of survival and recruitment on the annual rate of population change.
The survival axis is 28-day survival of hares older than 30 days (i.e., postweaning juvenile and
adult survival combined). Juvenile recruitment is measured as number of leverets alive at 30
days per mother per year. The rates of population increase were calculated using a simple Leslie
matrix model, and they indicate that population growth basically occurs only when monthly
survival is >0.90.

whether populations will increase or decline. This interpretation is supported by a recent analysis of the 1995–1996 increase data, which suggests that the numeric changes in the cycle predominantly result from changes in juvenile and adult survival, followed by leveret survival and reproductive output (Haydon et al. 1999).

8.3.5   Causes of Death

In most years, >90% of the adult snowshoe hares that died were killed by predators (table 8.6). On control areas, some hares died of nonpredation causes (e.g., starvation, injury) during the peak and decline years, but no control hares died of nonpredation causes during the low phase. Similar patterns of predation mortality were observed on all other treatments except predator exclosure + food; this treatment had some nonpredation deaths in all years except one year of the low phase, and the predation deaths are consequently 5–10% lower than on control sites. The fences were permeable to hares; throughout the cycle, 31.2% of radio-collared predator exclosure hares and 35.4% of predator exclosure + food hares died outside the fences. We excluded these deaths in calculating survival rates and causes of death for these treatments.

Mammalian predators, mainly coyotes and lynx, were responsible for 65–75% of the predator-caused deaths of adult hares on unfenced treatments (figure 8.8), whereas raptors, mainly great horned owls and goshawks, killed 70–80% of the predator-killed hares inside the fences. Although about 15% of deaths were identifiable only as predation and another 15% were completely unknown as to the cause of death, we expect that the causes of these deaths were distributed in a similar fashion. It is possible that great horned owl kills were less likely to be identified than were kills by other predators because owl kills have fewer diagnostic characteristics. Across all treatments and all years, only 18 kills were positively identified as due to other predators: 4 hares were killed by wolves, 2 by marten, 2 by wolverine, 2 by weasels, 1 by a bald eagle, 1 by a Harlan's hawk, and 6 by hawk owls. Five of the 6 hawk owl deaths occurred in February and March 1991 on control 1. Additionally, the fences were not completely effective at keeping out mammalian predators; over all the years 12.4% and 7.8% of deaths of radio-collared hares on predator exclosure and predator exclosure + food, respectively, were due to lynx before the lynx left the fences. The predator exclosure + food treatment had a coyote for about 3 months during winter 1995–1996, and more than half of the hare deaths on this treatment in the entire year were due to this one animal.

Adult hares on control sites had seasonal sources of mortality (figure 8.9). Most of the coyote kills (65%) were in October and November, and of these, 55% were cached whole rather than eaten immediately. Lynx, goshawks, and great horned owls made 70–75% of their kills of adult hares between December and May. Most hares (56%) that died of nonpredation causes died in February and March; <20% of nonpredation deaths occurred between June and November. These patterns may be slightly biased because 51% of the hares for which we could not identify a cause of death died during the summer. This bias is unlikely to be large, because only 26% of hares died during the summer, and we successfully identified the cause of death for 34% of these hares.

Like adults, postweaning juvenile hares mainly died of predation (table 8.7). On control sites, mammalian predators killed more juveniles than did raptors (12 and 3, respectively), and no nonpredation deaths were observed. On food sites, mammalian predators

Table 8.6 Causes of deaths of adult snowshoe hares.

Grid/Year	Deaths from known predators						Deaths from all causes				
	Coyote	Lynx	Goshawk	Owl	Mammal	Raptor	Unidentified predator	Nonpredation	Unknown	% Predation	No. Dead
Control											
1988–1989	16.7	0	11.1	0	22.2	27.8	11.1	22.2	16.7	73.3	18
1989–1990	19.4	9.7	9.7	3.2	29.0	16.1	16.1	9.7	29.0	86.4	31
1990–1991	4.3	10.0	14.3	11.4	15.7	42.9	7.1	11.4	22.9	85.2	70
1991–1992	43.3	14.2	3.9	10.2	58.3	17.3	9.5	0	15.0	100.0	127
1992–1993	13.9	19.4	11.1	8.3	41.7	22.2	13.9	0	22.2	100.0	36
1993–1994	42.1	15.8	5.3	0	57.9	10.5	5.3	0	26.3	100.0	19
1994–1995	40.8	6.1	8.2	2.0	55.1	14.3	16.3	0	14.3	100.0	49
1995–1996	22.5	15.0	12.5	5.0	50.0	22.5	12.5	2.5	12.5	97.1	40
Fertilizer											
1988–1989	0	15.8	5.3	5.3	15.8	26.3	26.3	5.3	26.3	92.9	19
1989–1990	8.3	8.3	4.2	4.2	33.3	16.7	29.2	0	20.8	100.0	24
1990–1991	8.3	30.6	22.2	12.5	41.7	38.9	5.6	6.9	6.9	92.5	72
1991–1992	22.9	20.0	4.3	8.6	55.7	14.3	14.3	1.4	14.3	98.5	70
1992–1993	7.7	30.8	15.4	0	38.5	15.4	23.1	0	23.1	100.0	13
1993–1994	22.2	11.1	11.1	22.2	33.3	33.3	0	0	33.3	100.0	9
1994–1995	6.3	6.3	6.3	6.3	18.8	12.5	37.5	0	31.3	100.0	16
1995–1996	11.1	27.8	0	0	44.4	11.1	11.1	5.6	27.8	92.3	18
Food											
1988–1989	16.7	8.3	0	8.3	33.3	8.3	16.7	8.3	33.3	87.5	12
1989–1990	4.5	0	4.5	4.5	18.2	31.8	31.8	4.6	13.6	94.7	22

(continued)

Table 8.6 (Continued)

Grid/Year	Deaths from known predators						Deaths from all causes				
	Coyote	Lynx	Goshawk	Owl	Mammal	Raptor	Unidentified predator	Nonpredation	Unknown	% Predation	No. Dead
1990–1991	24.2	15.2	9.1	9.1	45.5	18.2	15.2	3.0	18.2	96.3	33
1991–1992	26.4	12.5	12.5	1.4	48.6	13.9	15.3	1.4	20.8	98.2	72
1992–1993	28.6	14.3	3.6	7.1	50.0	10.7	21.4	3.6	14.3	95.8	28
1993–1994	38.7	22.6	3.2	3.2	61.3	6.5	16.1	0	16.1	100.0	31
1994–1995	17.6	29.4	0	0	52.9	0	11.8	0	35.3	100.0	17
1995–1996	10.0	30.0	10.0	0	45.0	10.0	25.0	0	20.0	100.0	20
Predator exclosure[a]											
1988–1989	0	21.4	0	0	21.4	0	35.7	7.1	35.7	88.9	14
1989–1990	0	11.1	11.1	22.2	11.1	33.3	22.2	11.1	22.2	85.7	9
1990–1991	0	7.4	14.8	14.8	7.4	59.3	11.1	3.7	18.5	95.5	27
1991–1992	0	17.0	13.2	39.6	17.0	56.6	7.6	7.6	11.3	91.5	53
1992–1993	0	20.0	20.0	20.0	20.0	70.0	10.0	0	0	100.0	10
1993–1994	0	0	20.0	60.0	0	80.0	0	0	20.0	100.0	5
1994–1995	0	0	12.5	50.0	12.5	62.5	25	0	0	100.0	8
1995–1996	0	0	27.3	9.1	0	45.5	36.4	9.1	9.1	90.0	11

Predator Exclosure + Food[a]

Year										
1988–1989	0	30.0	0	0	30.0	0	50.0	10.0	10.0	88.9	10
1989–1990	0	15.4	0	0	23.1	15.4	15.4	7.7	38.5	87.5	13
1990–1991	0	7.1	35.7	0	7.1	57.1	14.3	14.3	7.1	84.6	14
1991–1992	0	9.7	32.3	9.7	9.7	51.6	16.1	9.7	12.9	88.9	31
1992–1993	0	6.9	13.8	20.7	6.9	62.1	17.2	6.9	6.9	92.6	29
1993–1994	0	0	17.6	11.8	0	52.9	23.5	5.9	17.7	92.9	17
1994–1995	0	9.1	18.2	27.3	9.1	72.7	9.1	0	9.1	100.0	11
1995–1996	53.6	0	3.6	0	53.6	10.7	14.3	14.3	7.1	84.6	28

Each time span includes deaths from 1 April through 31 March. Values are percentages of hares dead of each cause. "Owl" refers to great horned owls. Hares killed by marten, weasels, wolves, eagles, hawk-owls, and Harlan's hawks are included in the "mammal" and "raptor" categories as appropriate. Percent predation is calculated out of deaths for which cause of death was positively determined (i.e., excluding "unknown" deaths).

[a]On several occasions, mammalian predators were inside the fences for time periods of hours to days. Lynx were the most common.

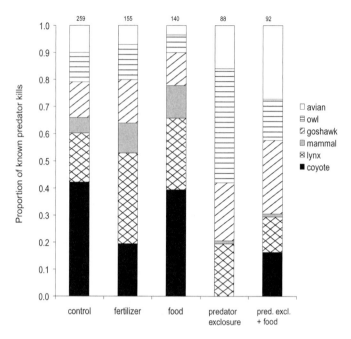

Figure 8.8 Predation deaths of adult snowshoe hares. All predator-caused deaths from 1 April 1988 to 31 March 1996 are included. Sample sizes for each treatment are at the top of each bar. *Avian* kills include kills made by bald eagles, hawk owls, and Harlan's hawks, plus hares killed by a raptor but for which the species of raptor could not be determined. *Mammal* kills include kills made by martens, weasels, wolverines, and wolves, plus kills for which the species of mammalian predator could not be determined. The mammalian kills on the two fenced grids are due to individual lynx and coyote that were inside the fences.

and raptors killed 7 and 6 juveniles, respectively, and 2 hares died of natural nonpredation causes. Juvenile mortality patterns were also seasonal, with most of the kills by raptors and coyotes occurring before November (Gillis 1999).

Leverets were killed by a different suite of predators (table 8.8). We observed no lynx or coyote kills of leverets. Red and ground squirrels were responsible for up to a third of leveret mortalities on control areas and food addition sites. On the predator exclosure + food grid during the decline, great horned owls and ground squirrels were the main predators (each responsible for 13% of all deaths), and we suspect that the high loss rate of leverets (27%) was due to great horned owls or other raptors carrying the animals off the grid and out of telemetry range. Most nonpredation deaths (abandoned/exposed) of leverets occurred during the decline phase (84% and 25% of deaths in 1991 and 1992, respectively; the next highest proportion of non-predation deaths on control areas was 3% in 1995).

Habitat structure and spatial patterning of predators affected the kill patterns that we observed. Coyotes hunted more frequently at either end of the Shakwak Trench than in the middle, affecting the kill patterns of adult snowshoe hares. The predator exclosure + food treatment had less dense spruce cover, scattered old trees, and exceptionally high densities of ground squirrels but comparatively low densities of red squirrels. For the lev-

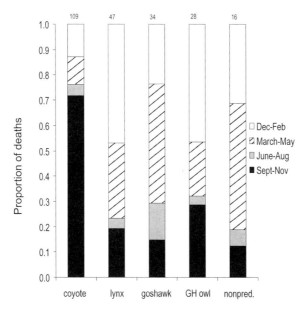

Figure 8.9 Seasonality of hare mortalities. Each bar indicates all the deaths of control hares for which cause of death could be determined for the period 1 April 1988 to 31 March 1996. Sample sizes for each mortality type are given above the bars.

Table 8.7 Causes of mortality for postweaning juvenile hares.

Cause of Death	Control	Food
Lynx	23.1 (6)	11.8 (2)
Coyote	7.7 (2)	23.5 (4)
Great horned owl	0 (0)	11.8 (2)
Goshawk	11.5 (3)	11.8 (2)
Unidentified mammal	15.4 (4)	5.9 (1)
Unidentified raptor	0 (0)	11.8 (2)
Unidentified predator	26.9 (7)	11.8 (2)
Total predation	84.6 (22)	88.2 (15)
Nonpredation	0 (0)	11.8 (2)
Unknown	15.4 (4)	0 (0)
No. of mortalities	26	18[a]
No. radio collared	49	35

These data were collected from radio-collared animals in 1995–1996, during the population increase. Animals were radio collared shortly after weaning and monitored until the following spring. Causes of mortality are presented as percentages of total deaths.

[a]One hare killed by a hunter was excluded from the analyses; percentages are based on 17 mortalities.

Table 8.8 Causes of deaths of leverets.

Cause of Death	Control							Food			Predator Exclosure + Food
	1989	1990	1991	1992	1994	1995	1996	1989	1990	1995	1992
Red squirrel	20.0 (6)	20.4 (11)	4.0 (1)	12.5 (1)	35.6 (16)	14.3 (5)	30.5 (22)	15.6 (5)	23.1 (12)	0 (0)	3.7 (2)
Arctic ground squirrel	13.4 (4)	13.0 (7)	0 (0)	0 (0)	0 (0)	17.1 (6)	8.5 (6)	9.4 (3)	7.7 (4)	15.4 (6)	13.0 (7)
Short-tailed weasel	0 (0)	0 (0)	0 (0)	0 (0)	20.0 (9)	0 (0)	0 (0)	0 (0)	0 (0)	0 (0)	0 (0)
Unidentified small mammal	0 (0)	0 (0)	0 (0)	25.0 (2)	6.7 (3)	5.7 (2)	6.9 (5)	0 (0)	0 (0)	5.1 (2)	3.7 (2)
Great horned owl	3.3 (1)	0 (0)	0 (0)	0 (0)	2.2 (1)	20.0 (7)	0 (0)	0 (0)	1.9 (1)	18.0 (7)	13.0 (7)
Goshawk	3.3 (1)	5.6 (3)	0 (0)	0 (0)	0 (0)	2.9 (1)	5.6 (4)	0 (0)	3.8 (2)	5.1 (2)	1.8 (1)
Boreal owl	0 (0)	0 (0)	0 (0)	0 (0)	6.7 (3)	0 (0)	1.4 (1)	0 (0)	0 (0)	0 (0)	0 (0)
Harlan's hawk	0 (0)	0 (0)	0 (0)	0 (0)	2.2 (1)	0 (0)	0 (0)	3.1 (1)	0 (0)	0 (0)	0 (0)
American kestrel	0 (0)	0 (0)	0 (0)	0 (0)	0 (0)	0 (0)	0 (0)	0 (0)	0 (0)	0 (0)	1.8 (1)
Unidentified raptor	0 (0)	0 (0)	0 (0)	12.5 (1)	13.3 (6)	2.9 (1)	4.2 (3)	0 (0)	0 (0)	12.8 (5)	18.5 (10)
Carcass or radio in tree	40.0 (12)	27.8 (15)	12.0 (3)	0 (0)	4.4 (2)	22.8 (8)	26.3 (19)	34.4 (11)	15.4 (8)	12.8 (5)	18.5 (10)
Unidentified predator	0 (0)	25.9 (14)	16.0 (4)	12.5 (1)	0 (0)	5.7 (2)	4.2 (3)	12.5 (4)	23.1 (12)	20.5 (8)	24.2 (13)
Total predation	80.0 (24)	92.6 (50)	16.0 (4)	62.5 (5)	91.1 (41)	91.4 (32)	87.6 (63)	75.0 (24)	75.0 (39)	89.7 (35)	98.2 (53)
Unknown	20.0 (6)	5.6 (3)	0 (0)	12.5 (1)	8.9 (4)	5.7 (2)	9.6 (7)	25.0 (8)	23.1 (12)	0 (0)	0 (0)
Abandoned/exposure	0 (0)	1.8 (1)	84.0 (21)	25.0 (2)	0 (0)	2.9 (1)	2.8 (2)	0 (0)	1.9 (1)	10.3 (4)	1.8 (1)
No. lost	4	4	8	1	15	11	10	6	10	9	27
No. of mortalities	30	54	25	8	45	35	72	32	52	39	55
No. radio tagged	46	76	38	13	123	99	166	42	93	64	101

The values are percentages of all mortalities of hares from birth until 30 days old. "Unidentified small mammal" deaths are deaths due to red squirrels, ground squirrels, or weasels, but for which the exact predator could not be determined. These data summarize kills for litters 1–3 in each year, except for 1991 and 1992, when hares had only two litters. Many of the "carcass or radio in tree" cases are probably due to red squirrel predation.

erets on this treatment, the low amount of red squirrel predation is probably due to this spatial pattern. Similarly, the nesting locations of raptors may have affected their contribution to hare mortality patterns. Of the raptorial kills of leverets that we observed, goshawk kills occurred only on controls 3 and 4, and the great horned owl kills occurred only on the two food grids and one off-grid site (Stefan 1998).

8.3.6 Immigration and Emigration

We assessed immigration rates by calculating the proportion of hares caught on a grid for the first time in each spring trapping session (including animals that had been caught previously on other grids). This index is problematic because it does not allow us to differentiate among resident hares from the area that were caught for the first time, hares that had immigrated since the previous autumn, and hares that were transient at the time of trapping. The food addition treatment had higher proportions of newly caught animals each spring than did all the other treatments, except during the low and increase phases when the fertilizer treatments had a higher proportion of newly caught hares (figure 8.10a). This pattern suggests that hares are attracted to the food. The fenced treatments had lower proportions of newly caught hares, possibly indicating that the fence or the relatively high hare densities within the fences acted as a deterrent to immigration.

Emigration was also difficult to assess because the experiment was not set up to study dispersal. As a very crude index, we consider the proportion of hares caught during only one trapping session. Hares on food addition areas were less likely to be caught multiple times than were hares on other treatments (figure 8.10b). On control sites, 54% of hares were caught in only one session, compared with 66% of the hares on food addition sites.

For the two fenced treatments, we could use one additional assessment of emigration: the proportion of radio-collared hares dying outside the fences. A greater proportion of hares left the fences as densities declined (figure 8.10c). This pattern could indicate two things—that more hares dispersed during this period, or that hares whose home ranges included both sides of the fence were more likely to be killed outside the fence during the decline phase. During the low phase, many of the animals that were later located outside fences had clearly left home ranges inside the fences (K. E. Hodges, unpublished data), which suggests that the decline pattern may similarly be due to dispersal. Furthermore, on the predator exclosure, the population decline from spring 1995 to spring 1996 is partially attributable to dispersal: of 19 hares that were radio collared in autumn 1995, 4 went through the fence (2 were trapped elsewhere, 2 died), and 3 hares moved from the trapping grid to a far corner of the entire fenced area; 7 radio-collared hares died on the trapping grid during the same period.

8.4 Impacts of Experimental Treatments

All treatments increased hare densities, although the effect of fertilization was minimal in relation to the large effects fertilization had on plant growth (see chapters 5 and 6). Food addition and excluding predators had their largest impacts during the decline with 3.8-fold and 2.4-fold increases in density, respectively. The simultaneous manipulation of predator exclosure + food also had the largest impact during the decline, with densities 14.4-fold higher than control densities. These treatment patterns are also observed when

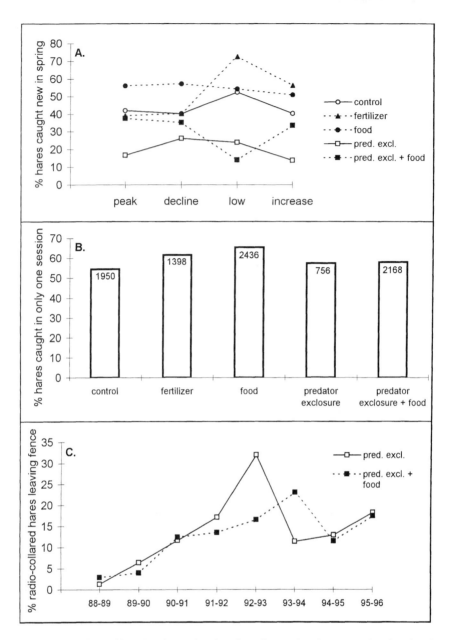

Figure 8.10 Indices of immigration and emigration of snowshoe hares. (A) Immigration, indexed as the proportion of hares caught each spring that were not previously captured on that
grid. Proportions therefore include new hares and hares previously caught elsewhere. The increase phase includes the years 1994–1996; the trapping history was not deemed adequate to
assess immigration during the first period of increase. (B) Proportion of hares caught during
only one trapping session. The number of individual hares caught on each treatment are given
in the bars. (C) Percentage of radio-collared hares leaving the fences within each year, 1 April–
31 March. Through the entire cycle 31.2% or 62/199 hares left the predator exclosure, and
35.4% or 84/237 hares left the predator exclosure + food area. In the figure, the percentages
are lower because the values are calculated as (n dead outside)/(n radio collared per year), and
many radio-collared hares lived more than one year.

impacts are integrated across the entire cycle or examined in the other phases (CD-ROM frame 52).

This large impact indicates that food and predation interact in their effects on hare densities, because this magnitude is higher than the combined effects shown from the single-factor manipulations. The one caveat is that the manipulations were established during the increase phase, and hence for each phase the treatments had been in effect for a different length of time. The relative densities in each phase may therefore partly result from the length of time that each manipulation had been imposed.

The population rates of change also suggest that food and predation interacted in their impacts on hares. Hares on control areas, the fertilizer treatments, and predator exclosure had 2 years of severe decline and an additional 1 or 2 years of mild decline, whereas hares on food areas had only 2 years of severe decline. The hare population on predator exclosure + food had 2 years of mild decline, followed by only 1 year of severe decline. These patterns suggest that the predator exclosure + food manipulation altered the timing and pattern of the cycle, in addition to affecting hare densities throughout.

Clearly, the numeric changes are composed of the changes in reproduction, survival, and dispersal that occur during the cycle (table 8.9). Of these, juvenile survival, adult survival, and reproductive output are thought to be the major parameters contributing to cyclic numeric changes, both because of the sensitivity of the rate of increase to changes in these rates (figure 8.7; Haydon et al. 1999) and because of the observed magnitude of change in each of these parameters (Krebs et al. 1996, Stefan 1998). Understanding the cycle is therefore mostly an exercise in explaining why these demographic rates show the regular changes they do. Determining the effects of the experimental treatments on these parameters therefore allows a more sensitive test of cyclic mechanisms than does just looking at density and rate of population change. If an experimental manipulation changed either the magnitude or the timing of the changes in survival, reproduction, or dispersal, that would provide evidence that the factor manipulated is necessary for hare populations to cycle.

8.4.1 Impacts on Hare Demography

Hares on food addition sites had a total reproductive output similar to that of control hares, even though the onset of reproduction was slightly earlier in some years. Stillborn rates were higher for hares on food addition sites, which compensated for slightly higher litter sizes. Food addition did not improve the survival of leverets or juvenile hares, but adult hares on food addition sites survived slightly better than hares on control sites in peak years and slightly worse in decline years. Most of the increase in density on food addition sites relative to control areas resulted from movement of hares onto food sites, rather than from reproductive and survival rates in situ (see also Boutin 1984, Krebs et al. 1986b). Fertilization resulted in increased densities of natural food but did not change patterns of hare survival or rates of increase; there were small positive effects on hare density.

Hares inside the predator exclosure had higher survival rates than hares on control sites, but the cyclic patterns of change in survival and in density still manifested themselves. Unfortunately, we have no reproductive information for this treatment, but we suspect that reproduction on this treatment was similar to that on control areas. The predator exclosure treatment showed patterns of numeric change similar to those on the control areas,

Table 8.9 Summary of the effects the experimental treatments had on snowshoe hare demography during the cycle, for 1988–1996.

	Fertilizer	Food	Predator Exclosure	Predator Exclosure + Food
Attributes of Cycle				
Peak year	1990	1991	1990	1990
Low year	1993	1993, 1994	1993	1994
Spring density (minimum/maximum amount higher than controls)	1.0–1.9	2.0–3.8	1.3–2.4	1.2–14.4
Cyclic amplitude	46.9	24.7	8.8	6.2
Survival and Causes of Death				
Adult survival	Similar	Similar/lower	Higher	Higher
Juvenile survival	ND	Same in increase[a]	ND	ND
Leveret survival	ND	Same or lower[b]	ND	Higher in decline[c]
Predation deaths of adults (%)	Similar	Similar	Similar	Lower
Reproduction and Age Structure				
Litter 1 birth date	ND	Earlier by 4–12 days[b]	ND	Earlier by ~10 days in decline[c]
Litter number	ND	Same[b]	ND	Higher by one litter in decline[c]
Litter size	ND	Same[b]	ND	Higher in decline[c]
Total annual reproduction	ND	Same[b]	ND	Higher in decline[c]
Age structure	Same	Same	Older	Older
Dispersal				
Emigration (% single captures)	Higher	Highest	Higher	Higher
Immigration (% new in spring)	Same/higher	Higher	Lower	Lower

ND indicates the parameter was not detrimental. Values are relative to the control. Multiple values indicate the range from replicate grids. Note that reproductive attributes and juvenile survival were only measured in some years.

[a]Juvenile survival was assessed in 1995–1996.

[b]Reproductive parameters were measured on food grids in only 1989, 1990, and 1995.

[c]Reproductive parameters for predator exclosure + food were measured in 1991 and 1992 only.

whereas the predator exclosure + food treatment did not. The two fenced treatments had similar survival rates, which suggests that the different reproduction on the predator exclosure + food treatment led to the changes in the cyclic dynamics. The similar dynamics between the predator exclosure and controls therefore indicates no difference in reproductive output.

The combined manipulation of food and predators altered both the magnitude and timing of the changes in reproduction and survival. Hares on the predator exclosure + food treatment did not show the collapse in reproductive output typical of the decline phase. Survival rates were uniformly high on the predator exclosure + food treatment throughout the cycle and did not show the reduction in survival seen on the other treatments (except predator exclosure). During the decline years on control sites, hares on this treatment remained at comparatively very high densities, and many hares went through the fences. This movement appears to be partially responsible for the decline in numbers on this treatment because reproduction and survival would have maintained a higher population in the absence of emigration. The decline was also partially due to low leveret survival in 1992 because of high predation rates by great horned owls.

8.4.2 Causation of the Hare Cycle

Our results show that the combined manipulation of food and predators had a much stronger impact on hare dynamics than did manipulation of either factor alone. The implication is that the food–hare (Bryant 1981, Fox and Bryant 1984) and hare–predator (Trostel et al. 1987) models are inadequate explanations of hare cyclicity. Instead, explanations of the hare cycle should focus on the analysis of the interactive effects of food and predation on hare dynamics. Studies at Kluane and elsewhere have indicated several promising approaches for analyzing the interactive effects of food and predation on the hare cycle (Wolff 1980, 1981, Sievert and Keith 1985, Sovell 1993, Hik 1994, Murray et al. 1997, 1998, Boonstra et al. 1998a, Hodges 1998). These approaches focus on hare behavior and physiology, thus emphasizing the ways individual hares are affected by food and predation.

The boreal forest is a mosaic of habitat types, and it is possible that modifications in hare habitat use patterns, movements, or foraging behavior in response to predation pressure could affect their diets, nutrition, survival, and perhaps fecundity (Dolbeer and Clark 1975, Wolff 1981, Hik 1994, Hodges 1998). Additionally, predators can cause hares stress through unsuccessful chases or by causing hares to change their behavior. These stress patterns change through the cycle as the number of predators varies, and the stress patterns lead to a number of physiological changes that might affect fecundity of hares or, through maternal effects, survival of the leverets (Boonstra et al. 1998a,b). Finally, levels of parasite infestation vary through the cycle, and parasites may affect hare behavior and physiology, leading to changes in fecundity or survival (Sovell 1993, Sovell and Holmes 1996, Murray et al. 1997, 1998).

8.4.3 Efficacy of the Experimental Treatments

The reliability of our demographic results depends partly on how effectively the treatments created food-rich and predator-poor areas. Food addition was popular among the fauna: moose, grizzly bears, ground squirrels, corvids, and microtines all ate the rabbit chow. For most of the year, we succeeded in spreading enough chow that hares had constant access to it, although we cannot be certain how much chow each hare obtained. During the spring thaw (~4 weeks) and periods of rain, the chow that was available became waterlogged and unpalatable. During the thaw, we were also unable to feed as regularly as during the rest of the year, which is problematic because that may be the time of year when hares are most nutritionally stressed due to the cumulative effects of overwinter browsing, the lack of new growth, and reproduction (mating chases and pregnancy). We made more effort to keep the predator exclosure + food addition treatment regularly fed during this period. Despite these problems, hares were routinely observed eating the rabbit chow, and there is evidence that it positively affected their nutritional intake and body condition (Hik 1994, Hodges 1998, Hodges et al. 1999b).

The fences were only partially effective. Raptors still had access to hares on the predator exclosure and predator exclosure + food grids. The monofilament on predator exclosure was not very effective at protecting hares: snow and trees damaged sections of monofilament, raptors were observed hunting beneath the monofilament, hares were killed underneath the monofilament, and many hares had home ranges that extended beyond the confines of the monofilament. Furthermore, the actual fences were not completely effective at excluding terrestrial predators, and there were some mammalian kills of hares on these treatments. Additionally, hares were able to go through the fences, and roughly one in three hares originally radio collared inside the fences died outside the fences. Despite these problems, these treatments did dramatically affect the hare populations within the fences: densities, rates of population change, survival rates, and causes of death were all different from those on the control sites. We therefore conclude that although the treatments were imperfect, they still reduced predation pressure enough to consider them effective.

8.4.4   Experimental Scale and Methodological Concerns

Hares' movement patterns may have differentially affected our density estimates for the various treatments because it is unclear what area the traps on each grid actually sampled. Although hare home ranges average about 10 ha, home range size and degree of overlap vary through the cycle and with the experimental treatments (Allcock 1994, Hik 1994, Hodges 1998). Hares with well-defined home ranges may also undertake forays of up to 1–2 km before returning to their home ranges (Hodges 1998). On the predator exclosure + food treatment, our density estimates were potentially underestimated because hares on the predator exclosure + food treatment had small home ranges (Hik 1994, Hodges 1998), so fewer hares living near the edge of the trapping grid were likely to be caught. In contrast, the food grids appeared to attract hares, and hares that spent most of their time off the grids may have made forays onto the food grids (K. E. Hodges unpublished data; see also Boutin 1984); this movement pattern could lead to overestimates of density, as hares that were only ephemerally present might have been trapped.

Understanding patterns of dispersal would also help in the intepretation of our results. Approximately 60% of captured hares were caught during only one trapping session, and it is difficult to know whether they were residents that avoided traps or transients that lived elsewhere. Additionally, our spring trapping sessions were just before or during the breeding season, which may have amplified hares' movements. The movement patterns of hares on the two fenced sites indicates that dispersal can affect the population dynamics of areas at least 1 km² in size.

Our survival estimates are likely to be underestimates (Haydon et al. 1999, C. J. Krebs and W. Hochachka unpublished data). The models of hare dynamics suggest that survival must be higher for the population growth rates seen during the increase phase. Empirically, the low survival estimates for the decline phase would yield much lower hare populations than we observed. There is some suggestion that trapping hares and perhaps radio collaring hares both affect survival negatively (C. J. Krebs and W. Hochachka unpublished data). Hare survival estimates from trapping data are much lower than radio telemetry estimates because of low trappability (Boutin and Krebs 1986), thus preventing the use of this alternative methodology.

In terms of the predator–prey interactions, the physical location of our hare grids with respect to raptor nests and coyote distribution may have affected both the survival rates and causes of mortality of hares on our trapping areas (Rohner and Krebs 1998, M. O'Donoghue unpublished data). In this case, our treatments were smaller than the patchiness of predator distributions.

8.5   Interactions with Other Species

Researchers have postulated that the cyclic decline is initiated because of food limitation (Pease et al. 1979, Keith 1990). Observations on shrubs and trees at Kluane indicate that sufficient food was available to hares throughout the cycle (see chapter 6; Smith et al. 1988). Additionally, few hares died of starvation (table 8.6), and patterns of hare body condition do not support the view that hares were nutritionally stressed during peak populations (Hik 1995, Hodges et al. 1999b). Hares on sites where food was added appeared to have heavy impacts on the vegetation, in that more twigs were browsed and more shrubs and spruce trees had bark stripped from them (see chapter 6; Hodges 1998), but the hares did not appear to be undernourished. Our estimates of available browse did not detect absolute food limitation at any time, but it is impossible to determine relative food limitation from examination of the plants (Hik 1995, Hodges 1998).

The herbivore community interactions were unexpected. Both ground squirrels and red squirrels exerted considerable predation pressure on leverets, possibly with a type-3 functional response (Stefan 1998). If squirrel-caused mortality is additive to other sources of mortality, leveret survival could easily have doubled without the presence of squirrels. Competitive interactions between hares and squirrels were either nonexistent or not obvious. The only food overlaps appeared to be in summer, when both hares and ground squirrels eat forbs. There potentially was moose–hare competition, in that moose and snowshoe hares both rely on woody browse through the winter, but moose densities were relatively low in the Shakwak Trench. Moose may have had an impact inside the predator exclosure fences; moose commonly broke through the fences and sometimes were inside for periods of several months. The size of the enclosures meant that this represented an unusually high density of moose for those time periods. Our indices of food abundance did not indicate food shortage on these treatments, however.

8.6   Conclusions

We conclude that the snowshoe hare cycle is caused by the simultaneous and interacting impacts of food supply and predation pressure on hares. There are clear changes in annual reproduction and leveret, juvenile, and adult survival; dispersal rates may change through the cycle as well. The single-factor manipulations resulted in changed densities and had some impacts on survival and perhaps reproductive output, but the overall cyclic pattern remained.

The combined manipulation of predator exclosure + food addition altered the timing and magnitude of changes in hare survival and reproduction, which, in turn, led to changes in the pattern of the numeric cycle. The extreme numeric decline was delayed for several years, perhaps because of high reproduction, until one year of particularly low survival coupled with high dispersal led to a dramatic numeric decline in this population. Future research should therefore focus on the ways in which food distribution and predation pressure interact in their effects on these demographic patterns. There are several potential behavioral and physiological pathways through which the impacts of food and predation pressure may affect hare densities.

Literature Cited

Allcock, K. 1994. Do predation risk and food affect the home ranges of snowshoe hares? BSc thesis, University of British Columbia, Vancouver.

Boonstra, R., D. Hik, G. R. Singleton, and A. Tinnikov. 1998a. The impact of predator-induced stress on the snowshoe hare cycle. Ecological Monographs **79**:371–394.

Boonstra, R., C. J. Krebs, and N. C. Stenseth. 1998b. Population cycles in small mammals: the problem of explaining the low phase. Ecology **79**:1479–1488.

Boutin, S. 1984. Effect of late winter food addition on numbers and movements of snowshoe hares. Oecologia **62**:393–400.

Boutin, S., B. S. Gilbert, C. J. Krebs, A. R. E. Sinclair, and J. N. M. Smith. 1985. The role of dispersal in the population dynamics of snowshoe hares. Canadian Journal of Zoology **63**:106–115.

Boutin, S., and C. J. Krebs. 1986. Estimating survival rates of snowshoe hares. Journal of Wildlife Management **50**:592–594.

Boutin, S., C. J. Krebs, R. Boonstra, M. R. T. Dale, S. J. Hannon, K. Martin, A. R. E. Sinclair, J. N. M. Smith, R. Turkington, M. Blower, A. Byrom, F. I. Doyle, C. Doyle, D. Hik, L. Hofer, A. Hubbs, T. Karels, D. L. Murray, M. O'Donoghue, C. Rohner, and S. Schweiger. 1995. Population changes of the vertebrate community during a snowshoe hare cycle in Canada's boreal forest. Oikos **74**:69–80.

Boutin, S., C. J. Krebs, A. R. E. Sinclair, and J. N. M. Smith. 1986. Proximate causes of losses in a snowshoe hare population. Canadian Journal of Zoology **64**:606–610.

Bryant, J. P. 1981. The regulation of snowshoe hare feeding behaviour during winter by plant antiherbivore chemistry. *in* K. Myers and C. D. MacInnes (eds). Proceedings of the World Lagomorph Conference, pages 720–731. University of Guelph, Guelph, Ontario.

Cary, J. R., and L. B. Keith. 1979. Reproductive change in the 10-year cycle of snowshoe hares. Canadian Journal of Zoology **57**:375–390.

Dolbeer, R. A., and W. R. Clark. 1975. Population ecology of snowshoe hares in the central Rocky Mountains. Journal of Wildlife Management **39**:535–549.

Elton, C. S. 1924. Periodic fluctuations in the numbers of animals: their causes and effects. British Journal of Experimental Biology **2**:119–163.

Ernest, J. 1974. Snowshoe hare studies. Final Report, Alaska Department of Fish and Game, Juneau.

Finerty, J. P. 1980. The population ecology of cycles in small mammals. Yale University Press, New Haven, Connecticut.

Fox, J. F., and J. P. Bryant. 1984. Instability of the snowshoe hare and woody plant interaction. Oecologia **63**:128–135.

Gillis, E. A. 1999. Survival of juvenile snowshoe hares during a cyclic population increase. Canadian Journal of Zoology **76**:1949–1956.

Gillis, E. A., and C. J. Krebs. 1999. Natal dispersal of snowshoe hares during a cyclic population increase. Journal of Mammalogy **80**:933–939.

Haydon, D. T., E. A. Gillis, C. I. Stefan, and C. J. Krebs. 1999. Biases in the estimation of the demographic parameters of a snowshoe hare population. Journal of Animal Ecology **68**:501–512.

Hik, D. S. 1994. Predation risk and the 10-year snowshoe hare cycle. PhD dissertation, University of British Columbia, Vancouver.

Hik, D. S. 1995. Does risk of predation influence population dynamics? Evidence from the cyclic decline of snowshoe hares. Wildlife Research **22**:115–129.

Hodges, K. E. 1998. Snowshoe hare demography and behaviour during a cyclic population low phase. PhD dissertation, University of British Columbia, Vancouver.

Hodges, K. E., C. J. Krebs, and A. R. E. Sinclair. 1999a. Snowshoe hare demography during a cyclic population low. Journal of Animal Ecology **58**:581–594.

Hodges, K. E., C. I. Stefan, and E. A. Gillis. 1999b. Does body condition affect fecundity in a cyclic population of snowshoe hares? Canadian Journal of Zoology **77**:1–6.

Keith, L. B. 1974. Population dynamics of mammals. Proceedings of the International Congress of Game Biologists **11**:17–58.

Keith, L. B. 1981. Population dynamics of hares. *in* K. Myers and C. D. MacInnes (eds). Proceedings of the World Lagomorph Conference, pages 395–440. University of Guelph, Guelph, Ontario.

Keith, L. B. 1983. Role of food in hare population cycles. Oikos **40**:385–395.

Keith, L. B. 1990. Dynamics of snowshoe hare populations. *in* H. H. Genoways (ed). Current mammalogy, pages 119–195. Plenum Press, New York.

Keith, L. B., J. R. Cary, O. J. Rongstad, and M. C. Brittingham. 1984. Demography and ecology of a declining snowshoe hare population. Wildlife Monographs **90**:1–43.

Keith, L. B., A. W. Todd, C. J. Brand, R. S. Adamcik, and D. H. Rusch. 1977. An analysis of predation during a cyclic fluctuation of snowshoe hares. Proceedings of the International Congress of Game Biologists **13**:151–175.

Keith, L. B., and L. A. Windberg. 1978. A demographic analysis of the snowshoe hare cycle. Wildlife Monographs **58**:1–70.

Krebs, C. J., R. Boonstra, S. Boutin, M. Dale, S. Hannon, K. Martin, A. R. E. Sinclair, J. N. M. Smith, and R. Turkington. 1992. What drives the snowshoe hare cycle in Canada's Yukon? *in* D. R. McCullough and R. H. Barrett (eds). Wildlife 2001: Populations, pages 886–896. Elsevier Applied Science, London.

Krebs, C. J., S. Boutin, R. Boonstra, A. R. E. Sinclair, J. N. M. Smith, M. R. T. Dale, K. Martin, and R. Turkington. 1995. Impact of food and predation on the snowshoe hare cycle. Science **269**:1112–1115.

Krebs, C. J., S. Boutin, and B. S. Gilbert. 1986a. A natural feeding experiment on a declining snowshoe hare population. Oecologia **70**:194–197.

Krebs, C. J., B. S. Gilbert, S. Boutin, A. R. E. Sinclair, and J. N. M. Smith. 1986b. Population biology of snowshoe hares. I. Demography of food-supplemented populations in the southern Yukon, 1976–84. Journal of Animal Ecology **55**:963–982.

Krebs, C. J., A. R. E. Sinclair, and S. Boutin. 1996. Vertebrate community dynamics in the boreal forest of north-western Canada. *in* R. B. Floyd, A. W. Sheppard, and P. J. DeBarro (eds). Frontiers of population ecology, pages 155–161. CSIRO Publishing, Melbourne.

Murray, D. L., J. R. Cary, and L. B. Keith. 1997. Interactive effects of sublethal nematodes and nutritional status on snowshoe hare vulnerability to predation. Journal of Animal Ecology **66**:250–264.

Murray, D. L., L. B. Keith, and J. R. Cary. 1998. Do parasitism and nutritional status interact to affect production in snowshoe hares? Ecology **79**:1209–1222.

O'Donoghue, M. 1994. Early survival of juvenile snowshoe hares. Ecology **75**:1582–1592.

O'Donoghue, M., and C. M. Bergman. 1992. Early movements and dispersal of juvenile snowshoe hares. Canadian Journal of Zoology **70**:1787–1791.

O'Donoghue, M., S. Boutin, C. J. Krebs, and E. J. Hofer. 1997. Numerical responses of coyotes and lynx to the snowshoe hare cycle. Oikos **80**:150–162.

O'Donoghue, M., S. Boutin, C. J. Krebs, G. Zuleta, D. L. Murray, and E. J. Hofer. 1998. Functional responses of coyotes and lynx to the snowshoe hare cycle. Ecology **79**:1193–1208.

O'Donoghue, M., and C. J. Krebs. 1992. Effects of supplemental food on snowshoe hare reproduction and juvenile growth at a cyclic population peak. Journal of Animal Ecology **61**:631–641.

Pease, J. L., R. H. Vowles, and L. B. Keith. 1979. Interaction of snowshoe hares and woody vegetation. Journal of Wildlife Management **43**:43–60.

Pollock, K. H., S. R. Winterstein, C. M. Bunck, and P. D. Curtis. 1989. Survival analysis in telemetry studies: the staggered entry design. Journal of Wildlife Management **53**:7–15.

Rohner, C., and C. J. Krebs. 1998. Response of great horned owls to experimental "hot spots" of snowshoe hare density. Auk **115**:694–705.

Royama, T. 1992. Analytical population dynamics. Chapman & Hall, London.

Sievert, P. R., and L. B. Keith. 1985. Survival of snowshoe hares at a geographic range boundary. Journal of Wildlife Management **49**:854–866.

Sinclair, A. R. E., C. J. Krebs, J. N. M. Smith, and S. Boutin. 1988. Population biology of snowshoe hares. III. Nutrition, plant secondary compounds and food limitation. Journal of Animal Ecology **57**:787–806.

Smith, J. N. M., C. J. Krebs, A. R. E. Sinclair, and R. Boonstra. 1988. Population biology of snowshoe hares. II. Interactions with winter food plants. Journal of Animal Ecology **57**:269–286.

Sovell, J. R. 1993. Attempt to determine the influence of parasitism on a snowshoe hare population during the peak and initial decline phases of a hare cycle. MSc thesis. University of Alberta, Edmonton.

Sovell, J. R., and J. C. Holmes. 1996. Efficacy of ivermectin against nematodes infecting field populations of snowshoe hares (*Lepus americanus*) in Yukon, Canada. Journal of Wildlife Diseases **32**:23–30.

Stefan, C. I. 1998. Reproduction and pre-weaning juvenile survival in a cyclic population of snowshoe hares. MSc thesis, University of British Columbia, Vancouver.

Trostel, K., A. R. E. Sinclair, C. J. Walters, and C. J. Krebs. 1987. Can predation cause the 10-year hare cycle? Oecologia **74**:185–192.

Vaughan, M. R., and L. B. Keith. 1981. Demographic response of experimental snowshoe hare populations to overwinter food shortage. Journal of Wildlife Management **45**:354–380.

Windberg, L. A., and L. B. Keith. 1976. Experimental analyses of dispersal in snowshoe hare populations. Canadian Journal of Zoology **54**:2061–2081.

Wolff, J. O. 1980. The role of habitat patchiness in the population dynamics of snowshoe hares. Ecological Monographs **50**:111–130.

Wolff, J. O. 1981. Refugia, dispersal, predation, and geographic variation in snowshoe hare cycles. *in* K. Myers and C. D. MacInnes (eds). Proceedings of the World Lagomorph Conference, pages 441–449. University of Guelph, Guelph, Ontario.

9

The Role of Red Squirrels and Arctic Ground Squirrels

RUDY BOONSTRA, STAN BOUTIN, ANDREA BYROM,
TIM KARELS, ANNE HUBBS, KARI STUART-SMITH,
MICHAEL BLOWER, & SUSAN ANTPOEHLER

A consistent feature throughout the boreal forest of North America is the rattle call of the red squirrel, *Tamiasciurus hudsonicus,* as it advertises its whereabouts to conspecifics. Like the snowshoe hare, the red squirrel's distribution encompasses the entire boreal forest (see figure 2.7). The "keek keek" call of a second squirrel species, the arctic ground squirrel (*Spermophilus parryii,* the *siksik* of the Inuit), is also heard in the boreal forests of northwestern North America (Banfield 1974). In terms of biomass of herbivores in these forests, these two squirrels are the second and third most important, respectively, after snowshoe hares (see figure 1.2). Both squirrel species could serve as alternate food sources for the many predators who eat primarily snowshoe hares. However, before our study, no one had investigated experimentally the possible linkages between populations of these squirrels and the snowshoe hare population cycle. The conventional wisdom is that any link would be a secondary one, as predators switch from hares to squirrels during the hare decline. Though both squirrels are active during the summer and thus potentially available to predators, only red squirrels remain active during the long boreal winter and are one of the few alternate prey available to hare predators.

9.1   Natural History

9.1.1   Red Squirrels

Red Squirrels are medium-sized (ca. 250 g) granivores. They defend individual, non-overlapping territories year-round (C. C. Smith 1968, Price et al. 1986, 1990), which are maintained by means of specialized calls, scent marking, and aggressive expulsion of intruders. Possession of a territory (0.2–0.5 ha) is essential for overwinter survival (Larsen and Boutin 1994). Territory owners rarely lose their territory to another squirrel, but breeding females sometimes bequeath part or all of their territory to one or more of their offspring. If this happens, females seldom move more than one or two territories away. The beauty of this social system for the researcher is that when a squirrel disappears from a territory, one can be almost certain that the squirrel has died. This territoriality, coupled with the fact that squirrels are easily trapped and highly visible, makes for accurate measures of density and survival.

Females can breed when 2 years old and the norm is to give birth to a single litter of 1–6 offspring (mean 2.6 young/female), usually during April and May (Boutin and Larsen 1993). Births are not synchronous, with the first and last litters in a year being separated by 55–82 days (O'Donoghue and Boutin 1995). Young begin to spend time out of the nest at about 42 days and are weaned at about 70 days. Dispersal occurs shortly after weaning, but it is usually to a site within one to two territories of where they were born (Larsen and Boutin 1994). Each territory has a central food cache or midden where 12,000–16,000 unopened conifer cones (white spruce in Kluane) are stored for winter consumption (M. C. Smith 1968, Gurnell 1984, Obbard 1987). Squirrels feed on a wide variety of foods, such as conifer buds and flowers, lichens, berries, mushrooms, insects, bird eggs, young birds, and hare leverets, but their staple food is conifer seed obtained by husking cones. Squirrels begin to consume the seed in July or August when cones are still green. Before this they rely on cones stored from previous years. Cones are not clipped until late August or early September, and harvesting is complete by late September. Cones are stored in a series of shallow underground tunnels, primarily in middens. Once underground, the cones remain closed and can last several years. The cones are also protected from pilfering by mice and voles because these species cannot extract the seed efficiently. The only potential thieves are other conspecifics, and this leads to a strict territorial social system.

9.1.2   Arctic Ground Squirrels

Arctic ground squirrels are medium-sized (ca. 500 g), diurnal, burrowing herbivores adapted for life close to the ground: they are short, stocky, and cylindrical, with strong, powerful claws for digging through soil. The burrow is critical to survival, and possession of a burrow gives ground squirrels protection from both predators and the elements and a place in which to hibernate through the long northern winter. Most of the range of the arctic ground squirrel is in arctic and alpine tundra regions of Canada, Alaska, and Siberia (Nadler et al. 1974, Nadler and Hoffmann 1977). Consequently, most of the research on ground squirrels has been done either there (Carl 1971, Green 1977, Batzli and Sobaski 1980) or in open meadows in the boreal forest (McLean 1981, Lacey 1991). These researchers have concluded that the Arctic ground squirrel populations remain relatively stable over time in these open habitats and that food, spacing behavior, and burrow availability, but not predation, limit their numbers.

Arctic ground squirrels live in small groups of related females (female home ranges of 2.5 ha, densities of 1–3/ha, Hubbs and Boonstra 1997), with no more than one territorial male per group. They are obligate hibernators, emerging above ground from a 7- to 8-month hibernation in early to mid-April, with males appearing about 1–2 weeks before females (Carl 1971, Lacey 1991). Mating occurs within several days of female emergence and is generally highly synchronous. Yearlings are reproductively mature, and thus the entire population breeds each year. Litters of 4–7 (mean 4.9; Hubbs and Boonstra 1997) are born 25 days later and appear above ground in late June. Weaning takes place about 1 week later, and juveniles (primarily males) disperse in July. Adult females usually enter hibernation in late July to mid-August, adult males shortly thereafter, and juveniles as late as September or early October. Arctic ground squirrels eat a large variety of foods, ranging from seeds (*Arctostaphylos* spp.), leaves, and flowers (e.g., *Anemone* spp.) of forbs (especially legume species), mushrooms, and, to a lesser extent, willows leaves and catkins, grasses, and sage (*Artemisia* spp.) (McLean 1985). Like red squirrels, they readily eat animal matter and may kill to get it (e.g., microtines: Boonstra et al. 1990; hare leverets: O'Donoghue 1994). They also readily ate the rabbit chow we distributed on the food supplementation treatments.

9.2   Community Interactions and Factors Affecting Population Dynamics

The populations of these two squirrel species could be responding in one of two ways to the hare cycle. First, they could be driven by forces unrelated to the hare cycle (e.g., stochastic factors such as weather which affect the squirrel populations either directly or indirectly through their effects on food supply) and thus could be operating largely independently of the hare cycle. Indeed, for red squirrels, the general conclusion has been that the strict territoriality in prime habitat maintains population stability, while regional populations fluctuate markedly as a direct consequence of fluctuations in the critical food supply, the conifer cone crop (Kemp and Keith 1970, Rusch and Reeder 1978, Sullivan 1990, Klenner and Krebs 1991). Alternatively, squirrels could be tightly linked to the hare cycle through two major ecological processes, competition and predation. Competition is unlikely to play a significant role, as the niches of squirrels and hares are so different and overlap little. Predation, on the other hand, is more probable, possibly occurring directly as both squirrel species kill and eat hare leverets, particularly at certain times of the cycle (O'Donoghue 1994), and indirectly as predators switch from hares to squirrels during the hare decline. The current evidence is unclear. Both squirrel species are readily preyed upon by all of the predators that eat hares (see chapters 13–16). Anecdotal evidence from the aspen–coniferous forests of Alberta suggests that red squirrels decline after a hare decline (Keith and Cary 1991) and that this is possibly caused by predator switching. The evidence that declines in hares are echoed by declines in ground squirrels is slightly better in those species living along the southern border of the boreal forest (Adamcik et al. 1979, Erlien and Tester 1984, Keith and Cary 1991). Our objectives were to determine (1) whether food, predators, or an interaction between food and predators limited populations of these two squirrel species and (2) whether squirrel survival and density were affected by the snowshoe hare cycle. We deal with each species separately in the Methods and Results and then have a combined Discussion section.

9.3 Methods

9.3.1 Red Squirrels

Population and Survival Estimation   Population changes from 1987 to 1996 were determined by a 3- to 5-day trapping session in late May–early June (spring) and another session in late July–early August (summer) on 10×10 grids with stations 30 m apart (7.3 ha) on all experimental treatments. We prebaited 50 traps with peanut butter 3–5 days before trapping, placed traps at every other trapping station, and set them at 0700 h. Traps were checked twice and locked open by 1230 h. All new squirrels were tagged in both ears, and on each capture we recorded tag number, location, mass, sex, and breeding condition (males, scrotal or abdominal testes; females, lactating or not). Newly captured squirrels were classified as adult if they weighed more than 200 g. In addition to this trapping, we periodically trapped at middens in autumn (September–October) to obtain a cohort of 20–30 adult squirrels to monitor survival rates over the winter. In some cases these individuals were also radio collared. Middens were retrapped in March and May to determine which squirrels had survived. We calculated survival as the proportion of squirrels still alive at the end of a season.

Intensive Studies on Control and Experimental Sites   As part of a long-term study of the behavioral ecology of red squirrels, we monitored squirrels intensively on four grids. This work began in 1987 on control 1, fertilizer 1, and a new grid, control E (figure 9.1). Control 4 was added in 1988. We also followed the predator exclosure grid intensively from summer 1991 through summer 1993 (Stuart-Smith and Boutin 1995). The objective

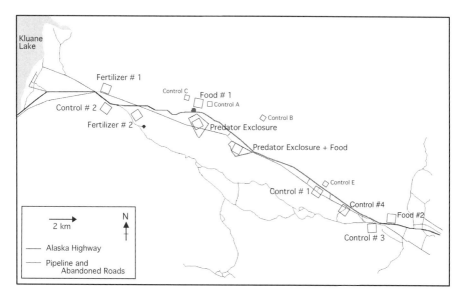

Figure 9.1 Spatial arrangement of treatments and controls at Kluane. Red squirrels were monitored on all experimental treatments and on three controls (1,4, and E). Ground squirrels were monitored on all experimental treatments and on four controls (A–D). Controls C and D were monitored from 1993 onward. Control D is located within Control 3.

was to enumerate all squirrels with individually numbered tags and color marks, determine the middens owned by each, and follow the reproductive condition of females. All females were monitored closely by trapping and observation during the breeding season (April–August). We located and entered nests to enumerate the young (CD-ROM frame 28). A cohort of squirrels was often radio collared in late August or early September, and their fates were monitored weekly throughout the winter. Details of methods can be found in Boutin and Larsen (1993). Our efforts were largely successful, and we were able to obtain the birth date, midden ownership, and date of disappearance of most squirrels from the grid.

The size of the area on which we were able to follow squirrels intensively varied somewhat from year to year. To be consistent, we determined midden ownership changes on a 270 m $\times$ 270 m (7.3 ha) grid, which was similar in size and location to the area trapped for population estimation. However, from 1989 onward, we monitored an area at least as large as the entire hare grid (35 ha). We included individuals caught on this area in assessments of adult survival, juvenile production, and juvenile survival.

Our intensive research allowed us to determine the reliability of information obtained from the trapping sessions. We estimated population size using the program CAPTURE (Otis et al. 1978). Confidence limits on the capture estimates were narrow (often 0), and estimates were always within four individuals of the total number of different animals caught each session. In all cases we report the total number. We also estimated population size by recording the number of squirrels that owned middens within the trapping grid. Midden owners and the number caught were highly correlated ($r^2 = .53$; $y = 8.11 +$

0.98x). Estimates of actual density as determined by midden ownership indicated that the 7.3-ha trapping grid effectively trapped an area closer to 10 ha. Thus, we report all estimates of population density as number per 10 ha. The extensive research also revealed that squirrels that disappeared from their middens could be considered dead (Larsen and Boutin 1994).

Finally, the extensive trapping was largely ineffective as a means of enumerating or indexing juvenile production and early survival. Juvenile squirrels are difficult to catch for the first time, so it takes a considerable amount of exposure to traps. Also, the date of emergence of juveniles can vary by as much as 6 weeks from year to year. For example, in 1994, many young were old enough to be trapped in the May session, whereas in most years they are not present until late July, and in some cases (e.g., 1990) many have not emerged until mid August. Our work schedule at Kluane did not allow us to vary the timing of trapping, so relatively few juveniles were captured in the late summer session. As a consequence, we only report measures of juvenile production from the intensively studied grids.

9.3.2   Arctic Ground Squirrels

Population and Survival Estimation   Ground squirrels were monitored intensively for 7 consecutive years (1990–1996) on 10 × 10 grids with stations 30 m apart (see Hubbs and Boonstra 1997 for details). Because the experimental treatments effectively ceased for arctic ground squirrels in late winter 1995–1996, the presentation of the results of the treatment effects concentrates on the period when the treatments were in place (1990–1995). Over this period, we captured 8758 ground squirrels on all control grids and experimental grids combined. Though minimal trapping occurred from 1987 to 1989, the intensity was not sufficient to assess population parameters, and thus we ignore these years. All treatment and control populations were censused twice per year using Tomahawk live traps, and additional trapping occurred throughout the active season on all grids except on the fertilizer treatments. From 1990 to 1992, two control grids (A and B) were trapped, and from 1993 to 1996, two additional control grids (C and D) were added (figure 9.1). On the experimental treatments, the trapping grids were nested within the snowshoe hare trapping grids. We tagged all new squirrels in both ears and on each capture recorded tag number, location, mass, sex, and breeding condition (males, scrotal or abdominal testes; females, lactating or not). All populations were censused once in spring (May 1–15) to assess spring densities and overwinter survival and once in summer (July 14–31) to assess densities before hibernation but after reproduction and juvenile dispersal. To assess reproductive parameters, an additional intensive trapping targeting juveniles occurred at natal burrows in June just after juvenile emergence.

Two main trapping methods were used:

1. Burrow trapping. From 1990 to 1992, adult squirrels were trapped at their burrows. One trap was placed at each burrow entrance, and two to three traps were placed at large, perennial burrow systems with multiple entrances. Traps were baited with peanut butter, set at 0800 h, checked twice per morning at 2-h intervals, and closed at 1200 h (Hubbs and Boonstra 1997). We trapped squirrels for 2 consecutive days (to give a total of four trapping sessions) and estimated population densities with the program CAPTURE (Otis et al. 1978).

2. Stake trapping. From 1993 to 1996 we trapped squirrels at alternate grid stakes both to maximize our efficiency as density increased on some treatments and to simultaneously assess red squirrel numbers. Traps were set at 0700 h, checked three times at 1.5-h intervals, and closed by 1230 h. Captures from the three daily checks were pooled to give an overall daily capture rate. This procedure was repeated for at least 3 consecutive days until the population estimate had an associated standard error of $\leq 10\%$. We calculated estimates using the mark–recapture heterogeneity (jack knife) model (Pollock et al. 1990) of CAPTURE (Otis et al. 1978), as recommended by Menkens and Anderson (1988) and Boulanger and Krebs (1994). We are confident that this trapping method produced robust demographic information for two reasons: first, trappability of arctic ground squirrels is high (77% for adults and 88% for juveniles; Hubbs and Boonstra 1997); and second, there was no bias in estimates as the density increased (i.e., the relationship between the estimate and the numbers actually caught was linear over a range of 3–356 squirrels ($r^2 = .995$, $n = 76$, $p < .001$). We tested whether the change from burrow trapping to stake trapping affected estimates by carrying out both techniques sequentially (separated by 3 days) in 1993 on the predator exclosure. There was no difference between methods (burrow trapping: 12 ± 0 [95% CL] vs. stake trapping: 13 ± 0). Finally, on treatments with very high densities (food and predator exclosure + food), traps became saturated when only stakes were trapped and thus trapping occurred at both stakes and burrows to reduce trap competition.

To obtain estimates of survival and home range size, we radio collared squirrels. During the active season, adults were radio collared with 5-g radio collars (PD-2C transmitters, Holohill Systems Limited, Carp, Ontario) and juveniles with 6-g expandable collar radios (SS-2 transmitters with Hg-675 batteries; Biotrack, Dorset, England). Survival of adults (from emergence in April to immergence in August or September) was obtained on a sample of squirrels on each of the treatments ($n = 13–30$/year/treatment from 1992 to 1995; $n = 13–25$/year on all four of the control grids combined from 1993 to 1995). Survival of juveniles was obtained on a sample of squirrels from each of the treatments from June to August 1993–1995 ($n = 195$; Byrom 1997). Adults were located twice per week, and juvenile squirrels were located every 2 days with a hand-held antenna. We estimated survival of adults and juveniles using formulas in Pollock et al. (1989). All survival estimates were compared using a log-rank test (Pollock et al. 1989). For all radio-collared squirrels killed by predators, we identified the predator based on characteristics at the kill site, such as feathers, pellets, whitewash, plucking, or scats.

We calculated overwinter survival of adults and juveniles using radio-collared squirrels from 1993 to 1996 fitted with the same radio transmitter type as used on adults during the summer. Radios with fresh batteries were placed on adults in late July and on juveniles after mid-August. The location of each squirrel was determined by telemetry during mid-February when the squirrels were still in hibernation. From early April, each squirrel was relocated and its fate determined by cross-referencing spring positions to winter hibernacula. If the radio signal remained in fixed location until June and the squirrel was never captured again, the squirrel was recorded not to have survived the winter. Because this method was not used before 1993, we calculated overwinter survival of adult and juvenile females using trapping data to include previous years. Squirrels that were absent in the spring after being recorded during autumn trapping were considered not to have survived the winter.

Reproduction We trapped juveniles as soon as they emerged in June on all experimental treatments and controls; thus we do not know the number born, only the number recruiting to the traps on emergence. Juveniles typically remain close to the natal burrow for the first week (Lacey 1991). However, some mothers move their young immediately upon emergence (McLean 1981) to their adjacent resident burrow system, which may be shared with two to four females in high-density populations. Natal burrows of adult females were identified from a combination of live trapping, visual observations, and radio telemetry. A wire mesh, multiple-capture trap designed specifically to capture juveniles (Waterman 1986) was placed at each natal burrow and adjacent resident burrow, along with two to six Tomahawk traps. On control grids and on the predator exclosure grid, most juveniles at a particular burrow system could be assigned to an individual female. However, an accurate measure of litter size for an individual female was not possible for most females from food-supplemented treatments owing to the high degree of mixed litters because of high density. When this occurred, we estimated litter size by dividing the total number of juveniles caught at these targeted burrow systems by the total number of breeding females that lived within that burrow system.

9.4 Results

9.4.1 Red Squirrels

Population Demography on Control Sites Spring populations of red squirrels from 1987 to 1996 were not correlated with snowshoe hare populations ($r = .14, p = .69, n = 10$), remaining relatively constant at between 1.5 and 2.8 squirrels/ha (figure 9.2). Population estimates based on midden ownership on the intensively studied control areas indicated that between-site variation was as great as between-year variation (figure 9.3). Densities varied roughly twofold between control 4 (mean = 1.72/ha) and control 1 (mean = 2.74/ha; figure 9.3). These between-site differences were partly a consequence of the amount of usable habitat on each site (50% spruce cover on control 4 as compared to 90% on other controls). All grids showed a slight decline to relatively low values in 1991–1993. This was followed by increases to maximum levels in 1994. Control E showed the largest change, where numbers doubled between 1991 and 1994. All grids showed a brief decline overwinter in 1994–1995. To summarize, red squirrel densities at Kluane were best characterized by their remarkable constancy over the course of the hare cycle.

Like population density, adult survival (per 6 months) showed little variation. It was consistently near 80% in both summer and winter during the study (figure 9.4). The most notable exception was during the summer of 1992, when almost half of the adults disappeared. Survival also dipped to just under 70% over summer 1995. Juvenile survival was more variable, and we examined it by looking first at overall changes from birth to the next spring and then by separating survival into two periods: from birth to emergence from the nest and from emergence to the next spring. Juveniles had an average survival rate from birth to the next spring of 23%, with 1989, 1992, and 1994 being years of particularly poor survival (<15%), and 1991 and 1993 being years of particularly good survival (38–43%; figure 9.5a). Nestling survival was consistently near 70% except in 1992 and 1995, when it dropped below 40% (figure 9.5b). Postemergence survival was extremely

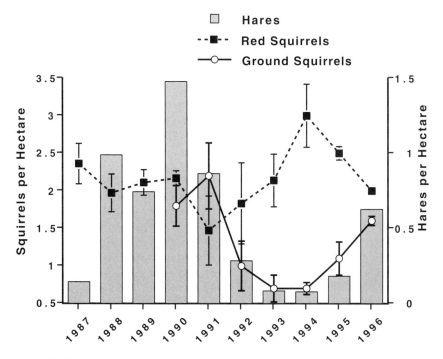

Figure 9.2 Changes in spring abundance of snowshoe hares, arctic ground squirrels, and red squirrels. Ground squirrel estimates from 1990 to 1992 were based on two control areas. Two additional control areas were added in 1993 to increase the precision of our estimates. Red squirrel abundances were averages of three control grids. Error bars (± 1 SE) are given for both squirrel species.

low in 1994 (5%), relatively low in 1989 and 1990 (<28%), and high in 1991, 1993, and 1995 (>55%; figure 9.5c).

To summarize, with one exception (1992, the second summer of the hare decline), adult survival was consistently high. In contrast, juvenile survival from birth to the next spring varied nine-fold from 5% to 45%, and most of this variation occurred in the period between emergence from the nest and the following spring. Recruitment tended to be higher from 1991 to 1995 as compared to 1988–1990.

Reproduction on control areas showed two phases during the study (table 9.1). From 1988 to 1991, less than 70% of the females bred per year, and litter size averaged less than 2.7. Thereafter, the proportion breeding increased to almost 100%, and litter size was well above three. The net result was a substantial increase (two- to threefold) in the number of young produced per hectare in these years. The only exception was 1994, in which the proportion breeding dropped, as did number of young produced per hectare.

Treatment Effects

FOOD ADDITION   Red squirrels were frequently observed eating the rabbit chow during both winter and summer. However, it is unlikely that the chow represented high-quality

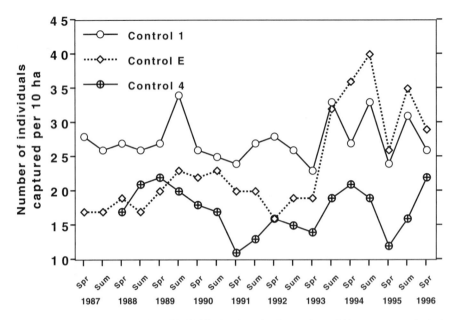

Figure 9.3 Densities per 10 ha of individual red squirrels owning middens on controls 1, 4, and E.

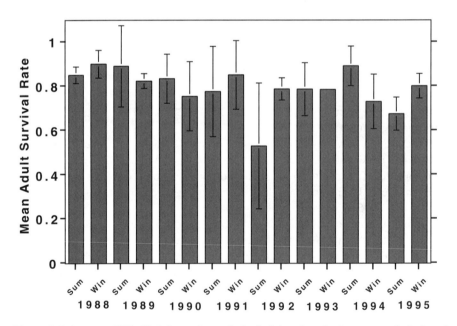

Figure 9.4 Average (95% CLs) 6-month survival of adult red squirrels on controls 1, 4, and E. Summer period covered April through August and winter period covered September through March. Sample sizes of number of individuals followed per grid for each period ranged from 35 to 140 and averaged near 70.

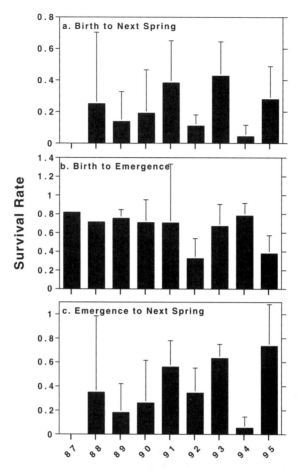

Figure 9.5 Average (95% CLs) survival of young red squirrels on controls 1, 4, and E. Survival is given for three periods: (a) from birth to the next spring; (b) from birth to emergence from the nest (ca. 42 days), and (c) from emergence to the next spring. Sample size of number of young born ranged from 11 to 137 per grid per year and was <30 in only 3 cases.

food given the differences in composition and digestibility of chow compared to conifer seed. The ratio of the number of squirrels on the food grids relative to the control grids was highly variable but tended to be >1 from 1990 onward (figure 9.6). Ratios were never >1.5. As another measure of whether food addition increased squirrel densities, we conducted call surveys on the food grids and adjacent to the grids after spring and summer trapping periods (1991–1995). Calls were counted over a 5-min period on 2 consecutive mornings at each of 16 stations distributed over 35 ha. Call rates on and off of the grids were virtually identical (mean = 7.4 and 8.2 for on vs. off grid), indicating that densities were not higher on the food grids relative to surrounding habitat. Finally, we compared survival of adult squirrels on control and food-supplemented grids over winters following cone crop failures (1989–1990, 1991–1992, and 1994–1995). In all cases, survival

Table 9.1 The average proportion of female red squirrels breeding, the average litter size, and the average number of young produced per hectare on controls 1, 4, and E (sample sizes in parentheses).

Year	% Breeding	Litter Size	Number Produced/ha
1988	55 (80)	2.70 (31)	1.16
1989	67 (99)	2.69 (48)	1.75
1990	39 (104)	2.45 (37)	1.13
1991	65 (100)	2.56 (60)	1.53
1992	81 (162)	3.26 (101)	3.6
1993	82 (150)	3.19 (109)	3.77
1994	66 (173)	3.05 (108)	2.9
1995	98 (134)	3.34 (105)	3.6
1996	93 (122)	3.13 (103)	3.85

on the food supplemented areas was less than that on the control areas (table 9.2; sample sizes of 16–36 on the three food-supplemented grids: food 1, food 2, and predator exclosure + food). Thus, addition of rabbit chow had no positive effect on red squirrel numbers or survival.

PREDATOR EXCLUSION   Exclusion of predators did not lead to higher squirrel densities. Figure 9.6 shows that the density ratio on the predator exclosure was always <1, and it remained constant over the course of the hare cycle. We hypothesized that predation would have its greatest impact on squirrels when snowshoe hares had declined and the predators would be forced to consume alternate prey (1991–1993). As a test, we intensified our efforts on the predator exclosure just before and during the hare decline. Survival rates of adults were significantly higher on the predator exclosure than on the control grids in the winter of 1991–1992 and summer of 1992, but not in summer of 1993 (table 9.2; Stuart-Smith and Boutin 1995). In the summer of 1992, all mortalities were a result of predation (Stuart-Smith and Boutin 1995). The predator exclosure eliminated only large mammalian predators, which suggests that mammalian predators had a short-term impact on adult survival during the hare decline. The particularly poor survival of juveniles from birth to emergence in 1992 (figure 9.5) may in part have been due to death of their mothers in that period (figure 9.4). However, this poor survival did not carry over into the period after emergence, as survival from emergence to the following spring was average. Thus, predator-induced low survival rates in both adults and young were short-lived, and the impact on the red squirrel populations had little long-term effect.

INTERACTIVE EFFECTS OF FOOD AND PREDATION   Given the results of the food addition and predator exclosure treatments, we did not expect to see any effect of the combined treatment on red squirrel densities. The density ratios on this treatment were almost always <1 (figure 9.6), and survival on the predator exclosure + food grid was never substantially greater than that observed on the control areas (table 9.2).

FERTILIZATION   We do not think that the fertilizer treatment had any direct relevance to red squirrels. The only potential effect may have been on cone crop, and the fertilizer had

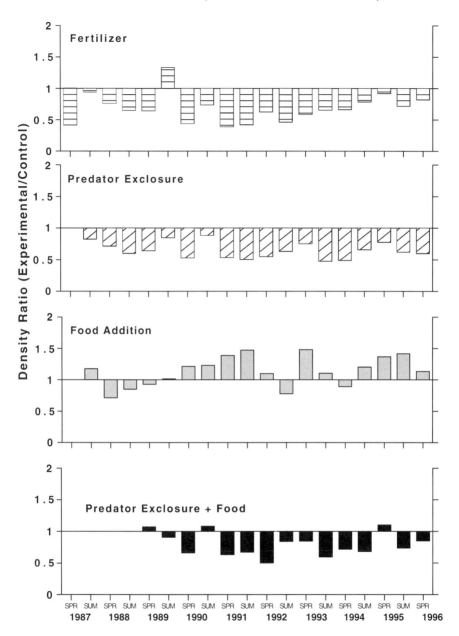

Figure 9.6 Ratios of treatment densities to average control population densities of red squirrels. Ratios of 1 indicate no difference with control densities. Control densities as in figure 9.2. Values for the fertilizer and food-addition grids were averages of two grids in each case.

Table 9.2 Six-month winter survival rates (2 SE) for red squirrel adults on control and experimental grids.

Grid	1989–90	1990–91	1991–92	1992–93	1993–94	1994–95
Control	0.84 (0.02)	0.79 (0.06)	0.87 (0.06)	0.89 (0.08)	0.82 (0.01)	0.77 (0.06)
Food addition	0.77 (0.16)	0.76 (0.13)	0.74 (0.18)	0.88 (0.12)	0.91 (0.08)	0.66 (0.08)
Predator exclosure	0.63	0.42	0.96	0.82	0.84	0.67
Predator exclosure + food	0.72	0.83	0.70	0.71	0.79	0.68

Rates were calculated as the proportion of individuals caught during summer or autumn that were recaptured the following May. Time periods between sessions varied from 24 to 42 weeks. All rates were standardized to a 6-month period.

no effect on this (see figure 7.7). We monitored fertilizer 1 intensively from 1987 through 1993. The changes in numbers on this grid are notable, as they were the most drastic of any of the sites studied. Numbers dropped almost threefold between autumn 1989 and spring 1990. They remained low through the early 1990s before recovering in 1993. The decline in 1989–1990 is even more drastic when survival rates are examined. Only 15% ($n = 60$) of the adults and 0% of the juveniles ($n = 27$) survived that winter. We regularly found mummified carcasses of squirrels in nests while we were searching for young in the spring of 1990. It appears that most of the mortality was due to starvation. Fertilizer 1 appeared to represent low-quality habitat. It was the driest site, and cone crops on this area were the lowest of all the study sites.

Role of Stochastic Factors   Squirrels rely heavily on conifer seed, which is highly variable from year to year. All study grids showed similar year-to-year patterns, but the average number of cones per tree varied between sites. Over the course of the study, there were cone failures in 1989, 1991, and 1994, above-average cone years in 1992 and 1995, and a mast year in 1993 (see 7.1.3 for details). We have limited data for years before 1987, but 1983 was a mast year and 1985 and 1986 were average or above average. We hypothesized that changes in density would be related to cone availability. To test this idea, we standardized squirrel numbers caught on all grids by calculating standardized deviates for each grid and compared this with our cone index. The cone index was significantly correlated with the standardized deviate of the number of squirrels caught in the following spring ($r = .48, p < .0001, n = 78$) and autumn ($r = .55, p < .0001, n = 68$; figure 9.7). Similarly, finite population growth from spring to spring was related to cone crop. When the cone index was <10 per tree, populations increased in only 8 of 29 cases, whereas when it was >10 per tree, numbers increased in 26 of 34 cases.

We also examined the survival rates of squirrels in relation to cone supply and predicted low overwinter survival in years of cone failure. This prediction was not supported for adults, as survival was only slightly lower in one (1994–1995) of the 3 years of cone failure. The lowest survival rates occurred in summer 1992, which followed the cone crop failure of 1991. However, similar failures in 1989 and 1994 did not result in low survival the next summer (1990 and 1995, respectively). Thus, it is unlikely that the absence of cones from the previous autumn has anything to do with poor adult survival in the following summer. Juvenile survival did not show any clear relation to cone crop. Survival from birth to emergence (figure 9.5b) was low in 2 of the 3 years following a cone fail-

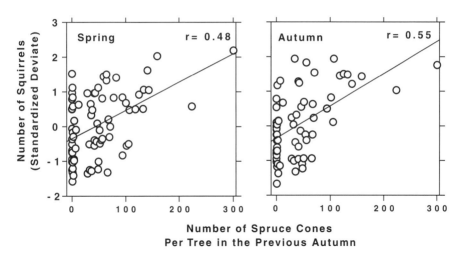

Figure 9.7 The relationship between the production of spruce cones per tree on trapping grids in late summer and the standardized deviate of red squirrel numbers on these trapping grids in the following spring and the following autumn.

ure. The summer of 1990 was the exception. Survival from emergence to spring was both low (<10% in 1994) and high (>50% in 1991) in years of cone failure (figure 9.5c).

Finally, production of juveniles appeared to be related to the cone crop in the current year rather than to that in the preceding year. This is surprising given that all young are conceived before the current year's cone crop appearing on the trees. Table 9.1 shows that juvenile production doubled in 1992, a year that followed a cone failure but preceded a mast year. Production declined somewhat in 1994 even though the previous year was a mast year. Adult females appear to anticipate (possibly through consumption of cone buds that are produced in the previous summer and are available to squirrels during the winter before a mast crop) and adjust their investment accordingly.

9.4.2 Arctic Ground Squirrels

Population Demography on Control Sites   Spring populations from 1990 to 1996 fluctuated in close synchrony with that of the snowshoe hares populations ($r = .89, n = 7$; figure 9.2). Ground squirrels were still increasing from 1990 to 1991, reaching an average of 2.2/ha in 1991, but as hares rapidly declined from 1991 to 1992, finite rates of increase for squirrels dropped to <0.5 (figure 9.8). Ground squirrel populations declined 54% by spring 1992 and another 30% by spring 1993, to reach constant levels of 0.7/ha from 1993 to 1994. When hare populations started increasing in the summer of 1994 (chapter 8), so did ground squirrels, with some controls doubling their numbers each year (finite rates of increase of 2; figure 9.8).

From 1990 to 1993, survival of adult ground squirrels declined as total snowshoe hare biomass declined (figure 9.9a, c). Hubbs and Boonstra (1997) calculated an index of survival of adult squirrels on controls based on trapping records and showed that survival

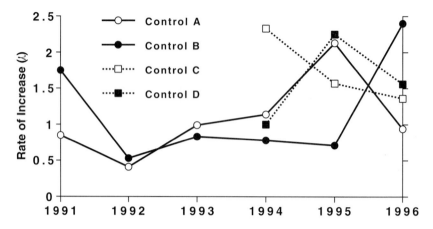

Figure 9.8 Finite rates of increase (λ) of ground squirrel populations on each of the control sites monitored from 1990 to 1996. The value for each year represents the rate of change in abundance from the previous year.

over the 3-month active season declined by more than half from 70% in 1991 to 31% in 1992 (table 2 in Hubbs and Bonstra 1997). Survival estimates from trapping records underestimate survival by 35–39% but still provide a comparable relative index of survival (Hubbs and Boonstra 1997). Further estimates of survival based on radio telemetry showed that adult ground squirrel survival increased markedly from a 3-month survival of 36% in 1993 to 91% in 1994 (table 9.2), when both total mammalian predator and total avian predator biomass were lowest (figure 9.9b).

During the active season, predation was the greatest proximate cause of mortality of both radio-collared juveniles and adults from 1992 to 1995. Of the adults radio collared from 1992 to 1995 and juveniles radio collared from 1993 to 1995, predation accounted

---

Figure 9.9 (a) Total biomass (kg/ha) for snowshoe hares (1989–1995) and for arctic ground squirrels (1990–1995) on control areas. Methods for estimating snowshoe hare density and biomass are described in chapter 8. Biomass for ground squirrels was calculated by multiplying the average weight of a ground squirrel (0.47 kg) by population density. (b) Total biomass for mammalian predators (lynx and coyotes; kg/350 km^2) and for avian predators (great horned owls, red-tailed hawks, and goshawks; kg/100 km^2) from 1989 to 1995. Methods for estimating mammalian predator densities and biomass are described in chapter 13. Methods for estimating avian predator densities and biomass are described in chapter 15. (c) Twenty-eight-day summer season survival and overwinter survival of adult and juvenile ground squirrels on control sites. Summer survival of juvenile ground squirrels was measured from 1993 to 1995. Active season survival in 1990 and 1991 was estimated from trapping data and in 1992 from radio telemetry collected by Hubbs and Boonstra (1997). Overwinter survival in females was recorded from trapping records obtained in late autumn and early spring. (d) Litter sizes and weaning rates of adult female ground squirrels on control sites. Litter size was measured from 1991 to 1995. Weaning rate, which is the percentage of female ground squirrels that were successful in weaning a litter, was measured from 1993 to 1995.

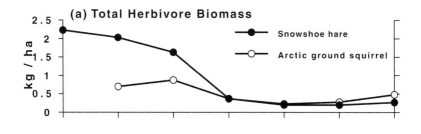

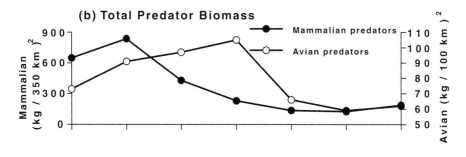

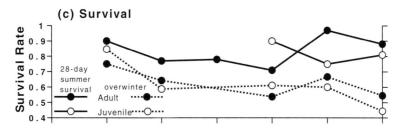

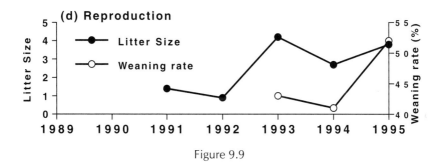

Figure 9.9

for 96% (73 of 76) and 96% (52 of 54), respectively. During the second year of the hare decline (1992), avian predators accounted for 62% of known predator kills (Hubbs and Boonstra 1997). From 1993 to 1995, when snowshoe hare densities (figure 9.2) and total biomass (figure 9.9a) were at their lowest, and avian predator biomass at its highest (figure 9.9b), avian predators accounted for most of the known kills (75% of adults, $N = 24$ and 79% of juveniles, $N = 31$) of known predators.

Overwinter survival of adult and juvenile females was determined from trapping data (1990–1996). We present only overwinter survival of females (figure 9.9c), as it is difficult to separate dispersal from overwinter mortality in males because we were not present when possible dispersal could have occurred from late September to October before hibernation or from late March–April just after hibernation. Overwinter survival averaged 63% ($N = 72$) for juvenile females and 61% ($N = 72$) for adult females from 1990 to 1996. Survival of both age classes combined was greatest during the winter of 1990–1991 (81%, $N = 21$), declined to approximately 60% and remained at this level until the winter of 1995–1996 when survival decreased to 52% ($N = 31$).

Reproduction on control sites (figure 9.9d) varied considerably from 1991 to 1995. Recruited litter sizes were small during 1991 and 1992 (approximately 1.2 pups per litter) when predators were numerous and hares declining (figure 9.9b). When hares and predators were at low densities from 1993 to 1995, litter sizes for ground squirrels tripled to an average of 3.6 pups per litter. We only measured weaning rate from 1993 to 1995 (figure 9.9d). During this period from 42% to 51% of females weaned litters.

Treatment Effects on Demography

DENSITY   Our experimental treatments had major effects on ground squirrel populations, and, to assess the impact of the treatments, we present the changes relative to those occurring on the control grids (figure 9.10). The fertilizer addition had a progressively negative effect on squirrel densities. Densities were either similar to (fertilizer 2) or higher (fertilizer 1) than those on controls in 1990, but thereafter (1992–1996), they averaged only 40% of that on controls (figure 9.10). Fertilizer 2 declined to such an extent that ground squirrels were absent during the last 2 years of the project. On the predator exclosure spring population densities averaged 1.9 times those of controls (average spring density over the study was 3.2/ha) and reached a maximum of 2.6 times control densities by spring of 1995. The doubling of density on the predator exclosure was already evident by 1991, but the population was not able to sustain this and declined drastically from 1991 to 1992 when hares were declining. This treatment may have converged back to control levels as a result of great horned owls, which continued to increase until 1992 (chapter 15), and these were not excluded from the predator exclosure. However, we were not able to detect any difference from control levels in mortality during this period (either during the active season or overwinter). After 1992 densities again reached values more than double control levels (figure 9.10).

On the two food grids, densities averaged five times that of the controls, and there was little change over time (spring densities averaged 6.1/ha on the food 1 and 3.4/ha on food 2 over the study, reaching a maximum of 7.3 times on food 1 during the spring of 1996 [figure 9.10]). Food 1 and food 2 fluctuated together during 1993 and 1994 but increased loss due to overwinter mortality on food 2 during the winter of 1994/1995 caused them to diverge. On the predator exclosure + food grid densities averaged 10.8 times that of

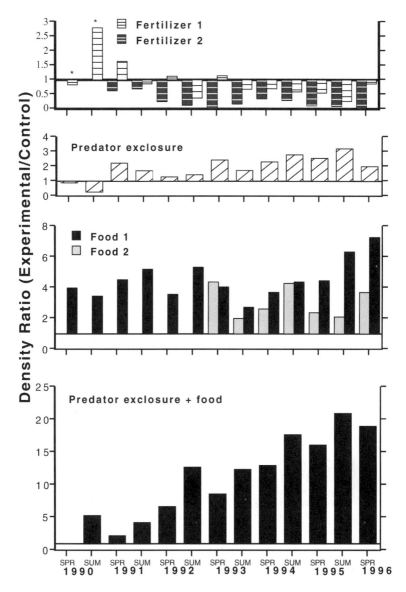

Figure 9.10 Ratios of treatment densities to average control population densities of arctic ground squirrels. Ratios of 1 indicate no difference with control densities. Control densities as in figure 9.9a. *Minimum number alive used for 1990 density on fertilizer treatment.

controls and diverged progressively from 1991 to 1996, reaching a maximum spring density of 31/ha in the spring of 1996 (19 times that of controls). Thus, whereas the effect of nutrient addition was clearly negative, the exclusion of predators was positive (2-fold increase over control levels), the addition of food was positive (5-fold increase), and, when the treatments were combined, they had a multiplicative effect (11-fold increase on the predator exclosure + food).

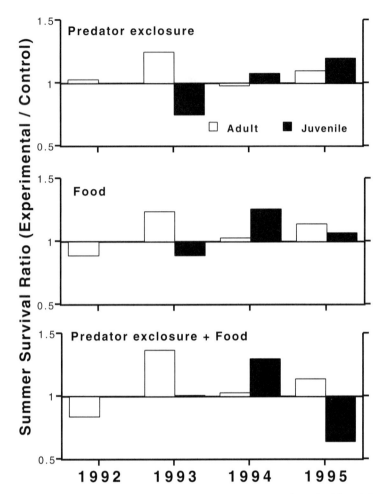

Figure 9.11 Ratios of treatment summer survival to average control survival for adult (1992–1995) and juvenile (1993–1995) arctic ground squirrels. Values >1 represent positive treatment effects (i.e., increased survival relative to controls); values <1 represent negative treatment effects (lower survival relative to controls).

SURVIVAL   Active season survival of radio-collared adults or juveniles was unaffected by the experimental treatments from 1992 to 1995 (figure 9.11; (adults: $\chi^2 = 5.88$, df = 3, $p \geq .21$ for all years, juveniles: $\chi^2 = 6.25$, df =3, $p \geq .10$ for all years). Survival of both adults (1992–1995) and juveniles (1993–1995) was affected by year (adults: $\chi^2 = 159$, df = 3, $p < .0001$; juveniles: $\chi^2 = 5.6$, df = 2, $p = .06$). Thus, changes in survival on the controls from one year to the next were reflected by similar changes in survival on the treatments. Average juvenile 28-day survival over 1993–1995 ranged from 2% to 14% lower than the survival of adults on the experimental treatments (table 9.3).

Overwinter survival, calculated from trapping records, averaged 74% ($N = 903$) over

Table 9.3 Twenty-eight-day active season survival (with 95% confidence limits) of radio-collared adult and juvenile ground squirrels on experimental sites and controls from 1992 to 1995.

	Year	Controls	Predator Exclosure	Food	Predator Exclosure + Food
Adults	1992	0.79 (0.74–0.83)	0.81 (0.76–0.87)	0.70 (0.66–0.74)	0.66 (0.61–0.71)
	1993	0.71 (0–0.92)	0.89 (0.78–0.97)	0.88 (0.77–0.96)	0.97 (0.94–1.0)
	1994	0.97 (0.92–1.0)	0.95 (0.85–1.0)	1	1
	1995	0.88 (0.73–0.98)	0.97 (0.93–1.0)	1	1
	All	0.84 (0.59–1.0)	0.90 (0.71–1.0)	0.90 (0.82–0.97)	0.91 (0.86–0.95)
Juveniles	1993	0.91 (0.71–1.0)	0.68 (0.42–0.83)	0.81 (0.63–0.94)	0.92 (0.74–1.0)
	1994	0.74 (0.47–0.91)	0.80 (0.62–0.93)	0.93 (0.85–1.0)	0.96 (0.87–1.0)
	1995	0.81 (0.58–0.97)	0.97 (0.9–1.0)	0.87 (0.73–0.98)	0.52 (0–0.76)
	All	0.82 (0.48–0.9)	0.80 (0.58–0.87)	0.87 (0.76–0.93)	0.77 (0–0.88)

all treatments and years. Both year and treatment were significant in explaining overwinter survival of ground squirrels. However, their effect was interactive (logistic regression: $\chi^2 = 28$, df $= 12, p = .006$), meaning that the effects of the treatments on overwinter survival changed from year to year. Squirrels on food-supplemented grids had worse survival than controls in the winters of 1991–1992 and 1992–1993, but better survival than controls from 1993 to 1995 (figure 9.12). Predator exclusion had no consistent effect on the overwinter survival of females (figure 9.12). High overwinter survival from 1993 to 1995 was echoed by our estimates obtained using radio-collared females. Overwinter survival from 1993 to 1996 (all treatments combined for radio-collared females) was 97% (92 of 97 individuals). Comparisons among treatments or years were not possible because so few died.

REPRODUCTION   The size of litters at emergence showed the same pattern of variation on both controls and treatments from 1991 to 1995, with 1992 having the smallest litter sizes (figure 9.13). However, litters were larger on all treatments in at least 4 of the 5 years, but when pooled, treatment effects were not significant (figure 9.13). Thus, these differences were not sufficient to explain the differences in density among treatments.

Weaning success differed markedly among controls and treatments. For the period from 1993 to 1995, we examined the proportion of females that were successful in weaning a litter by logistic regression (year × treatment). Weaning success was significantly better on the treatments, but the effect was not independent of the effect of year (interaction effect: 21.3, df $= 11, p = .03$). Weaning success (figure 9.14) varied as follows: predator exclosure + food (83%) > food (77%) > predator exclosure (72%) > controls (45%). Thus, the presence of food or the absence of predators had a similar effect and increased the weaning rate by 60% over control levels.

First emergence of young from the natal burrow was strongly affected by the food addition. Because estrus occurs within 3–4 days of emergence, and both gestation and the length of time the young spend in the burrow are constant (25 days and 27 days, respec-

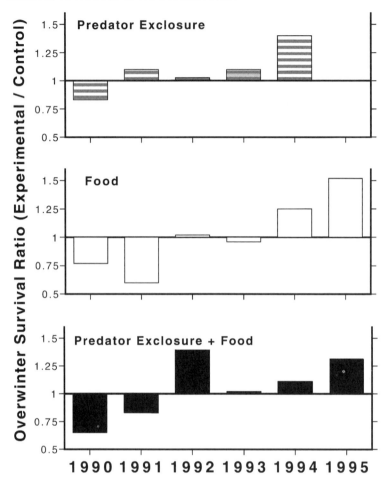

Figure 9.12 Ratios of treatment overwinter survival to average control survival for adult female arctic ground squirrels (1990–1995). Values >1 represent positive treatment effects (i.e., increased survival relative to controls); values <1 represent negative treatment effects (lower survival relative to controls).

tively; Lacey 1991), earlier appearance of the young must mean earlier breeding. Thus the young have a longer period of time to grow before hibernation. On the controls and on the predator exclosure, emergence date ranged from June 13 to June 23 for all years except 1992 (table 9.4). Severe spring weather in 1992 delayed breeding, and hence juvenile emergence occurred in early July, about 3 weeks later than normal. On the food grids (including the predator exclosure + food), emergence was generally 1–2 weeks earlier than in control areas, even in 1992. Thus, food addition permitted earlier initiation and completion of reproduction.

WEIGHT DYNAMICS   Body mass of females just after emergence from hibernation is critical for reproductive success. We examined how our treatments affected body mass with

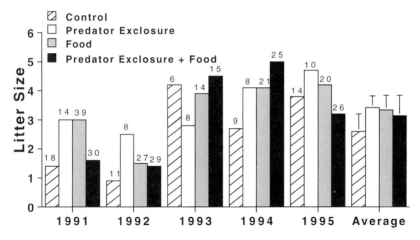

Figure 9.13 Litter sizes for arctic ground squirrels on treatments and controls (1991–1995). Sample size for adult females is indicated above bar. Average litter sizes (all years pooled) are shown with SEs.

a two-way ANOVA (treatment × year; figure 9.15). All variables were significant (treatment: $F = 74.0$, df = 3, $p < .0001$; year: $F = 8.6$, df = 6, $p < .0001$; interaction: $F = 4.6$, df = 18, $p < .0001$). Body mass was lower on all sites during the spring of 1992 (figure 9.15) when weather conditions were poor, and this contributed to the year effect. There was overlap in body mass between food-addition grids and non–food-addition grids in some years, which accounts for the interaction effect that we detected. Spring female body mass over the entire study was ranked as follows: predator exclosure + food = food > predator exclosure > controls. Average spring body mass was found to be positively correlated with average litter size produced by the females that year ($r = .51$, $N = 5$ years) and average weaning success ($r = .68$, $N = 3$).

9.5 Discussion

The responses of the two squirrel species, red squirrels and arctic ground squirrels, are at two ends of the spectrum. Red squirrel populations showed virtually no response to the snowshoe hare cycle, modest responses to the fluctuations in the cone crop, and virtually

Table 9.4 Earliest date of juvenile arctic ground squirrel emergence from natal burrows for treatment and control areas, 1991–1995.

Treatment	1991	1992	1993	1994	1995
Controls	21 June	4 July	14 June	13 June	20 June
Predator exclosure	23 June	5 July	18 June	13 June	20 June
Food	17 June	26 June	14 June	6 June	5 June
Predator exclosure + food	13 June	27 June	7 June	6 June	6 June

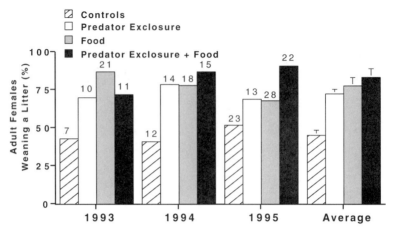

Figure 9.14 Weaning rate for arctic ground squirrels on treatments and controls (1993–1995). Weaning rate is the percentage of females that successfully wean their litter. Average weaning rates (all years pooled) are shown with SEs.

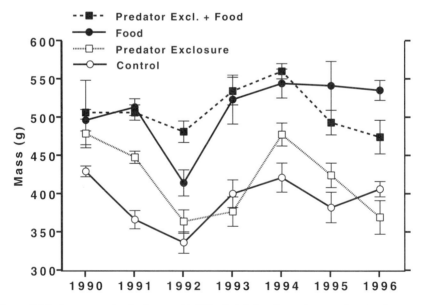

Figure 9.15 Average spring body mass ($\pm$ SE) of adult female arctic ground squirrels on treatments and controls from 1990 to 1996.

no response to any of the experimental treatments. In contrast, arctic ground squirrels fluctuated in tandem with the snowshoe hare cycle, showed dramatic responses to changes in food supply (both positive [food addition treatments] and negative [fertilizer addition treatment]) and to removal of predators. Thus, in the boreal forest ecosystem of the southern Yukon, red squirrel populations are characterized by constancy and arctic ground squirrel populations by long-term fluctuations. We examine the pattern first with respect to changes in demography relative to the snowshoe hare cycle and to stochastic changes in the environment and then in relation to the experimental treatments.

9.5.1   Role of Stochastic Events

Populations rarely remain constant over time because short-term environmental stochasticity, particularly acting though weather, can affect such factors as food supply (Sinclair and Pech 1996), which then affect the population demography. We found that the size of the spruce cone crop was a good predictor of red squirrel population growth and density the next spring and autumn. This begs the question of the causes of the fluctuations in the cone crop. In the deciduous forests of Europe, outbreaks of microtines are directly related to mast production by the major seed-producing trees, and these mast years are directly related to weather patterns of the previous year (Pucek et al. 1993). In red squirrel populations of Alberta, Kemp and Keith (1970) found that drier summers one year produced larger cone crops or mast years the next. When we examined correlations between temperature and precipitation from May to August of one year, (chapter 2; months examined individually or summation of these months) and the size of the spruce seed crop the next year ($N = 8$), no correlation was significant (the highest was the average May temperature, $r = -.64, p = .09$). Thus, the peak spruce cone and seed mast year of 1993 (chapter 7, figure 10.4), the highest in our 10-year study, appeared unrelated to delayed weather patterns in any obvious way, and some other explanation must account for this mast year. Both this peak mast year and the previous one (1983) followed the peak of the snowshoe hare cycle by about 3 years, and in the nutrient section below, we suggest that this may not simply be chance.

Variation in weather can have major effects on ground squirrel populations. Though the annual cycle of hibernation and activity in ground squirrels is an endogenous, circannual one (Blake 1972, Davis 1976), timing of emergence and reproduction may be controlled by exogenous factors linked to weather (Morton 1975, Davis 1976). Delayed snow melt in spring reduces the availability of food, causing ground squirrels to rely more on their remaining fat reserves, thereby affecting their survival and reproduction. Ground squirrels depend heavily on remaining fat reserves for reproduction after hibernation, but those remaining fat reserves are insufficient without available forage to support a litter to weaning (Kiell and Millar 1980). Michener and Michener (1977) found that Columbian ground squirrels (*S. columbianus*) and Richardson's ground squirrels (*S. richardsonii*) emerged from hibernation 12 and 21 days earlier, respectively, during a warm spring than during a cool one. After a particularly severe and long winter, Phillips (1984) found pregnancy loss in golden-mantled ground squirrels (*S. saturatus*). After a month-long snow storm in spring, Morton and Sherman (1978) reported poor survival, delayed breeding, and reduced reproduction in Belding's ground squirrels (*S. beldingi*). Finally, after a drought and a prolonged winter, Van Horne et al. (1997) reported lower adult survival, delayed emergence and reproduction, and low mass and complete loss of juveniles in Townsend's ground squirrels (*S. townsendii*).

In our study, 1992 had a delayed spring in which snow cover persisted for almost 3 weeks longer than normal, and this was associated with major population declines in all populations (Hubbs and Boonstra 1997). Overwinter survival dropped from an average of 74% in all other years to 50% during the winter of 1991–1992. Reproduction was delayed by 17–19 days across all treatments and controls as indicated by timing of juvenile emergence (table 9.4). Combined with small litter sizes during 1992 (figures 9.9d, 9.13), these effects may have also contributed to the further decline in ground squirrel densities on control sites from 1992 to 1993 (figure 9.2). Food addition ameliorated the severity of this spring only on the predator exclosure + food grid, where we could deliver the food by hand in spite of the snow. Food addition was not effective on the food-addition grid at that time because of food consumption by grizzly bears and moose (Hubbs and Boonstra 1997). This severe spring was coupled to intense predator pressure during 1992 and served to converge all the trapping areas more than they otherwise might have been.

9.5.2   Treatment Effects

Role of Nutrients   The boreal forest is a nutrient-limited environment, and we had expected that the addition of fertilizer would increase squirrel densities by stimulating plant productivity (spruce seed production in the case of red squirrels; herbs and grasses in the case of ground squirrels). Spruce seeds increased slightly, and perennial graminoides (*Festuca altaica* and *Calamagrostis lapponica*) and two herbs (*Epilobium angustifolium* and *Achillea millefolium*) (Nams 1993) increased markedly, but legumes did not (chapter 5, John and Turkington 1995). Both squirrel species decreased markedly. Why? Red squirrels may have declined because of the negative effects of fertilization on mushroom crops (S. Boutin personal observation), an alternative food which is readily harvested and stored (M. C. Smith 1968) and which may be one of the few food sources if the cone crop fails. However, our evidence for this is weak. Thus, intensive fertilization such as we applied on the fertilizer treatments may be partly responsible for the red squirrel decline. However, three lines of evidence suggest that widespread, low-level fertilization related to nutrient release associated with the hare peak (decaying hare pellets produced at the hare peak and possibly decaying vegetation clipped from shrubs at the peak) may affect red squirrel populations through its affects on the cone crop. First, our biggest mast year, 1993, was not related in any obvious way to weather patterns and followed the hare peak by about 3 years. Another mast year, 1983, followed the hare peak by about 3 years. Second, peaks in northern red-backed voles also occurred about 2–3 years after the hare peak (chapter 10). Third, shrub biomass reached its highest level at the peak (1990) and in the next 2 years (see figures 6.2, 6.5).

Ground squirrels may have declined because certain plants (and the essential chemicals they contain) critical to successful hibernation largely disappeared with the addition of fertilizer. Addition of fertilizer shifted the competitive balance from those plant species that grow well in a nutrient-limited environment (particularly the nitrogen fixers such as the lupines) to those that grow well in a nutrient-sufficient environment (particularly the grasses, chapter 5, Turkington et al. 1998). For successful hibernation, ground squirrels require a diet high in polyunsaturated fatty acids (Geiser and Kenagy 1987, 1991, Frank

1992, 1994). Legumes (such as lupines and vetches) are good sources of these fatty acids; grasses are not (Harwood and Geyer 1964, Lehninger 1982). Polyunsaturated fatty acids lower the melting point of depot fats (Mead et al. 1986), increasing the proportion of individuals capable of hibernating, increasing torpor bout duration, and decreasing body temperature during hibernation. Thus, our evidence suggests that ground squirrels declined markedly because they were not able to hibernate successfully on the fertilized grids. The corollary of this is that ground squirrels would not be able to inhabit the boreal forest if it were not nutrient limited.

Role of Food   Red squirrel populations were relatively constant over time in the face of major changes in two key factors, cone supply and predation pressure. Although local areas experienced changes in numbers as great as fourfold, average numbers remained remarkably constant. Cone availability seemed more important than predation in causing year-to-year changes. It was only when cones were scarce (cone crop failure in 1991) and predation pressure near its maximum (summer 1992) that predation seemed to have any noticeable effect on survival. Despite low adult and early juvenile survival, populations on many areas increased from 1992 to 1993. This was due to a substantial increase in litter size, in the proportion of females breeding, and in the number of multiple litter attempts in that year. Females anticipated the abundant cone crops of 1992 and 1993 by increasing their reproductive effort. This anticipatory reproductive response was first observed by Svärdson (1957), who proposed that increases in clutch size of Finnish game birds in years of peak seed production was a consequence of the peak in nutritionally superior reproductive buds the previous winter. This phenomenon has also been observed in a number of red squirrel studies (C. C. Smith 1968, Kemp and Keith 1970). The ability of squirrels to cache food and use it over multiple years tends to smooth out the major fluctuations in cone supply. Individuals that acquire a territory are capable of surviving complete cone failures.

Although cone availability provided a reasonable explanation for some of the red squirrel dynamics observed, there were some perplexing inconsistencies. Juvenile survival to emergence was low in two of the three summers following a cone failure in the previous autumn. The exception to this pattern occurred in 1990, the summer of maximum hare numbers. Our research on survival of leverets (chapter 8) indicated that squirrels were killing baby hares at a high rate during this period (O'Donoghue 1994). Relatively few female squirrels bred that year, but those who did may have been able to get enough energy from eating baby hares to keep their offspring alive to emergence. Similarly, survival from emergence to the next spring was good in 1991–1992 (the winter of hare decline) despite the cone crop failure of 1991. This high survival may have been due to squirrels scavenging hares that had been killed and cached by predators.

Though both species ate the rabbit chow on the food-supplementation grids, rabbit chow had no effect on the demography of red squirrels but marked effects on that of the arctic ground squirrels, and thus we only address its effects on the latter. Food addition played a greater role in the population dynamics of ground squirrels than did predation. Increases in density on food-addition treatments averaged fivefold over control densities, twice that of the predator exclosure. The typical response to food addition in vertebrate populations has been a doubling of density, but few studies have supplied food for more than one year (Boutin 1990). In Columbian ground squirrels (*S. columbianus*), Dobson and Kjelgaard (1985) supplemented two populations at different elevations over a 2-year period, resulting in a 1.5–2.3 increase in density. One of the problems with short-term food-addition studies is that they do not incorporate possible pretreatment maternal effects (Bernardo 1996) that may reduce the response to the experimental manipulation. Pretreatment maternal effects should have been minimal in our study, as the length of our study greatly exceeded their average life span (1.5 years ± 0.05 SE) and even exceeded the longest lived individual (6 years on the predator exclosure + food grid). Our study spanned 10 generations of squirrels, and food addition resulted in a maximum sevenfold increase. Our intensive study on ground squirrels began 3 years after food addition started, and at that time densities were already about fourfold that of controls. Thus we were too late to investigate population regulatory processes, but rather could only make conclusions about limitation, and food was clearly a major limiting factor.

Although food addition had little or no effect on summer (figure 9.11, table 9.3) or winter survival (figure 9.12) or on litter sizes at emergence (figure 9.13), in arctic ground squirrels, it had marked positive effects on adult female body mass (figure 9.15), on the percentage of females weaning a litter (a 32% increase over that on controls; figure 9.13), and on earlier emergence of juveniles from the natal burrow (by 1–2 weeks; table 9.4). Similar findings have been reported for other ground squirrel species, though some have also emphasized the positive effects of food on overwinter survival. Timing of breeding, which determines the date of juvenile emergence, can affect both body size and survival of offspring as well as litter size. Parturition dates of Richardson's ground squirrels were positively related to litter size at emergence (earlier litters were larger, Dobson and Michener 1995). Rieger (1996) also found the same relationship between weaning date and litter size in Uinta ground squirrels (*S. armatus*), but also found that offspring weaned later in the season were larger than those weaned earlier. Juveniles weaned when larger in mass may have an advantage to those weaned smaller in mass; however, the greater investment by the mother may potentially affect her survival. Trombulak (1991) found in Belding's ground squirrels that food-supplemented females did not have larger litters, but weaned offspring that were 28% heavier than controls. Juveniles weaned earlier also have the potential to forage longer to accumulate the necessary fat reserves to survive hibernation. Rieger (1996) found that survival to the next spring decreased with later weaning dates in female *S. armatus*. Armitage et al. (1976) found that earlier juvenile emergence in yellow-bellied marmots (*Marmota flaviventris*) was associated with increased overwinter survival. The positive relationship between prehibernatory mass and increased overwinter survival of juvenile ground squirrels has been documented for several species of ground squirrels (*S. armatus*: Slade and Balph 1974, *S. columbianus*: Murie and Boag 1984, *S. richardsonii*: Michener 1974, Michener and Locklear 1990). However, we found no effect of food addition on survival and conclude that the major mechanism by which food addition operated in arctic ground squirrels in the boreal forest was through reproduction, not survival.

Role of Predation   Predators had negative effects on aspects of the demography of both squirrel species, and reduction of predation (exclusion of large mammalian predators and reduction of avian predators under the monofilament area) ameliorated these negative effects. In red squirrels, the impact of predators was only transitory during the winter of 1991–1992 and the summer of 1992, when survival outside the predator exclosure was significantly lower than that inside it. This coincided with the highest predator–hare ratio of the decline (see chapters 13–16; Stuart-Smith and Boutin 1994). However, this increased red squirrel mortality did not translate into a marked reduction in population density, with the population easily compensating for any losses and the density remaining constant.

We were surprised that hare predators did not have a greater effect on red squirrels during the hare decline. Some reasons for the minor impact may be as follows. First, lynx were the only predators to show a major shift to squirrels, but by the time this occurred lynx numbers were already so low that they had little effect on squirrel populations. Goshawks were scarce and great horned owls tended to prey on hares throughout. Squirrels appear to be relatively safe from predation during winter. This may be due to the fact that their activity is considerably reduced during this time. With limited daylight and cold temperatures, they may only be active for a few hours. When in a nest they are simply unavailable. Extreme temperatures, such as those experienced in late January 1992, may reduce activity even further. Thus, despite constant densities, red squirrels are unreliable alternate prey because their activity patterns render them unavailable during periods when predators would need them most. It appears that predation is more important during the breeding season than it is in winter. In retrospect this is not surprising given the increased amount of time that squirrels are active and exposed to predators. It appears that food supply and predation can interact to lower survival as in summer 1992 when squirrels were without cones and predator numbers were lagging behind the hare crash.

The influence of the cyclic decline of snowshoe hares on arctic ground squirrel populations from 1990 to 1995 was clearly demonstrated by the negative population growth rates (figure 9.8) and decline in adult active season survival on all areas (including the predator exclosure) from 1991 to 1993. This is the period when predation intensity was at its highest (figure 9.9). Active season survival of adult arctic ground squirrels was lowest in 1992 and 1993, when predators were using arctic ground squirrels as alternative prey during summer (chapters 13, 15; Rohner 1994). In addition, adult survival was consistently high in 1994 and 1995, when predator biomass (and possibly predation intensity) was lowest (figure 9.9, table 9.2). Thus, predation played a major role in limiting ground squirrel populations, and much of its impact appeared to operate through changes in survival.

The ground squirrel population on the predator exclosure reached a maximum of three times the density of the control populations (figure 9.10). The predator exclosure population experienced small but nonsignificant increases in survival of adults (figure 9.11) but not of juveniles. The population within the predator exclosure and underneath the monofilament declined synchronously with that on control areas, so the monofilament was not effective at preventing the decline, and avian predators must have had access to the squirrels and compensated to some extent for the lack of large mammalian predators. On control areas in 1992 (Hubbs and Boonstra 1997), avian predators and mammalian predators caused 45% and 55%, respectively, of the identifiable kills, but on the predator exclosure (and on the predator exclosure + food) avian predators were responsible for all squirrels killed by predators.

Other studies have found increased predation on alternate prey species after declines of the principal prey. Keith and Cary (1991) observed that increased predation occurred on alternative prey (including Franklin's ground squirrels, *S. franklinii*) during a snowshoe hare decline in Alberta. Steenhof and Kochert (1988) reported that Townsend's ground squirrels (*S. townsendii*) were alternative prey for golden eagles (*Aquila chrysaetos*) during a jackrabbit (*L. californicus*) decline in Idaho. Marcstrom et al. (1989) concluded that arctic hare (*L. timidus*) densities and growth rates declined in response to increased predation by red foxes *(Vulpes vulpes)* and martens *(Martes martes)* when *Microtus* and *Clethrionomys* populations declined on two islands in the northern Baltic sea. Sutherland (1988) found that changes in survival rates of Brent geese may be due to predators such as arctic foxes *(Alopex lagopus)* switching to goose eggs and goslings as a consequence of the cyclic decline of lemmings on the Taimyr Peninsula in Siberia.

   Reproduction in ground squirrels, but not in red squirrels, may also have been negatively affected by the presence of predators. More females weaned young on the predator exclosure (72%) than on the controls (45%), and litter sizes on the predator exclosure were greater than on the controls in 4 of the 5 years of the study, but juvenile emergence times were the same. We have no evidence to suggest that this was the result of food differences in the quality or quantity of food on these areas. However, the same trend in weaning rate between the predator exclosure and the controls was observed between the predator exclosure + food and the food addition site, but to a lesser degree (83% vs. 77%, respectively). In some species, the indirect effects of predators on their prey may be more important in determining changes in population growth than the direct effect of predation. Antipredatory behaviors (Holmes 1984, Carey and Moore 1986, Lima 1987) can result in less time spent foraging (Holmes 1991, Kieffer 1991) or in more time spent in protective habitats with poor-quality forage causing body condition to decline and fecundity to decrease (Hik 1995). Stress induced by the presence of predators may also have repercussions on physiological condition, causing a decline in reproductive performance (Boonstra and Singleton 1993, Hik 1995). Recent findings by Boonstra et al. (1998) indicated that snowshoe hares were chronically stressed and had low reproduction during the population decline when predation risk was high and that stress physiology and reproduction did not improve until predation risk declined. They also suggested that predator-induced stress effects may carry over into future generations, delaying population recovery after a decline. We do not know which of these processes were operating in the ground squirrels.

Interaction between Food and Predation   If food and predators are influencing the dynamics of the squirrel species, was the effect additive or multiplicative? We were not able to answer this question for red squirrels, as the rabbit chow was inappropriate, we had no monofilament grid nested within this exclosure (and thus could not even partially remove the impact of avian predators), and the habitat on this treatment was more open and thus less appropriate than in prime spruce habitat. Red squirrel densities on this treatment were largely unaffected by the experimental treatment (figure 9.6). However, for ground squirrels, this treatment had the greatest, cumulative effect on densities and we conclude the effect was multiplicative. By 1995, this treatment had 60 squirrels/ha, which far exceeded density estimates on the other areas in our study (figure 9.10) and in prime natural habitat (in alpine, 5.8/ha: Green 1977, in meadows, 3–10/ha: McLean 1983). Thus, when predators are removed and food is in abundance, ground squirrel populations appear to increase without asymptoting, indicating that neither food nor predators alone can explain ground squirrel populations but that the limits of ground squirrel population densities are determined by the interaction between these two important factors.

There were three aspects of demography, which, though small, operated to increase densities in successive years. First, juvenile emergence was 1 week earlier in 2 of 5 years on the predator exclosure + food when compared with the food addition and 2 weeks earlier than that on the predator exclosure (table 9.4). Second, more adult females weaned litters on this grid (83%) than on either the food addition (77%) or predator exclosure (72%) grids (figure 9.14). Third, overwinter survival after the winter of 1993–1994 increased on food-supplemented treatments relative to unfed sites (figure 9.12). Thus, during the decline and low of the snowshoe hare cycle, ground squirrel populations are limited by an interaction between food and predation.

9.5.3   Role of Squirrels in the Boreal Forest Community

Would the dynamics of the snowshoe hare and its associated predators change if either or both of these squirrel species were absent? We cannot answer this question directly for red squirrels because the range of hares and red squirrels overlap completely. However, the arctic ground squirrel is absent over most of the boreal forest, yet the hare cycle continues, and thus ground squirrels are clearly unnecessary for its continuance. Given the constancy in numbers of red squirrels over our study, changes in squirrel densities are not necessary for the hare cycle. Nevertheless, we think that the microdynamics of the system may be affected significantly by squirrel presence. Both squirrels kill and eat baby hares (of 254 leverets radio collared during the peak of the cycle, 67% died within about 4 weeks, and of that, three-quarters were either killed or possibly killed by squirrels; O'Donoghue 1994). Thus, the presence of these squirrels provides one explanation for why conventional predators may be able to catch up to hare populations.

Squirrels may be very efficient type III predators, switching to hares at peak densities and contributing to the reduction in recruitment at the peak. This may help to explain the differences in processes that we have observed in the Yukon from what Keith (1974, 1990) observed in Alberta. Keith found that the hare decline was initiated by winter food shortage at the peak, leading to poor nutrition, some starvation losses, and a reduction in fecundity, and that the decline was maintained by predation acting in a delayed, density-dependent manner. In contrast, we have never observed absolute food shortage for hares at Kluane (chapter 6, Smith et al. 1988) and propose that predation alone is sufficient to account for the decline (Krebs et al. 1986, Trostel et al. 1987, Sinclair et al. 1988). Winter food shortage at the peak may not be seen in the Yukon where squirrels are present but may be a reality at Keith's study site where the impact of squirrels is less (arctic ground squirrels are not present, and the habitat is less suitable for red squirrels). Thus squirrels may shift the mechanisms acting during the hare decline from a starvation–predation scenario to a pure predation scenario.

Red squirrels and northern red-backed voles may be possible competitors for access to spruce seed. Given the large numbers of cones harvested by the red squirrels (up to 16,000 per territory in nearby Alaska; M. C. Smith 1968), the seeds from these cones are thus not available for the voles. However, we review evidence in chapter 10 that spruce seed is a poor food for these voles and present evidence that vole population dynamics appear to operate largely independent of the spruce seed crop. Given that the dynamics of red-backed vole populations in deciduous forests elsewhere are tightly linked to that of tree seed production (Jensen 1982, Pucek et al. 1993, Wolff 1996), what is the explanation for our results? One possibility is that red squirrels are such efficient harvesters of cones in boreal forests that spruce seeds do not become available for secondary consumers such as voles often enough and that mast years are too rare to specialize on.

9.6   Conclusions

Red squirrel populations showed little variation over the hare cycle. Larger cone crops in one year were positively related to squirrel density in the next year. Adult survival was not related to cone supply, whereas reproduction was, with more young being produced in the last half of the study when cone crops were above average.

The experimental treatments were largely ineffective in influencing red squirrel population dynamics: the food addition had no effect on populations; the exclusion of predators had an effect, but only during one summer when hare predators were high and spruce cone crops were low. The interaction treatment (food addition and exclusion of predators) had no effect, and the fertilization may have had a small negative effect, but the reason is not clear.

Populations of arctic ground squirrels declined and increased in tandem with the decline and increase of snowshoe hares from 1990 to 1996. Increased predation and weather contributed to lower survival and reproduction, causing a decline of control populations from 1991 to 1993. Populations did not show positive growth until predators were at their lowest densities from 1994 to 1996.

Exclusion of predators caused a twofold increase in ground squirrel densities, addition of food a fivefold increase, and both exclusion of predators and food addition an 11-fold increase. Predators were largely responsible for the decline of ground squirrel populations, independent of treatment. Absence of predators improved weaning rate. The benefits of food acted primarily by improving aspects of ground squirrel reproduction (improved weaning success, earlier emergence of young from the natal burrow, and higher adult female body weight). We conclude that ground squirrel populations are predominately limited by the multiplicative interaction between food and predation through changes in reproduction of adult females. Fertilization had strong negative effects on arctic ground squirrel populations, and we suggest that ground squirrels would not be present in the boreal forest if it were not nutrient limited.

Red squirrel and arctic ground squirrel predation on hare leverets during the peak of the hare cycle may be key contributors to slowing down hare population growth sufficiently to allow predators time to catch up. They thus prevent hares from overgrazing their food supply.

Literature Cited

Adamcik, R. S., A. W. Todd, and L. B. Keith. 1979. Demographic and dietary responses of red-tailed hawks during a snowshoe hare fluctuation. Canadian Field-Naturalist **93**:16–27.

Armitage, K. B., J. F. Downhower, and G. E. Svendsen. 1976. Seasonal changes in weights of marmots. American Midland Naturalist **96**:36–51.

Banfield, A. W. F. 1974. The mammals of Canada. National Museum of Canada and University of Toronto Press, Toronto, Ontario.

Batzli, G. O., and S. T. Sobaski. 1980. Distribution, abundance, and foraging patterns of ground squirrels near Atkasook, Alaska. Arctic and Alpine Research **12**:501–510.

Bernardo, J. 1996. Maternal effects in animal ecology. American Zoologist **36**:83–105.

Blake, B. H. 1972. The annual cycle and fat storage in two populations of golden-mantled ground squirrels. Journal of Mammalogy **53**:157–167.

Boonstra, R., D. Hik, G. R. Singleton, and A. Tinnikov. 1998. The impact of predator-induced stress on the snowshoe hare cycle. Ecological Monographs **68**:371–394.

Boonstra, R., C. J. Krebs, and M. Kanter. 1990. Arctic ground squirrel predation on collared lemmings. Canadian Journal of Zoology **68**:757–760.

Boonstra, R., and G. R. Singleton. 1993. Population declines in the snowshoe hare and the role of stress. General and Comparative Endocrinology **91**:129–143.

Boulanger, J., and C. J. Krebs. 1994. Comparison of capture-recapture estimators of snowshoe hare populations. Canadian Journal of Zoology **72**:1800–1807.

Boutin, S. 1990. Food supplementation experiments with terrestrial vertebrates: patterns, problems, and the future. Canadian Journal of Zoology **68**:203–220.

Boutin, S., and K. W. Larsen. 1993. Does food availability affect growth and survival of males and females differently in a promiscuous small mammal, *Tamiasciurus hudsonicus?* Journal of Animal Ecology **62**:364–370.

Byrom, A. E. 1997. Population ecology of arctic ground squirrels in the boreal forest during the decline and low phases of a snowshoe hare cycle. PhD dissertation. University of British Columbia, Vancouver.

Carey, H. V., and P. Moore. 1986. Foraging and predation risk in yellow-bellied marmots. American Midland Naturalist **116**:267–275.

Carl, E. A. 1971. Population control in arctic ground squirrels. Ecology **52**:395–413.

Davis, D. E. 1976. Hibernation and circannual rhythms of food consumption in marmots and ground squirrels. The Quarterly Review of Biology **51**:477–514.

Dobson, F. S., and J. D. Kjelgaard. 1985. The influence of food resources on population dynamics in Columbian ground squirrels. Canadian Journal of Zoology **63**:2095–2104.

Dobson, F. S., and G. R. Michener. 1995. Maternal traits and reproduction in Richardson's ground squirrels. Ecology **76**:851–862.

Erlien, D. A., and J. R. Tester. 1984. Population ecology of sciurids in northwestern Minnesota. Canadian Field-Naturalist **98**:1–6.

Frank, C. L. 1992. The influence of dietary fatty acids on hibernation by Golden-mantled ground squirrels (*Spermophilus lateralis*). Physiological Zoology **65**:906–920.

Frank, C. L. 1994. Polyunsaturate content and diet selection by ground squirrels *(Spermophilus lateralis).* Ecology **75**:458–463.

Geiser, F., and G. J. Kenagy. 1987. Polyunsaturated lipid diet lengthens torpor and reduces body temperature in a hibernator. American Journal of Physiology **252**:R897–901.

Geiser, F., and G. J. Kenagy. 1991. Dietary fats and torpor patterns in hibernating ground squirrels. Canadian Journal of Zoology **71**:1182–1186.

Green, J. E. 1977. Population regulation and annual cycles of activity and dispersal in the Arctic ground squirrel. MSc thesis. University of British Columbia, Vancouver.

Gurnell, J. 1984. Home range, territoriality, caching behaviour and food supply of the red squirrel *(Tamiasciurus hudsonicus fremonti)* in a subalpine forest. Animal Behavior **32**:1119–1131.

Harwood, H. J., and R. P. Geyer. 1964. Biology data book. Federation of American Societies for Experimental Biology, Washington, DC.

Hik, D. S. 1995. Does risk of predation influence population dynamics? Evidence from the cyclic decline of snowshoe hares. Wildlife Research **22**:115–129.

Holmes, W. G. 1984. Predation risk and foraging behavior of the hoary marmot in Alaska. Behavioral Ecology and Sociobiology **15**:293–301.

Holmes, W. G. 1991. Predator risk affects foraging behaviour of pikas: observational and experimental evidence. Animal Behavior **42**:111–119.

Hubbs, A. H., and R. Boonstra. 1997. Population limitation in arctic ground squirrels: effects of food and predation. Journal of Animal Ecology **66**:527–541.

Jensen, T. S. 1982. Seed production and outbreaks of non-cyclic rodent populations in deciduous forests. Oecologia **54**:184–192.

John, E., and R. Turkington. 1995. Herbaceous vegetation in the understorey of the boreal forest: does nutrient supply or snowshoe hare herbivory regulate species composition and abundance? Journal of Ecology **83**:581–590.

Keith, L. B. 1974. Some features of population dynamics in mammals. Proceedings of the 11th International Congress Game Biology, Stockholm, pages 17–58.

Keith, L. B. 1990. Dynamics of snowshoe hare populations. *in* H. H. Genoways (ed). Current mammalogy, pages 119–195. Plenum Press, New York.

Keith, L. B., and J. R. Cary. 1991. Mustelid, squirrel, and porcupine population trends during a snowshoe hare cycle. Journal of Mammalogy **72**:373–378.

Kemp, G. A., and L. B. Keith. 1970. Dynamics and regulation of red squirrel *(Tamiasciurus hudsonicus)* populations. Ecology **51**:763–779.

Kieffer, J. D. 1991. The influence of apparent predation risk on the foraging behaviour of eastern chipmunks *(Tamias striatus)*. Canadian Journal of Zoology **69**:2349–2351.

Kiell, D. J., and J. S. Millar. 1980. Reproduction and nutrient reserves of arctic ground squirrels. Canadian Journal of Zoology **58**:416–421.

Klenner, W., and C. J. Krebs 1991. Red squirrel population dynamics: I. The effect of supplemental food on demography. Journal of Animal Ecology **60**:961–978.

Krebs, C. J., B. S. Gilbert, S. Boutin, A. R. E. Sinclair, and J. N. M. Smith. 1986. Population biology of snowshoe hares. I. Demography of food-supplemented populations in the southern Yukon, 1976–84. Journal of Animal Ecology **55**:963–982.

Lacey, E. A. 1991. Reproductive and dispersal strategies of male Arctic ground squirrels (*Spermophilus parryii plesius*). PhD thesis. University of Michigan, Ann Arbor.

Larsen, K. W., and S. Boutin. 1994. Movements, survival, and settlement of red squirrel *(Tamiasciurus hudsonicus)* offspring. Ecology **75**:214–223.

Lehninger, A. L. 1982. Principles of biochemistry. Worth, New York.

Lima, S. L. 1987. Vigilance while feeding and its relation to the risk of predation. Journal of Theoretical Biology **124**:303–316.

Marcstrom, F., L. B. Keith, E. Engren, and J. R. Cary. 1989. Demographic responses of arctic hares *(Lepus timidus)* to experimental reductions of red foxes *(Vulpes vulpes)* and martins *(Martes martes)*. Canadian Journal of Zoology **67**:658–668.

McLean, I. G. 1981. Social ecology of the Arctic ground squirrel *Spermophilis parryii*. PhD thesis. University of Alberta, Edmonton, Alberta, Canada.

McLean, I. G. 1983. Paternal behaviour and killing of young in arctic ground squirrels. Animal Behavior **31**:32–44.

McLean, I. G. 1985. Seasonal patterns and sexual differences in the feeding ecology of arctic ground squirrel (*Spermophilus parryii plesius*). Canadian Journal of Zoology **63**:1298–1301.

Mead, J. F., R. Alfin-Slater, D. Howton, and G. Popjak. 1986. Lipids: chemistry, biochemistry, and nutrition. Plenum, New York.

Menkens, G. E., Jr., and S. H. Anderson. 1988. Estimation of small-mammal population size. Ecology **69**:1952–1959.

Michener, D. R. 1974. Annual cycle of activity and weight changes in Richardson's ground squirrel, *Spermophilus richardsonii*. Canadian Field Naturalist **88**:409–413.

Michener, G. R., and L. Locklear. 1990. Over-winter weight loss by Richardson's ground squirrels in relation to sexual differences in mating effort. Journal of Mammalogy **71**:489–499.

Michener, G. R., and D. R. Michener. 1977. Population structure and dispersal in Richardson's ground squirrels. Ecology **58**:359–368.

Morton, M. L. 1975. Seasonal cycles of body weights and lipids in Belding ground squirrels. Bulletin of the Southern California Academy of Sciences **74**:128–143.

Morton, M. L., and P. W. Sherman. 1978. Effects of a spring snowstorm on behavior, reproduction, and survival of Belding's ground squirrels. Canadian Journal of Zoology **56**: 2578–2590.

Murie, J. O., and D. A. Boag. 1984. The relationship of body weight to overwinter survival in Columbian ground squirrels. Journal of Mammalogy **65**:688–690.

Nadler, C. F., and R. S. Hoffmann. 1977. Patterns of evolution and migration in the arctic ground squirrel, *Spermophilus parryii* (Richardson). Canadian Journal of Zoology **55**: 748–758.

Nadler, C. F., R. I. Sukernil, R. S. Hoffmann, N. N. Vorontsov, C. F. Nadler, Jr., and I. I. Formichova. 1974. Evolution in ground squirrels I. Transferrins in holarctic populations of *Spermopilus*. Comparative Biochemistry and Physiology **47A**:663–681.

Nams, V. O. 1993. Effects of nitrogen fertilization on several woody and nonwoody boreal forest species. Canadian Journal of Botany **71**:93–97.

Obbard, M. E. 1987. Red squirrel. *in* M. Novak, M. E. Obbard, and B. Malloch (eds). Wild furbearers management and conservation in North America, pages 265–281. Ontario Ministry of Natural Resources, Toronto.

O'Donoghue, M. 1994. Early survival of juvenile snowshoe hares. Ecology **75**:1582–1592.

O'Donoghue, M., and S. Boutin. 1995. Does reproductive synchrony affect juvenile survival rates of northern mammals? Oikos **74**:115–121.

Otis, D. L., K. P. Burnham, G. C. White, and D. R. Anderson. 1978. Statistical inference for capture data from closed populations. Wildlife Monographs **62**:1–135.

Phillips, J. A. 1984. Environmental influences on reproduction in the Golden-mantled ground squirrel. *in* J. O. Murie and G. R. Michener (eds). The biology of ground-dwelling squirrels: annual cycles, behavioral ecology, and sociality, pages 108–141. University of Nebraska, Lincoln.

Pollock, K. H., J. D. Nichols, C. Brownie, and J. E. Hines. 1990. Statistical inference for capture-recapture experiments. Wildlife Monographs **107**:1–97.

Pollock, K. H., S. R. Winterstein, C. M. Bunck, and P. D. Curtis. 1989. Survival analysis in telemetry studies: the staggered entry design. J. Wildlife Management **53**:7–15.

Price, K., K. Broughton, S. Boutin, and A. R. E. Sinclair. 1986. Territory size and ownership in red squirrels: response to removals. Canadian Journal of Zoology **64**:1144–1147.

Price, K., S. Boutin, and R. Ydenberg. 1990. Intensity of territorial defense in red squirrels: an experimental test of the asymmetric war of attrition. Behavioral Ecology and Sociobiology **27**:217–222.

Pucek, Z., W. Jedrzejewski, B. Jedrzejewska, and M. Pucek. 1993. Rodent population dynamics in a primeval deciduous forest (Bialowieza National Park) in relation to climate, seed crop and predation. Acta Theriologica **38**:199–232.

Rieger, J. F. 1996. Body size, litter size, timing of reproduction, and juvenile survival in the Uinta ground squirrel, *Spermophilus armatus*. Oecologia **107**:463–468.

Rohner, C. 1994. The numerical response of great-horned owls to the snowshoe hare cycle in the boreal forest. PhD thesis. University of British Columbia, Vancouver.

Rusch, S. A. and W. G. Reeder. 1978. Population ecology of Alberta red squirrels. Ecology **59**:400–420.

Sinclair, A. R. E., C. J. Krebs, J. M. N. Smith, and S. Boutin. 1988. Population biology of snowshoe hares. III. Nutrition, plant secondary compounds and food limitation. Journal of Animal Ecology **57**:787–806.

Sinclair, A. R. E., and R. P. Pech. 1996. Density dependence, stochasticity, compensation and predator regulation. Oikos **75**:164–173.

Slade, N. A., and D. F. Balph. 1974. Population ecology of Uinta ground squirrels. Ecology **55**:989–1003.

Smith, C. C. 1968. The adaptive nature of social organization in the genus of three squirrels *Tamiasciurus*. Ecological Monographs **38**:31–63.

Smith, J. N. M., C. J. Krebs, A. R. E. Sinclair, and R. Boonstra. 1988. Population biology of snowshoe hares. II. Interactions with winter food plants. Journal of Animal Ecology **57**:269–286.

Smith, M. C. 1968. Red squirrel responses to spruce cone failure in interior Alaska. Journal of Mammalogy **32**:305–317.

Steenhof, K., and M. N. Kochert. 1988. Dietary responses of three raptor species to changing prey densities in a natural environment. Journal of Animal Ecology **57**:37–48.

Stuart-Smith, A. K., and S. Boutin. 1995. Predation on red squirrels during a snowshoe hare decline. Canadian Journal of Zoology **73**:713–722.

Sullivan, T. P. 1990. Responses of red squirrel *(Tamiasciurus hudsonicus)* populations to supplemental food. Journal of Mammalogy **71**:579–590.

Sutherland, W. J. 1988. Predation may link the cycles of lemmings and birds. Trends in Ecology and Evolution **3**:29–30.

Svärdson, G. 1957. The invasion type of bird migration. British Birds **50**:314–343.

Trombulak, S. C. 1991. Maternal influence on juvenile growth rates in Belding's ground squirrel *(Spermophilus beldingi)*. Canadian Journal of Zoology **69**:2140–2145.

Trostel, K., A. R. E. Sinclair, C. Walters, and C. J. Krebs. 1987. Can predation cause the 10-year hare cycle? Oecologia **74**:185–193.

Turkington, R., C. J. Krebs, R. E. John, M. Dale, V. Nams, R. Boonstra, S. Boutin, K. Martin, A. R. E. Sinclair, and J. N. M. Smith. 1998. The effects of NPK fertilization for nine years on the vegetation of the boreal forest in northwestern Canada. Journal of Vegetation Science **9**:333–346.

Van Horne, B., G. S. Olson, R. L. Schooley, J. G. Corn, and K. P. Burnham. 1997. Effects of drought and prolonged winter on Townsend's ground squirrel demography in shrub steppe habitats. Ecological Monographs **67**:295–315.

Waterman, J. M. 1986. Behaviour and use of space by juvenile Columbian ground squirrels *(Spermophilus columbianus)*. Canadian Journal of Zoology **64**:1121–1127.

Wolff, J. O. 1996. Population fluctuations of mast-eating rodents are correlated with production of acorns. Journal of Mammalogy **77**:850–856.

10

Voles and Mice

RUDY BOONSTRA, CHARLES J. KREBS, SCOTT GILBERT,
& SABINE SCHWEIGER

Small mammals are a ubiquitous, but less obvious, component of the herbivore community in the boreal forest. Small mammals are defined as those <100 g and generally represent <4% of the herbivore biomass in the Kluane Lake ecosystem. Across the boreal forest of North America, there are three main genera of cricetids. There are two species of the genus *Clethrionomys,* with the northern red-backed vole (*C. rutilus*) occupying the forests approximately north of 60° latitude (Martell and Fuller 1979, West 1982, Gilbert and Krebs 1991) and the southern red-backed vole (*C. gapperi*), occupying the rest (Grant 1976, Fuller 1985, Vickery et al. 1989). The deer mouse, *Peromyscus maniculatus,* is present throughout most of the boreal forest region (though in the Kluane area it is nearing the northern limit of its range). The *Microtus* voles (in order of decreasing abundance: the meadow vole, *M. pennsylvanicus;* the root vole, *M. oeconomus;* the singing vole, *M. miurus;* and the long-tailed vole, *M. longicaudus*) all occupy primarily grassy regions within the forest (meadows, marshes, and forest openings). In addition, the Kluane area has a rich diversity of rarer species (Krebs and Wingate 1976, 1985).

10.1   Natural History and Food Web Links

In this chapter we focus primarily on the northern red-backed vole and on the *Microtus* voles. We do not discuss the deer mouse, as it has been rare over the last 10 years, though formerly it made up about half of the small mammals captured (Krebs and Wingate 1985, Gilbert and Krebs 1991). For both groups of microtines, the breeding season typically lasts from May to the end of August or mid-September. Both have typical litter sizes of five or six. *Clethrionomys* females can produce 28 young per season, while *Microtus* females can produce 19–23 young per season. However, more realistic rates of increase are half or less than this (i.e., sevenfold for *Clethrionomys;* four- to sixfold for *Microtus;* Krebs and Wingate 1985).

The northern red-backed vole eats a wide variety of foods (documented in studies from Fairbanks, Alaska: Grodzinski 1971, West 1982; and from Great Slave Lake, Northwest Territories: Dyke 1971). It prefers to eat seeds (particularly of *Arctostaphylos* spp., *Vaccinium* spp., *Geocaulon lividum,* spruce seeds, but also *Empetrum nigrum*), leaves of some trees and shrubs (especially trembling aspen in spring but also a modest amount of *Shepherdia, Vaccinium,* and *Empetrum*), leaves of a number of herbs (particularly *Equisetum pratense* in spring, but also *Geocaulon* and *Cornus*), lichens (especially during winter: *Alectoria jubata, Usnea* spp. *Cladonia* spp., *Parmelia* spp., and *Peltigera aphtosa*), fungi (especially *Laccinum scaber, Russula* spp., *Clavaria* spp., *Hygrocybe punciea,* and *Hydnellum* spp.) and occasionally mosses. It will also readily eat arthropods. Because of its consumption of both epigeous and hypogeous fungi (Maser et al. 1978), it may play a significant ecosystem role in the boreal forest by dispersing spores of decomposer and ectomycorrhizal fungi (Johnson 1996, Pastor et al. 1996), particularly after fires (West 1982).

Members of the genus *Microtus* are primarily herbivorous, eating both dicotyledon and monocotyledon leaves, stems, and roots, with the proportion varying depending on season and site (Batzli 1985, Bergeron and Jodoin 1989). Seeds generally make up a small proportion of the diet (particularly in root voles; Grodzinski 1971), and arthropods are generally insignificant in the diet (Tast 1974). In Alaska, root voles also consume leaves of *Salix, Shepherdia, Vaccinium,* and *Viburnum,* as well as the herbs *Equisetum, Linnaea,*

Cornus, and *Mertensia* (Grodzinski 1971). Fungi, lichens, and mosses may be eaten, but appear to be minor components of the diet (Grodzinski 1971, Maser et al. 1978, Pastor et al. 1996). Finally, virtually all predators eat *Microtus* spp. and *Clethrionomys,* and thus small rodents may play a significant ecosystem role in influencing the dynamics of the predators.

10.2 Community Interactions and Factors Affecting Population Dynamics

Two major peaks in *Clethrionomys* have occurred in the last 20 years, one in 1973 and one in 1984, and with additional modest vole peaks in 1975 and 1987 (Krebs and Wingate 1985, Gilbert et al. 1986, Gilbert and Krebs 1991, Boutin et al. 1995). Both major peaks coincided with the late decline or low phase of the snowshoe hare cycle. Four possible explanations can be proposed for this inverse relationship: competition, predation, stochastic processes, and nutrient release. First, red-backed voles may be acting as a buffer prey, such that when hares are abundant, so are their predators, which also eat voles. These predators may be eating voles to such an extent during the hare peak and early decline that vole numbers are depressed (Erlinge et al. 1984), and only when the predator populations collapse after the hare decline can vole populations grow (Hansson and Henttonen 1985). In northern Fennoscandia, numbers of small mammal species show 3- to 4-year cycles, and predators have been implicated in causing population declines (Henttonen et al. 1987, Hanski et al. 1991, Korpimäki and Norrdahl 1991). Second, competition for food between hares and red-backed voles during the peak summers of hare abundance may seriously depress food supply for the latter. Third, stochastic processes, operating principally through weather, may indirectly affect vole populations through their effect on seed or mushroom crops (Kalela 1962, Dyke 1971, West 1982) or may directly affect vole populations though negative effects on survival (e.g., low temperatures caused by insufficient snow cover) (Fuller 1969, Martell and Fuller 1979, Whitney and Feist 1984). In northern Europe, *C. glareolus* outbreaks followed years with a superabundant production of seeds from deciduous trees, and these mast years were directly related to weather patterns (Jensen 1982, Pucek et al. 1993). Fourth, a nutrient release hypothesis proposes that high hare densities at the peak cause high production of hare pellets, which then fertilize the forest. Plants critical to the diet of small mammals, particularly seed-bearing forbs, capture nutrients from the decay of these pellets. Forbs bear large crops of seeds when weather conditions are suitable, and this results in a vole peak (variants of this hypothesis have been put forward for lemmings in the arctic by Pieper [1964] and Schultz [1969]). With the large-scale experiments discussed in chapter 4, we were able to examine each of these hypotheses.

Though densities of meadow and root voles vary from one year to the next in the Kluane area, there is no clear evidence of a 3- to 4-year cycle in *Microtus* voles. However, Whitney (1976) found evidence of a 3-year cycle in root voles in Fairbanks, Alaska. Because *Microtus* species occur primarily in grassland habitat (which makes up about 7% of the valley; see chapter 2) and our experiments were directed principally at the boreal forest, we did not expect any major response from these voles to most of our manipulations. The fertilization experiment, however, was the exception because the boreal forest is nutrient limited (Chapin and Shaver 1985, Bonan and Shugart 1989), and grass growth was stimulated by fertilization (chapter 5, Nams et al. 1993, John and Turkington 1995, Turk-

ington et al. 1998). We thus predicted that *Microtus* species should benefit from this manipulation and that the effects would be particularly pronounced on grassland meadows
which occurred throughout the study area. In addition, in Fennoscandia, *Microtus* has been
postulated to be more vulnerable to predation than *Clethrionomys* (Hansson 1987, Henttonen et al. 1987), and if this relationship holds for the boreal forest in North America, we
predicted that dynamics of *Clethrionomys* should be closely tied to that of *Microtus*.

10.3 Methods

10.3.1 Small Mammal Trapping

We trapped small mammals from 1987 to 1996. Additional information on the small
mammals for this area covers the period from 1973 to 1989 (Krebs and Wingate 1976,
Gilbert and Krebs 1991). We set up 12 trapping grids to assess the responses of small mammals to the experimental manipulations. Each grid covered 2.81 ha (including a buffer
strip around the perimeter) and had 100 trap points spaced 15.24 m apart and arranged in
a 10 × 10 pattern. Traps were placed at every other trap point for a total of 50 traps. All
small mammal grids were either nested within or immediately adjacent to the hare trapping grid or in the middle of the experimental treatment in comparable open spruce habitat with well-developed shrub, herb, and grass layers. We had two replicate small mammal grids for the treatments. In 1989, we set up three additional grids to cover meadow
habitats (grass and shrub) that were present in the valley but which we were not trapping.
Each covered 1.49 ha and had 49 trap points in a 7 × 7 pattern.

Trapping sessions were carried out twice per year, with the first occurring in mid- to
late May just after snow melt to assess overwinter survival and the second occurring in
mid- to late August to assess summer production. Longworth livetraps were left permanently on the grids at all times and covered with protective boards to protect the traps
against sun and rain. All grids on the study area were trapped within 2 weeks of each other.
One week before trapping, we prebaited all traps with oats. During a trapping session,
traps were set in the evening of the first day and checked three times: by 0900 h the next
morning, by 2000 h the next evening, and by 0900 h the second morning, when all traps
were locked open. Traps were baited with oats and a slice of apple; cotton stuffing was
provided for warmth. When first caught, animals were ear tagged; tag number, species, location, weight, sex, and sexual condition (males, scrotal or not; females, vagina perforate
or not, lactating or not, and pregnant or not) were recorded, and the animals were immediately released (CD-ROM frame 29). Because *Microtus* species were difficult to tell
apart, especially the female root voles from the female meadow voles, we pooled them
into a single *Microtus* category.

Before giving the results, two caveats must be born in mind, which may have affected
the quality of the data. First, in the trapping sessions of 1987 and spring 1988, traps on
many of the grids were disturbed by arctic ground squirrels and red squirrels. By the summer of 1988 we eliminated this problem by putting each trap inside a protective wire mesh
cage. Second, an outfitter allowed his horses to range freely on the study area from 1987
to the spring 1990. The horses were found almost exclusively on fertilizer 1.

10.3.2 Data Analysis

We calculated the minimum number of small mammals alive (MNA) for comparison among grids. Because we trapped only three times per trapping session and only twice per year, estimators using the Jolly-Seber method or the CAPTURE program (White et al. 1982, Menkens and Anderson 1988) were inappropriate. For each trapping grid we calculated the finite rate of summer population growth standardized to a 4-month period; no rates were calculated for times when zero animals were captured. These data were analyzed with a one-way ANOVA plus Tukey-Kramer multiple range test using SuperANOVA (Gagnon et al. 1990). For the overall northern red-backed vole data set, we calculated an index of overwintering survival as spring density in year $t + 1$/late summer density in year t. We were not able to get estimates of overwinter survival based on recovery of tagged animals because few were recaught in spring that had been tagged in summer. Variation in population size among treatments was statistically analyzed using a randomization test (e.g., Manly 1991).

10.3.3 Robustness of Data

To assess the efficiency of our trapping technique in accurately assessing the small mammal populations, we examined a population of red-backed voles trapped in a spruce forest in our valley for 7 years (grid S+M, trapped by S. Gilbert from 1983 to 1989, except for 1987). This population was trapped in most years at 2- to 3-week intervals, and we used it to compare our estimation method (MNA) with a robust estimation method (Jolly-Seber). Two points can be made from this analysis. First, these voles were highly trappable (mean Jolly estimate of trappability/year = 92.1% $\pm$ 2.76 (SE), range 81.3–100%; Krebs and Boonstra 1984). Second, the Jolly estimate and MNA were highly correlated ($r^2 = .99$, $n = 44$), with MNA underestimating the Jolly estimate by 4% at a population size of 5 and by 9% at 20. Thus, our trapping method should accurately reflect demographic parameters.

10.4 Impacts of the Manipulations

Over the entire study and on all areas combined, we caught 2780 different animals (1813 northern red-backed voles, 732 voles of the genus *Microtus,* 66 deermice, 38 chipmunks, 11 heather voles, and 126 shrews) in 24,650 trap nights. In addition, 9 short-tailed weasels (*Mustela erminea*) were caught in live traps (2 in 1988, 5 in 1993, and 2 in 1994). To assess general population changes over the study area, the data were divided into the major species and pooled for all the grids we trapped continuously from 1987–1996. Red-backed voles were generally the most abundant small mammal in the valley. Their population density was moderately high in 1987, low from 1988 to spring 1991, growing rapidly over the summer of 1991, at high densities in 1992, moderately high densities in 1993 and 1994, and finally declining to extremely low levels in 1995 (figure 10.1). During the years of high snowshoe hare numbers (1988–1990), the northern red-backed voles remained low, whereas after the hares declined, these voles increased dramatically in density. Spring densities of these two species were inversely correlated ($r = -.76$, $N = 9$, $p = .02$).

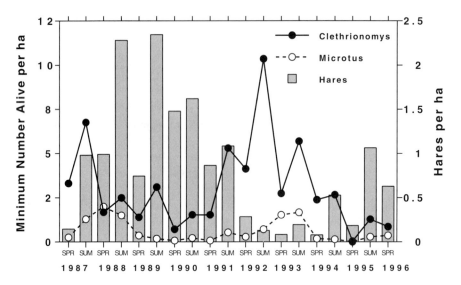

Figure 10.1 Population changes in small mammals at Kluane. All live-trapping grids were pooled. For the small mammal estimates (scale on the left y axis), only those forest grids trapped from the start of the study were included. Changes in snowshoe hare density on control grids were plotted for comparison.

Microtus voles were found at high densities in 1988 and 1993 but at low densities in all other years on the main trapping grids (figure 10.1). There was no relationship between *Microtus* density and either *Clethrionomys* density ($r = .32$) or snowshoe hare density ($r = .01$). *Microtus* density was unrelated to our index of weasel abundance (vole density in spring vs. weasel abundance the previous winter [$r = .37$] or vole density in autumn vs. weasel abundance the next winter [$r = .03$]). Both chipmunks and deermice made up minor parts of the small mammal community, being consistently low throughout the study.

Shrews (*Sorex cinereus* and *S. vagrans*) were only periodically abundant and almost never were caught in spring (only 4% shrews were caught then). The year of peak shrew abundance, 1995, may have been associated with an outbreak of spruce bark beetle, whose effects were becoming particularly pronounced at that time. Shrews were patchily distributed on the grids. In the sample in which all the grids were pooled, there was no obvious correlation between shrews and any of the other species (e.g., shrews vs. late-summer density of *Clethrionomys*, $r = .41$; shrews vs. late-summer density of hares, $r = .14$).

10.4.1 Northern Red-Backed Vole

Treatment Effects Variation in red-backed vole densities among plots within the same treatment was great, but the general changes in population abundance over time were correlated. Thus, there was a significant relationship between population changes on control 1 with that on control 2 ($r = .74$, $N = 19$, $p = .0003$). We obtained an average treatment density within each trapping session and compared treatments by two methods: graphically with a density effect comparison and statistically with a randomization test compar-

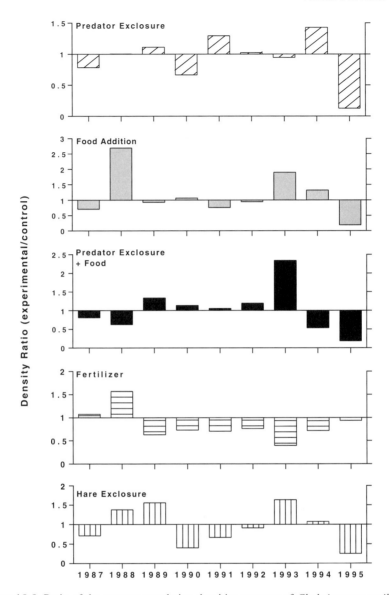

Figure 10.2 Ratio of the average population densities per year of *Clethrionomys rutilus* for the major treatments to the average control population densities at the same time. If there is no treatment effect, we expect a ratio of 1.0.

ison. The density effect of the treatments was compared as the ratio of the average effect of the treatment (replicates pooled) to average control densities (figure 10.2). Initially only the data from 1988 onward were examined, as the experiments were being set up in 1987. For most treatments there appeared to be no consistent effect, with the density ratio being approximately 1 (i.e., averaged density ratio was 1.3 for food addition, 1.0 for predator exclosure, 1.1 for predator exclosure + food addition, 1.0 for hare exclosure, and 1.0 for hare exclosure + fertilization). In contrast, the predator exclosure + monofilament had consistently lower average densities (0.7), but those lower densities were present from the start of the experiment, and thus site-specific effects may explain these effects. The fertilizer density ratio averaged 0.78 from 1988 onward and 0.66 from 1990 onward. Fertilization actually reduced vole numbers. From 1987 to 1988, populations on the fertilized grids were either similar or higher than those on control grids, and thus it is unlikely that there were initial site-specific peculiarities limiting numbers. Independent evidence from other long-term experiments at Kluane on the effects of fertilizer indicates that the negative effects on a number of forest herbs and dwarf shrubs take 3 years to become evident (Nams et al. 1993; John and Turkington 1995).

Three major conclusions result from the randomization analysis. First, the control treatment was not significantly different from any other treatment except for the predator exclosure + monofilament. Second, this latter grid had the lowest average densities per trapping session over the 10-year period (mean $\pm$ SE, 2.21 $\pm$ 0.55) of any treatment (e.g., controls 3.27 $\pm$ 0.64), including the other small mammal grid also within the predator exclosure, but not under the monofilament (3.28 $\pm$ 0.67). We do not know why this occurred. Third, the fertilizer treatment (2.74 $\pm$ 0.60) had lower average densities than the food treatment (3.53 $\pm$ 0.61; figure 10.2), and we interpret this to be a real effect. In the first summer (1987), average densities of red-backed voles were higher on the fertilizer treatments than on the food treatments, but thereafter they were almost always lower. This result could be caused by a slight positive effect of food addition on red-backed vole populations, a negative effect of fertilization on critical plant foods, or both. Because of changes in the vegetation with fertilization (chapter 5), we believe that a negative effect of fertilization on certain herbs was probably the cause of these patterns in density.

Large-Scale Effects   In no case did the treatments prevent large fluctuations in numbers over time, and all numbers fluctuated in relative synchrony (e.g., correlation for control grids given above; changes in average density on fertilizer vs. food grids: $r = .62, p = .004$). We interpret this to mean that the treatment effects, for the most part, were not dramatic and that many of the differences we observed were due to site-specific variation in vegetation or in site-quality factors. To examine which factors may have caused the large-scale fluctuations in numbers (figure 10.1) and to remove the noise caused by site-specific differences among grids, we pooled the entire data set and treated our 12 trapping grids as one large grid. In attempting to tease out the relationships, we focus the analysis around two questions: what determined population growth in summer, and what determined the rate of population decline over winter?

Initial spring density was a good predictor of late-summer density ($r = .88, N = 9$), but late-summer density did not predict densities the next spring ($r = -.26$). Spring densities and overwinter survival were unrelated either to the track index of weasel (see 14.5) abundance ($r = .30$ and $r = -.16$, respectively) or to the average production of spruce seeds (see 7.1.3) in our valley ($r = .03$ and $r = .04$, respectively). Overwinter survival was positively related to the finite growth rate the next summer ($r = .75$). Thus, conditions that produced good overwintering survival appear to carry over into the next summer, resulting in higher densities in late summer.

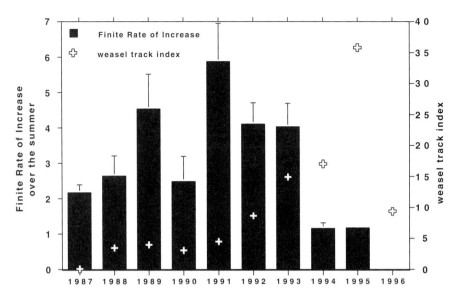

Figure 10.3 Variation among years (+ SE) in the finite rate of population growth of *Clethrionomys rutilus* over the summer (May–August) and an index of weasel winter abundance (number of weasel tracks per 100 km of snow machine traveled). The weasel index as plotted refers to the previous winter (i.e., index for 1991 refers the winter of 1990–1991). Average finite rates of growth were calculated for each grid separately (only those with non-zero captures in spring and late summer were included). All grids but 1 had zero spring captures in 1995 and there were no late summer trapping in 1996. The year 1991 differed significantly from 1994, 1987, and 1988.

There were pronounced differences in summer population growth among years ($F = 3.86$, df $= 8,77$, $p = .0007$), with 1991 having the highest rate and 1994 the lowest (figure 10.3). Summer population growth bore no relationship to spruce seed fall the previous autumn ($r = -.20$) or to snowshoe hare spring abundance ($r = -.26$). Weasel abundance the previous winter was unrelated to growth the next summer ($r = -.47, p = .20$).

Spring body mass varied significantly over the study (figure 10.4; two-way ANOVA, sex × year), with males being significantly smaller than females (mean ± SE, 24.77 g ± 0.20, $N = 315$ vs. 26.62 g ± 0.38, $N = 214$, respectively; $F = 7.75$, df $= 1, 510, p = .006$). In some years animals were much heavier than in other years ($F = 7.66$, df $= 9$, $510, p < .0001$). There was no interaction between sex and year ($p = .16$). Male and female spring mass were correlated ($r = .87, p < .001$) and thus reflected the same processes. Body mass of females in spring did not predict the rate of population growth (figure 10.4; $r = .38$). Female mass bore no relationship to the index of weasel abundance the previous winter ($r = .04$). Thus, patterns of population growth and body mass were unrelated to levels of one possible winter food (spruce seeds) and to the abundance of a specialist small mammal predator.

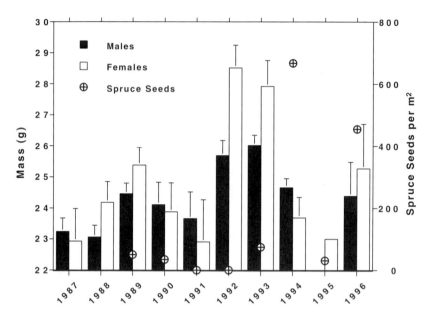

Figure 10.4 Variation (+ SE) in the spring body mass of *Clethrionomys rutilus* and spruce seed fall, 1990–1996. Spruce seed production refers to the previous winter (e.g., numbers for 1991 refer to the winter of 1990–1991). Obviously pregnant females were excluded from the averages.

10.4.2 Microtus *Voles*

Of 776 individual *Microtus* voles captured over the study, 37% (286) were caught on the two grassland meadow grids over an 8-year period (1989–1996 = 17.9/grid/year), and the rest were caught on the other 13 grids (about 3.9/grid/year). Because capture on these latter grids was sporadic and bore no relation to the treatments, our primary focus is on the meadow grids (figure 10.5). Though *Microtus* populations on the shrub meadow grid were correlated with those on the unfertilized meadow grid ($r = .72, p = .002$), average densities were much lower on the former (1.52 ± 0.50 per ha) than on the latter (6.41 ± 1.10), and so we ignore the shrub meadow grid in the following analysis. We ask three questions. First, were changes in the grassland meadow correlated? We would expect this if they were responding to some major environmental variable, even if fertilizer caused an overall increase on the fertilized meadow. Second, were changes in *Microtus* populations in the grassland meadow correlated with changes in forest *Clethrionomys* populations? It has been argued that predators such as weasels find it easier to capture *Microtus* than *Clethrionomys*. Thus, if predators were causing *Clethrionomys* populations to crash or remain low, *Microtus* populations should also crash or remain low simultaneously. The meadows were small and interspersed throughout the spruce forest. Third, as the soils of the boreal forest are nutrient limited, did fertilization stimulate grass growth in meadows and in more open forests, causing *Microtus* populations to increase? We address these questions in turn.

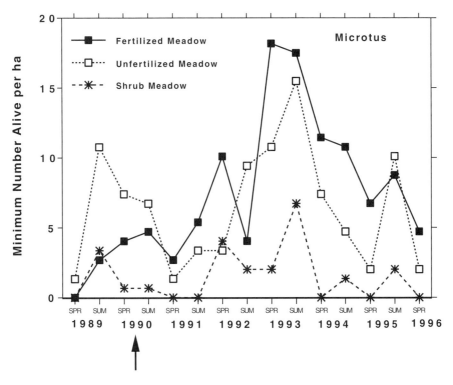

Figure 10.5 Population changes in *Microtus* spp. on the meadow grids over the study. Horses were selectively grazing the fertilized meadow until the summer of 1990 when they were removed, as indicated by the arrow.

We examined only the data collected from 1991 onward because the horses had major effects on the vegetation before that time. *Microtus* populations in the two grassland meadows were correlated ($r = .71, N = 11, p = .01$). We thus pooled these data and compared it with the pooled *Clethrionomys* sample from the two control grids (population changes on these were also correlated; see above). There was no correlation between changes in these two species, either when we examined only the spring periods ($r = .15$) or when we included both periods per year ($r = .14$). Thus, the species of small mammals living in the grasslands were not fluctuating in synchrony with those living in the forest. Neither was there any correlation between the weasel track index and either the pooled *Microtus* meadow spring populations (1991–1996, $r = .10$) or the unfertilized *Microtus* spring populations (1989–1996, $r = .37$). Thus, even when overwintering weasel indices reached their second highest level (1994–1995; figure 10.3), spring *Microtus* numbers in meadows remained moderately high (figure 10.5).

Finally, fertilization appeared to have positive effects on the growth of *Microtus* populations in grassland (figure 10.5). From 1991 to 1996, densities averaged 9.2 ± 1.6 per ha on fertilized grassland meadow versus 6.4 ± 1.4 on the unfertilized meadow ($p = .01$; a modest 44% increase). In addition, a comparison of the changes in densities on the forest fertilizer 1 and nearby control 2 grids (two grids that were the most similar in vegetation; figure 10.5), indicated that *Microtus* numbers were similar on both grids in the first 3 years (1987–1989: fertilizer 1, 1.1 ± 0.4 vs. control 2, 1.2 ± 0.5). Thereafter, *Microtus* populations on the fertilizer 1 grid were virtually always higher than on control 2 (1990–1996: fertilizer 1, 2.2 ± 0.8 vs. control 2, 1.0 ± 0.5). In contrast, the *Clethrionomys* populations showed the opposite pattern. Thus, we conclude that fertilization had a modest positive effect on *Microtus* numbers.

10.5   What Limits Mice and Vole Populations at Kluane?

In general, except for a minor fertilizer effect, the treatments had no marked impact on any of the small mammal species in our study area. There are two possible explanations for this. First, most of our treatments were directed toward the principal actors in the boreal forest, the snowshoe hares and their predators. Thus, the treatments were largely inappropriate for manipulating the abundance of the small mammals in this system and for teasing apart their contribution to it. Though weasels have been regarded as major predators of small mammals in other boreal forest systems (e.g., Fennoscandia), we could not exclude them from our exclosures. Though food addition is known to affect small mammal populations markedly (reviewed in Boutin 1990), we used food that was more appropriate for hares and ground squirrels. Thus, our manipulations may have had no major effect because we designed the experiments incorrectly from the small mammal viewpoint. Second, our manipulations should have affected small mammals as well, but did not because other factors were more important. The responses to the fertilization treatment (which was directed at all players in this system), the hare exclosures (which were directed specifically at the small mammals), and the other treatments give insight into the factors influencing this group.

Small mammal populations did not remain constant over time on the main forest grids. On the contrary, both *Clethrionomys* and *Microtus* populations reached two peaks (not simultaneously). Over the 10 years of our study, *Clethrionomys* populations appeared to be inversely related to hare populations (figure 10.1). Was our 10-year window of time long enough to reveal an invariant pattern? The answer is no. However, others have observed striking similarity to the pattern we observed. Gilbert et al. (1986) found two peaks 10 years apart, and both followed hare peaks by about 2 years. They did not find a vole peak just before the hare peak as we did (also reported in Gilbert and Krebs, 1991). Grant (1976) found peaks in the abundance of *C. gapperi* at 10-year intervals in southern Quebec, but did not have population data on hares. Previous studies in this and adjacent areas on both the northern red-backed vole and the hare allowed us to extend the window of time back another 15 years (17 years for hares). Combined data have to be interpreted with caution because of differences among the studies in trapping protocol and because of site-specific differences in productivity. However, the general pattern over 25 years is shown and reveals three hare peaks (1971, 1980–1981, 1989–1990), with each one being followed by a pronounced vole peak 2–3 years later. In addition, minor vole peaks were observed at some sites in 1975 (by Krebs and Wingate 1985, and Gilbert and Krebs 1991, but not by Gilbert et al. 1986; figure 10.4) and in 1987 (Gilbert and Krebs 1991). Combining the vole data into one continuous series (1973–1975 from Gilbert et al. 1986; 1976–1986 from Gilbert and Krebs, 1991; 1987–1996, our data) and correlating it to a similar series in hare

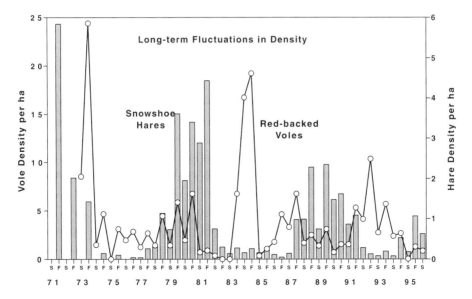

Figure 10.6 Long-term fluctuations in the density of northern red-backed voles and snowshoe hares. All data come from the Kluane Lake, Yukon, area except for the 1971–1975 hare data, which come from Fairbanks, Alaska. The vole data come from the following sources: 1973–1975, Gilbert and Krebs (1981); 1976–1986, Gilbert and Krebs (1991); 1987–1996, present study. The hare data come from the following sources: 1971–1975, Keith (1990); 1976–1986, Krebs et al. (1986) and Boutin et al. (1995); 1987–1996, Krebs et al. (1995) and present study.

data (see figure 10.6) indicated no correlation between vole and hare densities either in spring ($r = .23$, $N = 19$, $p = .34$) or in autumn ($r = .19$, $N = 20$, $p = .43$; $r = .08$, $N = 23$, $p = .72$ if the Alaska hare data are included; see also Boutin et al. 1995). This suggests that hares and northern red-backed voles do not compete directly for resources. However, a phase shift is suggested between hare and major *Clethrionomys* peaks of about 2 years (between autumn hare densities and autumn *Clethrionomys* densities 1 year later, $r = .08$, $N = 23$, $p = 0.72$, and 2 years later is $r = .37$, $N = 23$, $p = .09$). Below we argue that the coupling between hare and vole peaks is real and propose a plausible mechanism that does not preclude secondary vole peaks occurring at other times.

10.5.1   Northern Red-Backed Vole

Role of Predation   The northern red-backed vole did not respond to the predator removal experiments, whether or not food was added (figure 10.2). However, our manipulations of predators was limited to the large mammal predators and to a possible reduction of avian predation underneath the monofilament. Diet information indicates that virtually all predators eat this vole, but none specialize on it. There are three possible reasons for our findings. First, some other compensatory mortality (e.g., socially induced mortality or disease) replaced mortality normally accounted for by these predators. Second, weasel predators, which we could not keep out of the exclosures, were the critical predator component limiting these voles. Or, third, predation was neither limiting nor regulating. We discuss the first and second explanations here and the third in later sections.

Compensatory mortality is unlikely given our low densities. Gilbert et al. (1986) found that although high density (in peak years) did suppress maturation in juveniles, it did not increase mortality. They concluded that social control of density could not explain population fluctuations. Because we found no relationship between weasel abundance and *Clethrionomys* numbers, we conclude that weasels had no major impact on our vole populations (figure 10.3). The most common weasel in our area is the short-tailed weasel. Compared to this weasel species in Europe (208–320 g; King 1983), ours are about half as large (45–106 g; Banfield 1974), being more similar in size to large least weasels in Europe (range 30–60 g for females, 40–100 g for males; Güttinger 1995). Thus, our short-tailed weasels probably fill an intermediate niche relative to the two European species, and some of the findings of the role of the least weasels in Europe may be relevant to our situation here. Our conclusion of no impact is relatively weak. We would have liked to correlate the weasel index to actual density estimates, but this would have required radio collaring plus tracking (Jedrzejewski et al. 1995), and because the larger carnivores were deemed more significant, this is where we put our resources.

There must still be a minimum number of weasels in the area to permit them to have a measurable impact on the prey, and we think that for most of the time, vole densities are too low. A prolonged period of higher vole densities may permit weasel numbers to increase. One such period occurred from 1991 to 1993 when both *Clethrionomys* in the forest (figure 10.1) and *Microtus* in meadows (1992–1993; figure 5.5) were high and this was followed by a high weasel index in the winters of 1993–1994 and 1994–1995 (figure 10.3; see also chapter 16). The lowest rate of *Clethrionomys* population growth was observed in the summers of 1994 and 1995 (figure 10.3), which is consistent with weasels continuing to exert a negative impact then. The *Clethrionomys* data for 1995 present a problem. In spring 1995, only 1 *Clethrionomys* was caught on 12 grids, whereas in late summer 1995, 60 were caught. Vole numbers cannot increase at this rate, and we suggest that the low numbers in spring 1995 were a trapping artifact induced by trap avoidance related to high weasel abundance in the winter of 1994–1995 (Stoddart 1976, Jedrzejewski et al. 1993).

Henttonen (1987) reviewed the evidence in Fennoscandia and proposed that least weasel populations could increase only if high densities of *Microtus* (which are thought to be easier to capture than *Clethrionomys*) were present in meadows and open habitats. Once the weasel population was high, it could then depress both species. If these arguments apply to our area, the *Microtus* in meadows should have been driven to extremely low levels when the weasels peaked in the winter of 1994–1995. This was not observed (figure 10.5), but *Microtus* did decline in 1994 and again in 1995 from their peak in 1993. This evidence is consistent with weasels occasionally having a negative impact on both populations of both species. However, we suggest that other factors (probably overwintering food supplies) are more likely to explain the dynamics of these two species most of the time.

Role of Food   The northern red-backed vole did not respond to the food addition experiments, whether or not predators were present (figure 10.2). Though we have limited evidence that these voles will occasionally eat rabbit chow (unpublished data), food addition never resulted in earlier reproduction or higher densities as it did in hares (chapter 8) and ground squirrels (chapter 9). Relative to boreal forest communities in Fairbanks or Fennoscandia, our forest was unproductive for voles, with densities usually being about 1–3/ha in spring and 2–4/ha in autumn (occasionally reaching maximum densities of 10–20/ha; figures 10.1, 10.6). Comparable low values were found by Fuller (1985) in the southwestern Northwest Territories in *C. rutilus,* with spring densities being <1/ha and autumn densities being 2–3/ha in most years and 9/ha in peak years; parapatric populations of *C. gapperi* were slightly higher. In contrast, the Fairbanks, Alaska, area appears more productive; both West (1982) and Whitney and Feist (1984) reported *C. rutilus* densities varying from about 5/ha in spring to 40–60/ha in autumn. However, neither study found evidence of a 3- to 4-year cycle. West (1982) attributed this to chronically poor overwintering survival.

In Pallisjärvi, Finland (68°N), Henttonen et al. (1987) found densities similar to those in Alaska when combining the numbers of three different species of *Clethrionomys,* with densities typically being 2–3 times ours (spring densities in peak years being about 10/ha and autumn densities 50/ha). In northern Sweden, Lofgrën (1995) found slightly lower densities. However, in contrast to the North American situation, both these Fennoscandian areas have shown evidence of the 3–4 year population cycles. The reason for this difference between northern Europe and North America may be related to better overwintering survival and to higher productivity of the ground layer in Europe.

Jedrzejewski and Jedrzejewska (1996) presented evidence indicating that these microtine cycles can only occur in ecosystems where the standing crop of ground vegetation is >4000 kg/ha, as is found in the boreal forest of northern Europe. In a white spruce forest near Fairbanks (a site likely to be more productive than ours), Grodzinski (1971) calculated that the standing crop of ground vegetation averaged 2250 kg/ha, of which only 237 kg was herbs and dwarf shrubs and the rest was mosses and lichens. Productivity of the ground layer in Kluane appears even lower than in Alaska, and thus our low vole densities are understandable. In contrast, the standing crop of edible birch and willow (the main foods for the cycling snowshoe hare at Kluane) varied from 43–168 kg/ha during the hare peak to 2200–19,700 kg/ha during the hare low (Smith et al. 1988). Thus, productivity of the ground layer, as opposed to the shrub layer, may be the major reason that cycles are not seen for voles that rely on the ground layer, but are seen in hares that rely on the shrub layer.

Populations of small mammals living in forests of Europe and eastern North America with low ground cover do not cycle, but do fluctuate, and these are directly linked to years with superabundant production of tree seeds (Jensen 1982, Pucek et al. 1993, Wolff 1996). However, we found no correlation between spruce seed production and population fluctuations in *Clethrionomys* (figures 10.1, 10.4). Years in which virtually no spruce seeds were produced (1991 and 1992) were followed by springs in which voles were heavy and populations grew to reach high densities. Conversely, the year with maximal seed production (1993) was followed by a spring in which voles were smaller and populations grew poorly. Though the northern red-backed vole eats spruce seeds, these apparently are an incomplete diet, and voles will lose weight if forced to only eat spruce seeds, but not if a supplement of lichens is also available (Dyke 1971, Grodzinski 1971). Thus, the absence of a good spruce seed crop did not prevent high vole densities, nor did its presence permit them. Some other food must explain fluctuations in our populations.

The overall pattern of demography suggests that not only must food supply permit good population growth in summer, it must also permit good survival overwinter and good survival into the next breeding season. Both we and Gilbert and Krebs (1991) found that the size of the vole population in the spring was a good predictor of the size of the autumn population. Gilbert and Krebs (1991) thus postulated that the key demographic factor leading to population peaks was high winter survival leading to high spring densities. However, because the size of the autumn population can be increased by up to three times by providing high-quality supplementary food in spring and summer (sunflower seeds, but not oats; Gilbert and Krebs 1981), typical summer food levels must constrain population growth (especially through survival of juveniles). In addition, Gilbert and Krebs (1981) found that production of higher densities in the autumn did not carry over to the next spring, and thus some aspect of winter conditions must also constrain growth. Schweiger and Boutin (1995) tested this idea by adding food during the autumn and winter and observed higher spring densities. However, when they stopped feeding the next spring, these higher spring densities did not result in higher autumn densities. Thus, some aspect of spring or summer food was inadequate to sustain these populations.

The mushroom crop (both epigeous and hypogeous fungi) may be a critical component of the diet, but mushrooms are available only in summer and early autumn and do not carry through to the next year. The size of the mushroom crop depends on the amount of summer rain (Dyke 1971). *Clethrionomys* spp. readily eat mushrooms, and they may be a major summer and early autumn dietary item. Variation in this resource could be expected to affect summer vole populations. However, Dyke (1971) found fungi in *C. rutilus* stomachs only in summer and early autumn, not in winter, and concluded that fungi were unimportant thereafter. Merritt and Merritt (1978) found a similar pattern in *C. gapperi* in Colorado.

Another potentially critical food item is seed produced by dwarf shrubs and herbs, some of which overwinter (*Arctostaphylos uva-ursi, Vaccinium* spp., *Rosa* spp., *Juniperus* spp.). Evidence suggests that the critical food permitting good survival in both late summer, winter, and into the next growing season is the size of the berry crop produced by dwarf shrubs (Dyke 1971, Grodzinksi 1971, West 1982). What needs to be understood are the factors that cause variation in the size of this crop and the crop's role in supporting overwinter survival of *Clethrionomys*.

Role of Stochastic Events   Stochastic events, primarily variation in weather from one year to the next, may affect small mammal populations. In general, variation in winter weather does not explain variation in densities. Fuller (1977, 1985) postulated that more severe winters, particularly those with unfavorable autumns before stabilization of snow cover and those with unfavorable springs with delayed snow melt and accumulation of melt water, should cause poor survival and depress *Clethrionomys* populations. However, he rejected this hypothesis and concluded that a population's ability to withstand the stresses of winter varied over time and that some unknown property (possibly related to food) was responsible for better survival in years when winter conditions were more severe. West (1982) argued that winter survival in Alaska was largely contingent on the size of the autumn berry crop but that the benefits of a large overwintering berry crop could be eliminated by an early spring with rapid melt off. The early spring and rapid melt would cause remaining berries to disappear due to foraging birds and insects and decomposition.

The result would be vole populations that would decline rapidly in spring. A late spring snow melt would delay this loss and bridge the period until other foods became available.

The spring of 1992 appears to be an example of this. In that year, approximately twice as much snow fell in April and May as normal, and snowmelt was delayed almost 3 weeks (until the end of May) (Hubbs and Boonstra 1997). These conditions were detrimental for arctic ground squirrel populations, which declined markedly that spring. In contrast, 1992 stood out as one of the best years of our study for *Clethrionomys:* populations reached their highest spring densities (figure 10.1), females were their heaviest (figure 10.4), and the population had a high growth rate that summer (figure 10.3). Dyke (1971) experienced a similar set of environmental conditions, but interestingly, it only benefitted *C. rutilus,* not *C. gapperi or P. maniculatus,* both of which declined. Thus, spring conditions may accentuate the amplitude of a vole peak and prolong it if a bountiful berry crop occurred the previous summer.

Does the size of the berry crop vary from year to year, and, if so, what is the cause? Virtually no effort has been made to measure the size of the berry crop over a long time and correlate it with environmental variables. Dyke (1971) found extreme variation from one year to the next in overwintering berry crops in the Northwest Territories. *Arctostaphylos uva-ursi* varied from 72.4 berries/m^2 in 1966 to 0.6 in 1967 and to 0.5 in 1968 in jackpine–juniper stands; *Vaccinium vitis-idea* at the same time varied from 13.3 berries/m^2, 1.4, and 5.6, respectively, in jackpine–juniper stands and 51.4 berries/m^2, 5.9, and 0.6, respectively, in black spruce. High seed production in the autumn of 1966 was associated with high *C. rutilus* density throughout the next summer (1967); conversely, low seed abundance in 1967 was associated with low density in 1968. Dyke attributed this variation to an early, warm growing season in 1965 and 1966 and a delayed, cool, and wet season in 1967 and 1968. West (1982) found modest variation in the berry crop over 3 years in Fairbanks (29.0–63.7 berries/m^2), but this variation had no effect on vole density the next year.

Kalela (1962) proposed that microtine density fluctuations were coupled to variation in plant production caused by variation in weather. Tast and Kalela (1971) fine-tuned this hypothesis, proposing that favorable climatic conditions in northern latitudes, particularly warm temperatures, over several years were needed to create a synchronous pulse of seed and vegetative production, after which plants had low production as they rebuilt their energy levels. Microtine populations were proposed to track these years of good and bad production, resulting in cycles. However, tests of this hypothesis have been equivocal or contradictory (Laine and Hentonnen 1983, Järvinen 1987, Oksanen and Ericson 1987), and the evidence argues in favor of microtine grazing being the cause, not the consequence, of the variation in vegetation performance.

We attempted to relate vole density in spring to a variety of environmental variables in the growing seasons 1 and 2 years before (i.e., average temperatures and precipitation in each of the months May to August and summed combinations of these). Only one variable was significant: the precipitation in July in one year was positively correlated to spring vole densities the next ($r^2 = .28, N = 20, p = .02$). Thus, in our area, which is a relatively dry boreal forest site (see chapter 2), moisture during the height of the growing season, not temperature, may partly limit productivity. Given suitable conditions several years in a row, productivity could remain high (e.g., 1992–1994) and result in an extended period of high numbers. However, much of the variation remains unexplained, and below we propose a hypothesis that includes a critical variable, nutrients, not included in the various weather hypotheses.

Role of Nutrients   We propose that the above evidence is consistent with the following hypothesis: population peaks in *Clethrionomys* follow population peaks in hares by 2–3 years. Foods required by these voles (particularly seeds and berries produced by herbs and dwarf shrubs) are stimulated by the flush of nutrients released by the decomposition of large quantities of pellets produced by hares at the population peak, particularly those produced during peak winters. Though the nutrients are released primarily in the summer after their winter production (D. S. Hik personal communication), lags are introduced into this system because it takes 2 years to produce a good berry crop. This delay is due to the development of flower primordia in the first summer and then berries in the next summer. The high berry production stimulates population growth of voles in the third year. Stochasticity in weather can affect the exact timing of the seed crop (e.g., severe late frosts in spring may kill the flowers, or drought in summer may prevent adequate seed maturation). If weather conditions permit, high seed abundance in the second autumn results in high overwintering survival of the voles and high spring densities in the third year. The bumper crop of seeds may last long enough into the spring and summer of the third year (especially in years with late snowmelt) to enable the animals to survive until the new vegetation starts growing. The importance of decaying pellets as stimulators of plant growth through mineral recycling and herbivore population responses have also been postulated for lemmings in the arctic tundra by Pieper (1964) and by Schultz (1969). Outbreaks of forest lepidoptera in northern ecosystems result in large-scale defoliation, which then inject pulses of nutrients into these forests (Nilsson 1978; Swank et al. 1981). In northern Fennoscandia, Neuvonen (1988) proposed that a nitrogen pulse caused by defoliation of birch forests by a geometrid moth cause an increase in microtine populations. However, definitive evidence on this interaction has yet to be collected.

Evidence for the effect of fertilization on berry production is limited, but it appears to depend on the amount of fertilizer used. In Fennoscandia (reviewed in Raatikainen and Niemelä 1994), fertilization generally increases berry production, often with a lag of 1–2 years, but too much fertilizer will decrease berry production as competitors grow more rapidly than the dwarf shrubs. In the arctic tundra, growth of evergreen dwarf shrubs is significantly stimulated by low levels of fertilization (Chapin and Shaver 1985, Henry et al. 1986), but higher levels lead to mortality. At the levels of fertilization used in our long-term experiments at Kluane (John and Turkington 1995, Turkington et al. 1998), declines occurred in dwarf shrubs. The decline in *Clethrionomys* populations (figure 10.2) on the fertilizer 1 grid 2 years after fertilizer addition started is consistent with this interpretation.

Are the amounts of nutrients released from hare pellets significant to stimulate plant growth? From a study to predict snowshoe hare densities based on hare pellet counts (Krebs et al. 1987), we were able to estimate how many pellets were produced at peak densities. In the autumn of 1980, we estimate that peak hare densites of 3.4/ha produced 71.2 pellets/m^2 and in the autumn of 1989, peak hare densities of 2.34/ha produced 46.0 pellets/m^2. Low levels of nutrients likely to come from these large numbers of decaying pellets may be sufficient to stimulate berry and seed production. Shrub biomass reaches it highest level at the peak (1990) and in the next 2 years (figures 6.2, 6.5), particularly in willow, but possibly also in birch (see Turkington et al. 1998). In addition, high levels of browsing by hares on birch and willow in peak years may make them less competitive relative to the dwarf shrubs and may increase the amount of illumination reaching the dwarf shrubs. Dwarf shrubs may then be able to sequester larger amounts of nutrients than they would when the taller shrubs are growing vigorously. In addition, we know from our fertilizer experiment that plants immediately incorporate the nutrients when they are added artificially. Our evidence indicates that the shrub growth is responding dramatically at and just after the peak, and this is consistent with the argument that there has been a stimulation of growth resulting from release of nutrients from hare pellets.

An implication of this hypothesis is that if snowshoe hare peaks vary in amplitude, peaks with higher hare densities should result in greater browsing pressure, greater production of hare pellets, increased release of nutrients, greater production of seeds and berries, and ultimately in higher *Clethrionomys* densities 2 years later. Figure 10.6 suggests that the hare densities in the peak of 1981 may have been twice as high as densities in 1990. Indeed, the mortality rate of tagged twigs of birch and willow tend to corroborate this. The mortality rate was much higher in the hare peak of 1981 (100% and 57%, respectively, of twigs of the two shrub species that were clipped 0–100 cm above the ground; Smith et al. 1988) than in the hare peak of 1990 (71–90% and 8–16%). Figure 10.6 indicates that vole densities were up to two times higher in 1984 than in 1992. Interestingly, figure 10.6 also indicates that hare densities were about three times higher in 1971 (from Fairbanks) than in 1992, and vole densities in 1973 were also about three times higher. If the hare peak in Kluane in the early 1970s was also about three times as high as that in 1990, the correlation would be tight. However, given differences in methods of data collection and given higher productivity in Alaska (chapter 2), this correlation must be treated with caution. An alternative approach would be to examine the size of the vole peaks within treatments that increased hare densities (through food addition, exclusion of predators, or both). We predict that this should result in high production of hare pellets, the subsequent release of a large amount of fertilizer, and the production of a large berry crop. However, all of these treatments were confounded by increased summer foraging on herbs and dwarf shrubs due to high hare and ground squirrel densities (chapters 8 and 9), and we saw no effect on vole densities.

If a hare peak is followed by a *Clethrionomys* peak, preventing the first should prevent the second. Within the hare exclosure (unfertilized), we not only prevented a hare peak, we eliminated hares altogether, and this should have prevented a *Clethrionomys* peak within the exclosure. It did not (figure 10.2). We were not able to detect any difference between the controls and the hare exclosure. Part of the problem may be one of scale. Relative to the home ranges of these voles, our 4-ha hare exclosure was too small. More intensive monitoring of demography (survival and reproduction) may have permitted us to detect lower indices of demographic performance within the exclosure.

Other boreal forest sites also show irregular *Clethrionomys* fluctuations, and some may be associated with the hare cycle, though the nature of the hare cycle is unknown for these areas. Fuller (1977, 1985) trapped voles in the southwestern Northwest Territories from 1961 to 1979 and found peaks in *C. gapperi* in 1961–1962, 1974, and 1976, but not at other times. *C. rutilus* was trapped intensively from 1970 to 1979, and peaks similar to those in *C. gapperi* were found in 1974 and 1976. Whitney and Feist (1984) summarized

C. rutilus data from 1970 to 1981 (the data after 1972 were collected intermittently and were missing for 1977–1978) for live-trapping grids near Fairbanks and found increasing populations in 1971, peak populations in 1972, and low populations at all other times. For the same general area, West (1982) live trapped voles from 1972 to 1976 and found peak populations in 1976. Mihok et al. (1985) snap-trapped *C. gapperi* in the springs from 1969 to 1981 in spruce forests in southern Manitoba and found high vole populations only from 1978 to 1980. Thus, there seems to be some evidence of peaks in *Clethrionomys* being 10 or more years apart, but the link to the snowshoe hare cycle is less clear in these areas than for our area where we have good trapping data on both species. Because our site is more arid than most boreal forest sites and thus productivity is lower, it may be more sensitive to fluctuations in nutrients than other sites.

We propose that for the boreal forest in the Kluane area, the primary 10-year cycle in snowshoe hares results in a secondary cycle in northern red-backed voles (cf. Angelstam et al. 1985). This occurs through a fertilization effect on the plants eaten by voles as a consequence of the production and decay of large quantities of hare pellets produced during the hare peak. Other vole peaks can occur independent of the hare cycle, but we suggest that these are more related to stochastic weather events which cause variation in size of the berry crop of the dwarf shrubs and mushroom production.

10.5.2   Microtus *Species*

All our major experiments, except for fertilization, had no effect on *Microtus* populations, and this is understandable given that these voles primarily eat grass and our experiments were directed at boreal forest sites. On the forest grids, *Microtus* populations reached higher numbers (mean of about 2/ha) twice, both after *Clethrionomys* peaks (figure 10.1). We do not know why this occurred. From the standpoint of the dynamics of the vertebrate predators in the boreal forest, populations of *Microtus* in meadows, marshes, forest clearings, and stream margins may be particularly significant, especially during decline and low phases of the hare cycle. Diet information (see chapters 13–18) indicates that all predators consume *Microtus,* and some (especially coyotes) appear to preferentially hunt for them during hare declines. *Microtus* voles even feature prominently in the diet of great horned owls and boreal owls, both species that are regarded as typical forest predators (chapter 16). Thus, the importance of *Microtus* to sustaining the predators in the boreal forest ecosystem during hare declines and lows may not be proportional to their numerical abundance.

Fertilization stimulated grass growth (Nams et al. 1993, John and Turkington 1995, Turkington et al. 1998) and *Microtus* populations in meadows responded with a modest population increase (44%; figure 10.5). Passerine birds also showed a modest increase (46%) in response to fertilization (Folkard and Smith 1995; see chapter 12). Our results argue that growth of vegetation in these meadows is also severely constrained by low nutrient abundance characteristic of the boreal forest. It is possible that predators cued into these hot spots of higher vole density, and, if we had been able to remove predators, the differences would have been even more marked than we observed. In summary, only two groups of vertebrates responded to fertilization of this nutrient limited ecosystem: the microtine herbivores and the passerine insectivores.

10.6   Summary

There are some unanswered questions that remain. How does the size of the principal foods of the northern red-backed vole, particularly the berry crop, vary over time, and what are the causes for this? Does nutrient release following snowshoe hare peaks play any role in this variation? As all predators heavily eat *Microtus* voles, what is the role of predation in limiting the population size of this group?

We propose that the key variable limiting northern red-backed vole populations in the boreal forest of North America is overwintering survival and that this is a function of over-wintering food, not predation. Evidence from other research indicates that the key food necessary for good overwintering survival is overwintering berries from dwarf shrubs, not white spruce seeds, and that this may vary as a function of weather. We propose that pro-duction of overwintering berry crop is usually low because primary productivity in the shrub layer in North American boreal forests is concentrated in the tall shrubs, not in the dwarf shrubs, as it is in Fennoscandia. Because of this, vole densities are usually low (1–3/ha), though occasionally they increase dramatically. Most of the major factors we ma-nipulated with our large-scale treatments (large mammalian predators, snowshoe hares, and food in the form of rabbit chow) did not affect northern red-backed voles. Large-scale fertilization had a slight negative effect on these vole populations, probably acting through a reduction in dwarf shrubs (and hence the food produced by them) caused by competi-tion from grasses that were stimulated by fertilizer. Specialist weasel predators are irrel-evant as limiting factors most of the time because vole densities usually do not reach den-sities high enough to sustain weasel populations, and the other boreal forest predators do not focus on these voles. However, if the red-backed vole population were to increase over a series of years, this may permit weasel populations to increase and then depress these voles in the winter. Our long-term trapping data, in which red-backed vole peaks always follow hare peaks by 2–3 years, suggest that periodic fertilization pulses produced by large quantities of hare pellets at hare peaks stimulates dwarf shrub berry crops and pro-duces high overwintering survival and good summer reproduction of voles.

We propose that one of the key factors limiting *Microtus* populations in the meadows of the boreal forest is the low productivity of the grasslands. Both grassland productivity and *Microtus* population density can be increased by fertilization. Predation by both spe-cialist and nonspecialist predators following the hare decline may also periodically de-press *Microtus* in meadows. The impact of weasels on these voles is uncertain because the patchiness of meadows within the forest and their small size could make it difficult for weasels to use *Microtus* efficiently.

Literature Cited

Angelstam, P., E. Lindstrom, and P. Widen. 1985. Synchronous short-term population fluctua-tions of some birds and mammals in Fennoscandia—occurrence and distribution. Hol-arctic Ecology **8**:285–298.

Banfield, A. W. F. 1974. The mammals of Canada. National Museum of Canada and Univer-sity of Toronto Press, Toronto.

Batzli, G. O. 1985. Nutrition. *in* R. H. Tamarin (ed). Biology of the New World *Microtus,* pages 779–811. Special Publication No. 8. American Society of Mammalogists, Pennsylvania.

Bergeron, J., and L. Jodoin. 1989. Patterns of resource use, food quality, and health status of voles *(Microtus pennsylvanicus)* trapped from fluctuating populations. Oecologia **79**:306–314.

Bonan, G. B., and H. H. Shugart. 1989. Environmental factors and ecological processes in boreal forests. Annual Review of Ecology and Systematics **20**:1–28.

Boutin, S. 1990. Food supplementation experiments with terrestrial vertebrates: patterns, problems, and the future. Canadian Journal of Zoology **68**:203–220.

Boutin, S., C. J. Krebs, R. Boonstra, M. R. T. Dale, S. J. Hannon, K. Martin, A. R. E. Sinclair, J. N. M. Smith, R. Turkington, M. Blower, A. Byrom, F. I. Doyle, C. Doyle, D. Hik, L. Hofer, A. Hubbs, T. Karels, D. L. Murray, V. Nams, M. O'Donoghue, C. Rohner, and S. Schweiger. 1995. Population changes of the vertebrate community during a snowshoe hare cycle in Canada's boreal forest. Oikos **74**:69–80.

Chapin III, F. S., and G. R. Shaver. 1985. Individualistic growth responses of tundra plant species to environmental manipulations in the field. Ecology **66**:564–576.

Erlinge, S., G. Göransson, G. Högstedt, G. Jansson, O. Liberg, J. Loman, I. Nilsson, T. von Schantz, and M.Sylvén. 1984. Can vertebrate predators regulate their prey? American Naturalist **123**:125–133.

Dyke, G. R. 1971. Food and cover of fluctuating populations of northern cricetids. PhD thesis. University of Alberta.

Folkard, N. F. G., and J. N. M. Smith. 1995. Evidence for bottom-up effects in the boreal forest: do passerine birds respond to large-scale experimental fertilization? Canadian Journal of Zoology **73**:2231–2237.

Fuller, W. A. 1969. Changes in numbers of three species of small rodents near Great Slave Lake, N.W.T. Canada, 1964–1967, and their significance for population theory. Annales Zoologici Fennici **6**:113–144.

Fuller, W. A. 1977. Demography of a subarctic population of *Clethrionomys gapperi:* numbers and survival. Canadian Journal of Zoology **55**:42–51.

Fuller, W. A. 1985. Demography of *Clethrionomys gapperi,* parapataric *C. rutilus,* and sympatric *Peromyscus maniculatus* in northern Canada. Annales Zoologici Fennici **22**:229–241.

Gagnon, J., J. Roth, B. Finzer, R. Hofmann, K. Haycock, J. Simpson, and D. Feldman. 1990. SuperANOVA—accessible general linear modelling. Abacus Concepts, Inc., Berkeley, California.

Gilbert, B. S., and C. J. Krebs. 1981. Effects of extra food on *Peromyscus* and *Clethrionomys* populations in the southern Yukon. Oecologia **51**:326–331.

Gilbert, B. S., and C. J. Krebs. 1991. Population dynamics of *Clethrionomys* and *Peromyscus* in southwestern Yukon, 1973–1989. Holarctic Ecology **14**:250–259.

Gilbert, B. S., C. J. Krebs, D. Talarcio, and D. B. Cichowski. 1986. Do *Clethrionomys rutilus* females suppress maturation of juvenile females? Journal of Animal Ecology **55**:543–552.

Grant, P. R. 1976. An 11-year study of small mammal populations at Mont St. Hilaire, Quebec. Canadian Journal of Zoology **54**:2156–2173.

Grodzinski, W. 1971. Energy flow through populations of small mammals in the Alaskan taiga forest. Acta Theriologia **16**:231–275.

Güttinger, R. 1995. *Mustela nivalis* L., 1766. *in* J. Hausser (ed). Saugetiere der Schweiz, pages 383–388. Birkhäuser Verlag, Basel, Switzerland.

Hanski, I., L. Hansson, and H. Henttonen. 1991. Specialist predators, generalist predators, and the microtine rodent cycle. Journal of Animal Ecology **60**:353–367.

Hansson, L. 1987. An interpretation of rodent dynamics as due to tropic interactions. Oikos **50**:308–318.

Hansson, L., and H. Henttonen. 1985. Gradients in density variations of small rodents: the importance of latitude and snow cover. Oecologia **67**:394–402.

Henry, G. H. R., B. Freedman, and J. Svoboda. 1986. Effects of fertilization on three tundra plant communities of a polar desert oasis. Canadian Journal of Botany **64**:2502–2507.

Henttonen, H. 1987. The impact of spacing behavior in microtine rodents on the dynamics of least weasels *Mustela nivalis*—a hypothesis. Oikos **50**:366–370.

Henttonen, H., T. Oksanen, A. Jortikka, and V. Haukisalmi. 1987. How much do weasels shape microtine cycles in the northern Fennoscandian taiga? Oikos **50**:353–365.

Hubbs, A. H., and R. Boonstra. 1997. Population limitation in arctic ground squirrels: effects of food and predation. Journal of Animal Ecology **66**:527–541.

Järvinen, A. 1987. Microtine cycles and plant production: what is cause and effect? Oikos **49**:352–357.

Jedrzejewski, W., and B. Jedrzejewska. 1996. Rodent cycles in relation to biomass and productivity of ground vegetation and predation in the Palearctic. Acta Theriologica **41**:1–34.

Jedrzejewski, W., B. Jedrzejewska, and L. Szymura 1995. Weasel population response, home range, and predation on rodents in a deciduous forest in Poland. Ecology **76**:179–195.

Jedrzejewski, W., L. Rychlik, and B. Jedrzejewska. 1993. Responses of bank voles to odours of seven species of predators: experimental data and their relevance to natural predator-vole relationships. Oikos **68**:251–257.

Jensen, T. S. 1982. Seed production and outbreaks of non-cyclic rodent populations in deciduous forests. Oecologia **54**:184–192.

John, E., and R. Turkington. 1995. Herbaceous vegetation in the understorey of the boreal forest: does nutrient supply or snowshoe hare herbivory regulate species composition and abundance? Journal of Ecology **83**:581–590.

Johnson, C. N. 1996. Interactions between mammals and ectomycorrhizal fungi. Trends in Ecology and Evolution **11**:503–507.

Kalela, O. 1962. On the fluctuations in the numbers of arctic and boreal small rodents as a problem of production biology. Annales Academiae Scientiarum Fennicae Series A IV **66**:1–38.

Keith, L. B. 1990. Dynamics of snowshoe hare populations. Current Mammalogy **2**:119–195.

King, C. M. 1983. *Mustela erminea*. Mammalian Species **195**:1–8.

Korpimäki, E., and K. Norrdahl. 1991. Numerical and functional responses of kestrels, short-eared owls, and long-eared owls to vole densities. Ecology **72**:814–826.

Krebs, C. J., and R. Boonstra. 1984. Trappability estimates for mark-recapture data. Canadian Journal of Zoology **62**:2440–2444.

Krebs, C. J., S. Boutin, R. Boonstra, A. R. E. Sinclair, J. N. M. Smith, M. R. T. Dale, K. Martin, and R. Turkington. 1995. Impact of food and predation on the snowshoe hare cycle. Science **269**:1112–1115.

Krebs, C. J., B. S. Gilbert, S. Boutin, A. R. E. Sinclair, and J. N. M. Smith. 1986. Population biology of snowshoe hares. I. Demography of food-supplemented populations in the southern Yukon, 1976–84. Journal of Animal Ecology **55**:963–982.

Krebs, C. J., B. S. Gilbert, S. Boutin, and R. Boonstra. 1987. Estimation of snowshoe hare population density from turd transects. Canadian Journal of Zoology **65**:565–567.

Krebs, C. J., and I. Wingate. 1976. Small mammal communities of the Kluane region, Yukon Territory. Canadian Field-Naturalist **90**:379–389.

Krebs, C. J., and I. Wingate. 1985. Population fluctuations in the small mammals of the Kluane Region, Yukon Territory. Canadian Field-Naturalist **99**:51–61.

Laine, K., and H. Henttonen. 1983. The role of plant production in microtine cycles in northern Fennoscandia. Oikos **40**:407–418.

Lofgrën, O. 1995. Spatial organization of cyclic *Clethrionomys* females: occupancy of all available space at peak densities? Oikos **72**:29–35.

Manly, B. F. J. 1991. Randomization and Monte Carlo methods in biology. Chapman and Hall, London.

Martell, A. M., and W. A. Fuller. 1979. Comparative demography of *Clethrionomys rutilus* in taiga and tundra in the low Arctic. Canadian Journal of Zoology **57**:2106–2120.

Maser, C., J. M. Trappe, and R. A. Nussbaum. 1978. Fungal-small mammal interrelationships with emphasis on Oregon coniferous forests. Ecology **59**:799–809.

Menkens, G. E., Jr., and S. H. Anderson. 1988. Estimation of small-mammal population size. Ecology **69**:1952–1959.

Merritt, J. F., and J. M. Merritt. 1978. Population ecology and energy relationships of *Clethrionomys gapperi* in a Colorado subalpine forest. Journal of Mammalogy **59**:576–598.

Mihok, S., B. N. Turner, and S. L. Iverson. 1985. The characterization of vole population dynamics. Ecological Monographs **55**:399–420.

Nams, V. O., N. F. G. Folkard, and J. N. M. Smith. 1993. Effects of nitrogen fertilization on several woody and nonwoody boreal forest species. Canadian Journal of Botany **71**:93–97.

Neuvonen, S. 1988. Interactions between geometrid and microtine cycles in northern Fennoscandia. Oikos **51**:393–397.

Nilsson, I. 1978. The influence of *Dasychira pudibunda* (Lepidoptera) on plant nutrient transports and tree growth in a beech *Fagus sylvatica* forest in southern Sweden. Oikos **30**:133–148.

Oksanen, L., and L. Ericson. 1987. Dynamics of tundra and taiga populations of herbaceous plants in relation to the Tihomirov-Fretwell and Kalela-Tast hypotheses. Oikos **50**:381–388.

Pastor, J., B. Dewey, and D. P. Christian. 1996. Carbon and nutrient mineralization and fungal spore composition of fecal pellets from voles in Minnesota. Ecography **19**:52–61.

Pieper, R. D. 1964. Production and chemical composition of arctic tundra vegetation and their relation to the lemming cycle. PhD thesis. University of California, Berkeley.

Pucek, Z., W. Jedrzejewski, B. Jedrzejewska, and M. Pucek. 1993. Rodent population dynamics in a primeval decidous forest (Bialowieza National Park) in relation to climate, seed crop and predation. Acta Theriologica **38**:199–232.

Raatikainen, M., and M. Niemelä. 1994. The effect of fertilization on the yield of wild forest berries. Biological Research Reports (University of Jyväskylä). **38**:123–129.

Schultz, A. M. 1969. A study of an ecosystem: the arctic tundra. *in* G. M. van Dyne (ed). The ecosystem concept in natural resource management, pages 77–93. Academic Press, New York.

Schweiger, S., and S. Boutin. 1995. The effects of winter food addition on the population dynamics of *Clethrionomys rutilus*. Canadian Journal of Zoology **73**:419–426.

Smith, J. N. M., C. J. Krebs, A. R. E. Sinclair, and R. Boonstra. 1988. Population biology of snowshoe hares. II. Interactions with winter food plants. Journal of Animal Ecology **57**:269–286.

Stoddart, M. 1976. Effect of the odour of weasels (*Mustela nivalis* L.) on trapped samples of their prey. Oecologia **22**:439–441.

Swank, W. T., J. B. Waide, D. A. Crossley, Jr., and R. Todd. 1981. Insect defoliation enhances nitrate export from forest ecosystems. Oecologia **51**:297–299.

Tast, J. 1974. The food and feeding habits of the root vole, *Microtus oeconomus*, in Finnish Lapland. Aquilo Series Zoologica **15**:25–32.

Tast, J., and O. Kalela. 1971. Comparison between rodent cycles and plant production in Finnish Lapland. Annales Academiae Scientiarum Fennicae Series A IV **186**:1–14.

Turkington, R., C. J. Krebs, E. John, M. R. T. Dale, V. Nams, R. Boonstra, S. Boutin, K. Martin, A. R. E. Sinclair, and J. N. M. Smith. 1998. The effects of NPK fertilization for nine years on the vegetation of the boreal forest in northwestern Canada. Journal of Vegetation Science **9**:333–346.

Vickery, W. L., S. L. Iverson, S. Mihok, and B. Schwartz. 1989. Environmental variation and habitat separation among small mammals. Canadian Journal of Zoology **67**:8–13.

West, S. D. 1982. Dynamics of colonization and abundance in central Alaskan populations of the northern red-backed vole, *Clethrionomys rutilus*. Journal of Mammalogy **63**:128–143.

White, B. D., D. R. Anderson, K. P. Burnham, and D. L. Otis. Capture-recapture and removal methods for sampling closed populations. Los Alamos National Laboratory LA-8787-NERP, 235 pp.

Whitney, P. 1976. Population ecology of two sympatric species of subarctic microtine rodents. Ecological Monographs **46**:85–104.

Whitney, P., and D. Feist. 1984. Abundance and survival of *Clethrionomys rutilus* in relation to snow cover in a forested habitat near College, Alaska. Special Publication of the Carnegie Museum of Natural History **10**:113–120.

Wolff, J. O. 1996. Population fluctuations of mast-eating rodents are correlated with production of acorns. Journal of Mammalogy **77**:850–856.

Forest Grouse and Ptarmigan

KATHY MARTIN, CATHY DOYLE, SUSAN HANNON, & FRITZ MUELLER

11.1 The Ecological Role of Forest and Alpine Grouse

11.1.1 Trophic Position

The Kluane system harbors five species of grouse. Two species, spruce grouse (*Falcipennis canadensis*) and ruffed grouse (*Bonasa umbellus*), inhabit forest year-round. Three species of tundra grouse, willow ptarmigan (*Lagopus lagopus*), white-tailed ptarmigan (*L. leucurus*), and rock ptarmigan (*L. mutus*), inhabit the alpine and subalpine above treeline in the Shakwak Valley most of the year. Willow ptarmigan descend into the open forest from October to March. All grouse are almost exclusively herbivores; females during egg laying and young chicks also eat a variety of invertebrates. Spruce grouse are conifer specialists, foraging principally on spruce and pine in winter and on forbs and leaves, fruits, and seeds of shrubs in summer and autumn (Boag and Schroeder 1992). Spruce grouse spend significant amounts of time foraging on ground vegetation in summer and thus might compete with hares for food during the breeding season. Ruffed grouse feed principally on buds and leaves of deciduous trees (aspen, bog birch) and shrubs and forbs seasonally (Rusch et al. 2000). Ptarmigan feed on willow buds and leaves throughout the year and on flowers, leaves, seeds and berries of sedges, ericaceous shrubs, and mosses in season (Braun et al. 1993, Hannon et al. 1998). Both forest grouse and ptarmigan form part of the herbivore trophic level year-round in Kluane, as their wide-ranging predators (goshawk, golden eagle, harrier, coyote, lynx, fox, wolverine, wolf) hunt in both forest and adjacent alpine areas.

11.1.2 Life History

Forest grouse have a promiscuous mating system and are assumed to have an equal sex ratio (Johnsgard 1983, Boag and Schroeder 1992). Breeding densities for the *F. canadensis* race of spruce grouse range from <0.1 to 83 birds/km^2 (Boag and Schroeder 1987, 1992). Given their dispersal and migration behavior, a year-round study of spruce grouse will consist of year-round residents, breeding birds, winter residents, and some birds encountered only during spring and autumn migration (Schroeder 1985). Spruce grouse tend to be relatively long lived compared to alpine grouse, with above-average nesting success for ground-nesting birds (Ricklefs 1969). Annual mortality averaged 30.1% (range 22–49%), with the longevity record being 13 years in natural populations (Boag and Schroeder 1992).

Ruffed grouse, a close relative of spruce grouse (Ellsworth et al. 1995), are more sedentary and extend farther south into the northern United States in the Appalachian, Cascades, and Rocky Mountain cordilleras than do spruce grouse. Male ruffed grouse have traditional drumming display stations that make males easy to count. Females of both species are difficult to census because they range over large areas in spring and autumn and have secretive behavior most of the year. Both spruce grouse and ruffed grouse choose nest sites in willow or slash, often at the base of trees in deciduous or mixed forests. Overbrowsing by hares at the peak might reduce nesting cover for forest grouse. Hatching success (proportion of nests hatching at least one chick) for *F. canadensis* populations ranges from 40% to 81% in southern Canada and the northern United States, and 81% in Alaska.

Ptarmigan are monogamous and territorial and breed in open tundra habitats. Ptarmigan are unique in the avian world because they change their plumage from fully white in winter to a cryptic brownish and reddish grey plumage in other remaining seasons. Male willow ptarmigan participate in brood rearing, whereas white-tailed and rock ptarmigan females supply all of the parental care, as is typical for most grouse species. Annual survival of willow ptarmigan ranges from 30% to 64% and varies with sex and age class (Hannon et al. 1998). A long-term population dynamics study of willow ptarmigan conducted in the Chilkat Pass, about 200 km south of Kluane, showed that densities of ptarmigan were strongly correlated with the snowshoe hare cycle (Boutin et al. 1995, Hannon and Martin 1996).

11.1.3   Habitats

Spruce grouse range across the northern boreal coniferous forest from Alaska to Labrador and southward into New England and the northern United States (Boag and Schroeder 1992). Highest spruce grouse densities in North America are in 15- to 25-year-old pine plantations in central Ontario (Boag and Schroeder 1992). Spruce grouse choose microhabitats within forests that vary between the sexes and across seasons. In spring, males choose territorial sites in mature or young forests with sparse canopy cover and openings on the ground for display. Breeding females use areas where food is most available such as open wet and dry meadows with interspersed trees and a well-developed shrub and herb layer. Hens with broods may use forest with much deadfall. The selection of microhabitats for both sexes appears to be a compromise between food acquisition and predator avoidance. Ruffed grouse range throughout northern North America from temperate coniferous rainforest to relatively arid deciduous forest types in both young and mature stands (Johnsgard 1983). The unifying criterion is that all forest stages include deciduous trees, especially *Betula* or *Populus* species.

A winter study (1989–1992) showed that spruce grouse at Kluane feed exclusively on needles of white spruce (*Picea glauca*), and certain individual trees were fed on extensively (Mueller 1993). Captive spruce grouse had a strong preference for young needles from "feeding trees" over control trees of similar age, size, and location. Feeding trees had lower concentrations of two monoterpene antifeedants, camphor and bornyl acetate, and lower resin than control trees. Feeding trees showed increased growth of lateral twigs, longer needles, and decreased cone production.

Willow and rock ptarmigan have a holarctic distribution that includes arctic, subarctic, and subalpine regions (Hannon et al. 1998), while white-tailed ptarmigan are restricted to alpine regions of western North America from Alaska to New Mexico (Braun et al. 1993). The ptarmigan species in the southwestern Yukon are sympatric in a regional sense, but are stratified altitudinally (Weeden 1963, 1964). In the breeding season, willow ptarmigan occur in the subalpine with clumps of *Salix* and *Betula* (0.2–2.5 m high) interspersed with sedge and grass hummocks or deciduous shrubs. In winter, they descend into open forest with more cover than the subalpine but still with exposed forage (Weeden 1964, Gruys 1993, Mossop 1994). White-tailed ptarmigan remain in the alpine most of the year and descend in winter to willow-dominated basins with considerable snow accumulation (Braun et al. 1993).

11.1.4   Current Understanding of Population Dynamics

For all grouse species occurring in Kluane, there are long-term population studies, ranging from 10 to 30 years, in multiple locations in North America that provide detailed information on life history, population dynamics, and behavior (Mercer 1967, Keppie 1982, Keith and Rusch 1986, Bergerud and Gratson 1988, Martin et al. 1989, Boag and Schroeder 1987, 1992, Braun et al. 1993, Hannon and Martin 1996, Rusch et al. 2000). In general, data for ruffed grouse support top-down regulation of densities by predators that switch from declining hare or vole populations to grouse (Keith and Rusch 1986). Ruffed grouse show 10-year cycles with peaks and declines 1–2 years before hares in the northern part of their range (Keith and Rusch 1986). Population density and recruitment of other forest grouse decreases with increasing age of the forest (Bendell and Elliott 1967, Zwickel and Bendell 1972, 1985). Willow ptarmigan numbers fluctuate on 8- to 11-year cycles in North America (Mossop 1994, Hannon et al. 1998), with populations declining in synchrony with snowshoe hare cycles where these species co-exist (Keith 1963, Boutin et al. 1995). Rock ptarmigan in Iceland function as keystone herbivores, precipitating both functional and numerical responses in gyrfalcon populations (Nielsen 1999).

Despite numerous experiments and long-term correlative studies, there is no overall agreement on mechanisms of population change in grouse. Angelstam et al. (1985), Hannon (1986), and Bergerud (1988) summarized hypotheses proposed to explain population changes in grouse. Regulation may occur through (1) predation or starvation in winter, (2) territorial or spacing behavior in spring or autumn, or (3) variation in chick production during the breeding season. Poor chick production might be caused by poor maternal quality related to food supply or by heavy parasite loads. Declines in ruffed grouse are associated with high predation on juvenile grouse in summer in Alberta (Keith and Rusch 1986), or on all birds in autumn and winter in Wisconsin (Small et al. 1991). In northern Fennoscandia, where forest and tundra grouse show synchronous 4-year cycles with voles and their predators (Angelstam et al. 1985), it is hypothesized that generalist predators such as mustelids switch to grouse nests when vole numbers decline. In summary, forest and tundra grouse are considered an important component of population cycles in southern Alberta, Iceland, United Kingdom, and Fennoscandia. They have functioned as the keystone herbivore in Iceland, but more commonly as the alternative herbivore. The timing of the declines and the life-history stage showing the most dramatic impact (survival of adults or juveniles in spring, summer, or winter, or in nesting success) differed across these studies.

11.1.5   Predicted Responses to Experimental Treatments

Given previous studies, we expected all grouse in the Yukon to decline in synchrony with the hares. We designed the Kluane grouse censusing to describe numerical and reproductive responses of forest and alpine grouse to the hare cycle and to examine potential mechanisms for population change. We tested whether forest and alpine grouse numbers were correlated with changes in hare densities. Given other studies, we expected to see increased predation on grouse by great horned owls, goshawks, lynx, and coyotes as hares declined. Because ptarmigan live above the forest most of the year and thus vary in relative availability seasonally, we expected them to be a seasonal component of the boreal forest trophic structure. Their contribution to prey biomass in winter might improve survival of resident predators and might also improve survival of other forest herbivores through a buffering effect (reduction of predation risk).

We did not expect to see strong responses to our experimental treatments when it became clear that forest grouse were patchily distributed and at much lower density than hares on our plots. Thus, we present a limited number of comparisons among the treatment plots, and given the small number of grouse per plot we make only qualitative inferences for two secondary predictions. First, if predators regulate grouse populations, we expected spruce grouse numbers to remain stable on predator exclosure plots. Second, if hares and grouse have an overlap in summer food supply, then hares might compete with grouse by changing food supplies for grouse due to overbrowsing at the peak. If so, spruce grouse should decline less on the fertilizer addition plots and decline more on food addition treatments after hares severely overbrowse the natural vegetation.

11.2 Methods

11.2.1 Duration of Grouse Studies

Data on population trends for forest and alpine grouse at a valley-wide scale are available for the duration of the Kluane study. During the 1990–1997 breeding seasons, we conducted specific studies on spruce grouse on six plots, with less emphasis on ruffed grouse. There are data on spruce grouse abundance for two of these plots in 1987 and 1988. Personnel recorded grouse mortalities encountered during field work. A population study of willow ptarmigan in Chilkat Pass, 200 km southeast of Kluane, from 1979 to 1992 provided concurrent population data (Hannon and Gruys 1987, Boutin et al. 1995, Hannon and Martin 1996).

11.2.2 Shakwak Valley Population Trends

The Kluane Seen-Sheet Data   Field staff recorded all encounters of 25 avian and mammalian species (including spruce grouse, ruffed grouse, and ptarmigan), the hours afield and mode of transportation used (foot, snowmobile) for the extensive Kluane study area, (Hochachka et al. 2000). A visual detection during daylight hours was required to score a sighting. We recorded trips where individuals saw no animals. Radio-tagged individuals were not included unless encountered accidentally. Observers were trained in species identification. Data were collected year-round from May 1988 through April 1996 by 12–45 observers annually (total of 212 observers and 98,058 field h of observations during the study). The Kluane seen-sheet data is a coarse-grained data set that provides information on relative abundance/activity on multispecies population trends in the valley.

11.2.3 Numerical and Reproductive Parameters on Treatment Grids

We measured abundance and reproductive parameters of forest grouse during breeding season on hare grids in response to the experimental manipulations using dawn census data and grouse search and capture data from 1990 to 1997 (CD-ROM frame 30).

Dawn Census In 1990 we began counts of total breeding densities of spruce grouse in relation to experimental treatments over a 10-day period in late April to early May. Counts of breeding displays of both male and female spruce grouse can be used to estimate populations in spring (Herzog and Boag 1977, Schroeder and Boag 1989, Keppie 1992). Females give a cantus call considered to have a spacing or territorial function during prelaying and laying periods (Nugent and Boag 1982). This call broadcasts up to 500 m, and other females answer and may approach the call, occasionally to within meters. Females are most responsive to other females calling around dawn. Males also approach and perform flutter flights (nonvocal displays; Boag and Schroeder 1992) in response to these female calls. We used standardized playback techniques for six grids (control 1, control 4, fertilizer 1, food addition 1, predator exclosure, predator exclosure + food) from 0430 h to 0645 h (about 75 min before sunrise to about 60 min after sunrise). In 1991 and 1993, we tried playback calls at intervals from early April through late May to determine the optimal timing for census.

On the G and N transect lines (CD-ROM frame 24) on each 36-ha grid, we played calls at a total of seven stations per transect, each station separated by 100 m. At each playback station, we first listened for 30 sec, then played two cantus calls (15 sec) twice, each followed by a 30-sec period to listen for responses (total of 2 min/station). Tape recorder volume was adjusted to broadcast about 200 m. We recorded all responses by females or males and the time and the approximate locations of each, counting only those birds on the plot. Individuals recorded at multiple stations or between lines were only included once in each survey-morning. We also recorded male ruffed grouse heard during the dawn census. Because the prime response period was only about 45–50 min, and responses dropped off quickly after sunrise, each observer was able to conduct a maximum of two transects (one grid) per morning, which took about 75–90 min. Each plot was censused three times (rounds) per season, and the population size was calculated as double the number of the maximum grouse of either sex counted during one of the surveys. There were two main observers (K.M. from 1990 to 1997 and C.D. from 1990 to 1996), and a third individual differed each year. Observers and starting points were rotated for each round. The total of 18 surveys was completed in 6–9 mornings. We did not sample on days with falling snow or moderate to high winds.

Grouse Search and Capture Two to three observers searched for grouse on six grids after the dawn census at intervals from 0630h to 2200 h to monitor population changes across the hare cycle on six treatment plots in spring. Radio-marked birds were included only when encountered accidentally. During each search we used a dog (1987–1994), and occasionally we used female cantus calls. We did not play calls on plots we planned to survey the next morning. Grouse search data are available for fertilizer 1 and control 1 in 1987 and 1988. We recorded all ruffed grouse encountered. During grouse search periods, we captured and banded all spruce grouse located (41 males, 31 of the 46 females banded were also radio tagged), and we measured local survival, body condition, and reproductive parameters such as clutch size, nesting success, and chick survival.

11.2.4 Data Robustness and Limitations

The main field observers and dog contributing to our year-round and spring abundance estimates of grouse populations over the hare cycle worked throughout much of the Kluane study. Thus, we have the advantage of continuity, normally difficult to achieve across a long-term study. The seen sheet, grouse search and capture, and dawn census use different spatial scales, observers, and methods to estimate abundance and thus are independent data sets to examine population trends. The concordance across these independent data sets suggests that our general demographic trends are robust. The radio tagging provided data on reproductive parameters over the hare cycle, but the sample is insufficient to assess responses to treatments. Spruce grouse, the most abundant grouse accounting for the highest biomass in Kluane, is the only species for which we can assess responses to the experimental treatments, and here power to detect responses in abundance or reproductive parameters is low, given their low density and patchy distribution.

We have less confidence in our estimates of ptarmigan densities over the hare cycle because we expended limited census effort on them. In winter, ptarmigan occur in flocks, and the seen-sheet data are highly skewed (zero observations for many observer-days) that conform to neither normal nor Poisson distributions (Hochachka et al. 2000). We conducted aerial censusing from 1990 to 1996 to measure ptarmigan breeding densities, but it was difficult to achieve comparable count conditions across years given annual variation in snow melt, weather, and flight conditions (birds are harder to see when snow melt is advanced). An MSc student, Luc Pelletier, conducted replicate sets of aerial spring surveys in 1995 and 1996 and compared these with ground counts for ptarmigan. The aerial census did not provide reliable indices for comparing ptarmigan abundance across years (Pelletier 1996, Pelletier and Krebs 1997) and are thus not presented here.

11.3 Demography of Grouse

11.3.1 Population Trends in the Shakwak Valley

The Kluane seen-sheet data indicate that densities of spruce grouse (both breeding and winter) increased from 1987 to 1988, peaked in the summer of 1989 and the winter of 1989–1990, began to decline during the 1990 breeding season, declined markedly in the 1990–1991 winter, and bottomed out in the 1991 breeding season (figure 11.1). Densities remained low through 1993. In 1994, numbers increased and continued to increase through the end of data collection in April 1996 (figure 11.1). Summer densities of spruce grouse started declining again in 1996 (figure 11.2), but spruce grouse numbers appeared to remain high over the 1997–1998 winter (E. Hofer, personal communication).

The start of the grouse decline was in late winter or early spring of 1990. Data available from the seen-sheet observations show encounters of spruce grouse greatly decreased early in 1990. In November and December 1989, we observed 10.9 grouse/100 h ($n = 830$ hours afield) compared with only 1.9 grouse/100 h ($n = 859$ h) for January and February of 1990. The following year, 1.9 grouse/100 h were seen in November and December 1990 ($n = 859$ h), compared with 1.4 grouse/100 h ($n = 1071$ h) in January and February 1991. The spruce grouse decline appears to be caused by elevated predation on adult grouse. Sixteen grouse mortalities were recorded in the spring of 1990 (when data collection began), and 16 mortalities over the autumn, winter, and spring of 1990–1991. For the same period in 1992, we recorded only two dead grouse. During the spring of 1991, we located four recently killed grouse and only five live birds during 60 h of grouse search (table 11.1).

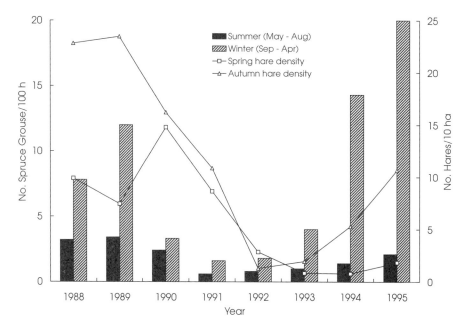

Figure 11.1 Average number of spruce grouse seen by multiple observers on foot in the Shakwak Valley during the breeding season (summer) and the nonbreeding season (winter). Summer 1988 is May–August 1988; winter 1988 is September 1988 through April 1989, and so on. The lines indicate spring and autumn hare densities.

Ruffed grouse at Kluane were confined to patchily distributed aspen (*Populus tremuloides*)-dominated stands and thus occurred at lower overall densities than spruce grouse. Ruffed grouse were found mostly in larger aspen stands, but they occurred also in smaller deciduous stands when population densities were high. Ruffed grouse accounted for 9% of forest grouse detections during grouse search. Ruffed grouse declined and remained low in synchrony with spruce grouse (figure 11.2). Virtually no ruffed grouse were heard in the Shakwak Valley by project personnel from 1991 to 1993 (F. I. Doyle, personal communication).

Ptarmigan were present in the valley only in winter. Seen-sheet data indicate that willow ptarmigan numbers at Kluane peaked in the winter of 1989–1990 and started declining the next winter simultaneous with declines in hare winter densities ($r = .76$, 8 years; figure 11.3). Long-term field studies in the Chilkat Pass showed numbers of willow ptarmigan pairs peaked in 1980–1981 and again in 1990; in both cases, declines began synchronously with changes in hare densities ($r = .84$, $n = 12$ years, $p < .001$; Boutin et al. 1995).

11.3.2 Reproductive Parameters

To address mechanisms relating to the predation and food competition hypotheses, we measured several survival, body condition, and reproductive variables. Birds banded in

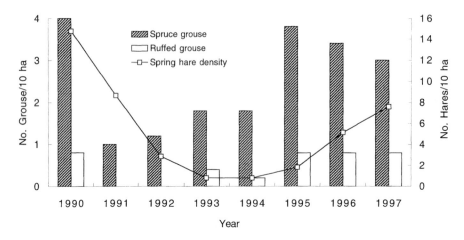

Figure 11.2 Total numbers of spruce and ruffed grouse estimated from the dawn census using playbacks of female cantus calls in late April to early May on six treatment plots (36 ha each, total = 216 ha) in relation to snowshoe hare spring densities (lines).

our study ranged from 1 to 6 years of age. Local survival, as measured by birds returning to the study plots in subsequent years, varied significantly across the cycle from a low of 27% in the hare decline years to a high of 54% at the hare low (table 11.2). As survival of radio-tagged birds during the breeding season remained high throughout the study (80% of 31 females), it appears that spruce grouse declined due to predation over the winter during the hare decline years of 1990 and 1991. At the hare low, local survival of the few remaining birds was high (table 11.2). The proportion of clutches that hatched chicks and the proportion of hatched chicks that survived to late summer remained normal to high for spruce grouse across the study, but data are limited (table 11.2). Although production of independent offspring appeared normal to high during the hare low, we recorded limited recruitment to the breeding population in subsequent years (table 11.1), suggesting that young produced during the hare low did not survive well over winter.

Body condition and reproductive effort indices run counter to the food competition hypotheses as body mass for females was lowest during the hare decline and highest during the hare low ($F = 3.33, p = .03, n = 35$ females; table 11.2). In addition, clutch size increased through the decline and hare low ($F = 5.82, p = .004$; table 11.2). Body mass of male spruce grouse did not vary across the hare cycle ($F = 1.36, p = .27$; table 11.2).

11.3.3 Population Trends on Control Plots

On the control plots, forest grouse varied from 0.28 to 3.33 birds/10 ha over the hare cycle. Spruce grouse varied from 0.28 to 2.5 birds/10 ha, and ruffed grouse varied from 0 to 2.2 birds/10ha on control 4 (figure 11.4). The lowest densities were recorded in 1991, two years before the hares reached their lowest spring densities (figure 11.4). Recovery started in 1992 on control 1, but not until 1993 for control 4. The temporal trends on control 4 may be somewhat complicated by the presence of two grouse species. Spruce grouse

Table 11.1 Numbers of spruce grouse (SG) and ruffed grouse (RG) encountered on grouse search in Kluane.

		Grouse Located Alive			Spruce grouse		Spruce Grouse Recruitment		
Year	Total Search Time (h)	Total Grouse per h	No. RG	No. SG	No. Found Dead	% Dead of Total Located	Total No. Individuals Caught	No. Returns (Banded)	No. New Recruits
1990	52.2	0.38	3	17	3	15	11	3	8
1991	60.6	0.08	0	5	4	44	5	2	3
1992	48.8	0.15	0	7	2	22	5	2	3
1993	57.1	0.19	0	11	0	0	8	4	4
1994	84.0	0.21	3	15	1	6	9	3	6
1995	68.3	0.50	4	30	0	0	18	6	12
1996	64.7	0.42	4	23	0	0	13	9	4
1997	89.5	0.23	6	15	1	6	9	3	6

Total search time is the number of hours searched across 4–6 plots annually and calculated by tallying total person hours in the field. A dog was used from 1990 to 1994. Time required to capture and band spruce grouse was subtracted from the total search time. Grouse/h includes total number of grouse (SG + RG) encountered/total search hours.

numbers were greatest in 1993 on control 4, and thereafter remained at or below 1990 levels, but ruffed grouse increased from none in 1991 and 1992 to eight birds in 1997 (figure 11.4). Habitat suitability for each species and sex varied between control plots. Control 1 had a substantial amount of suitable habitat for both male and female spruce grouse, but not for ruffed grouse. Control 4 was basically the reverse, with a large area of aspen in the center comprising the most suitable ruffed grouse habitat of all the census plots, while the periphery had suitable coniferous habitat for female spruce grouse. Ruffed grouse were present only on control 4 grid in 1990 (the first year of dawn census); they were not encountered for 2 years, reappeared in 1993, increased until 1995, and remained high through 1997 (figure 11.4). We recorded one ruffed grouse male on fertilizer 1 in 1995 and 1996 but none in 1997.

11.3.4   Relationship of Changes to the Snowshoe Hare Cycle

Across the valley, spruce grouse breeding densities peaked in 1989 and the winter of 1989–1990 (figure 11.1), about one season ahead of the hare peak in the winter of 1989–1990 and peak hare breeding densities in 1990 (chapter 8). Using breeding season counts for both grouse and hares, serial correlations were strongest for spruce grouse declining 1 year ahead of the hares ($r = .89$, 1988–1995) using seen-sheet data, and also when using dawn census data for spruce grouse ($r = .87$, 1990–1997) or for all forest grouse ($r = .86$, 1990–1997). Using seen-sheet data, correlations of spring grouse populations with autumn hare densities were similar for both the preceding year where $r = .90$ (i.e., lag = -1), and for the subsequent winter ($r = .89$, lag = -2, 1988–1996). Using dawn census data, the strongest correlations of summer grouse numbers were with autumn hare densities in the preceding winter for spruce grouse ($r = .82$) and for all forest grouse ($r = .82$,

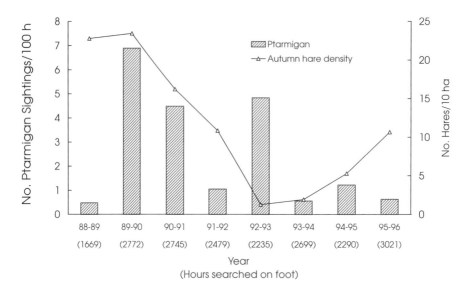

Figure 11.3 Numbers of ptarmigan encountered per 100 h afield from October to March by multiple observers (seen-sheet observation data) in the Shakwak Valley. Total observation hours for each winter are included in parentheses under each year. Autumn hare densities presented as lines.

1989–1997). Thus, forest grouse began to decline one season in advance of the hares, their numbers collapsed quickly, and they began their recovery while hares were still declining. Willow ptarmigan started declining in the winter of 1990–1991, 1 year after the forest grouse decline (figure 11.3). The seen-sheet data counts suggest that ptarmigan declined with a 1-year lag with autumn hare densities ($r = .76$, 8 years) or a 2-year lag with spring hare densities ($r = .76$, 8 years). The trends in Kluane ptarmigan counts accord well with patterns in the Chilkat Pass, where ptarmigan densities peaked in 1990, and were still declining in 1992 (Boutin et al. 1995).

11.4   Response of Forest Grouse to Experimental Treatments

Although the size of the manipulated areas for fertilizer addition, food addition 1, predator exclosure, and food addition + predator exclosure was 1 km², for consistency, we restricted our grouse census to a 36-ha hare grid area. Here we discuss whether numbers, encounter rates, or timing changed on experimental treatments relative to controls and within plots across the cycle.

11.4.1   Fertilizer Addition

For spruce grouse, the quantity of forage (spruce needles, ericaceous shrubs) was enhanced by the fertilizer treatment (chapters 5 and 6). On fertilizer 1, spruce grouse declined to a low in 1991, numbers increased in 1992, and by 1993 the plot had recovered

Table 11.2 Body size, reproductive parameters, and annual survival for spruce grouse across the snowshoe hare cycle stages at Kluane.

	Stage of hare cycle											
	Peak (1987–90)			Decline (1991–92)			Low (1993–94)			Increase (1995–97)		
Parameter	Mean	SD	N	Mean	SD	N	Mean	SD	N	Mean	SD	N
Morphometrics and Condition[a]												
Males[b]												
Body mass (g)	570	50	17	620	7	2	601	31	5	586	38	11
Wing chord (mm)	187	4	17	194	1	3	193	3	5	193	2	11
Females[c]												
Body mass (g)	626	39	10	571	58	4	649	26	6	637	43	15
Wing chord (mm)	186	5	10	189	2	4	190	3	6	190	4	15
Survival (%)[d]	—		—	27		15	54		13	45		40
Reproductive Success												
Clutch size[e]	7.6	1.3	5	8.5	0.7	2	9.8	0.8	5	8.7	0.6	14
N = nests												
% nests hatched	80			50			100			79		
Chicks fledged (%)	50		2	44		1	76		2	57		7
N = broods												

[a]Body mass and wing chord measurements were taken from birds captured between 25 April and 15 May 1987–1997.
[b]Male metrics over hare cycle: body mass: $F = 1.36, p = .27$; wing chord: $F = 8.61, p < .001$.
[c]Female metrics over hare cycle: body mass: $F = 3.33, p = .03$; wing chord: $F = 2.74, p < .06$.
[d]Survival is local survival as measured by the return in a subsequent year of birds known to be alive in spring on treatment grids, 1990–1997. Both sexes pooled.
[e]Clutch size across the hare cycle: $F = 5.82, p = .004$. Clutch size did not differ for adult (mean $= 8.73, n = 15$), and yearling hens (mean $= 8.78$ eggs, $n = 9$) pooled across the study.

to peak densities and remained high through 1997 (figure 11.4). This recovery was 2 years earlier than on the other treatment plots and faster than on control 1. The data are consistent with the fertilization treatment resulting in an earlier recovery of spruce grouse from the low than control 4 and more rapidly than control 1.

11.4.2 Food Addition

It is not clear what, if anything, was manipulated for spruce grouse on this plot. We did not observe spruce grouse eating the rabbit chow. The food-addition plot had average grouse densities in 1990, after which no birds were detected for 3 years and only two were found in 1994 (figure 11.4). The habitat appeared suitable for spruce grouse, but numbers did not recover to 1990 levels by 1997. Spruce grouse may compete with ground squirrels and hares for herbaceous shrubs, forbs, and grasses in spring. Spruce grouse may have experienced increased predation risk because more grouse predators were observed on food 1 during the low and increase phases.

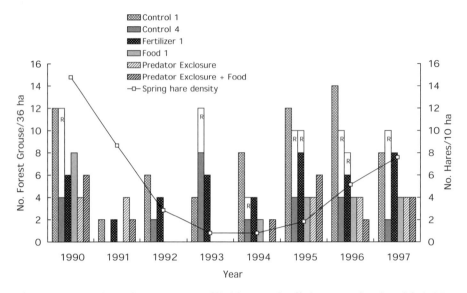

Figure 11.4 Numbers of spruce grouse (filled bars) and ruffed grouse (clear bars labeled R) estimated on control and treatment plots (36 ha each) from the dawn census using playbacks of female cantus calls in late April to early May, 1990–1997, across the hare cycle (spring hare densities as lines). Population size was estimated by doubling the count of the maximum number of birds of one sex (either sex) on the grid in one of three survey mornings.

11.4.3   Predator Exclosure

Risk from mammalian predators was largely removed and the monofilament cover over part of the plot reduced the risk from avian predators, but this would have affected only a few spruce grouse individuals when they were on or close to the ground. Although the predator exclosure had the lowest abundance of grouse in 1990, it was the only plot that remained similar in density in 1991 (figure 11.4). For the next 3 years, we found no spruce grouse on this plot. Grouse reoccupied the plot in 1995, and densities remained at 1990 levels for the next 3 years. The predator exclosure treatment possibly delayed the spruce grouse decline by 1 year. In 1991 we encountered only five spruce grouse in 61 h of search across all plots, and the only female we located was under the monofilament in the predator exclosure plot. Positive effects of the monofilament might have been offset by two pairs of great horned owls nesting near the area (see chapter 15).

11.4.4   Predator Exclosure + Food

Numbers of spruce grouse declined from 1990 to none or two birds for the next 4 years (figure 11.4). Numbers recovered by 1995, but decreased subsequently. Hares remained at or above control peak levels until 1993, and ground squirrels remained high through 1996 (Krebs et al. 1995; see figure 9.10). Although risk from mammalian predators was removed, extensive overbrowsing by hares reduced ground cover in some sections such that risk from avian predators might have increased. In addition, grouse may have continued to experience high predation risk here if avian predators concentrated activities on this plot. Natural vegetation was strongly affected, despite food addition, as both hares and ground squirrels were maintained at high densities for 4 years on this plot (Krebs et al. 1995; see chapters 8 and 9). Thus, spruce grouse possibly competed with ground squirrels and hares for the remaining herbaceous shrubs and forbs in spring.

11.5.   Discussion

11.5.1   Hypotheses Related to Grouse Population Trends

Our results suggest that densities of all grouse in the valley were strongly influenced by the hare cycle. We found the strongest support for the predator regulation hypothesis because grouse body condition and fecundity measures did not decrease across the hare decline. Grouse appeared to decline due to harvest bycatch by predators when hares were still abundant. The goshawk was the only predator observed switching to grouse during the hare decline (see chapter 16). At the low of the hare cycle, avian predators might switch during the winter to young-of-the-year grouse. This switch could explain the slow recovery of grouse, despite good summer production. Thus, grouse may require mainly a reduction in predation to permit density increase. We have little evidence for bottom-up regulation, but because little direct work was done on grouse foraging and competition with other herbivores, we incur the usual risk of biological error associated with nonequivalent testing of alternative hypotheses.

11.5.2   Did the Experiments Change the Dynamics
for Grouse?

The major treatments failed to change grouse population dynamics. However, our census grids were small given the spatial scale and patchy distribution of forest grouse. We observed a possible delay of 1 year in decline of spruce grouse on the predator exclosure (with monofilament) plot. Fertilizer 1 had the least severe decline, the most rapid recovery of grouse, and postfertilization numbers exceeded 1990 levels (figure 11.4). Given these qualitative patterns, the simplest model to predict density changes for grouse is a top-down predator control model that primarily influences survival of adult grouse just before the peak of hare populations.

11.5.3   Linkages on the Same Trophic Level

Except for females with broods, spruce grouse feed primarily on spruce needles and ruffed grouse on aspen, and competition with other major herbivores is not expected to be strong, especially since grouse can access the upper parts of trees that are unavailable to hares. Hares may compete with ptarmigan in winter as both species rely on willow, but we expect the magnitude of competitive interactions to be small because ptarmigan select more open areas than hares and willow is abundant.

The relative contribution of forest grouse to the overall herbivore biomass in the Kluane vertebrate community varied from <1% at the peak to 5–6% at the low (see figure

1.2). Hare demography just before the hare peak had a large impact, via predators, on the density of forest grouse in summer and likely all grouse in winter.

11.5.4 Linkages to Other Trophic Levels

Trophic Level Above   In the Kluane ecosystem, most predator species can integrate over a spatial scale from valley bottom to the alpine and thus can prey on grouse year-round. Most avian and mammalian predators present in Kluane were recorded feeding on adult or juvenile grouse (see figure 2.8). However, the magnitude of trophic interactions between grouse and their predators was asymmetrical. Predators usually have a large impact on population dynamics of most grouse (Braun et al. 1993, Hannon et al. 1998), and this appeared to be the case in Kluane. Predation on adult grouse in winter and early spring appeared to be responsible for the crash in grouse populations. Many grouse mortalities were found in 1990 and 1991, but no one predator species appeared to be responsible for the grouse declines.

We have no compelling evidence that grouse had a large impact on the population dynamics of any predator species, except during the low. Grouse were a minor diet item for most predators examined; they comprised 1% of 208 kills by coyotes and 2.5% of 572 kills by lynx, and furthermore 56% of 16 recorded grouse kills occurred when hares were still abundant (see chapter 14). Over the study, data on prey remains and pellets at raptor nests indicated that grouse composed 20% of diet items for goshawks and less for northern harriers, red-tailed hawks, and great horned owls (see chapters 15, 16). During the low, red squirrels were the most important alternative prey, but great horned owls, harriers, red-tailed hawks, and goshawks all preyed on grouse during this period. Thus, grouse may be most important to avian predators when both they and hares are at low densities, as they contributed to maintaining the few remaining individuals of several predator species. At the hare low, grouse comprised about 80% of the diet for the only breeding pair of goshawks left on the study (chapter 16).

Trophic Level Below   All grouse are primarily herbivorous. Spruce grouse foraging on new needles can have a significant impact on individual spruce trees and occasionally on patches of trees (Mueller 1993). The impact of spruce grouse herbivory on spruce trees was quite localized at wintering sites and may have been mitigated by feces deposition in this nutrient-limited system. During winter, ptarmigan flocks move irregularly over distances of up to 50 km (Hannon et al. 1998) and thus local impact on willow by ptarmigan is likely minimal.

Importance of Interactions for the Community   The association of ruffed grouse with aspen and spruce grouse with conifer may result in a rather even availability of grouse biomass to boreal forest predators. Because coniferous forest is predominant in Kluane, spruce grouse contributed about 85% of forest grouse biomass in summer (figure 11.2; 91% of grouse located/search hour). In winter, both species occur in flocks, and given their similarities in body mass and apparent vulnerability to predation at Kluane, spruce and ruffed grouse may be interchangeable prey for most mid-sized and larger predators. Ptarmigan augment the prey base year-round for predators because they winter in open willow areas at lower elevations and because goshawks and harriers hunt for them at high elevation in summer. The overlap of the forest and tundra grouse may increase the reliability of the winter food supply to predators. Winter concentrations of grouse may be vital to maintaining low densities of several avian predator species during the hare low.

11.6   Conclusions

11.6.1   Comparison with Other Studies

The timing of the forest grouse and hare peaks and declines at Kluane agree with previously documented fluctuations of ruffed grouse in southern populations where grouse decline in abundance 1 year ahead of hare declines (Keith and Rusch 1986). Spruce grouse and snowshoe hares are sympatric over much of their ranges, but despite several long-term grouse population studies, the Kluane study is the first to record a strong cycle in numbers of breeding and wintering spruce grouse in synchrony with hare densities. Annual harvest data in the Yukon support the hypothesis of spruce grouse cycling with hare densities. Peak numbers of spruce grouse were harvested in the autumn of 1989 (in agreement with peak abundance data for Kluane) and 11 years earlier in 1978, and lows were recorded in 1982 and 1992 (last year reported; Mossop 1994). Cyclicity has also been observed in Alaskan spruce grouse (D. Mossop, personal communication). In northern Ontario, where snowshoe hares cycle, and elsewhere, there is little evidence of spruce grouse and hare densities changing in synchrony (J. F. Bendell and D. M. Keppie personal communications). Other spruce grouse populations studied do not coexist with cyclic populations of snowshoe hares (Fritz 1979, Keppie 1987, Boag and Schroeder 1992).

Reproductive potential of Kluane spruce grouse females was exceptional. The mean clutch size of 8.7 + 0.2 S.E. eggs ($n = 26$ clutches, range 7–11 eggs) for Kluane spruce grouse was 3.1 eggs more than previously reported for the species (Keppie 1982). Furthermore, clutch size varied over the study with the largest clutches being laid during the low. Other researchers have not recorded density-dependent effects of clutch size for grouse or ptarmigan (D. M. Keppie unpublished data for three other spruce grouse populations in Canada; Hannon and Martin 1996).

Ptarmigan normally cycle with periods between peaks of 8–11 years in North America (Hannon et al. 1998), 4–5 years in England, 4–8 years in Scotland for red grouse (Jenkins et al. 1963), 3–4 years for willow grouse in Norway (Myrberget 1984), and 10 years for rock ptarmigan in Iceland (Nielsen 1999). Angelstam et al. (1985), Hannon (1986), and Bergerud (1988) summarized hypotheses to explain population changes in grouse. Regulation may occur through predation or starvation in winter, by territorial or spacing behavior in spring or autumn, by variation in chick production, or by internal parasite loads during the breeding season (Hudson 1992, Hudson et al. 1998, but see Moss et al. 1996). In most studies, alternative hypotheses were not tested simultaneously. Although there is no overall agreement on mechanisms of population change despite numerous experiments and long-term correlative studies, data for unmanaged grouse with natural population densities tend to support top-down regulation by predators often mediated by densities of keystone herbivores. The keystone herbivores are microtine mammals in Fennoscandia (Angelstam et al. 1985) and hares in the Yukon (this study), Alberta, and Wisconsin (Keith 1963, 1990). In Iceland, rock ptarmigan is the keystone herbivore for gyrfalcons (Nielsen 1999). The critical period for rock ptarmigan in Iceland is during autumn and winter, and during decline years, mortality in this period increases greatly, especially for juveniles (O. Nielsen personal communication). Our study provides indirect evidence that grouse in the Yukon are also vulnerable during the same period. Studies of managed populations of grouse in England and Scotland, with greatly elevated densities and predator control, have found more support for bottom-up limitations of populations through food quality, maternal condition, or intrinsic factors such as weather (Moss and Watson 1984, Watson et al. 1998), as well as through kin-selection social behaviors (Watson et al. 1994).

11.6.2   Unexpected Results

Despite five species of grouse in the Kluane system at reasonable densities, grouse did not appear to be important alternative prey for predators during the hare decline as reported for other studies which had the same predator species (Keith 1963, Nellis et al. 1972, Brand et al. 1976). However, grouse and red squirrels (chapter 9) were important alternative prey in the hare low during winter for maintaining the few predators remaining in the Kluane system.

The Kluane study is the first to record a cycle in spruce grouse in synchrony with the snowshoe hare cycle. There was at least a 75% decline in spring spruce grouse numbers as they had already begun their decline before the start of the dawn census counts in 1990. Also, there was a 10-fold reduction in encounter rates of spruce grouse in winter over a 2-year period from the peak in 1989–1990 (from the seen-sheet data). Density changes of this magnitude over 2–3 years have not been observed in other long-term population studies where the habitat remained unaltered (Boag and Schroeder 1992). After the population decline in 1991, we observed a dramatic behavioral shift, with the remaining spruce grouse being perceptibly wilder and more wary. This is a reasonable response by species undergoing intense predation pressure, but such behavioral shifts have not been recorded for forest or alpine grouse. However, Alden (1947) noted that surviving ruffed grouse during cyclic lows are wilder when being hunted than birds at peak population densities in Pennsylvania.

Despite high predator densities during the hare peak and decline, spruce grouse nesting and fledging success, which was comparable to other populations (Boag and Schroeder 1992), did not change. These results are in marked contrast to hares, where production of 15–19 leverets/summer/female in the increase/peak phase collapsed to about 6.9 in the decline phase (see chapter 8, figure 8.5). Reproduction of hares did not increase again until just after the low. Hares also appeared to show predation risk-induced stress (Boonstra et al. 1998), presumably because they can never reduce their exposure and vulnerability (Krebs et al. 1998). Grouse that survived into the low must have been significantly less vulnerable to predation than those during the decline, as we found high annual survival at the low. Grouse survivors in the low phase may have adopted risk-averse behaviors as suggested for red squirrels, and both possibly used spruce trees as shields against predators (chapter 9).

Despite the dependence of goshawks on grouse at the low phase, grouse maintained their breeding densities at this time. These enigmatic results for grouse may be explained by their migration patterns. There is a partial independence of populations in summer, winter, and during spring and autumn migration in the Kluane system (Boag and Schroeder 1992) that may allow sharing of the local predation risk load with individuals from a considerably larger geographic area. Since ptarmigan can disperse up to 90 km from breeding areas to winter sites, summer and winter populations do not have all the same individuals (Gruys 1993, Hannon et al. 1998). Extensive external dispersal and population rescue are documented for most ptarmigan and forest grouse populations (Small et al. 1991, Beaudette and Keppie 1992, Gruys 1993) and contribute increased stability to grouse breeding densities (Martin et al. 2000).

11.6.3   Unanswered Questions

Forest grouse in the southwestern Yukon have received little study and appear to differ in several life-history parameters from southern and eastern populations. Spruce grouse in Kluane are larger and heavier, have higher potential fecundity, and differ in male and possibly female displays from other populations (unpublished data). The behavioral responses to elevated predation risk in the low phase are fascinating and merit further study. Much remains unknown about migration patterns, scale of dispersal, and life-history differences for all Kluane grouse species—vital information for understanding population dynamics processes and for effective management (Martin 1998).

The Kluane grouse study has precipitated fascinating questions regarding the ecological factors that start and entrain cycles. Why do ruffed grouse and ptarmigan cycle regularly, whereas blue grouse and spruce grouse generally do not? Ruffed grouse do not exhibit cycles in all areas; regular cycles are less prevalent in southern and eastern North America (Keith 1963). We need to address questions about inducing species to exhibit cyclic behavior in locations where forest grouse are regionally sympatric with cyclic alpine grouse such as in Alaska and northwestern Canada to determine which ecological conditions must apply before spruce grouse exhibit cyclic population dynamics with coexisting hares. When forest and alpine grouse assemblages overlap in winter, it may allow interspecific population phenomena at a regional scale, which could result in the cyclicity of one species that might be noncyclic in most of its range.

Literature Cited

Alden, J. A. 1947. Ruffed grouse. A. A. Knight Knopf, New York.

Angelstam, P., E. Lindstrom, and P. Widen. 1985. Synchronous short-term population fluctuations of some birds and mammals in Fennoscandia—occurrence and distribution. Holarctic Ecology **8**:285–298.

Beaudette, P. D., and D. M. Keppie. 1992. Survival of dispersing spruce grouse. Canadian Journal of Zoology **70**:693–697.

Bendell, J. F., and P. W. Elliott. 1967. Behaviour and the regulation of numbers in blue grouse. Canadian Wildlife Service Report Series, no. 4. Department of Indian Affairs and Northern Development, Ottawa. Pages 1–76.

Bergerud, A. T. 1988. Mating systems in grouse. *in* A. T. Bergerud and M. W. Gratson (eds). Adaptive strategies and population ecology of northern grouse, pages 439–472. University of Minnesota Press, Minneapolis.

Bergerud, A. T., and M. W. Gratson (eds). 1988. Adaptive strategies and population ecology of northern grouse. University of Minnesota Press, Minneapolis.

Boag, D. A., and M. A. Schroeder. 1987. Population fluctuations in spruce grouse: what determines their number in spring? Canadian Journal of Zoology **65**:2430–2435.

Boag, D. A., and M. A. Schroeder. 1992. Spruce grouse (*Dendragapus canadensis*). in A. Poole, P. Stettenheim, and F. Gill (eds). The birds of North America, no. 5. American Ornithologists' Union, Washington, DC.

Boonstra, R., D. Hik, G. R. Singleton, and A. Tinnikov. 1998. The impact of predator-induced stress on the snowshoe hare cycle. Ecological Monographs **68**:371–394.

Boutin, S., C. J. Krebs, R. Boonstra, M. R. T. Dale, S. J. Hannon, K. Martin, A. R. E. Sinclair, J. N. M. Smith, R. Turkington, M. Blower, A. Byrom, F. I. Doyle, C. Doyle, D. S. Hik, L. Hofer, A. Hubbs, T. Karels, D. L. Murray, V. O. Nams, M. O'Donoghue, C. Rohner, and S. Schweiger. 1995. Population changes of the vertebrate community during a snowshoe hare cycle in Canada's boreal forest. Oikos **74**:69–80.

Brand, C. J., L. B. Keith, and C. A. Fischer. 1976. Lynx responses to changing hare densities in central Alberta. Journal of Wildlife Management **40**:416–428.

Braun, C. E., K. Martin, and L. A. Robb. 1993. White-tailed ptarmigan (*Lagopus leucurus*). in A. Poole and F. Gill (eds). The birds of North America, no. 62. American Ornithologists' Union, Washington, DC.

Ellsworth, D. L., R. L. Honeycutt, and N. J. Silvy. 1995. Phylogenetic relationships among North American grouse inferred from restriction endonuclease of mitochondrial DNA. Condor **97**:492–502.

Fritz, R. S. 1979. Consequences of insular population structure: distribution and extinction of spruce grouse populations. Oecologica **42**:57–65.

Gruys, R. C. 1993. Autumn and winter movements and sexual segregation of willow ptarmigan. Arctic **46**:228–239.

Hannon, S. J. 1986. Intrinsic mechanisms and population regulation in grouse—a critique. Acta Congressus Internationalis Ornithologici **19**:2478–2489.

Hannon, S. J., P. K. Eason, and K. Martin. 1998. Willow ptarmigan (*Lagopus lagopus*). in A. Poole and F. Gill (eds). The birds of North America, no. 369. American Ornithologists' Union, Washington, DC.

Hannon, S. J., and R. C. Gruys. 1987. Patterns of predation in a willow ptarmigan population in northern Canada. International Grouse Symposium **4**:44–50.

Hannon, S. J., and K. Martin. 1996. Mate fidelity and divorce in ptarmigan: polygyny avoidance on the tundra. in J. M. Black (ed). Partnerships in birds: the study of monogamy, pages 192–210. Oxford University Press, Oxford.

Herzog, P. W., and D. A. Boag. 1977. Seasonal changes in aggressive behavior of female spruce grouse. Canadian Journal of Zoology **55**:1734–1739.

Hochachka, W. M., K. Martin, F. Doyle, and C. J. Krebs. 2000. Monitoring vertebrate populations using observational data. Canadian Journal of Zoology **78**:521–529.

Hudson, P. J. 1992. Grouse in space and time. The population biology of a managed gamebird. Game Conservancy Ltd, Fordingbridge, UK.

Hudson, P. J., A. P. Dobson, and D. Newborn. 1998. Preventing population cycles by parasite removal. Science **282**:2256–2258.

Jenkins, D., A. Watson, and G. R. Miller. 1963. Population studies on Red Grouse, *Lagopus lagopus scoticus* (Lath.) in north-east Scotland. Journal of Animal Ecology **32**:317–376.

Johnsgard, P. A. 1983. The grouse of the world. University of Nebraska Press, Lincoln.

Keith, L. B. 1963. Wildlife's ten year cycle. University of Wisconsin Press, Madison.

Keith, L. B. 1990. Dynamics of snowshoe hare populations. in H. H. Genoways (ed). Current mammalogy, pages 119–195. Plenum, New York.

Keith, L. B., and D. H. Rusch. 1986. Predation's role in the cyclic fluctuations of ruffed grouse. Acta Congressus Internationalis Ornithologici **19**:699–732.

Keppie, D. M. 1982. A difference in production and associated events in two races of spruce grouse. Canadian Journal of Zoology **60**:2116–2123.

Keppie, D. M. 1987. Impact of demographic parameters upon a population of spruce grouse in New Brunswick. Journal of Wildlife Management **51**:771–777.

Keppie, D. M. 1992. An audio index for male spruce grouse. Canadian Journal of Zoology **70**:307–313.

Krebs, C. J., S. Boutin, R. Boonstra, A. R. E. Sinclair, J. N. M. Smith, M. R. T. Dale, K. Martin, and R. Turkington. 1995. Impact of food and predation on the snowshoe hare cycle. Science **269**:1112–1115.

Krebs, C. J., A. R. E. Sinclair, R. Boonstra, S. Boutin, K. Martin, and J. N. M. Smith. 1998. Community dynamics of vertebrate herbivores—how can we untangle the web? *in* H. Olff, V. K. Brown, and R. H. Drent (eds). Herbivores: between plants and predators, pages 447–473. Blackwell Scientific Publications, Oxford.

Martin, K. 1998. The role of animal behavior studies in wildlife science and management. Wildlife Society Bulletin **26**:911–920.

Martin, K., S. J. Hannon, and R. F. Rockwell. 1989. Clutch size variation and patterns of attrition in fecundity of willow ptarmigan. Ecology **70**:1788–1799.

Martin, K., P. B. Stacey, and C. E. Braun. 2000. Recruitment, dispersal and demographic rescue in spatially-structured white-tailed ptarmigan populations with local and regional stochasticity. Condor **102**:503–516.

Mercer, W. E. 1967. Ecology of an island population of Newfoundland willow ptarmigan. Technical Bulletin No. 2. Newfoundland and Labrador Wildlife Service, St. John's, Canada.

Moss, R., and A. Watson. 1984. Maternal nutrition and breeding success of Scottish ptarmigan *Lagopus mutus*. Ibis **126**:212–220.

Moss, R., A. Watson, and R. Parr. 1996. Experimental prevention of a population cycle in red grouse. Ecology **77**:1512–1530.

Mossop, D. H. 1994. Trends in Yukon upland gamebird populations from long-term harvest analysis. Transactions of the 59th North American Wildlife and Natural Resources Conference, pp. 449–456.

Mueller, F. P. 1993. Herbivore-plant-soil interactions in the boreal forest: selective winter feeding by spruce grouse. MSc thesis. University of British Columbia, Vancouver.

Myrberget, S. 1984. Population cycles of willow grouse *Lagopus lagopus* on an island in northern Norway. Fauna Norv. Ser. C, Cinclus **7**:46–56.

Nellis, C. H., S. P. Wetmore, and L. B. Keith. 1972. Lynx-prey interactions in central Alberta. Journal of Wildlife Management **36**:320–329.

Nielsen, O. K. 1999. Gyrfalcon predation on ptarmigan: numerical and functional responses. Journal of Animal Ecology **68**:1034–1050.

Nugent, D. P., and D. A. Boag. 1982. Communication among territorial female spruce grouse. Canadian Journal of Zoology **60**:2624–2632.

Pelletier, L. 1996. Evaluation of aerial surveys of ptarmigan (*Lagopus spp.*). MSc thesis. University of British Columbia, Vancouver.

Pelletier, L., and C. J. Krebs. 1997. Line-transect sampling for estimating ptarmigan (*Lagopus spp.*) density. Canadian Journal of Zoology **75**:1185–1192.

Ricklefs, R. E. 1969. An analysis of nesting mortality in birds. Smithsonian Contributions to Zoology **9**:1–48.

Rusch, D. H., S. DeStefano, and D. Lauten. 2000. Ruffed grouse (*Bonasa umbellus*). *in* A. Poole and F. Gill (eds). The birds of North America, no. 515. American Ornithologists' Society, Washington, DC.

Schroeder, M. A. 1985. Behavioural differences of female spruce grouse undertaking short and long migrations. Condor **85**:281–286.

Schroeder, M. A., and D. A. Boag. 1989. Evaluation of a density index for territorial male spruce grouse. Journal of Wildlife Management **53**:475–478.

Small, R. J., J. C. Holzwart, and D. H. Rusch. 1991. Predation and hunting mortality of ruffed grouse in central Wisconsin. Journal of Wildlife Management **55**:513–521.

Watson, A., R. Moss, R. Parr, M. D. Mountford, and P. Rothery. 1994. Kin landownership, differential aggression between kin and non-kin, and population fluctuations in red grouse. Journal of Animal Ecology **53**:639–662.

Watson, A., R. Moss, and S. Rae. 1998. Population dynamic of Scottish Rock ptarmigan cycles. Ecology **79**:1174–1192.

Weeden, R. B. 1963. Management of ptarmigan in North America. Journal of Wildlife Management **27**:673–683.

Weeden, R. B. 1964. Spatial separation of sexes of rock and willow ptarmigan in winter. Auk **81**:534–541.

Zwickel, F. C., and J. F. Bendell. 1972. Blue grouse, habitat, and populations. International Ornithological Congress, Proceedings **15**:150–169.

Zwickel, F. C., and J. F. Bendell. 1985. Blue grouse—effects on, and influences of a changing forest. Forestry Chronicle **6**:185–188.

12

Other Herbivores and Small Predators

Arthropods, Birds, and Mammals

JAMES N. M. SMITH & NICHOLAS F. G. FOLKARD

In this chapter, we discuss the responses of three groups of animals in the center of the Kluane food web (figure 2.8) to the snowshoe hare cycle, to the experimental addition of NPK fertilizer, and to an unexpected natural event, a large outbreak of bark beetles. None of these animals was studied intensively every year, either because we did not expect them to be central players in the drama of the snowshoe hare cycle or because they were rare.

The first group, arthropods, are usually the most diverse and abundant animals in terrestrial food webs (Daly et al. 1998), including the boreal forest (Danks and Footit 1989). Lepidopteran larvae can be dominant consumers of vegetation in temperate forests, removing foliage from almost the entire forest canopy (Myers 1993), and severe defoliation can have large impacts on ecological performance of trees (Kaitaniemi et al. 1999). Defoliating species may generate strong interactions in boreal food webs, including effects on passerine birds (Enemar et al. 1984, Neuvonen 1989). Wood-boring insects, particularly bark beetles (order Coleoptera, family Scolytidae), can kill many species of forest trees (Daly et al. 1998; see below).

Arthropod numbers in boreal systems are strongly constrained by a short warm season. Even in May and August, temperatures are often too low for feeding, mating, and development. Arthropods are the principal foods of a variety of other consumers, including spiders, wasps, passerine birds, woodpeckers, and even small raptors such as American kestrels (*Falco sparverius;* see chapter 16). Adding nutrients to the ground layer, as we did at Kluane by fertilization, may elevate the abundance of herbivorous insects and result in increases in predaceous insectivores (Chen and Wise 1999).

The second group included insectivorous songbirds (order Passeriformes) and woodpeckers (order Piciformes). Songbirds at Kluane are nearly all migratory or nomadic, the only permanent residents being three species of corvids and two species of chickadees. In the boreal forest, distinctly different assemblages of passerines are found in the valley bottoms, compared with the neighboring alpine and subalpine areas (Theberge 1976, Folkard and Smith 1995). The dominant songbird species in the white spruce forests of the valley are the yellow-rumped warbler (*Dendroica coronata*), dark-eyed junco (*Junco hyemalis*), and Swainson's thrush (*Catharus ustulatus*), whereas the subalpine willow shrub zones are dominated by American tree sparrows (*Spizella arborea*) and Wilson's warblers (*Wilsonia pusilla*). Woodpeckers are conspicuous birds at Kluane and serve an important role in avian community dynamics (Martin and Eadie 1999) by excavating cavities that are used by other vertebrates such as boreal owls (*Aegolius acadius*) and northern flying squirrels (*Glaucomys sabrinus*).

The final group examined here included larger mammalian herbivores, such as moose (*Alces alces*) and bears (order Carnivora), and some smaller herbivores including the riparian-associated muskrat (*Ondatra zibethica*) and beaver (*Castor canadensis*), and the arboreal porcupine (*Erithezon dorsatum*). During the study, the largest herbivores provided us with both excitement (especially when cantankerous grizzly bears [*Ursus arctos horribilis*] decided they "owned" our food grids), and repeated frustration, when moose and bears repeatedly damaged our fenced exclosures.

12.1   Methods

12.1.1   Arthropods

We sampled arthropods between 1988 and 1989. Samples were collected (by N.F.) on four plots: control 1, control 2, fertilizer 1, and fertilizer 2. In 1988, 40 sticky traps per plot were used to sample flying insects, and 20 pitfall traps (8 cm in diameter, 12 cm deep) per plot were used to sample ground arthropods. Both types of sampling were conducted at 30-m intervals along a 600-m transect. In 1989, we used 30 pitfall traps dispersed non-randomly across open grassy areas on each plot to sample ground-dwelling species. Foliage-dwelling arthropods were sampled by beating a nonrandom sample of thirty 1×1.5 m willow bushes dispersed widely across each plot. The bush was beaten with a 1 m stick and falling arthropods were collected on a 1-m square tray or sheet below the bush. Similarly, arthropods were sampled from 30 white spruce trees by beating foliage up to 3 m in height against a wire mesh screen (details in Folkard 1990). We sampled trees and bushes three to four times between 1330 h and 1800 h, the time of day when arthropods are most active, from mid-May to early August.

Arthropods were identified to order or family and sorted by size. An index of biomass trapped was calculated by cubing the length of each individual and summing these values per sample.

12.1.2   Songbirds and Woodpeckers

We estimated relative abundances and species richness of songbirds and woodpeckers using point counts with an unlimited radius in 1988 and 1989 and a 100-m radius thereafter. We sampled 11 stations per plot on each of the same four plots where arthropods were sampled (details in Folkard and Smith 1995). Birds flying overhead or through the plot within 100 m of the observer were included in the samples.

Counts were conducted from late May to early July, and the numbers of cycles of counts varied from year to year. In 1988, we censused both fertilizer plots four times and both control plots three times. In 1989, fertilizer 1 was censused five times and the other three plots four times. In 1990, 1991, 1995, and 1996, there were two censuses per plot, and in 1992 only one. Counts were conducted by N.F. from 1988 to 1989, by J.S. from 1990 to 1992, and by M. Evans from 1995 to 1996. The only obvious difference among observers was that N.F. and M.E. did not recognize the thin and insectlike song of the blackpoll warbler, *Dendroica striata*. Consequently, this species was only recorded by J.S. in 1990–1992, when it was moderately common.

In addition to the plot-based surveys, Delehanty (1995) studied the detailed use of space by gray jays (*Perisoreus canadensis*) from 1991 to 1993. Birds from nine groups were trapped and color banded on three control plots and three fed plots.

12.1.3   Other Herbivores

Numbers of other herbivores were estimated from opportunistic sightings on the seen-sheet records (see chapter 4.3.3; Hochachka et al. 2000). Muskrat dens (pushups) were counted from a low-flying aircraft in 8–11 lakes and ponds in early spring in 1988 and from 1990 to 1996.

12.2 Predicted Responses to Experimental Treatments

We expected herbivorous insects and vertebrates to increase in response to fertilization through increased foliage volume and plant quality (chapters 5–7). We also examined whether such a bottom-up effect might extend to passerine birds, because they are specialized predators on arthropods. We expected numbers of food-hoarding birds such as jays and chickadees to increase on food addition plots.

During the Kluane Project, there was a severe outbreak of spruce bark beetle (*Dendroctonus ruficollis*). The outbreak began in 1994 and, by late 1996, about half of the mature spruce trees in the valley had been killed, with particularly heavy mortality on south-facing slopes (CD-ROM frame 22, more photos button). For example, even on the north-facing control 1 in April 1997, beetles had recently killed 91 of 246 mature white spruce trees (37%), and a further 37%, were defoliated, but still alive. Only 26% of trees were free from noticeable beetle attack. Therefore, in addition to the effects of fertilization and food addition, we expected an increase in numbers of woodpeckers, which feed on wood-boring insects (Winkler et al. 1995), and perhaps a decline in the abundance of foliage-gleaning birds.

Larger herbivores operate on a spatial scale beyond that covered in our experiments, but we expected bears to respond to food addition by concentrating feeding activity on the food grids. Also, when large herbivores die, they provide carrion on which some of the focal species in our study depend.

12.3 Responses by Species Groups

12.3.1 Arthropods

The different sampling methods yielded different types of arthropods. We mainly captured dipterans and adult lepidopterans in sticky traps, predatory ground beetles (family Carabidae) and spiders (order Aranea, family Lycosidae) in pitfall traps, and lepidopteran and sawfly (order Hymenoptera, family Tenthredinidae) larvae from the foliage of bushes and trees.

Strong seasonal changes in arthropod biomass were detected (table 12.1). Flying insects increased sixfold in biomass from late May to late July in 1988. Arthropods on willow foliage increased nearly 90-fold from early May to late June 1989, and on spruce foliage they increased more than 16-fold from mid May to late June. In 1989, arthropod biomass in June–July was over 40 times greater on willow than on spruce foliage.

In contrast to these large seasonal effects, arthropod biomass was not enhanced by fertilization in either 1988 or 1989. The biomass of ground arthropods in pitfall traps on control and fertilized plots was similar in late summer 1988. The biomass of arthropods from willow foliage in June–July 1989 was over twice as high on control plots as on fertilized plots (figure 12.1). In contrast, the biomass dislodged from spruce foliage in June–July 1989 was about two times higher on fertilized plots (figure 12.1). Because arthropod bio-

Table 12.1 Seasonal changes in mean arthropod biomass on four 36-ha plots using three different sampling methods.

Group	Abundance Index in May (SE)	Abundance Index in late June–July (SE)
Flying insects 1988	188.3 (40.3)	1,126.3 (40.15)
Insects on willows 1989	105.3 (72.5)	9,387.5 (3,338.5)
Insects on spruce 1989	13.9 (3.6)	227.3 (57.1)

mass was much higher on willows, overall arthropod biomass sampled was still two times higher on control plots in 1989. Thus, fertilization reduced the abundance of herbivorous arthropods on foliage in 1989 but seasonal changes in arthropod abundance were much larger than effects of fertilization.

12.3.2 Birds

Responses of Passerine Birds to Fertilization Forty-five species of passerine birds have been found in the Kluane region (Theberge 1976, Folkard and Smith 1995). Twenty-eight of these species were encountered on the four study plots. Most species were present in all years, except for irruptive seed and fruit eaters (particularly white-winged crossbills, *Loxia leucoptera,* and Bohemian waxwings, *Bombycilla garrulus*), whose presence was irregular.

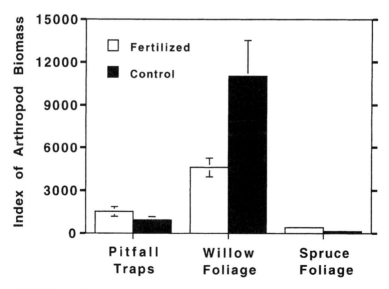

Figure 12.1 Mean indices of arthropod abundance (± SE) for three different sampling methods on two fertilized and two control plots in 1989.

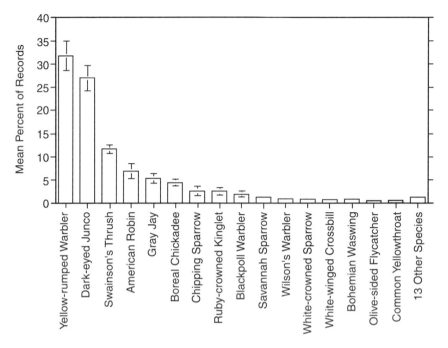

Figure 12.2 Relative abundances of songbirds at Kluane. Data from 1988 and 1989 are for all records, regardless of distance from the observer. Data for 1990–1995 are for records within 100 m of the observer. Plotted values are mean percentages of detections per species per year (±1 SE). Values for the blackpoll warbler are for 1990–1992 only, as observers did not count this species in other years.

The dominant species in all years and plots was the yellow-rumped warbler, followed by the dark-eyed junco and the Swainson's thrush. These three species made up an average of 70% of all records (figure 12.2). Six other species were moderately common: boreal chickadees (*Parus hudsonicus*), American robins (*Turdus migratorius*), gray jays, chipping sparrows (*Spizella passerina*), ruby-crowned kinglets (*Regulus calendula*), and blackpoll warblers (figure 12.2). The remaining species were uncommon or rare and tended to be restricted to habitats (e.g., riparian areas) that were uncommon on the plots.

Densities of passerines were low, averaging 122–155 males/km² for the 8 most common species on all plots in 1988–1989 (Folkard and Smith 1995). Given that these eight species made up 90% of all records, and assuming that (1) only singing males were detected in counts and (2) there was an even sex ratio, total adult densities in 1988–1989 were about 272–344 birds/km².

We assessed how passerine birds responded to fertilization by summing the abundances of the eight dominant species (see above) and calculating an average for the control and fertilized plots. These eight species clearly responded to fertilization, but only after a 3-year time lag (figure 12.3). Fertilizer was first applied in 1987; from 1990 onward, we averaged 41.6% more detections on the fertilized plots than on the control plots (figure 12.3). The effect of fertilization after 1990 was statistically significant (one-way AN-

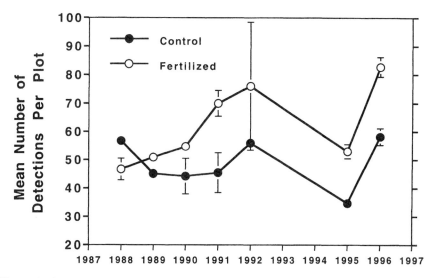

Figure 12.3 Responses of passerine birds to fertilization on two 36-ha control plots compared with two 36-ha fertilized plots. Bars denote standard errors.

COVA, $p < .001$). Although fertilization ceased in 1994, numbers on the two fertilized plots remained elevated in 1995 and 1996 (figure 12.3). Because some birds undoubtedly bred on the same plot from year to year, we also analyzed the data using a two-factor repeated-measures ANOVA; the result remained marginally significant ($p = .049$).

All eight dominant species showed positive responses to fertilization. The magnitudes of the responses of the three most dominant species were +36.9% for the yellow-rumped warbler, +70.8 % for the dark-eyed junco, and +96.0% for the Swainson's thrush averaged over all years after 1990.

Responses of Gray Jays to Supplemental Food Delehanty (1995) mapped the territories of nine groups of gray jays. Five groups on the control grids had a mean territory area of 23.2 ha (SE = 1.4), whereas four fed groups defended a mean area of 15.8 ha (SE = 1.9), a significant reduction ($p = .02$, t test). Jays on fed plots also maintained better body condition throughout the year and cached food at faster rates than jays on control plots (Delehanty 1995).

Effects of the Spruce Beetle Outbreak Two species of woodpeckers, the migratory northern flicker (*Colaptes auratus*) and the resident three-toed woodpecker (*Picoides tridactylus*) occurred commonly at Kluane. A third fire-dependent species, the black-backed woodpecker (*P. arcticus*), was also present, but we never encountered it in our censuses.

From 1990 to 1992, before the spruce beetle outbreak, we detected an average of 2.0 three-toed woodpeckers and 1.0 northern flickers per survey per plot. In 1995 and 1996, these averages rose to 7.5 three-toed woodpeckers and 7.0 flickers. This increase for flickers was significant (one-way ANOVA on log [$x + 1$] transformed numbers of detections per survey per plot per year, $p < .001$), but the response by the three-toed woodpecker was not (p = .30). However, 29 of 30 records of three-toed woodpeckers during the beetle outbreak were from the 2 plots with the oldest spruce trees (fertilizer 1 and control 1). Three-toed woodpeckers at Kluane forage mainly by scaling pieces of bark from large spruce trees. When only the two old-growth plots were considered, the increase in three-toed woodpeckers became significant (p = .03, one-way ANOVA). Because there were no bird surveys in 1994 when the beetle outbreak began, we cannot tell if the increase in abundance of woodpeckers was immediate or if it had a 1-year time lag.

Foliage-gleaning passerines were apparently not greatly affected by the beetle outbreak. Relative numbers of individuals of the eight dominant species were lower in 1995 than in earlier years (figure 12.3), but they rebounded to an all-time high at the peak of the beetle-induced foliage damage in 1996. The yellow-rumped warbler, a species that nests and forages mainly in spruce trees, remained the dominant species in 1995 and 1996.

12.3.3 Other Herbivores

Numbers of muskrat dens peaked in 1988, 2 years before the peak in the hare cycle, but dropped in concert with the hare decline from 1990 onward (figure 12.4). Overall, there was about a fivefold change in numbers of muskrat dens. Patterns of change in muskrat sightings in the seen-sheet records were less clear. Only one year (1989) revealed high numbers of muskrats (table 12.2).

Among the larger herbivores, moose were detected most commonly, followed by grizzly bear and American black bear (*Ursus americanus;* table 12.2). Grizzly bears were recorded about five times as often as black bears. Sightings for these species did not vary

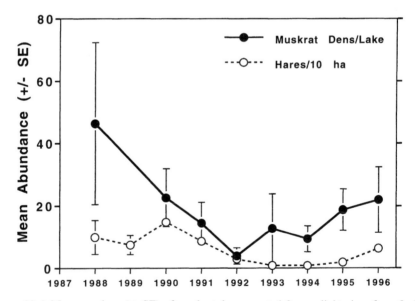

Figure 12.4 Mean numbers (± SE) of muskrat dens counted from a light aircraft on 8–11 lakes and ponds each spring, in relation to hare densities (hares/10 ha) on control grids in spring.

Table 12.2 Numbers of records of larger herbivores per 100 h from seen sheets (vehicle and foot records are combined).

Year	Moose	Grizzly Bear	Black Bear	Muskrat	Porcupine
1988	0.12	0.02	0.02	0.00	0.03
1989	0.25	0.09	0.00	0.08	0.00
1990	0.18	0.08	0.01	0.01	0.01
1991	0.32	0.08	0.02	0.02	0.06
1992	0.24	0.11	0.01	0.00	0.02
1993	0.24	0.18	0.01	0.01	0.01
1994	0.19	0.20	0.03	0.01	0.00
1995	0.28	0.08	0.09	0.03	0.00

predictably across years. Grizzly bears were attracted to the food addition grids, and this fact undoubtedly increased numbers of sightings. Grizzly bears were encountered particularly often in the summer of 1996, after the end of the seen-sheet records. We doubt if the seen-sheet data provide reliable indices of abundance for these large species. The pattern of sightings probably indicates the habits and degree of habituation of individual animals to humans.

Other species of larger herbivores detected included beaver, which were rarely recorded in the seen sheet data because of the distribution of wetlands relative to our study plots, and white-tailed deer (*Odocoileus virginianus,* one record).

12.4   Discussion

12.4.1   Links to the 10-Year Cycle

Only one of the species that we considered in this section exhibited a 10-year cycle, the muskrat. Muskrats exhibit 10-year cycles in boreal Canada, with more southern populations showing weaker cyclic dynamics (Elton and Nicholson 1942). The magnitude of the cycle suggested by the increases, about fivefold, is similar to that found in fur returns by Elton and Nicholson (1942). Presumably, cycles in muskrats are linked to cycles in snowshoe hares by their shared predators. Young muskrats are vulnerable to terrestrial predators such as lynx and coyote when they disperse in spring (Errington 1963).

Songbirds were commonly taken as prey by predatory birds (chapter 16), but the strength of this interaction was insufficient to affect songbird numbers markedly. The most specialized predator of small songbirds at Kluane, the sharp-shinned hawk (*Accipiter striatus*), was both secretive and uncommon (chapter 16), but the more abundant kestrels and northern harriers also commonly took songbirds. Goshawks took some larger songbirds as prey, but concentrated on mammals, grouse, and ptarmigan (table 16.2).

12.4.2   Effects of Fertilization

Arthropod abundance in 1988 and 1989 was not consistently higher on fertilized plots. In fact, the largest difference between control and fertilized plots (for insects from willow foliage) was a substantially lower biomass on the fertilized plots in 1989 (figure 12.1). Surveys of arthropod abundance after 1989 were clearly needed, and we could not, therefore, measure delayed effects of fertilization on the biomass of arthropods.

Despite the absence of a bottom-up effect on arthropod abundance in 1988 and 1989, we found a clear and consistent effect of fertilization on the numbers of birds after 1989. Fertilizing began in 1987 and, from 1990 onward, fertilized plots consistently supported about 40% more passerines than control plots.

Several mechanisms might explain these changes in numbers. First, arthropod standing crops may have been higher after 1989. Second, fertilization might have increased arthropod production, but songbirds and other small predators such as shrews could have consumed the increase, leaving the standing crop unchanged or even reduced, as seen in 1989. Third, birds might have responded directly to changes in vegetation promoted by fertilization, and not to the arthropods. For instance, fertilized plots were noticeably greener in spring, because of the increased growth by the dominant grass *Festuca altaica* (Turkington et al. 1998; chapter 5). Therefore, cover for nesting on fertilized plots might have been greater, thus encouraging birds to settle there.

12.4.3   Other Patterns in the Food Web

Gray jays responded to supplemental feeding by reducing the sizes of their territories, which presumably resulted in greater jay densities on food grids (Delehanty 1995). Gray jays are efficient predators on the eggs and nestlings of shrub-nesting birds (Pelech 1999), yet gray jays and other common passerine birds increased in response to fertilization. It is commonly believed that food supply and predation (Newton 1998) jointly regulate bird numbers. The above patterns suggest that the positive effects of food on numbers of songbirds are stronger than any negative effects of depredation on nests at Kluane.

The increase in the numbers of woodpeckers in 1995 and 1996 suggests that they increased in response to the spruce beetle outbreak that began in 1994 and continued to the end of the study. Point counts are not the ideal census method for woodpeckers (Martin and Eadie 1999), so there is some uncertainty over the strength of this response. It is not surprising that the resident three-toed woodpecker responded strongly, as it is specialized to feed on bark beetles in mature spruce trees (Winkler et al. 1995). The strong response by the migrant northern flicker was surprising, as flickers prefer open forest and are specialized to feed on ants and fruit (Winkler et al. 1995). Presumably, flickers had sufficient dietary flexibility to exploit the bonanza of bark beetles directly, or they benefited via increased ant numbers associated with the bark beetle outbreak. No marked response to the beetle outbreak was seen in the passerine birds, perhaps because the dominant songbirds at Kluane did not eat bark beetles, nor did they depend greatly on green spruce foliage for food or cover.

12.5   Summary

We examined the patterns of change in numbers of three groups of species that we expected to be largely uncoupled from the snowshoe hare cycle. Numbers of arthropods showed strong seasonal fluctuations, and insects decreased on willows on fertilized plots soon after fertilization began. Our studies of this important group were too short (1988 and 1989 only) to assess whether fertilization affected numbers of arthropods after 1989. Passerine birds showed a clear response to fertilization. Numbers of the eight dominant species were elevated by more than 40% on two fertilized plots, and all eight species showed a positive response to fertilization. Two species of woodpeckers increased in response to a severe attack by spruce bark beetles that began in 1994 and had killed nearly 40% of mature spruce trees on one plot by 1997. Muskrats exhibited a 10-year cycle with about a fivefold amplitude, but other larger mammalian herbivores showed no consistent patterns of change through the snowshoe hare cycle.

Literature Cited

Chen, B., and D. H. Wise. 1999. Bottom-up limitation of predaceous arthropods in a detritus-based terrestrial food web. Ecology **80**:761–772.

Daly, H. V., J. T. Doyen, and A. H. Purcell, III. 1998. Introduction to insect biology and biodiversity, 2nd ed. Oxford University Press, Oxford.

Danks, H. V., and R. G. Footit. 1989. Insects of the boreal zone of Canada. Canadian Entomologist **121**:625–690.

Delehanty, B. 1995. Effects of food addition on a population of gray jays. MSc thesis. University of British Columbia, Vancouver.

Elton, C., and M. Nicholson. 1942. Fluctuations in numbers of the muskrat (*Ondatra zibethica*) in Canada. Journal of Animal Ecology **11**:96–126.

Enemar, A., L. Nilsson, and B. Sjöstrand. 1984. The composition and dynamics of the passerine birds community in a subalpine birch forest, Swedish Lapland. Annales Zoologici Fennici **21**:321–338.

Errington, P. L. 1963. Muskrat populations. Iowa State University Press, Ames.

Folkard, N. F. G. 1990. An experimental study of the plant-arthropod-bird food chain in the southwestern Yukon. MSc thesis. University of British Columbia, Vancouver.

Folkard, N. F. G., and J. N. M. Smith. 1995. Evidence for bottom-up effects in the boreal forest: do passerine birds respond to large-scale experimental fertilization? Canadian Journal of Zoology. **73**:2231–2237.

Hochachka, W. M., K. Martin, F. Doyle, and C. J. Krebs. 2000. Monitoring vertebrate populations using observational data. Canadian Journal of Zoology **78**:521–529.

Kaitaniemi, P., S. Neuvonen, and T. Nyssönen. 1999. Effects of cumulative defoliations on growth, reproduction and insect resistance in mountain birch. Ecology **80**:524–532.

Martin, K., and J. M. Eadie. 1999. Nest webs: a community-wide approach to the management and conservation of cavity-nesting birds. Forest Ecology and Management **115**:243–257.

Myers, J. H. 1993. Population outbreaks in forest Lepidoptera. American Scientist **81**:240–251.

Neuvonen, S. 1989. Interactions between geometrid and microtine cycles in northern Fennoscandia. Oikos **51**:393–397.

Newton, I. 1998. Avian population ecology. Academic Press, London.

Pelech, S. A-M. 1999. Habitat use and searching success of red squirrels at a forest edge. MSc thesis. University of British Columbia, Vancouver.

Theberge, J. B. 1976. Bird populations in the Kluane Mountains, southwest Yukon, with special reference to vegetation and fire. Canadian Journal of Zoology **54**:1346–1356.

Turkington, R., E. John, C. J. Krebs, M. R. T. Dale, V. O. Nams, R. Boonstra, S. Boutin, K.

Martin, A. R. E. Sinclair, and J. N. M. Smith. 1998. The effects of NPK fertilization for nine years on boreal forest vegetation in northwest Canada. Journal of Vegetation Science **9**:333–346.

Winkler. H., D. A. Christie, and D. Nunery. 1995. Woodpeckers: an identification guide to the woodpeckers of the world. Houghton Mifflin, Boston.

PART IV

MAMMALIAN PREDATORS

Coyotes and Lynx

MARK O'DONOGHUE, STAN BOUTIN, DENNIS L. MURRAY,
CHARLES J. KREBS, ELIZABETH J. HOFER, URS BREITENMOSER,
CHRISTINE BREITENMOSER-WÜERSTEN, GUSTAVO ZULETA,
CATHY DOYLE, & VILIS O. NAMS

Coyotes and lynx are the two most important mammalian predators of snowshoe hares throughout much of the hare's geographic range. Between them, these two predators killed approximately 60% of all depredated radio-collared hares at Kluane from 1986 through 1996 (Krebs et al. 1995), and predation accounted for at least 78% of all mortalities during this same period.

The strong association of lynx numbers with those of snowshoe hares is well documented (Elton and Nicholson 1942, Moran 1953, Finerty 1980, Schaffer 1984, Royama 1992). The abundance of lynx typically follows the 10-year cyclic fluctuations of hares, and lynx–hare cycles have often been presented in ecology texts as classic predator–prey interactions. The geographic ranges of lynx and hares overlap almost exactly (Banfield 1974, figure 2.7), and all published studies of the food habits of lynx have found snowshoe hares to be the main dietary component (e.g., Saunders 1963a, Van Zyll de Jong 1966, Brand and Keith 1979, Parker et al. 1983). Morphologically, lynx are well adapted to hunting hares in the deep, powdery snows of the boreal forest because of their long legs and large snowshoe-like paws (Murray and Boutin 1991).

Whereas lynx may be considered prototypical specialists on hares, coyotes are generally viewed as opportunistic generalists (Bond 1939, Fichter et al. 1955, Van Vuren and Thompson 1982, but see MacCracken and Hansen 1987, Boutin and Cluff 1989), consuming prey items in the same relative proportions as are available. Indeed, across their large and expanding range in North America (figure 2.7), coyotes show a wide variety of feeding habits, habitat preferences, and social groupings (see Bekoff 1978). Coyotes are relatively recent colonizers of the boreal forest (Moore and Parker 1992); historical records suggest they arrived in the Yukon Territory between 1910 and 1920 (M. Jacquot and J. Joe, from unpublished transcripts of interviews by G. Lotenberg for Parks Canada). Unlike lynx, coyotes have fairly small paws, which restrict their ease of travel over soft snow (Murray and Boutin 1991).

There have been few studies of coyotes in the boreal forest, and only those in Alberta (Nellis and Keith 1968, 1976, Nellis et al. 1972, Brand et al. 1976, Keith et al. 1977, Brand and Keith 1979, Todd et al. 1981, Todd and Keith 1983) and the Yukon (Murray and Boutin 1991, Murray et al. 1994, 1995, O'Donoghue et al. 1995, 1997, 1998a,b) have addressed the relative roles of coyotes and lynx as predators of snowshoe hares and alternative prey. In the Alberta study, approximately one-third of the study area was agricultural land, and this area was used extensively by coyotes when hares were scarce. These Yukon studies therefore represent the first comparison of the dynamics and foraging behavior of coyotes and lynx in the contiguous northern forests.

Based on the generalist–specialist contrast between coyotes and lynx, we predicted that the two predators would respond dissimilarly to cyclic fluctuations of hares and, as a result, could affect populations of hares and other prey species differently as well. The direct effects of predation on prey populations are determined by the combined numerical (changes in rates of reproduction, survival, immigration, and emigration) and functional (changes in kill rates) responses of the predators to varying densities of prey (Solomon 1949). In Alberta, coyotes and lynx responded to a cyclic fluctuation in numbers of hares with similar four- to sixfold numerical responses; the highest numbers of coyotes occurred at the peak of the hare cycle, while abundance of lynx lagged 1 year behind (Keith et al.

1977). Coyotes, however, maintained densities roughly double those of lynx at all phases of the cycle (Brand et al. 1976, Keith et al. 1977, Todd et al. 1981). The calculated functional response of coyotes to changing densities of hares in Alberta was sigmoidal (type 3; Holling 1959b), whereas that of lynx was convex curvilinear (type 2) (Keith et al. 1977), which appears to support the generalist–specialist contrast between coyotes and lynx.

The estimated numerical and functional responses of coyotes, lynx, and avian predators during the cycle in abundance of hares in Alberta led Keith et al. (1977) to conclude that predation rates were not sufficient to account for the slowing of growth and eventual cyclic decline in hare populations. Rather, they considered the hare cycle to be primarily generated by an interaction between hares and their winter food supply (Keith and Windberg 1978, Keith et al. 1984, Keith 1990), with delayed density-dependent predation acting to increase mortality rates of hares once population growth had halted due to food shortage, thus deepening and hastening the cyclic declines. This view of the role of predation in generating population cycles corresponded with the conventional wisdom summarized by Finerty (1980). In addition to affecting cyclic amplitudes and the rapidity of population declines, predators could also act to increase the length of population cycles, due to the lag effect of predators persisting into periods of low prey populations, and possibly synchronize cycles over large geographic areas, due to the large-scale movements of northern predators (Finerty 1980).

The empirical results of the Kluane experiments presented in this book (see also Krebs et al. 1986, 1995, Sinclair et al. 1988, Smith et al. 1988) and the conclusions from recent models (Trostel et al. 1987, Akçakaya 1992, Royama 1992, Stenseth 1995) suggest that predation may play a larger role in generating cyclic fluctuations in numbers of snowshoe hares than previously thought. Likewise, researchers in Fennoscandia have suggested that differences in the degree and nature of predation can explain the geographic pattern of cyclicity in abundance of voles in that region. More diverse prey communities in the south support higher numbers of generalist predators, which can regulate vole numbers by strong functional responses (Erlinge 1987, Erlinge et al. 1983, 1984, 1988), while the resident specialists (weasels) of the north respond numerically, which results in a time lag and cyclic fluctuations (Henttonen et al. 1987, Hanski et al. 1991, Korpimäki et al. 1991, Korpimäki 1993, Lindström and Hörnfeldt 1994). This pattern is in agreement with the different theoretical effects of generalist and specialist predators on the stability of predator–prey interactions (Murdoch and Oaten 1975, Hassell and May 1986, Crawley 1992).

The aim of the research on predators that was conducted as a part of the Kluane Project was to examine in detail the changes in densities, demographics, predation rates, and behavior of the most important species during a cyclic fluctuation of hares. These studies would then complement the smaller-scale experimental manipulations of prey densities in elucidating the roles of different predators in the boreal food web (Boutin 1995). In particular, the main objectives of the studies of coyotes and lynx were to

1. Determine and contrast the numerical responses of coyotes and lynx during a cyclic fluctuation in numbers of snowshoe hares and the demographic mechanisms behind them,
2. Determine and contrast the functional responses of coyotes and lynx to changes in densities of hares,

3. Examine changes in foraging behavior (e.g., prey switching, changes in foraging tactics) of coyotes and lynx that contribute to their functional responses, and
4. Estimate the total effects of predation by coyotes and lynx on populations of snowshoe hares and the main alternative prey.

13.1 Methods

We used snow tracking (following and counting tracks) of coyotes and lynx during winter and radio telemetry as our main field methods during these studies. With the exception of some limited collection of scats, live trapping, and radio monitoring, field work was conducted during the winter months, from October to April.

13.1.1 Population Monitoring

Densities of large, mobile predators are notoriously difficult to estimate. We combined data from the movements of radio-collared coyotes and lynx with those from our snow tracking to make annual early-winter estimates of numbers of predators in our study area (O'Donoghue et al. 1997).

From 1986 through 1996, we live trapped and radio collared 21 individual coyotes (CD-ROM frame 35) and 56 lynx (CD-ROM frame 34) (techniques similar to Poole et al. 1993, Mowat et al. 1994). Most of the radio collars were equipped with mercury activity switches (Telonics, Mesa, Arizona), which allowed us to monitor the activity patterns of collared animals in addition to gathering data on their movements, home ranges, and survival rates (CD-ROM frame 41). We attempted to obtain at least one precise location per week for each collared coyote and lynx to calculate the boundaries of their home ranges; from 1990 through 1992, lynx were monitored more frequently, as a part of a study of their social organization (U. Breitenmoser, unpublished data). We plotted telemetry locations using LOCATE II (Nams 1990) and calculated home ranges as 95% minimum convex polygons (Mohr 1947) using CALHOME (Kie et al. 1994).

We counted the tracks of coyotes and lynx each winter (October through April) along a 25-km transect that traversed our study area, on days after fresh snowfalls while tracks were distinguishable (CD-ROM frame 43). These counts gave us indices of the numbers of predators present each winter and also information on the locations of animals that we did not radio collar (O'Donoghue et al. 1997). From 1986–1987 through 1996–1997, we counted tracks along 12,194 km of transect.

We also followed the tracks of coyotes (2134 km) and lynx (2500 km) during winter from 1987–1988 through 1996–1997 to gather data on foraging behavior (described below). Tracking began each season as soon as there was enough snow, usually in late October, and finished at the end of March. We selected fresh tracks crossing roads and trails on any days when weather and snow conditions were suitable. We attempted to distribute our tracking over the entire study area, throughout the winter, and among all group sizes of predators. During a tracking session, tracks were followed first backward (relative to the animal's direction of travel) as far as possible, and then forward from the starting point, to attain segments of trail of maximum length. The routes taken by predators we tracked provided us with additional information on the presence and locations of uncollared coyotes and lynx.

We estimated the total populations of coyotes and lynx in our study area by plotting the home ranges of radio-collared animals, and filling in missing animals based on our data from the track transects, snow tracking, and, in the case of coyotes, records of howling (O'Donoghue et al. 1997). We were not in the field enough during the first and last winters of this study (1986–1987 and 1996–1997) to estimate populations of predators accurately. We estimated survival and dispersal rates from our radio monitoring (Pollock et al. 1989) and calculated indices of recruitment by examining the distribution of group sizes of coyotes and lynx along the track transect (O'Donoghue et al. 1997).

13.1.2   Foraging Behavior

We gathered most of our detailed data on the foraging behavior of coyotes and lynx by snow tracking. Trackers, on snowshoes for most of the winter, followed the trails of predators and made a continuous record of the animals' hunting behavior, use of habitat and trails, and notable activities. Data were recorded on microcassettes in the field, and all events along the trails were noted at the tracker's step number, which was kept track of with a tally counter (steps were later converted to meters, using observer-specific conversion factors). Over the course of the study, 22 different observers collected data, of which 19 were full-time trackers.

We recorded all instances of kills and attempted kills by predators along trails. At each kill site, the prey species, the length of the chase, the percentage and parts of the carcass consumed, and any observations about feeding or caching behavior were noted. For attempted kills, we recorded the prey species and the length of the chase. Details of all scavenging by coyotes and lynx were also collected. We estimated the amount of food the predator had consumed when possible (from impressions in the snow), noted whether the prey had been previously cached, and looked for any evidence of the initial cause of death of the prey. We collected all scats along trails for later dietary analyses.

Data on the travel patterns of coyotes and lynx were also recorded. We classified all beds used by predators as "crouches" (or ambush beds), "short beds" (those in which the animal lay down, but did not stay long enough to melt the snow), or "resting beds." Instances of running, circling, or other hunting activities were noted. We recorded the amount of time that the trails we followed were on top of trails of other predators, moose, hares, snowmobiles, and snowshoes. (At high densities of hares, it became nearly impossible to record all instances of travel on hare trails, so we switched to noting whether predators were on or off of hare trails at 100-step intervals. For analyses, all records of travel on hare trails were converted to these 100-step measures; O'Donoghue et al. 1998a.)

We kept a continuous record of habitat used by coyotes and lynx by estimating the percent canopy cover (as <5%, 5–25%, 25–50%, 50–75%, or 75–100%) and dominant species and age (mature or immature) of the trees in the overstory. We also measured the sinking depth of the predator in the snow and snow depths and hardness on the trail and 1 m off the trail at 500-step (1987–1988 and 1988–1989) or 600-step (after 1988–1989) intervals (Murray and Boutin 1991). In 1996–1997, the last winter of tracking, we gathered data only on kills, attempted kills, scavenging, and beds used by predators, and discontinued our collection of data on habitat use, snow characteristics, and use of trails.

Our data on habitat use were collected continuously by many different observers, each of whom subjectively estimated canopy cover; these characteristics necessitated additional steps in the analyses of these data. We calculated the interval length (200 m) at which measurements of habitat should be independent based on patch sizes in our study area by using a boot-strapping technique with a set of patch lengths from transects run to classify habitat types (O'Donoghue et al. 1998a). We then sampled at 200-m intervals from each tracker's records separately to calculate frequencies of habitat types and mean overstory cover used by coyotes and lynx. Statistical analyses comparing habitat use between years and species were conducted separately by observer, and the results were combined using meta-analyses (Rosenthal 1978, Arnqvist and Wooster 1995) to control for observer-specific biases (O'Donoghue et al. 1998a).

The scats collected while tracking and during other field work were frozen and stored for later analysis. We subsampled scats by month and location in the study area each winter to maximize the temporal and spatial distributions of our samples. We identified the hairs in 10–30 random subsamples of each scat and summarized the data as relative frequency of occurrence by prey species (O'Donoghue et al. 1998b). Although prey species with small body sizes are often overrepresented in diets inferred by scat analyses (Floyd et al. 1978, O'Gara 1986), relative frequency of occurrence of prey in scats of coyotes was approximately equal to the percent biomass ingested in one study (Johnson and Hansen 1979). We also identified the species for any recognizable teeth and bone fragments found in scats.

In addition to snow tracking, we used several other field methods to supplement our data on foraging behavior of coyotes and lynx. While tracking, we made frequent observations of predators, particularly coyotes, scavenging prey from caches, but we were seldom able to tell how the prey had initially died, which predator had made the cache, or when the initial kill had been made. We were able to gather data on the frequency and timing of caching by predators by examining the patterns of caching of depredated, radio-collared hares. Beginning in 1992–1993, we monitored use of these caches by leaving all caches located while snow tracking or retrieving mortalities of radio-collared hares undisturbed, except for the attachment of a radio monitor (O'Donoghue et al. 1998b). A beacon radio was hung in a nearby tree at each cache site, and a second "cache collar" was hidden under the snow nearby with a steel wire running from its battery to the cache. When a cached carcass was disturbed, the battery from the cache collar became dislodged, and we could then return to the site and collect data on the return time, amount of carcass eaten, and details of cache use by predators.

Finally, to investigate the degree of use of prey hot spots by predators during the hare cycle, we made use of the experimental grids of the Kluane Project. Starting in 1988–1989, we counted tracks of predators around two food addition grids (which had roughly three times the density of hares as on control grids; Krebs et al. 1995; chapter 8) and one or two control grids after fresh snowfalls during the winter, when weather conditions were suitable. These counts were summarized as tracks per day for subsequent analyses (O'Donoghue et al. 1998a).

13.1.3   Functional Responses

The kill rates of coyotes and lynx from our snow-tracking data were expressed as kills/ distance of trail, but this was not a useful measure for calculating the functional responses or total impacts of predation by the two predators on densities of prey. We needed to convert our kill rates to kills/day for this purpose. Our annual tracking records consisted of measured distances that the predators were traveling and a variable number of events during which animals were inactive (beds) or stationary (kills, scavenging). To calculate the amount of time represented by our tracking records, we estimated the time spent by predators at each of the above activities.

We calculated travel rates of coyotes and lynx by opportunistically conducting time trials with our radio-collared animals from 1989–1990 through 1994–1995 (O'Donoghue et al. 1998b). When a collared animal was in an accessible location, we obtained an accurate location by walking toward the animal and making a visual observation or finding a recently left bed. If we were successful in obtaining this location without disturbing the animal, we then monitored the animal's activity until we could obtain a second location, usually one to several hours later, after which we followed and measured the trail in between these two points. We calculated travel rates from successful time trials by dividing the distance traveled by the time interval minus any inactive periods (CD-ROM frame 42). Breitenmoser et al. (1992) used this same technique to verify the use of monitoring via intensive radio telemetry for measuring movement rates of lynx.

We estimated the total amount of inactive time spent by coyotes and lynx (on beds) by monitoring the activity patterns of radio-collared animals. From 1990–1991 through 1995–1996, we used a Lotek SRX-400 programmable receiver with a data logger (Lotek, Newmarket, Ontario) to record pulse rate and signal strength every 30 sec (based on three signal pulses each reading) for selected collared predators (O'Donoghue et al. 1998b). From these records, we used a discriminant function, developed by simultaneously monitoring the activity of predators "by ear" (with observers judging periods of activity and inactivity based on the characteristics of the radio signal) and with the programmable receiver to calculate the percent time spent inactive each winter. Based on comparisons with activity patterns inferred from manual monitoring of predators, this remote technique allowed accurate calculation of inactive periods (O'Donoghue et al. 1998b).

With the calculation of travel rates and percentages of inactive time for coyotes and lynx, we could then convert our kill rates of hares to kills/day. We calculated the amount of active predator time represented by our track records each winter for coyotes and lynx separately by dividing the total track distances by the measured travel rates and adding standard times for each occurrence of a kill or scavenging event (O'Donoghue et al. 1998b). The total number of "coyote-days" and "lynx-days" were estimated by dividing the active times by the percent of time active during each winter to account for time spent on beds.

We plotted kills/day per coyote and lynx against the autumn, spring, and mean densities of hares each winter to calculate their functional responses (O'Donoghue et al. 1998b). We fitted linear and type-2 (Holling's disc equation; Holling 1959a) curves to our estimated kill rates for graphical description. Given that we estimated kill rates in only 10 winters, with no replication at given densities of hares, we had little statistical power to distinguish among shapes of the functional response curves (Trexler et al. 1988; Marshal and Boutin 1999) and did not attempt to do so.

Switching by predators among prey types, as the relative frequencies of prey species change, is one mechanism that may affect the shape and magnitude of a predator's functional response. *Switching* is defined as feeding on a prey species disproportionately less when its relative abundance to other prey species is low and disproportionately more when it is high (Murdoch 1969). We used a graphical analysis similar to that of Murdoch (1969) to analyze our data on prey use versus availability, using relative biomass of prey instead of relative frequencies due to large differences in body sizes of the main prey species at Kluane (O'Donoghue et al. 1998a). We first calculated null curves for coyotes and lynx preying on hares, red squirrels, and small mammals (to represent the predators' innate prey preferences: convex curves for preferred prey and concave for avoided; Murdoch 1969), using the ratios of mean Manly's alphas calculated each winter (Manly et al. 1972; these measure the probability that an individual prey item would be selected when all prey classes are equally available; Krebs 1999) as proportionality constants (Murdoch 1969, equation 2). When percent use of prey classes (in this case, hares, red squirrels, and small mammals) in predator diets are plotted against percent availability of the same prey, a pattern of points below the null curve at low relative availability ($<50\%$) and above it at high availability suggests that switching occurred.

13.2 Numerical Responses

13.2.1 Density

Populations of both coyotes and lynx fluctuated widely during this study, following the 26- to 44-fold cyclical change in numbers of snowshoe hares (chapter 8; Boutin et al. 1995). Densities of coyotes varied sixfold, with peak numbers (approximately $9/100 \text{ km}^2$) occurring in 1990–1991, a year after the cyclic high of hare populations (figure 13.1; O'Donoghue et al. 1997). This was followed by a rapid decline in densities of coyotes to a low of $1.4/100 \text{ km}^2$ in 1993–1994. Likewise, densities of lynx varied about 7.5-fold, and peak densities ($17/100 \text{ km}^2$) in 1990–1991 declined to a low of $2.3/100 \text{ km}^2$ in 1994–1995 (figure 13.1; O'Donoghue et al. 1997). The populations of both predators declined at similar rates: in 1992–1993, the second winter of declining predator numbers, coyote density was 30% ($r = -.60$ from 1990–1991 to 1992–1993) and lynx density 25% ($r = -.69$) of their respective peak densities. Our estimates of numbers of predators in our study area were highly correlated (coyotes: $r_s = .88$; lynx: $r_s = .95$) with the abundance of tracks along the 25-km transect (O'Donoghue et al. 1997).

When we plotted our estimates of densities of coyotes and lynx against densities of hares (means of autumn and late winter estimates), the numerical responses showed a counterclockwise pattern typical of delayed density-dependent responses (figure 13.2; O'Donoghue et al. 1997).

Coyotes in Alberta underwent a similar three- to sixfold fluctuation in abundance, but their numbers followed the cycle in numbers of hares with no lag (Keith et al. 1977, Todd et al. 1981). Densities of coyotes in that study were much higher than at Kluane, ranging from 8 to $44/100 \text{ km}^2$, and they declined at a slower rate from their peak; after 2 years of decline, densities were still 78% of peak numbers. Lynx in the Alberta study also varied three- to fourfold during the same cycle, their numbers peaking at $10/100 \text{ km}^2$ 1 year after the cyclic peak in numbers of hares (Brand et al. 1976, Keith et al. 1977). Other studies of lynx in northern Canada have found higher amplitude cycles of lynx: 10- to 17-fold in south-central Yukon (peak density $50/100 \text{ km}^2$; Slough and Mowat 1996), and 10-fold

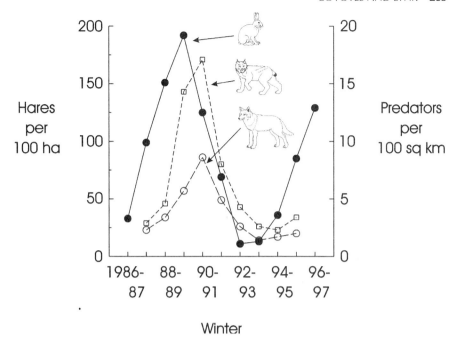

Figure 13.1 Estimated densities of snowshoe hares (means of autumn and late winter estimates) from 1986–1987 through 1996–1997, and coyotes and lynx during winter from 1987–1988 through 1995–1996 in the Kluane Project study area.

in the Northwest Territories (peak 30/100 km^2; Poole 1994), both with high densities 1 year after the highest densities of hares.

Numerical responses of predators may be associated with changes in survival, migration, and reproduction. We examined our data to determine which mechanisms were most important in causing the changes in densities that we observed.

13.2.2   Adult Survival

Survival of our radio-collared lynx varied considerably among phases of the hare cycle. Overwinter survival rates were relatively high (30-week Kaplan-Meier survival rate, 0.67–1.00) from the late cyclic increase (1987–1988) to the second year of declining hare numbers (figure 13.3; O'Donoghue et al. 1997). All of the lynx we had radio-collared at the beginning of 1992–1993 (the first year of the cyclic low; $n = 9$) either died or dispersed. The low survival calculated in 1993–1994 is largely due to starting that year with only two collared lynx, one of which died; three of the four lynx subsequently collared during the winter survived (O'Donoghue et al. 1997). Lynx that survived into the subsequent cyclic increase all survived to the end of the study.

The main causes of mortality of collared lynx during the study were human caused (fur trapping), except in the cyclic low (1992–1993 and 1993–1994), when predation and star-

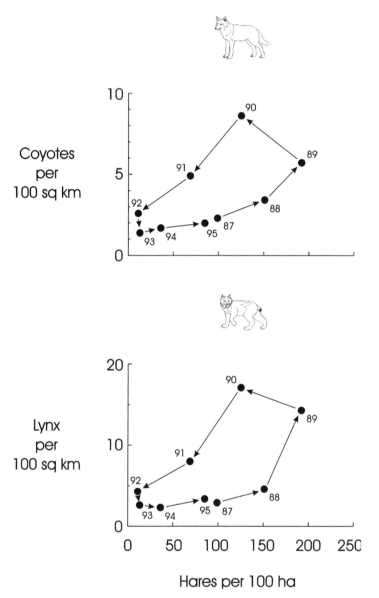

Figure 13.2 Numerical responses of coyotes and lynx to the cyclic fluctuation in densities of snowshoe hares at Kluane from 1987–1978 through 1995–1996. Densities of coyotes (r_s = 1.00) and lynx (r_s = .98) were highly correlated with densities of hares during the previous winter. Numbers next to the data points indicate years (e.g., 87 = winter of 1987–1988).

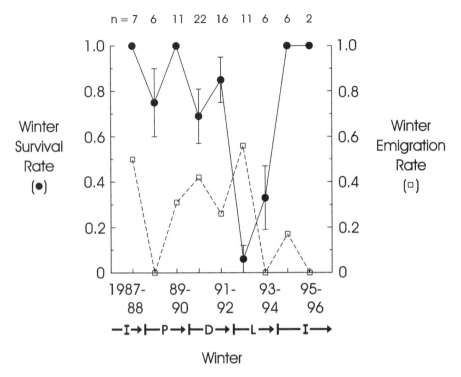

Figure 13.3 Kaplan-Meier survival (± SE) and emigration rates of lynx during winter (October through April) from 1987–1988 through 1995–1996 at Kluane. The number of radio-collared lynx monitored each winter is given at the top of the figure. The phases of the hare cycle (I = increase, P = peak, D = decline, L = low) are shown by the horizontal bar below the x-axis.

vation were more important (O'Donoghue et al. 1995, 1997). We confirmed three cases of predation on lynx by a wolf, a wolverine, and another lynx, and we also suspected predation by wolverines on two other lynx. Two lynx starved to death; we do not know if the depredated lynx would have survived, and suspect at least some were in poor condition.

Due to low sample sizes, we can say little about the survival rates of coyotes during our study. Of the 21 collared coyotes, 5 died, and 4 of these deaths were caused by humans. Most other animals emigrated from our study area, and we do not know if they survived.

Low survival of lynx following cyclic declines in densities of hares was noted in the two other northern studies of lynx as well. In the Northwest Territories, survival rates were high (0.90) during the peak and early decline phases, but decreased to 0.25 and 0.37 in the subsequent two winters (Poole 1994). Likewise, survival rates of lynx declined from near 0.90 at the cyclic peak to 0.40 in the second winter of declining abundance of hares in the south-central Yukon (Slough and Mowat 1996). Natural causes (starvation and predation) were the principal reasons for mortalities in both of these lightly-trapped study areas.

13.2.3   Emigration

Emigration was the main cause of loss of radio-collared coyotes from our study area. Thirteen of the 21 collared coyotes (62%) traveled beyond the range of our radio monitoring; this included 4 of 5 pups and 9 of 16 adults (O'Donoghue et al. 1997). We documented the distances moved by two coyotes that were trapped 23 and 40 km from their initial capture sites.

Radio-collared lynx emigrated from our study area during most winters, and rates were highest in 1987–1988 and from the peak to the early low phases of the hare cycle (figure 13.3; O'Donoghue et al. 1997). The calculated rate in 1992–1993 is inflated by the emigration of one of two surviving radio-collared lynx after most others had died. Over the whole study, 17 of 39 adults (44%) and 7 of 12 juveniles (58%) left our study area. We recovered the collars of 10 radio-collared lynx that had emigrated, mostly from fur trappers, and these were collected from 23 to 830 km from their capture sites. Five of six animals that moved long distances (>200 km) from their initial locations left our study area in 1991–1992; most of those leaving their home ranges later in the cyclic decline or low died within our study area (O'Donoghue et al. 1995, 1997).

High emigration rates of lynx during cyclic declines and lows have also been reported in the previous population cycle on our study area (Ward and Krebs 1985), in south-central Yukon (Slough and Mowat 1996), and in the Northwest Territories (Poole 1997). Lynx in all three of these other studies moved long distances (up to 1100 km) from their initial locations. Survival of dispersers was poor during the first winter of the cyclic low in the Northwest Territories (Poole 1997), and Ward and Krebs (1985) also noted that no animals made long-distance migrations once densities of hares had declined below 1/ha.

13.2.4   Recruitment

Although we did not directly measure reproductive output or juvenile survival of coyotes and lynx in our study area, we examined our observations from track counts for changes in the frequency and magnitude of group sizes. Family groups of lynx are easily recognized from tracks by the smaller foot-size of kittens, but this is not possible with coyotes, so we considered all groups of more than two coyotes to be family groups.

Both predators traveled in larger groups during periods with higher densities of hares. We observed family groups of coyotes only in the four winters from 1988–1989 through 1991–1992, and again in the last two winters of the study (figure 13.4). During the three winters of lowest abundance of hares (1992–1993 through 1994–1995), no groups larger than two were noted. Group sizes were largest in 1990–1991 and 1991–1992 (O'Donoghue et al. 1997).

Evidence of recruitment by lynx also followed the same trend. Few family groups were seen after the winter of 1990–1991, and none was seen in 1993–1994 and 1994–1995 (figure 13.4; no lynx kittens were noted in 1986–1987 either, but we cannot calculate percent family groups, as individual groups were not recorded separately). Family groups were largest from 1988–1989 through 1990–1991 (O'Donoghue et al. 1997).

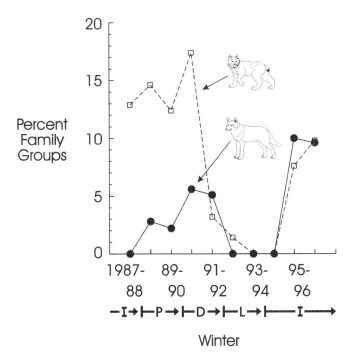

Winter

Figure 13.4 Percentages of observations of coyote and lynx tracks along a 25-km transect composed of family groups (groups of coyotes with more than two animals or groups of lynx with kittens) during winter from 1987–1988 through 1996–1997 at Kluane. The phases of the hare cycle (I = increase, P = peak, D = decline, L = low) are shown by the horizontal bar below the x-axis.

Large declines or cessation of reproduction by coyotes and lynx during cyclic declines of their main prey have been noted in many other studies. In Alberta, adult ovulation rates, pregnancy rates and litter sizes, and breeding by yearling female coyotes all declined as hare densities reached their cyclic low (Todd et al. 1981, Todd and Keith 1983). Reproductive output of coyotes in Utah was also reduced during a cyclic decline of black-tailed jackrabbits (*Lepus californicus*) (Clark 1972). Likewise, studies of lynx in south-central Yukon (Mowat et al. 1996, Slough and Mowat 1996), the Northwest Territories (Poole 1994), Alberta (Nellis et al. 1972, Brand et al. 1976, Brand and Keith 1979), Alaska (O'Connor 1986), and Nova Scotia (Parker et al. 1983) have all shown that recruitment by lynx virtually ceases during periods of low abundance of snowshoe hares. Based on examination of carcasses, several studies have suggested that postpartum mortality of kittens may be the largest factor contributing to the drop in recruitment (Brand and Keith 1979, Parker et al. 1983).

13.3   Social Organization

Our primary objectives in radio collaring coyotes and lynx were to determine their main areas of activity, measure their travel rates, determine their survival and emigration

rates, and monitor their activity patterns. Except for a more detailed study of spatial relationships among lynx from 1990 through 1992, we did not monitor predators frequently
enough, or with sufficient precision, to draw robust conclusions about the social organization of coyotes and lynx in our study area. We briefly discuss the limited data we gathered here, as they relate to those from other more comprehensive studies.

13.3.1 Home Ranges

We monitored few coyotes often enough to estimate the sizes or spatial arrangement
of their home ranges. The annual home ranges of five coyotes (two adult males and three
adult females) averaged 24.7 ± 10.6 (SD) km² in years of high densities of hares (1988–
1989 and 1990–1991), while those of two adult males were 75.4 ± 2.9 km² at the cyclic
low (1993–1994), suggesting that the sizes of home ranges may have increased with lower
prey abundance. We did not monitor any individual animals over the course of the cyclic
decline, though, so we do not have data on the persistence of family territories.

Our data on the home ranges of lynx in our study area are more complete after 1989.
In 1990–1991, the home ranges of adult males averaged 24.3 ± 9.3 km² ($n = 7$), while
those of females were 16.7 ± 13.7 km² ($n = 4$). There was a general trend of increasing
sizes of home ranges during the cyclic decline, and, by 1994–1995, the ranges of males
and females averaged 39.5 ± 11.4 km² ($n = 2$) and 45.2 ± 3.4 km² ($n = 2$), respectively.
The same trend was true for individual lynx that persisted in the study area between 1990–
1991 and 1992–1993. The sizes of home ranges of males were larger than those of females in all years of this study except 1994–1995, but these differences were not statistically significant (Mann-Whitney U-tests; $p > .12$) in any year.

The degree of intrasexual spatial overlap and fidelity to home ranges by lynx changed
considerably as densities of hares declined. In 1990–1991, adult lynx maintained essentially intrasexually exclusive territories (mean percent overlap = 7% among adult males
and 11% among adult females with >20 radio locations; calculated as in Poole 1995),
with the exception of two, probably younger, adult males, which had home ranges almost
completely within those of other males. In the next two winters, this territoriality apparently broke down. Established residents made longer forays outside and shifted the boundaries of their home ranges, and new adult males were captured with ranges that broadly
overlapped those of other consexuals (mean percent overlap = 75% and 28% for males
and females, respectively). During 1992–1993, all previously collared residents died or
emigrated, and new lynx captured in the subsequent two winters established home ranges
that were again intrasexually exclusive.

These observations about the spatial organization of lynx are consistent with those
made in other studies. Several lynx became nomadic, but persisted on our study area during the previous cycle in the early 1980s, once densities of hares had declined below 0.5/
ha (Ward and Krebs 1985). In the Northwest Territories, the home ranges of adult male
and female lynx averaged 36.3 km² and 20.8 km², respectively, at the cyclic peak, and
these ranges increased to 44.0 km² and 62.5 km² 2 years later when populations of hares
had crashed (Poole 1994). The ranges of males and females were mostly intrasexually exclusive and stable in the Northwest Territories study, but this spatial organization broke
down during the winter 2 years after the cyclic peak in abundance of hares, and all residents either died or dispersed (Poole 1995). The sizes of lynx home ranges also increased

>5-fold for males and >10-fold for females in the south-central Yukon study after the cyclic decline, but unlike our data and those of Poole (1994, 1995), most animals persisting into the cyclic low were known to have been born in the study area (Slough and Mowat 1996).

13.3.2 Social Groups

Both coyotes and lynx sometimes traveled in groups at Kluane. Coyotes were observed in groups of two to five, but groups larger than three animals were only noted in 1990–1991 and 1991–1992. These larger groups were likely composed of related animals, but we could not confirm this. The typical social organization of coyotes consists of exclusive family territories, with a mated pair and a variable number of their offspring that may delay or forgo dispersal, especially when food is abundant or dispersal opportunities limited (Andelt 1985, Bekoff and Wells 1986).

We observed lynx in groups of up to seven animals. Up until 1991–1992, all of the groups of lynx that we observed traveling together were composed of an adult (presumably, the mother) and one to six kittens. Lynx kittens typically remain with their mother for most of their first winter (Poole 1995, Mowat et al. 1996), and they hunt together as a group (Parker 1981). We noted lynx in adult groups of two or three animals for the first time in 1991–1992, the second winter of declining abundance of hares. These groups persisted throughout the winter and accounted for 51% of all group observations ($n = 81$) in 1991–1992, 90% in 1992–1993 ($n = 19$), and all groups in 1993–1994 ($n = 7$) and 1994–1995 ($n = 1$) on our track transects. We could not confirm the relationships among animals in groups, but we suspected one group consisted of an adult female and her grown female offspring. We have found only one other reference to adult groups of lynx (Barash 1971), although Poole (1995) also noted that the radio locations of two male–female pairs suggested positive attraction during the early part of the first winter of the cyclic low. We discuss evidence that hunting in adult groups increased the foraging efficiency of lynx below.

13.4 Foraging Behavior

13.4.1 Diets

We calculated the winter diets of coyotes and lynx from the kills we found along their trails and from analyses of their scats. During summer, we collected scats opportunistically and by searching along gravel roads; we found few scats of lynx during the summers.

We found 208 kills by coyotes during the 10 winters of tracking; 48% of these were snowshoe hares, 14% were red squirrels, and 36% were small mammals (mostly *Clethrionomys rutilus, Microtus pennsylvanicus,* and *M. oeconomus*). All of the kills were of hares during 1987–1988 and 1988–1989 (late cyclic increase to peak), followed by a period of increased use of small mammals and squirrels from 1989–1990 through 1991–1992 (cyclic peak to decline; figure 13.5). Coyotes killed mostly small mammals in 1992–1993 and 1993–1994, winters of low abundance of hares and high numbers of small mam-

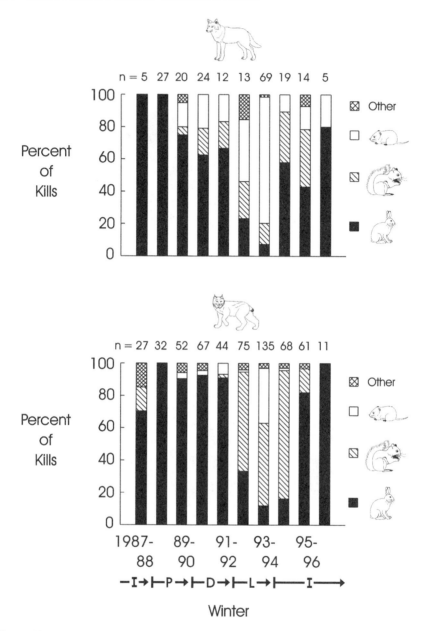

Figure 13.5 Kills by coyotes and lynx located along their tracks during winter from 1987–1988 through 1996–1997, at Kluane. Sample sizes are given above bars. "Other" kills for coyotes were 1 grouse, 2 flying squirrels, and 1 unknown prey, and for lynx were 14 grouse, 2 muskrats, 2 flying squirrels, 1 short-tailed weasel, 1 red fox, and 1 unknown prey. The phases of the hare cycle (I = increase, P = peak, D = decline, L = low) are shown by the horizontal bar below the x-axis.

mals. As densities of hares began increasing again from 1994–1995 to 1996–1997 and those of small mammals dropped, coyotes shifted their hunting to hares, with red squirrels more frequent as alternative prey in the first two of those winters (figure 13.5).

In terms of prey biomass, coyotes consumed >90% hares in all winters, based on our snow tracking, except in 1992–1993, 1993–1994, and 1995–1996, when red squirrels represented 12–20% of the biomass (figure 13.6). Small mammals represented <2% of biomass consumed in all years but 1993–1994. However, kills of small mammals are underrepresented from our tracking data because we often could not confirm whether an attempt at killing a vole had been successful. Scat analyses suggested a dietary pattern similar to that from our snow tracking, except that small mammals represented a larger proportion of the estimated biomass consumed (18–42% in winters of 1987–1988 through 1988–1989 and 1992–1993 through 1993–1994, based on their relative frequencies of occurrence), during periods of higher densities of small mammals (O'Donoghue et al. 1998b). Despite the fact that forested habitats were much more abundant in our study area than meadows (which were only about 7% of the area), coyotes killed many more *Microtus* ($n = 64$ identified jawbones in scats), which are mostly found in meadows, than *Clethrionomys* ($n = 9$). Hares made up the remaining bulk of scats in all winters; the relative frequency of occurrence of red squirrels was greater than 10% only in 1994–1995 (13%).

We collected an adequate number of coyote scats in summer only in 1988 ($n = 35$) and from 1990 through 1992 ($n = 22$–49). Analyses of these scats suggested that summer diets were more varied than winter diets, due to the availability of ground squirrels, which made up an estimated 15–21% of biomass consumed in all summers except 1990 (0%). Hares (33–73%; highest in 1990 and 1991) and small mammals (8–25%) made up the remaining bulk of the summer diets.

We located 572 kills by lynx during snow tracking, of which 55% were of hares, 32% of red squirrels, and 10% of small mammals. Lynx killed mostly hares in winters when hares were at densities >0.5/ha (1987–1988 through 1991–1992, 1995–1996 and 1996–1997), but more red squirrels than hares during the cyclic low (1992–1993 and 1993–1994) and early increase (1994–1995) winters (figure 13.5). Substantial numbers of small mammals were killed only in 1993–1994.

When considered in terms of biomass consumed, hares made up >90% of the diets of lynx, based on our snow tracking, in all winters except 1992–1993 through 1994–1995, when red squirrels made up an increasing proportion of prey (20–44%; figure 13.6). Analyses of winter scats suggested the same patterns, except that, as with coyotes, small mammals made up a larger proportion (23–26%) of scats in the winters of 1987–1988 and 1988–1989 (cyclic late increase to peak) and 1993–1994 (cyclic low) than indicated from our tracking (O'Donoghue et al. 1998b). Of identified small mammals in scats, more *Clethrionomys* ($n = 14$) were present than *Microtus* ($n = 11$), which may reflect the avoidance of open habitats by lynx (Murray et al. 1994). The relative frequency of occurrence of red squirrels ranged from 25% to 37% in scats from 1992–1993 through 1994–1995.

We gathered enough summer scats of lynx only in 1990 ($n = 13$) and 1991 ($n = 43$). As with coyotes, the diversity of summer diets was greater than in preceding winters due to the addition of ground squirrels (8–25% relative frequency of occurrence). The rest of the scats were composed mostly of hares (67–85%).

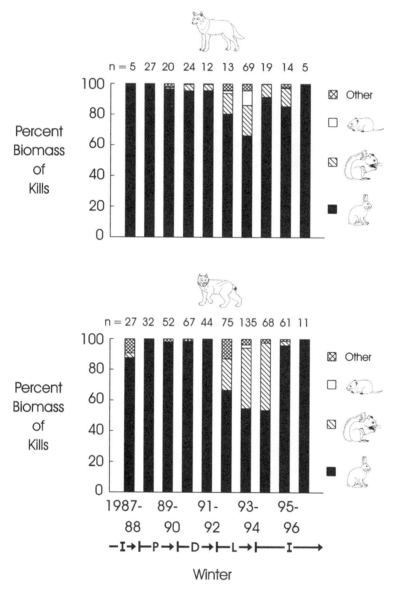

Figure 13.6 Percent biomass of kills by coyotes and lynx located along their tracks during winter from 1987–1988 through 1996–1997 at Kluane. Sample sizes (number of kills) are given above bars. The phases of the hare cycle (I = increase, P = peak, D = decline, L = low) are shown by the horizontal bar below the x-axis.

The dietary patterns that we observed for coyotes and lynx were consistent in all parts of our study area (O'Donoghue et al. 1998a). Therefore, although we were not able to distinguish among individual predators from our snow tracking, we suggest that the shifts in diets that we observed were due to most animals changing their patterns of hunting, rather than just a smaller proportion specializing in alternative prey during periods of low abundance of hares.

Changes in the diets of coyotes as numbers of snowshoe hares fluctuated were also observed in Keith's studies in Alberta. Based on analyses of scats and stomach contents, hares made up 67–77% of the estimated consumption of coyotes at the cyclic peak, but only 0–23% once densities of hares had declined (Nellis and Keith 1976, Todd et al. 1981, Todd and Keith 1983). Livestock carrion was the most important alternative food source in these studies.

Diets of lynx in Alberta were composed mostly of hares (97–100% at the cyclic peak, 65–81% during periods of low abundance of hares) at all phases of the hare cycle (Nellis et al. 1972, Brand et al. 1976, Brand and Keith 1979). Grouse and small mammals were the most important alternative prey in these studies. We are aware of only two other studies that have recorded squirrels as being more than a minor dietary component of lynx (24% of food items in scats in Washington; Koehler 1990; 28% of food items in Alaska; Staples 1995); both studies were conducted where densities of hares were low.

13.4.2   Scavenging Behavior

Signs of scavenging were much more frequent along the trails of coyotes than along those of lynx (figure 13.7; O'Donoghue et al. 1998b). Both predators also uncovered caches (40% of all opportunities to scavenge for coyotes, 25% for lynx) and encountered old kills (60% for coyotes, and 75% for lynx) most frequently during winters of highest densities of hares.

Data from monitoring the survival of radio-collared hares showed that coyotes cached the entire carcasses of 37.8% (82 of 217) of all kills they made from 1986 through 1996, and lynx cached the whole carcasses of only 0.5% (1 of 211) of their kills of hares (O'Donoghue et al. 1998b). This suggested that many of the caches dug up by coyotes were initially made by them as well. Observations of coyotes making side trips off of otherwise straight trails, sometimes for >0.5 km, to dig up caches also suggested that individual animals fed on their own caches. Caches made by coyotes were typically dug down through the snow to or into the ground, often at the bases of trees or shrubs. Cached carcasses were often pushed laterally under undisturbed snow, leaving little sign of food storage at the sites. Lynx, on the other hand, usually covered carcasses at the surface of the snow by pulling snow over them. While snow tracking, we recorded 17 observations of coyotes uncovering caches made by lynx, compared to 92 records of coyotes digging up typical coyote-made caches.

From 1992–1993 through 1994–1995, we monitored 30 caches made by coyotes (27 hares, 1 red squirrel, 1 flying squirrel, and 1 portion of a wolf-killed moose leg), and 7 caches made by lynx (all hares) that we had located while snow tracking and retrieving mortalities of radio-collared hares (O'Donoghue et al. 1998b). Coyotes cached mostly whole carcasses of hares (24 of 27 carcasses were entire; mean percent of carcass present

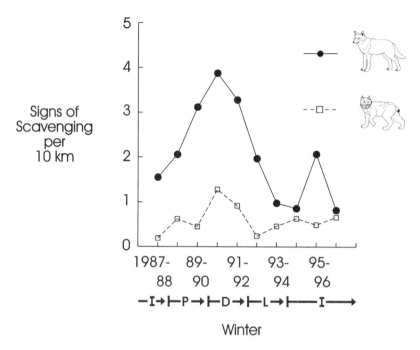

Figure 13.7 Frequency of scavenging (visiting old kills and caches) along the trails of coyotes and lynx during winter from 1987–1988 through 1996–1997 at Kluane. The phases of the hare cycle (I = increase, P = peak, D = decline, L = low) are shown by the horizontal bar below the x-axis.

= 99%, 95% CI = 97–100%), whereas lynx usually cached only portions (1 of 7 whole; mean = 61%, 95% CI = 33–86%). Most caches made by coyotes were also made early in winter (85% of the caches we monitored were made in October and November), whereas those of lynx were more evenly spread through the winter (43% in October and November). This reflects the relative temporal distributions of kills by coyotes and lynx. Of radio-collared hares killed by coyotes from 1986 through 1996 (n = 146, excluding hares killed on study grids with fencing or cut snowmobile trails), 77% of them were killed in October or November, compared to only 23% of hares killed by lynx (n = 132) in those months (figure 13.8).

Coyotes returned to 14 of the 27 (52%) caches of hares that we monitored and ate an estimated average of 74% (95% CI = 37–98%) of each carcass. They returned to caches an average of 56 ± 35 (SD) days (range 9–140 days) after making their kills. No caches made by coyotes were eaten by lynx, 2 were scavenged by red squirrels, and 11 were not used. We had evidence in two cases that coyotes approached these monitored caches but did not retrieve them, possibly because of our disturbance to the sites, so our use rates may be negatively biased.

Lynx retrieved six of their seven (86%) caches of hares and consumed an average of 99.5% (95% CI = 96–100%) of the portions cached. They returned to caches after an av-

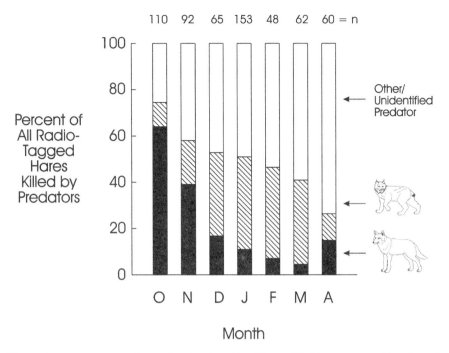

Figure 13.8 Percentages of depredated radio-collared hares killed by coyotes and lynx in each winter month from 1986 through 1995 at Kluane. Sample sizes (total number of mortalities) are given above bars. "Other" predators include red foxes, wolves, wolverines, avian predators, and cases in which the predator could not be identified.

erage of 0.9 ± 0.5 days (range 0–2 days). No lynx caches were eaten by coyotes, and one was not used. These results suggest that most of the scavenging at caches that we observed while snow tracking was likely of caches made by the predator being followed. Caching and use of caches by lynx has been noted in other studies (Nellis and Keith 1968, Parker 1981), but it does not appear to be a frequent behavior compared to consumption of freshly killed prey (Parker 1981).

13.4.3 Hunting Tactics

Coyotes usually pounced on hares from close range, with little sign of stalking, or flushed them out while walking or running through dense cover (Murray et al. 1995, O'Donoghue et al. 1998a). Coyotes seldom, if ever, used ambush beds to hunt hares (figure 13.9). During winters when coyotes actively hunted voles, they used a "mousing" hunting tactic in which they moved slowly, mostly through meadow areas, pausing frequently and leaping into the air to pin voles with their front paws. The frequency of mousing was low during winters of high and declining abundance of hares (0.00–0.05 attempts per 10 km trail in the winters of 1988–1989 through 1991–1992, 1995–1996, and 1996–1997; in all but 1991–1992, densities of voles were low during these winters). The fre-

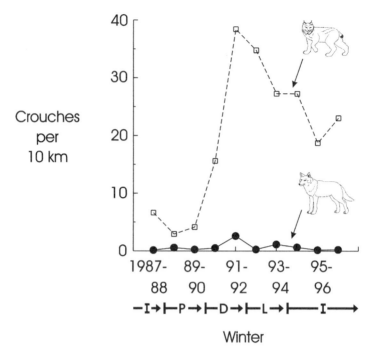

Figure 13.9 Frequency of crouches, or "ambush beds," along the trails of coyotes and lynx during winter from 1987–1988 through 1996–1997 at Kluane. The phases of the hare cycle (I = increase, P = peak, D = decline, L = low) are shown by the horizontal bar below the x-axis.

quency of mousing was higher in winters with densities of hares <0.5/ha and greater vole abundance (0.70–1.91 attempts/10 km in 1987–1988, 1992–1993, and 1994–1995), and very high in 1993–1994 (13.98 attempts/10 km).

Lynx hunted hares either by stalking them or ambushing them from crouches or ambush beds (Murray et al. 1995, O'Donoghue et al. 1998a). As densities of hares declined from the cyclic peak (1989–1990) to the low (1992–1993), the frequency of ambush beds increased more than ninefold along the trails of lynx (figure 13.9). Lynx also initiated progressively more chases of hares from hunting beds during the cyclic decline and low years (1990–1991 through 1993–1994), although hunting success from beds was about the same as that from stalks for hares, and slightly lower for red squirrels (O'Donoghue et al. 1998a). Use of beds declined after 1992–1993, but they were still frequently used until 1996–1997 (figure 13.9). This pattern was consistent in all parts of the study area, so the shift to greater use of an ambush hunting tactic during the cyclic decline was apparently widespread among individual lynx (O'Donoghue et al. 1998a).

Frequent use of ambush beds by lynx has been noted in other studies of lynx (Saunders 1963a,b, Haglund 1966), but in some studies, lynx seldom ambushed hares (Nellis and Keith 1968, Parker 1981). In Alberta, Brand et al. (1976) speculated that apparently lower daily movement rates during periods of low abundance of hares may have been related to increased use of beds, although this was not quantified.

13.4.4   Group Hunting

We observed evidence of group hunting by both coyotes and lynx. When hunting together, pairs or groups of coyotes typically traveled parallel to one another or spread out laterally relative to their direction of travel. Hares flushed by one animal were sometimes killed by another. We also observed coyotes running through dense cover, apparently with no prey initially sighted (based on distances run before hares were encountered) and killing hares flushed out. Ozoga and Harger (1966) described similar group hunting by coyotes in which one animal would run through a dense thicket and another would travel around its perimeter and chase hares flushed out.

Families of lynx hunted together in a similar manner. Although lynx seldom ran while not chasing prey, families generally spread out while moving through suitable hare habitat, and hares flushed out by one animal were often killed by another. This pattern of family hunting, with the mother typically flanked by kittens, was also described by Parker (1981).

We examined our tracking data to see if group hunting conferred advantages of higher food intake to predators. We calculated kill rates of hares by groups of coyotes and lynx (for all group sizes that we followed for >10 km in a given winter), and adjusted these to individual kill rates. There was a variable relationship between group size and individual kill rates for coyotes and no distinct trend of higher or lower food intake (O'Donoghue et al. 1998a). Larger groups of lynx killed more hares, but individual kill rates were lower with increasing group size from 1987–1988 through 1991–1992. In 1992–1993 and 1993–1994, when virtually all of the groups we tracked were adult groups, individuals in groups had similar or higher intake rates (O'Donoghue et al. 1998a). Our sample sizes are small in these years, but the limited data we gathered suggest that group hunting may confer some advantages in foraging success to lynx during periods of food scarcity.

13.4.5   Use of Trails

Coyotes traveled on top of the trails of other predators and those made by snowmobiles and snowshoes more often than did lynx at all phases of the hare cycle (figure 13.10; Murray and Boutin 1991, O'Donoghue et al. 1998a). Snow was harder on the trails of coyotes relative to those of lynx, and this likely eased their travel (Murray and Boutin 1991). Both predators often traveled along the trails of hares, especially when densities of hares were high, but there was no difference in the frequency of use between the two (figure 13.10). Keith et al. (1977) speculated that increased use of hare trails by coyotes at the cyclic peak led to higher kill rates.

13.4.6   Habitat Use

Coyotes and lynx showed a general trend of using habitats with progressively denser overstory cover (the densest habitats in our study area were dominated by young white spruce) from the winter of peak abundance of hares (1989–1990) through 1991–1992, the second winter of the cyclic decline (figure 13.11). After 1991–1992, there was an overall decline in density of cover until 1993–1994 (the cyclic low). This trend paralleled that of habitat use by snowshoe hares, which also selected progressively more dense cover as their numbers declined (Hik 1995, O'Donoghue et al. 1998a). Hares were consistently

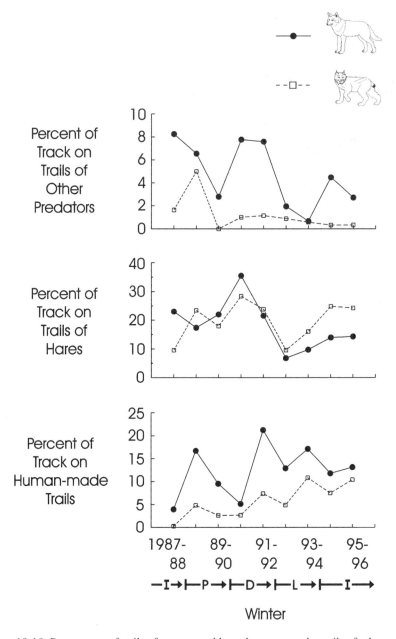

Figure 13.10 Percentages of trails of coyotes and lynx that were on the trails of other predators, snowshoe hares, and human-made (by snowmobile or snowshoes) trails during winter from 1987–1988 through 1995–1996 at Kluane. Coyotes used trails of other predators (Wilcoxon paired-sample test, $p = .004$) and humans ($p = .004$) more than did lynx; there was no difference between predators in the use of hare trails ($p = .660$). The phases of the hare cycle (I = increase, P = peak, D = decline, L = low) are shown by the horizontal bar below the x-axis.

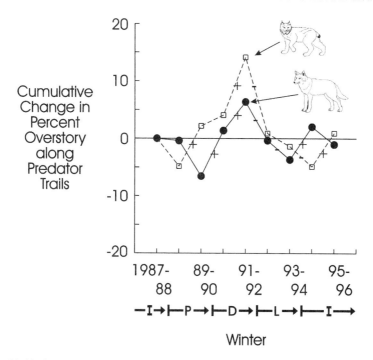

Figure 13.11 Cumulative changes between winters (starting at 0% in 1987–88) in percent overstory cover along the trails of coyotes and lynx during winter from 1987–1988 through 1995–1996 at Kluane. The + and − signs indicate statistically significant ($p < .05$) increases and decreases, respectively, in overstory cover between two years. The phases of the hare cycle (I = increase, P = peak, D = decline, L = low) are shown by the horizontal bar below the x-axis.

found in denser habitat than either predator at all phases of the cycle. Coyotes avoided open cover less than lynx, especially during winters at the cyclic low, when coyotes often hunted *Microtus* in meadow habitats (Murray et al. 1994, O'Donoghue et al. 1998a). During winters of high abundance of hares (1987–1988 through 1990–1991), hunting success of coyotes chasing hares was highest in dense habitats, whereas lynx were equally successful in all habitat types (Murray et al. 1994, 1995). Selection of dense and early successional coniferous habitats by lynx is well documented in other areas as well (Parker 1981, Koehler and Aubry 1994, Poole et al. 1996), but no changes in habitat selection by lynx were noted during a cyclic decline in the Northwest Territories (Poole et al. 1996).

At a larger spatial scale, coyotes may select topographical areas with shallower snow; they used areas of our study area at lower elevation more than they did the central plateau, which had deeper, softer snow (Murray and Boutin 1991). In Alberta, Todd et al. (1981) speculated that habitat selection by coyotes was less affected by snow depths at the cyclic peak due to trampling of snow by hares.

Both coyotes and lynx used the experimental food-addition grids, with higher densities of hares, more than the control grids during the cyclic decline (O'Donoghue et al. 1998a). This is further evidence that the predators concentrated their hunting activity in areas of locally high abundance of hares during this period, as was found for lynx during the previous cycle (Ward and Krebs 1985).

13.5 Functional Responses

Coyotes and lynx both showed strong functional responses to the cyclic changes in densities of hares. Estimated kill rates of coyotes varied approximately ninefold over the observed range of hare densities (figure 13.12). Linear and type-2 response curves fit the

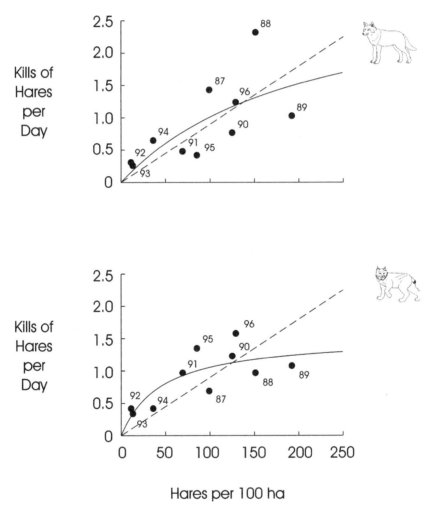

Figure 13.12 Functional responses of coyotes and lynx (kills per predator per day) to changing densities of snowshoe hares (means of autumn and late winter estimates of hare densities) during winter from 1987–1988 (labeled "87" on graphs) through 1996–1997 at Kluane. Type-2 (Holling's disk equation; Holling 1959b; coyote, $r^2 = .83$; lynx, $r^2 = .94$) and linear curves (coyote, $r^2 = .82$; lynx, $r^2 = .86$) are fitted to the estimated kill rates.

observed data equally well, so there was little evidence of satiation at higher densities of hares (which is expected theoretically, but seldom observed from field data; Boutin 1995). When calculated separately, using autumn densities of hares and kills by coyotes from October through December, the functional response was nearly linear, whereas the observed kill rates from January through March, plotted against late-winter densities of hares, showed a clear asymptote at about 1.65 kills/day (O'Donoghue et al. 1998b).

We observed peak kill rates of 2.3 hares per day per coyote in 1988–1989, one year before the peak in numbers of hares. With the exception of 1995–1996, kill rates were higher during winters of cyclic increase (1987–1988, 1988–1989, 1994–1995, and 1996–1997) than at comparable densities of hares during the peak, decline, and low phases of the cycle (figure 13.12).

Kill rates of hares by lynx varied four- to fivefold during the cyclic fluctuation of hares (figure 13.12). There was better evidence of an asymptotic kill rate over the range of observed hare densities for lynx than for coyotes, and the fit of the type-2 curve was slightly better than that of a linear functional response. Early-winter and late-winter functional responses were similar in shape, and, as with coyotes, the peak kill rates were higher at the beginning of the winter (asymptote = 1.8 hares/day) than at the end (1.0 hares/day) (O'Donoghue et al. 1998b).

During the first increase and decline in numbers of hares (1987–1988 through 1993–1994), kill rates by lynx were lower during cyclic increase years (1987–1988 and 1988–1989) than at comparable densities of hares during the decline (O'Donoghue et al. 1998b). The highest kill rate observed during this period was 1.2 kills/day, in 1990–1991, the year after the cyclic peak of hares. This pattern appeared to repeat itself with the low kill rates observed in 1994–1995, as densities of hares began to increase but lynx continued to hunt red squirrels. However, the kill rates estimated in 1995–1996 (1.35 kills/day) and 1996–1997 (1.58 kills/day) were the highest observed during the 10 winters of study, so it is questionable whether we would again observe even higher kill rates than these during the next cyclic decline.

The maximum kill rates of hares by coyotes and lynx that we observed were higher than those reported in the literature, and higher than estimates of the energetic requirements of these predators (coyotes: 0.9 hares/day, Litvaitis and Mautz 1980; lynx: 0.4 hares/day, Nellis et al. 1972). In Alberta, maximum daily kill rates were estimated to be 0.7 and 0.8 per day for coyotes and lynx, respectively (Brand et al. 1976, Keith et al. 1977). However, the functional responses in the Alberta study were estimated assuming that coyotes killed only as many hares as they required energetically (no surplus killing or wastage) and assuming that the distance between two resting beds of lynx represented 1 day of activity. Both of these assumptions are questionable. Surplus killing is relatively common among predators (Kruuk 1972), and use of two to three resting beds per day by lynx has been documented (Haglund 1966, Parker 1981, this study).

13.5.1 Components of Functional Responses

Functional responses by predators to changing densities of prey can result from changes in a number of attributes of the foraging behavior of the predator or from characteristics of the prey. Holling (1959a, 1966) considered that the basic components of functional responses are rate of successful search (determined by the reactive distance of the predator to its prey, the relative speeds of movement by predators and their prey, and the rate of successful capture by predators), foraging time by predators, and handling time (the amount of time needed to capture, consume, and digest a prey item). Although these parameters were considered constants in Holling's original equations, they may vary with changing prey densities (Abrams 1990). We examined our tracking data for evidence of changes in these components to clarify mechanisms of the functional responses.

Reactive Distances of Predators   The only index of the reactive distances of predators that we measured was the length of chases of hares by coyotes and lynx. Because hares usually fled before they were captured, chase lengths were invariably longer than the true reactive distances. Lengths of chases of hares by coyotes and lynx varied significantly over the hare cycle (coyotes: ranges of annual means for successful chases = 0.4–25.0 m, unsuccessful = 3.9–21.8 m; lynx: successful 1.8–19.3 m, unsuccessful 6.3–23.0 m), and were longest from 1992–1993 through 1994–1995, when densities of hares were lowest (O'Donoghue et al. 1998b).

Travel Rates of Predators   We successfully completed 7 time trials for coyotes and 16 for lynx to estimate their travel rates. The mean travel rates were 2.49 ± 0.39 (SD) km/h (range 1.77–3.06 km/h) and 1.09 ± 0.21 km/h (range 0.75–1.46 km/h) for coyotes and lynx, respectively (O'Donoghue et al. 1998b). Six of seven measures of travel rates for coyotes were made during winters of high densities of hares, so we could not examine these data for changes in travel rates. In winters with relatively high abundance of hares (1989–1990 through 1991–1992), lynx traveled at 1.02 ± 0.23 km/h ($n = 9$), while at the cyclic low (1992–1993 and 1993–1994), the comparable rate was 1.13 ± 0.16 km/h ($t = 1.63$, df = 14, $p = .13$), so we have no evidence of changing movement rates in response to prey density.

Hunting Success   Hunting success of coyotes chasing hares was relatively constant from 1989–1990 through 1993–1994, but considerably higher in the cyclic increase years of 1988–1989 and 1994–1995 through 1996–1997 (figure 13.13). The overall success rate was 38.1%. Coyotes chased a substantial number of squirrels only from 1992–1993 through 1995–1996, and their success rate during these winters was 36.5%.

The annual success rates of lynx chasing hares were less variable from 1987–1988 through 1994–1995 than those of coyotes, but we observed high success in 1995–1996 and 1996–1997 (figure 13.13). The overall success rate was 31.7%. During the three winters of lowest hare abundance (1992–1993 through 1994–1995) hunting success of lynx chasing hares was lowest, while there was a steady increase in their success at capturing red squirrels (figure 13.13).

The rates of hunting success that we observed were higher than published rates for coyotes chasing hares (6–10%; Ozoga and Harger 1966; Berg and Chesness 1978), but within the range of those reported for lynx (19–57%; Saunders 1963a, Haglund 1966, Brand et al. 1976, Parker 1981, Major 1989). Hunting success of lynx was not related to the density of hares in Alberta (Brand et al. 1976).

Foraging Time   We monitored the activity of radio-collared coyotes and lynx for 1342 and 4257 h, respectively, from 1990–1991 through 1994–1995 (O'Donoghue et al.

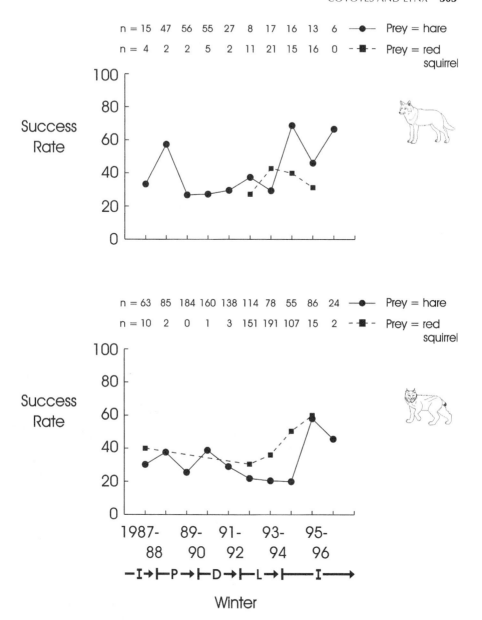

Figure 13.13 Hunting success (percentage of chases that were successful) for coyotes and lynx preying on snowshoe hares and red squirrels during winter from 1987–1988 through 1996–1997 at Kluane. Sample sizes (number of chases) are given at the top of each graph; data points are plotted only for winters with more than five chases. The phases of the hare cycle (I = increase, P = peak, D = decline, L = low) are shown by the horizontal bar below the x-axis.

1998b). Based on our verification trials, the automatic receiver with data logger was an accurate technique for remote measurement of the activity patterns of predators. The percent time active by coyotes varied little (range 44.3–49.7%) from 1990–1991, when the abundance of hares was still high, through 1993–1994, the cyclic low (O'Donoghue et al. 1998b). The frequency of resting beds along coyote trails also varied little among all winters (range 2.5–4.3 beds/10 km of trail), so we have little evidence that coyotes varied their foraging time in response to changes in the density of hares.

Likewise, the percent time active by lynx also varied little among winters (range 39.2–43.5%) and animals (mean annual CV = 4.8%) from 1990–1991 through 1994–1995. For lynx, though, the percent time active may not be a good measure of time they are exposed to prey because the use of ambush beds also increased greatly over that period (figure 13.9); during these "inactive" periods, lynx were still actively foraging. The frequency of resting beds also varied from 1.9 to 5.0 beds/10 km of trail during the 10 years of this study, with the most beds being used during winters of low abundance of hares (O'Donoghue et al. 1998b).

The activity rates of coyotes we measured were comparable to those measured elsewhere during winter (44–50%; Bekoff and Wells 1981, 1986, Bowen 1982). Several studies have found that lynx increased their daily movements as densities of hares declined (Ward and Krebs 1985, Poole 1994).

Handling Time   The only component of handling time that we directly recorded was the amount of prey eaten by coyotes and lynx at each kill site (which was presumably correlated with the amount of feeding time). Coyotes consumed an average of 82–96% of each carcass of hares in all winters except for 1993–1994 (62%) and 1996–1997 (68%) (sample sizes were small, $n = 5$ and 6, respectively, during these years) (O'Donoghue et al. 1998b). Lynx ate an estimated 78–95% of hare carcasses, with the lowest consumption rate during the cyclic peak. Although partial consumption of prey may decrease handling time (Abrams 1990), the time devoted to feeding on prey is likely not the major component of total handling time.

13.5.2   Prey Switching

We investigated prey switching by coyotes and lynx by comparing their relative use of snowshoe hares, red squirrels, and small mammals with the availability of these prey items in the autumn, late winter, and the mean of these two measures (O'Donoghue et al. 1998a). The results were the same regardless of which part of the winter was considered, so we present only the data considering mean densities of prey here.

Coyotes strongly preferred snowshoe hares in all winters (mean Manly's $\alpha = 0.85 \pm 0.12$ [SD], range 0.65–1.00), so the null curve for the switching test is a convex curve for hares and concave curves for squirrels and small mammals (figure 13.14). Based on these graphs, there is no evidence that coyotes switched among prey species (relative to their innate preference for hares) as the relative frequencies of these prey changed over the hare cycle.

Lynx also preferred hares in all winters ($\alpha = 0.91 \pm 0.13$, range 0.59–1.00), but this preference was lowest in 1993–1994 ($\alpha = 0.79$) and 1994–1995 ($\alpha = 0.59$), winters in which lynx more actively pursued red squirrels. The graphical tests suggest that lynx did

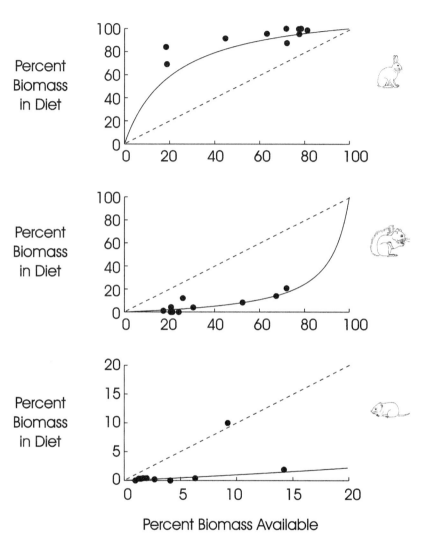

Figure 13.14 Graphical tests for prey switching by coyotes among snowshoe hares, red squirrels, and small mammals during winter from 1987–1988 through 1996–1997 at Kluane. The dashed lines represent equal availability and use of each prey, and the solid curves are the null hypotheses of no switching, taking into account the innate dietary preferences of coyotes. Evidence of switching is implied when data points lie below the null curve at relatively low (<50%) availability and above the curve at high availability.

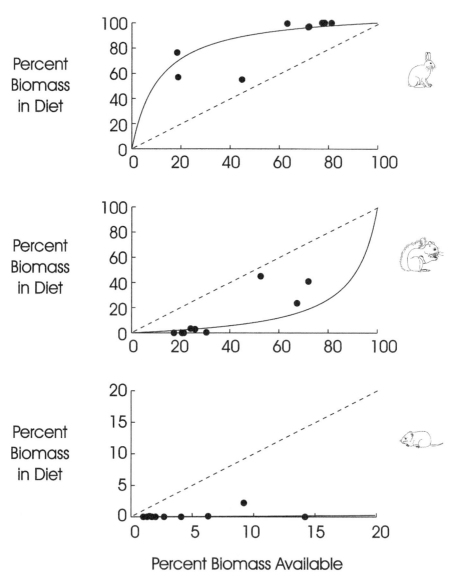

Figure 13.15 Graphical tests for prey switching by lynx among snowshoe hares, red squirrels, and small mammals during winter from 1987–1988 through 1996–1997 at Kluane. The dashed lines represent equal availability and use of each prey, and the solid curves are the null hypotheses of no switching, taking into account the innate dietary preferences of lynx. Evidence of switching is implied when data points lie below the null curve at relatively low (<50%) availability, and above the curve at high availability.

switch (sensu Murdoch 1969) to preying on red squirrels from 1992–1993 through 1994–1995, when red squirrels represented more than 50% of the prey biomass (figure 13.15). This conclusion is robust to two assumptions we made (O'Donoghue et al. 1998a). First, we assumed that the relative availabilities of prey equaled their relative densities. This is unlikely to be strictly true in any field situation, but our interpretation of switching remained the same if we considered that as little as 25% of red squirrels and small mammals were available due to arboreal and subnivean refuges, respectively. Second, we derived the null curves for these tests from our kill data, and calculated a 10-to-1 innate preference for hares. We would draw the same conclusion about switching even if this preference was as low as about 5 to 1. During the seven winters that red squirrels represented <50% of the prey biomass, the diets of lynx were composed of an average of 0.4 ± 0.7% (SD) squirrels, whereas 36.2 ± 1.5% of their diets were squirrels when squirrels composed >50% of the prey biomass present.

13.6   Synthesis and Conclusions

The main objectives of the Kluane research on mammalian predators were to determine and contrast the numerical and functional responses of coyotes and lynx to the changes in prey abundance associated with the 10-year cycle of snowshoe hares. Second, we wanted to investigate the behavioral mechanisms leading to changes in kill rates (i.e., the functional responses) of the two predators.

Coyotes and lynx responded to cyclic changes in the abundance of their main prey species, snowshoe hares, with major demographic and behavioral changes. Our main results were:

1. Coyotes and lynx responded numerically in much the same way to the 26- to 44-fold fluctuation in numbers of hares in our study area. Numbers of coyotes varied 6-fold and those of lynx 7.5-fold, and the abundances of both predators were maximal a year later than the peak in numbers of snowshoe hares.
2. Cyclic declines in numbers of coyotes were associated with lower reproductive output and high dispersal rates. Likewise, lynx produced few to no kittens after the second winter of declining numbers of hares. High dispersal rates were characteristic of lynx during the cyclic decline, and low *in situ* survival was observed later in the decline.
3. The sizes of home ranges of lynx, and possibly coyotes, increased as densities of hares declined, and the territorial organization of lynx apparently broke down after 2 years of declining abundance of prey.
4. Coyotes and lynx both fed mostly on hares during all winters except during cyclic lows. Coyotes killed more voles than hares during two winters when abundance of hares was lowest and numbers of small mammals were high. The main alternative prey of lynx was red squirrels, which they killed more than hares during the two winters of low hare numbers and during the first winter of the subsequent cyclic increase.
5. Coyotes and lynx both showed clear functional responses to changes in the densities of snowshoe hares. Coyotes responded with a ninefold change in kill rates, and their functional response is described equally well by linear and type-2 curves. Kill rates of hares by coyotes were generally lower during the cyclic decline in hare densities than during the increase. Kill rates of hares by lynx changed four- to fivefold

during the cyclic fluctuation. Their functional response is well described by a type-2 curve. Kill rates by lynx were lower during the increase phase of the hare cycle than they were during the hare decline.

6. Coyotes killed the maximum number of hares per coyote (2.3 hares/day) one year before the peak in hare abundance. During the first rise and fall of hare abundance, lynx killed the most hares (1.2 hares/day) one year after the peak in hare abundance, but as hare numbers reached high densities again, at the end of this study, lynx killed 1.6 hares/day. These kill rates are more than the estimated energetic requirements of the predators.

7. Coyotes killed more hares early in the winter and cached many of these for later retrieval.

8. Coyotes and lynx preferred hares to other prey species during all phases of the cycle. At low densities of hares, lynx switched to hunting more actively for, and preying on, red squirrels.

9. Habitat use by coyotes and lynx changed over the cycle and roughly paralleled trends in habitats used by hares. Both predators used the densest cover during the second winter of decline in hare abundance. Coyotes and lynx concentrated their activities in areas of high densities of hares during most winters. Coyotes hunted for voles in more open cover during the cyclic low in hare abundance, when voles were numerous.

10. Lynx increasingly used hunting beds for ambushing both hares and red squirrels during the cyclic decline and low. Coyotes switched their hunting tactics to active foraging for voles during only one winter. Both predators frequently traveled on the trails of hares, and coyotes also often used trails made by other predators and humans.

11. Lynx hunted in adult groups for the first time during the cyclic decline and low in hare numbers. Our data suggest the per-individual foraging success may have been higher for these groups.

Although the spatial and temporal scales of the Kluane Project were very large relative to most field studies, we could only work on the scale of the whole study area for these wide-ranging predators. We therefore have no replicates for examining variation in the demographic and behavioral parameters that we estimated. Furthermore, because we could not distinguish among individual predators while snow tracking, we could not measure variances in, for example, diets or kill rates within our study area either (although we did attempt to index variance by subdividing our data by time of winter and section of our study area). We therefore need to be conservative in the inferences we make from our analyses.

These data represent a fairly unique attempt to measure the specific parameters of predator–prey interactions among terrestrial vertebrates in the field, in a multispecies system with all of the major components quantified. Replicates will only accumulate over a period of decades. Below, we speculate on some of the implications suggested by our studies, and those of others, on the roles of coyotes and lynx in the boreal ecosystem. We see their value as much in suggesting hypotheses as in providing answers.

13.6.1 Numerical Responses

In theory, the 1-year lag in the numerical responses of coyotes and lynx should destabilize the predator–prey interaction, as is typical with the responses of many specialists, and contribute to its cyclic behavior. The delay of 1 year between peak densities of hares and lynx appears to be consistent among our study and other field studies (Keith et al. 1977, Poole 1994, Slough and Mowat 1996). But it is shorter than the 2- to 3-year lag, frequently used in models of the hare cycle, suggested by historical fur records (Royama 1992). Increased vulnerability of lynx to fur-trapping during periods of prey shortage may bias the accuracy of fur statistics as indices of population densities. For coyotes, the only other comparable field data show no lag in their numerical responses during the cyclic increase, but a 2-year lag in their cyclic decline (Keith et al. 1977). Modelers will need to incorporate shorter lags into their models to examine the effects of predation, as shorter time lags imply shorter cyclic periodicity (May 1981).

Densities of coyotes in our study area were about half those in Alberta over the course of the cycle (Keith et al. 1977, Todd et al. 1981), whereas our densities of lynx were about double (Brand et al. 1976, Keith et al. 1977). The estimated peak abundance of hares in Alberta was approximately five times that observed at Kluane (Keith 1990, Boutin et al. 1995). In other areas of their range, coyotes may reach densities as high as $50-100/100$ km^2 (Camenzind 1978, Andelt 1985) relative to the peak of $9/100$ km^2 we observed at Kluane, suggesting that the boreal forest is suboptimal habitat for coyotes. In the Alberta study, coyotes extensively used livestock carcasses as alternative prey during cyclic lows, which apparently allowed them to maintain higher densities. Peak densities of lynx were higher in our study and in the two other recent northern investigations (Poole 1994, Slough and Mowat 1996) than in Alberta. The lack of synchrony between population fluctuations of coyotes and lynx with those of small mammals in our study area (Boutin et al. 1995) underscores the importance of hares to the dynamics of these predators in the north.

Cyclic declines are characterized by long-distance emigration by lynx (Slough and Mowat 1996, Poole 1997, this study), and probably by coyotes as well. High mobility by predators may be important in synchronizing population cycles on a regional scale (Finerty 1980, Ims and Steen 1990, Korpimäki and Norrdahl 1991, Korpimäki and Krebs 1996). Influxes of lynx have been noted south of their usual range during cyclic declines to the north (Mech 1980), and these dispersers may establish new home ranges far from their points of origin (Poole 1997). Likewise, long-distance natal dispersal is typical of coyotes in some areas (Harrison 1992), and we observed high rates of emigration by adults from our study area. The role of these migratory movements in synchronizing the cycle on a continental scale has not been fully explored. Cyclic peaks in populations of introduced snowshoe hares on the island of Newfoundland have been synchronous with those on the mainland, despite the island's geographic isolation, which suggests that dispersal of terrestrial predators is not necessary for regional synchrony.

The role of emigration in maintaining the local persistence of populations of predators through cyclic lows is uncertain as well. Breitenmoser et al. (1993) proposed that locally born resident lynx occupied large, stable home ranges between periods of prey abundance and that these formed core populations for subsequent cyclic increases. Between the cyclic peak in the early 1980s, we found that at least one adult male persisted until the next cyclic peak in approximately the same home range. However, in both our study during the most recent cycle and in the Northwest Territories (Poole 1994, 1995), we observed a complete turnover of resident lynx during the cyclic declines. We could not confirm the origin of any of the residents that became established during the cyclic low. Slough and Mowat (1996), in contrast, noted that two of three resident animals at the cyclic low had been born in their study area. Local dynamics may vary consistently among source and sink areas (Pulliam 1988), or the patterns of persistence of resident lynx and settlement of dispersers may be random.

13.6.2 Functional Responses

The functional response of coyotes to changing densities of hares at Kluane showed no clear asymptote (figure 13.12), especially in the early winter (O'Donoghue et al. 1998b), whereas that of lynx had a more clearly decelerating slope. Our calculations suggest that both predators killed more hares than energetically required. Many of the hares killed by coyotes in the early winter were cached, and some of these caches were lost to scavengers or not retrieved. Storing excess food during times of plenty may guard against future periods of scarcity or a loss of caches (Vander Wall 1990). Surplus killing could then be the result, and this, in combination with some wastage (incomplete consumption) of carcasses by both coyotes and lynx (O'Donoghue et al. 1998b), apparently contributed to the high kill rates we observed. The higher proportion of juvenile hares early in winters could also have contributed to the higher kill rates. In this case, models of predator–hare interactions that assume functional responses based on energetic requirements of the predators would underestimate the impacts of predation.

We observed a 1-year lag in the functional responses of lynx to the changes in the densities of hares from the start of our study until the subsequent cyclic low. Kill rates by lynx were higher during the cyclic decline than during the increase, while the opposite was true for coyotes. Differences in kill rates between the two cyclic phases could be the result of changing age structures of predators—most offspring are born during cyclic increases and peaks, leading to progressively older mean ages (and more-experienced hunters) from the early cyclic decline to the subsequent cyclic increase (O'Connor 1986, Slough and Mowat 1996)—or result from lags in the prey preferences of predators relative to changing frequencies of prey. Phase dependency in kill rates could contribute to the asymmetry of populations cycles observed in historical time series and models (Royama 1992).

Of the basic behavioral components of functional responses (Holling 1959a, 1966), we observed large changes only in our index of the reactive distances of predators in response to changing densities of hares (O'Donoghue et al. 1998b). Reactive distances may be a function of hunger or environmental factors, such as density of habitat (Holling 1965, Abrams 1990, Bell 1991), and satiated predators have been shown to react to prey within smaller perceptual fields (e.g., Wood and Hand 1985). The increases in hunting success that we observed for coyotes and lynx during periods of cyclic increase (figure 13.13) may have been related to an older age structure in populations of animals at those times. Lack of change in travel rates and activity rates by predators relative to prey densities has been noted in several studies (Holling 1966, Schaller 1972), but changes in these parameters have been documented for other predators (Smith 1974, Bell 1991).

Our results suggest that lynx switched to hunting red squirrels during periods of low abundance of hares and that this switch persisted into the subsequent period of cyclic increase of hares. Switching is more characteristic of generalist predators, which typically have more plastic foraging behavior (MacNally 1995) than specialists. In northern boreal forests, however, the availability of alternative prey to hares is limited. Small mammals were the most frequently used alternative prey of coyotes in our study area, and their availability was limited by snow cover during winter (Wells and Bekoff 1982, Halpin and Bissonette 1988) and the habitat composition of our study area (*Microtus* seem more vulnerable to predation [Henttonen et al. 1987], and meadows only composed about 7% of our study area). Coyotes killed some red squirrels, but their hunting tactics (they seldom ambushed prey) and lack of ability to climb trees likely limited the availability of squirrels.

The main alternative prey of lynx were red squirrels, which were relatively abundant at all phases of the cycle (chapter 9; Boutin et al. 1995). The increasing use of ambush beds by lynx during the cyclic decline and progressively higher hunting success by lynx chasing squirrels during the cyclic low indicate that lynx surviving into this period became skilled at catching squirrels. Even when numbers of hares began to increase in 1994–1995 and coyotes shifted their hunting effort to hares again, these lynx continued hunting mostly squirrels. The foraging decisions of predators are typically more strongly influenced by recent feeding choices than those made over a longer time frame (Shettleworth et al. 1993). Assuming there is a cost to gaining enough information to accurately track resource availability, predators may maintain dietary preferences even as the relative profitabilities of those prey change (Lewis 1986, Dukas and Clark 1995, Dall and Cuthill 1997). Patterns of dietary choice and the functional responses of coyotes and lynx in northern boreal forests are likely closely related to the local availability of alternative prey species to hares.

Prey switching may result from a number of mechanisms, including the development of search images for specific prey (Tinbergen 1960, Lawrence and Allen 1983), switches among habitat types as the relative profitabilities of foraging in them change (Royama 1970, Murdoch and Oaten 1975), and changes in foraging tactics by predators (e.g., Lawton et al. 1974, Davies 1977, Akre and Johnson 1979, Formanowicz and Bradley 1987). We documented shifts in habitat use by coyotes and lynx during the hare cycle that paralleled those by hares (Wolff 1980, Hik 1995) but saw no major switches away from denser forested areas. The densest habitats may act as refuges for hares during cyclic lows (Wolff 1980, Akçakaya 1992), and hares were found in denser cover than their predators at all phases of the cycle (O'Donoghue et al. 1998a). Aggregation of predators in patches of good densities of prey can be considered a functional or numerical response, depending on scale, but the effect on the prey (increasing rates of predation) is the same (Solomon 1949, Murdoch and Oaten 1975, Hanski et al. 1991, Korpimäki and Krebs 1996).

Several other changes in the behavior of coyotes and lynx during our study may also have affected the pattern and magnitude of kill rates. The switch by lynx to hunting red squirrels during the cyclic low was accompanied by increased use of ambush beds. Although use of these beds did not seem to increase hunting success, ambush tactics may be more energy efficient or better for hunting alternative prey during periods of low abundance of hares. Hunting in groups may have led to increased foraging success for lynx during the cyclic low, as has been shown for other group-living carnivores (Gittleman 1989). Likewise, use of trails by predators for travel helps them conserve energy and may increase encounter rates with prey; this is especially important for coyotes because of their high foot-loads for travel in the soft snows of the northern forests (Murray and Boutin 1991).

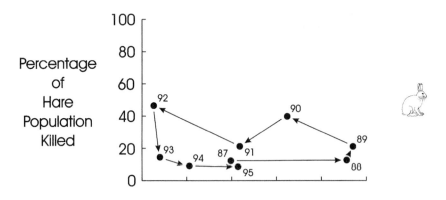

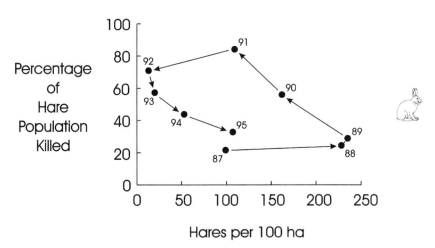

Figure 13.16 Estimated total impact of predation by coyotes and lynx during winter (October through April) on populations of snowshoe hares from 1987–1988 (labeled "87" on graph) through 1995–1996 at Kluane, as calculated from the combined numerical and functional responses of the predators and from survival rates of radio-collared hares.

13.6.3 Total Impact of Predation by Coyotes and Lynx on Hares

We estimated the total impact of predation by coyotes and lynx on populations of snowshoe hares by calculating the total number of hares killed by these predators each winter, from October through April, based on our measured functional and numerical responses. The percentage of the autumn population of hares killed by coyotes and lynx over the winter exhibits a counterclockwise temporal pattern typical of delayed density-dependent predation (Sinclair and Pech 1996; figure 13.16). However, the low predation rate calculated in 1991–1992, largely the result of rapidly declining populations of coyotes and lynx (figure 13.1), departs from the pattern expected with a simple time lag of the effects of predation.

We used data collected on the survival of radio-collared hares to make a second, independent estimate of the effects of predation by coyotes and lynx on hares. This analysis suggests that predation by these predators was 1.5–4 times higher than our estimates from this study (figure 13.16), particularly during the decline and low phases of the cycle. Rates of predation calculated from the hare survival data are likely overestimates, due to higher vulnerability of radio-collared hares to predation (Mech 1967), stress to hares caused by live trapping, or attraction of predators to trapping grids. Nonetheless, the data from the hare telemetry suggest that the effect of predation by coyotes and lynx was highest in 1991–1992, the winter we measured quite low rates from our predator data. Many radio-collared hares were killed by coyotes in October–November that winter, and we could have missed many kills because snow conditions were poor for snow tracking until November 1991.

Our estimates of the total impact of predation by coyotes and lynx ranged from 9–13% of the autumn population of hares during the cyclic increase, 21% at the peak, 21–40% during the decline, and 15–47% during the cyclic low in hare abundance (figure 13.16). These are approximately double those calculated in Alberta of 3–6% during the increase, and 9–20% during the decline of hare populations (Keith et al. 1977). We suggest that our estimates of kill rates by predators, particularly coyotes, are conservative (O'Donoghue 1998b) and that the true impact of predation by coyotes and lynx was somewhere between the two estimates calculated in figure 13.16.

The impact of all predators, in all seasons, needs to be assessed for a full evaluation of the effects of predation on the hare cycle. These data suggest that the delayed numerical (coyotes and lynx) and functional (lynx) responses of mammalian predators contributed to the cyclic dynamics of populations of hares (2- to 3-year lag in maximum predation rates) and that the magnitude of the effect of predation by coyotes and lynx is greater than previously measured. The functional and numerical responses that we have measured can be used to develop an accounting model of the whole system and to parameterize simulation models (Korpimäki and Krebs 1996).

13.6.4 Impact of Predation by Coyotes and Lynx on Alternative Prey

There is ample evidence from field studies that cyclic fluctuations of a preferred prey species may also occur in populations of alternative prey of predators (Henttonen 1985, Marcström et al. 1988, 1989, Lindström et al. 1994). In our study area, trends in populations of arctic ground squirrels, grouse, and ptarmigan were correlated with those of hares (Boutin et al. 1995), suggesting there may be a causal connection.

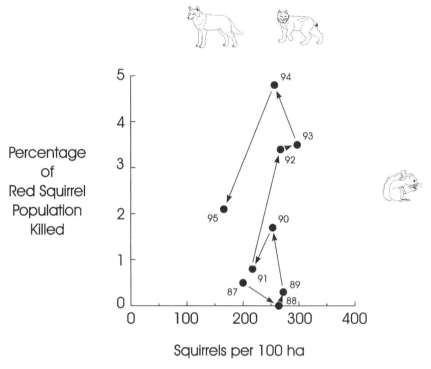

Figure 13.17 Estimated total impact of predation by coyotes and lynx during winter (October through April) on populations of red squirrels from 1987–1988 (labeled "87" on graph) through 1995–1996 at Kluane, as calculated from the combined numerical and functional responses of the predators.

We evaluated the effects of predation by coyotes and lynx during winter on populations of red squirrels using our measured numerical responses and kill rates calculated in the same manner as those for hares. We were unable to evaluate the effects of predation on numbers of small mammals because of the difficulty in judging whether attempted kills were successful, and we did not attempt to investigate predation on other prey species due to the low number of kills by coyotes and lynx. The estimated total effect of predation by coyotes and lynx on red squirrels was small (<5% of the autumn populations) in all winters and showed no pattern relative to densities of squirrels (figure 13.17). This agrees with the experimental results of Stuart-Smith and Boutin (1995), showing predation by mammals had little effect on population sizes of red squirrels.

13.6.5 Coexistence of Coyotes and Lynx

Competition between sympatric predators and how it affects partitioning of resources and breadth of resource use has been much explored and debated in the literature (Rosenzweig 1966, Schoener 1982, Wiens 1993). In general, overlap in resource use between sympatric species is less during lean times (high competition; Schoener 1982) than during periods of resource abundance. But this prediction may not hold due to complex rela-

tionships between resource abundance and species-specific thresholds of resource limitation (Wiens 1993) or different strategies of resource acquisition (Glasser and Price 1982,
MacNally 1995). Accordingly, studies of patterns of overlap among predators have shown
no change (Jaksic et al. 1993, Meserve et al. 1996), lower overlap (Korpimäki 1987), or
higher overlap (Wiens 1993) with lower resource abundance.

We did not conduct experiments that would enable us to add empirical evidence on
mechanisms permitting coexistence of coyotes and lynx. The question of how these two
predators, of similar size, survive using a very limited resource base is especially interesting, given that coyotes are relatively new immigrants in the north. We summarize our
observations here to suggest questions for further research.

We calculated Horn's index of overlap in diet (based on number of kills) and habitat
use between coyotes and lynx for each winter of our study. Overlap in diet was high
(>0.75) in all winters (figure 13.18). The lowest dietary overlap was in 1992–1993, when
coyotes killed mostly small mammals and lynx killed more red squirrels, at a time when

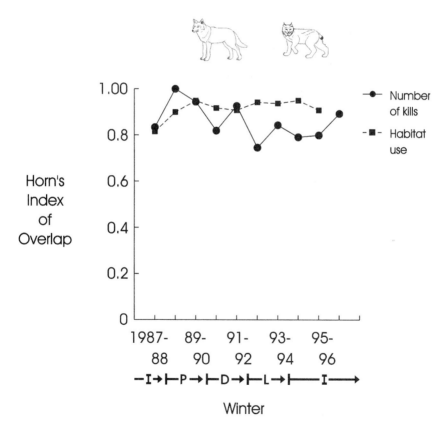

Figure 13.18 Overlap in diets (based on number of kills located during snow tracking) and
habitat use by coyotes and lynx during winter from 1987–1988 through 1996–1997 at Kluane.
The phases of the hare cycle (I = increase, P = peak, D = decline, L = low) are shown by the
horizontal bar below the x-axis.

numbers of hares were low. Overlap in habitat use was high (range 0.82–0.95) in all winters (figure 13.18).

These observations are consistent with those made of high overlap in diets between bobcats and coyotes in areas where coyotes were relatively recent colonizers (Witmer and DeCalesta 1986, Major and Sherburne 1987, Litvaitis and Harrison 1989). There is correlative evidence that coyotes may depress numbers of bobcats in the western and northeastern United States (Nunley 1978, Litvaitis and Harrison 1989), and both interspecific competition and intraguild predation have been suggested as mechanisms. We observed only one case of a coyote killing a young lynx (O'Donoghue et al. 1995), so the influence of predation by coyotes on lynx is not likely large in our area. Predation among predators (intraguild predation), however, has been recorded among many species of carnivores (e.g., Elsey 1954, Eaton 1979, Stephenson et al. 1991, O'Donoghue et al. 1995), and intraguild predation may have important consequences in some predator–prey systems (review in Polis et al. 1989).

The degree of overlap in the diets of coyotes and lynx during cyclic lows is likely highly dependent on the availability of alternative prey. In our study area, population fluctuations of red squirrels and voles, which are largely unrelated to cyclic hare dynamics (Boutin et al. 1995), would have a great influence on this overlap. Lynx appear better able than coyotes to take advantage of the most consistently available alternative prey species, red squirrels, and during periods when few voles or hares are available, lynx may persist in greater numbers than coyotes. In areas where other alternative food sources, such as the carrion of moose (e.g., Staples 1995) or livestock (e.g., Todd et al. 1981) are consistently available, coyotes may persist in higher numbers. Even with high or even complete overlap in resource use, though, recent models of competition have suggested that species may coexist for long periods of time (Hubbell and Foster 1986, MacNally 1995).

13.6.6 The Specialist–Generalist Contrast

The degree of versatility in resource use by any consumer is a function of species-specific constraints and external factors such as the relative availabilities of resources. Predators that are generalists, when considered over their whole geographic range, may be local specialists (MacNally 1995). Clearly, in a low-diversity environment such as the boreal forest in our study area, dietary options are limited for terrestrial predators. Given the limited availability of voles and other alternative prey, coyotes in our study area appeared to specialize on hares even during periods (1994–1995) when lynx were preying more heavily on red squirrels.

In fluctuating environments, facultative foraging strategies, in which predators concentrate their foraging efforts on the most profitable prey, which may vary in time, should be favored over obligate strategies (Glasser 1982). Over their geographic range, North American lynx can certainly be said to be more dependent on snowshoe hares than are coyotes—on this basis, they can be considered hare specialists. Locally, though, at least some individual lynx can specialize on preying on red squirrels during periods of low hare abundance, and so they may perhaps better be called facultative specialists. We do not know whether the changes that we observed during the hare cycle were due to all or only some individuals shifting their foraging behavior. Studies comparing the plasticities of individual lynx and coyotes in adjusting their patterns of resource use to changing availabilities of prey would be valuable for clarifying mechanisms behind shifts at the population level.

Literature Cited

Abrams, P. A. 1990. The effects of adaptive behavior on the type-2 functional response. Ecology **71**:877–885.

Akçakaya, H. R. 1992. Population cycles of mammals: evidence for a ratio-dependent predation hypothesis. Ecological Monographs **62**:119–142.

Akre, B. G., and D. M. Johnson. 1979. Switching and sigmoidal functional response curves by damselfly naiads with alternative prey available. Journal of Animal Ecology **48**:703–720.

Andelt, W. F. 1985. Behavioral ecology of coyotes in south Texas. Wildlife Monographs **94**:1–45.

Arnqvist, G., and D. Wooster. 1995. Meta-analysis: synthesizing research findings in ecology and evolution. Trends in Ecology and Evolution **10**:236–240.

Banfield, A. W. F. 1974. The mammals of Canada. University of Toronto Press, Toronto, Ontario.

Barash, D. P. 1971. Cooperative hunting in the lynx. Journal of Mammalogy **52**:480.

Bekoff, M. (ed). 1978. Coyotes: biology, behavior, and management. Academic Press, New York.

Bekoff, M., and M. C. Wells. 1981. Behavioural budgeting by wild coyotes: the influence of food resources and social organization. Animal Behaviour **29**:794–801.

Bekoff, M., and M. C. Wells. 1986. Social ecology and behavior of coyotes. Advances in the Study of Behavior **16**:251–338.

Bell, W. J. 1991. Searching behaviour. Chapman and Hall, London.

Berg, W. E., and R. A. Chesness. 1978. Ecology of coyotes in northern Minnesota. *in* M. Bekoff (ed). Coyotes: biology, behavior, and management, pages 229–247. Academic Press, New York.

Bond, R. M. 1939. Coyote food habits on the Lava Beds National Monument. Journal of Wildlife Management **3**:180–198.

Boutin, S. 1995. Testing predator-prey theory by studying fluctuating populations of small mammals. Wildlife Research **22**:89–100.

Boutin, S., and H. D. Cluff. 1989. Coyote prey choice: optimal or opportunistic foraging? A comment. Journal of Wildlife Management **53**:663–666.

Boutin, S., C. J. Krebs, R. Boonstra, M. R. T. Dale, S. J. Hannon, K. Martin, A. R. E. Sinclair, J. N. M. Smith, R. Turkington, M. Blower, A. Byrom, F. I. Doyle, C. Doyle, D. Hik, L. Hofer, A. Hubbs, T. Karels, D. L. Murray, V. Nams, M. O'Donoghue, C. Rohner, and S. Schweiger. 1995. Population changes of the vertebrate community during a snowshoe hare cycle in Canada's boreal forest. Oikos **74**:69–80.

Bowen, W. D. 1982. Home range and spatial organization of coyotes in Jasper National Park, Alberta. Journal of Wildlife Management **46**:201–216.

Brand, C. J., and L. B. Keith. 1979. Lynx demography during a snowshoe hare decline in Alberta. Journal of Wildlife Management **43**:827–849.

Brand, C. J., L. B. Keith, and C. A. Fischer. 1976. Lynx responses to changing snowshoe hare densities in Alberta. Journal of Wildlife Management **40**:416–428.

Breitenmoser, U., C. Breitenmoser-Würsten, G. A. Zuleta, F. Bernhart, and M. O'Donoghue.

1992. A method to estimate travel distances of fast-moving animals. Biotelemetry **12**:318–326.

Breitenmoser, U., B. G. Slough, and C. Breitenmoser-Würsten. 1993. Predators of cyclic prey: is the Canada lynx victim or profiteer of the snowshoe hare cycle? Oikos **66**:551–554.

Camenzind, F. J. 1978. Behavioral ecology of coyotes on the National Elk Refuge, Jackson, Wyoming. *in* M. Bekoff (ed). Coyotes: biology, behavior, and management, pages 267–294. Academic Press, New York.

Clark, F. W. 1972. Influence of jackrabbit density on coyote population change. Journal of Wildlife Management **36**:343–356.

Crawley, M. J. 1992. Overview. *in* M. J. Crawley (ed). Natural enemies, pages 476–489. Blackwell Scientific Publications, Oxford.

Dall, S. R. X., and I. C. Cuthill. 1997. The information costs of generalism. Oikos **80**:197–202.

Davies, N. B. 1977. Prey selection and search strategy of the spotted flycatcher (*Muscicapa striata*), a field study on optimal foraging. Animal Behaviour **25**:1016–1033.

Dukas, R., and C. W. Clark. 1995. Searching for cryptic prey: a dynamic model. Ecology **76**:1320–1326.

Eaton, R. L. 1979. Interference competition among carnivores: a model for the evolution of social behavior. Carnivore **2**:9–16.

Elsey, C. A. 1954. A case of cannibalism in Canada lynx (*Lynx canadensis*). Journal of Mammalogy **35**:129.

Elton, C., and M. Nicholson. 1942. The ten year cycle in numbers of lynx in Canada. Journal of Animal Ecology **11**:215–244.

Erlinge, S. 1987. Predation and non-cyclicity in a microtine population in southern Sweden. Oikos **50**:347–352.

Erlinge, S., G. Göransson, L. Hansson, G. Högstedt, O. Liberg, I. N. Nilsson, T. Nilsson, T. von Schantz, and M. Sylvén. 1983. Predation as a regulating factor on small mammal populations in southern Sweden. Oikos **40**:36–52.

Erlinge, S., G. Göransson, G. Högstedt, G. Jansson, O. Liberg, J. Loman, I. N. Nilsson, T. von Schantz, and M. Sylvén. 1984. Can vertebrate predators regulate their prey? American Naturalist **123**:125–133.

Erlinge, S., G. Göransson, G. Högstedt, G. Jansson, O. Liberg, J. Loman, I. N. Nilsson, T. von Schantz, and M. Sylvén. 1988. More thoughts on vertebrate regulation of prey. American Naturalist **132**:148–154.

Fichter, E., G. Schildman, and J. H. Sather. 1955. Some feeding patterns of coyotes in Nebraska. Ecological Monographs **25**:1–37.

Finerty, J. P. 1980. The population ecology of cycles in small mammals. Yale University Press, New Haven, Connecticut.

Floyd, T. J., L. D. Mech, and P. A. Jordan. 1978. Relating wolf scat content to prey consumed. Journal of Wildlife Management **42**:528–532.

Formanowicz, D. R., and P. J. Bradley. 1987. Fluctuations in prey density: effects on the foraging tactics of scolopendrid centipedes. Animal Behaviour **35**:453–461.

Gittleman, J. L. 1989. Carnivore group living: comparative trends. *in* J. L. Gittleman (ed). Carnivore behavior, ecology and evolution, pages 183–207. Cornell University Press, Ithaca, New York.

Glasser, J. W. 1982. A theory of trophic strategies: the evolution of facultative specialists. American Naturalist **119**:250–262.

Glasser, J. W., and H. J. Price. 1982. Niche theory: new insights from an old paradigm. Journal of Theoretical Biology **99**:437–460.

Haglund, B. 1966. Winter habits of the lynx (*Lynx lynx* L.) and wolverine (*Gulo gulo* L.) as revealed by tracking in the snow. (In Swedish; English summary.) Viltrevy **4**:81–310.

Halpin, M. A., and J. A. Bissonette. 1988. Influence of snow depth on prey availability and habitat use by red fox. Canadian Journal of Zoology **66**:587–592.

Hanski, I., L. Hansson, and H. Henttonen. 1991. Specialist predators, generalist predators, and the microtine rodent cycle. Journal of Animal Ecology **60**:353–367.

Harrison, D. J. 1992. Dispersal characteristics of juvenile coyotes in Maine. Journal of Wildlife Management **56**:128–138.

Hassell, M. P., and R. M. May. 1986. Generalist and specialist natural enemies in insect predator-prey interactions. Journal of Animal Ecology **55**:923–940.

Henttonen, H. 1985. Predation causing extended low densities in microtine cycles: further evidence from shrew dynamics. Oikos **45**:156–157.

Henttonen, H., T. Oksanen, A. Jortikka., and V. Haukisalmi. 1987. How much do weasels shape microtine cycles in the northern Fennoscandian taiga? Oikos **50**:353–365.

Hik, D. S. 1995. Does risk of predation influence population dynamics? Evidence from the cyclic decline of snowshoe hares. Wildlife Research **22**:115–129.

Holling, C. S. 1959a. Some characteristics of simple types of predation and parasitism. Canadian Entomologist **91**:385–398.

Holling, C. S. 1959b. The components of predation as revealed by a study of small-mammal predation of the European pine sawfly. Canadian Entomologist **91**:293–320.

Holling, C. S. 1965. The functional response of predators to prey density and its role in mimicry and population regulation. Memoirs of the Entomological Society of Canada **45**:1–60.

Holling, C. S. 1966. The functional response of vertebrate predators to prey density. Memoirs of the Entomological Society of Canada **48**:1–86.

Hubbell, S. P., and R. B. Foster. 1986. Biology, chance, and history and the structure of tropical rain forest tree communities. *in* J. Diamond and T. J. Case (eds). Community ecology, pages 314–329. Harper and Row, New York.

Ims, R. A., and H. Steen. 1990. Geographic synchrony in microtine population cycles: a theoretical evaluation of the role of nomadic avian predators. Oikos **57**:381–387.

Jaksic, F. M., P. Feinsinger, and J. E. Jiménez. 1993. A long-term study on the dynamics of guild structure among predatory vertebrates at a semi-arid Neotropical site. Oikos **67**:87–96.

Johnson, M. K., and R. M. Hansen. 1979. Estimating coyote food intake from undigested residues in scats. American Midland Naturalist **102**:363–367.

Keith, L. B. 1990. Dynamics of snowshoe hare populations. *in* H. H. Genoways (ed). Current mammalogy, pages 119–195. Plenum, New York.

Keith, L. B., J. R. Cary, O. J. Rongstad, and M. C. Brittingham. 1984. Demography and ecology of a declining snowshoe hare population. Wildlife Monographs **90**:1–43.

Keith, L. B., A. W. Todd, C. J. Brand, R. S. Adamcik, and D. H. Rusch. 1977. An analysis of predation during a cyclic fluctuation of snowshoe hares. Proceedings of the International Congress of Game Biologists **13**:151–175.

Keith, L. B., and L. A. Windberg. 1978. A demographic analysis of the snowshoe hare cycle. Wildlife Monographs **58**:1–70.

Kie, J. G., J. A. Baldwin, and C. J. Evans. 1994. CALHOME home range analysis program. Electronic user's manual. California Department of Fish and Game, Fresno.

Koehler, G. M. 1990. Population and habitat characteristics of lynx and snowshoe hares in north central Washington. Canadian Journal of Zoology **68**:845–851.

Koehler, G. M. and K. B. Aubry. 1994. Lynx. *in* L. F. Ruggiero, K. B. Aubry, S. W. Buskirk, L. J. Lyon, and W. J. Zielinski (eds). The scientific basis for conserving forest carnivores: American marten, fisher, lynx, and wolverine in the western United States, pages 74–98. General Technical Report RM-254, Fort Collins, Colorado. U.S. Department of Agriculture, Forest Service, Rocky Mountain Forest and Range Experiment Station.

Korpimäki, E. 1987. Dietary shifts, niche relationships and reproductive output of coexisting kestrels and long-eared owls. Oecologia **74**:277–285.

Korpimäki, E. 1993. Regulation of multiannual vole cycles by density-dependent avian and mammalian predation? Oikos **66**:359–363.

Korpimäki, E., and C. J. Krebs. 1996. Predation and population cycles of small mammals. BioScience **46**:754–764.

Korpimäki, E., and K. Norrdahl. 1991. Do breeding nomadic avian predators dampen population fluctuations of small mammals? Oikos **62**:195–208.

Korpimäki, E., K. Norrdahl, and T. Rinta-Jaskari. 1991. Responses of stoats and least weasels to fluctuating food abundances: is the low phase of the vole cycle due to mustelid predation? Oecologia **88**:552–561.

Krebs, C. J. 1999. Ecological methodology. Benjamin/Cummings, Menlo Park, California.

Krebs, C. J., S. Boutin, R. Boonstra, A. R. E. Sinclair, J. N. M. Smith, M. R. T. Dale, K. Martin, and R. Turkington. 1995. Impact of food and predation on the snowshoe hare cycle. Science **269**:1112–1115.

Krebs, C. J., B. S. Gilbert, S. Boutin, A. R. E. Sinclair, and J. N. M. Smith. 1986. Population biology of snowshoe hares. I. Demography of food-supplemented populations in the southern Yukon, 1976–84. Journal of Animal Ecology **55**:963–982.

Kruuk, H. 1972. Surplus killing by carnivores. Journal of Zoology **166**:233–244.

Lawrence, E. S., and J. A. Allen. 1983. On the term 'search image.' Oikos **40**:313–314.

Lawton, J. H., J. R. Beddington, and R. Bonser. 1974. Switching in invertebrate predators. *in* M. H. Williamson (ed). Ecological stability, pages 141–158. Chapman and Hall, London.

Lewis, A. C. 1986. Memory constraints and flower choice in *Pieris rapae*. Science **232**:863–865.

Lindström, E. R., H. Andrén, P. Angelstam, G. Cederlund, B. Hörnfeldt, L. Jäderberg, P.-A. Lemnell, B. Martinsson, K. Sköld, and J. E. Swenson. 1994. Disease reveals the predator: sarcoptic mange, red fox predation, and prey populations. Ecology **75**:1042–1049.

Lindström, E. R., and B. Hörnfeldt. 1994. Vole cycles, snow depth and fox predation. Oikos **70**:156–160.

Litvaitis, J. A., and D. J. Harrison. 1989. Bobcat-coyote niche relationships during a period of coyote population increase. Canadian Journal of Zoology **67**:1180–1188.

Litvaitis, J. A., and W. W. Mautz. 1980. Food and energy use by captive coyotes. Journal of Wildlife Management **44**:56–61.

MacCracken, J. G., and R. M. Hansen. 1987. Coyote feeding strategies in southeastern Idaho: optimal foraging by an opportunistic predator? Journal of Wildlife Management **51**:278–285.

MacNally, R. C. 1995. Ecological versatility and community ecology. Cambridge University Press, Cambridge.

Major, A. R. 1989. Lynx, *Lynx canadensis canadensis* (Kerr) predation patterns and habitat use in the Yukon Territory, Canada. MSc thesis. State University of New York, Syracuse.

Major, J. T., and J. A. Sherburne. 1987. Interspecific relationships of coyotes, bobcats, and red foxes in western Maine. Journal of Wildlife Management **51**:606–616.

Manly, B. F. J., P. Miller, and L. M. Cook. 1972. Analysis of a selective predation experiment. American Naturalist **106**:719–736.

Marcström, V., L. B. Keith, E. Engren, and J. R. Cary. 1989. Demographic responses of arctic hares (*Lepus timidus*) to experimental reductions of red foxes (*Vulpes vulpes*) and martens (*Martes martes*). Canadian Journal of Zoology **67**:658–668.

Marcström, V., R. E. Kenward, and E. Engren. 1988. The impact of predation on boreal tetraonids during vole cycles: an experimental study. Journal of Animal Ecology **57**:859–872.

Marshal, J. P., and S. Boutin. 1999. Power analysis of wolf-moose functional responses. Journal of Wildlife Management **63**:396–402.

May, R. M. 1981. Models for single populations. *in* R. M. May (ed). Theoretical ecology. Principles and applications, pages 5–29. Sinauer, Sunderland, Massachusetts.

Mech, L. D. 1967. Telemetry as a technique in the study of predation. Journal of Wildlife Management **31**:492–496.

Mech, L. D. 1980. Age, sex, reproduction, and spatial organization of lynxes colonizing northeastern Minnesota. Journal of Mammalogy **61**:262–267.

Meserve, P. L., J. R. Gutiérrez, J. A. Yunger, L. C. Contreras, and F. M. Jaksic. 1996. Role of biotic interactions in a small mammal assemblage in semiarid Chile. Ecology **77**:133–148.

Mohr, C. O. 1947. Table of equivalent populations of North American mammals. American Midland Naturalist **37**:223–249.

Moore, G. C., and G. R. Parker. 1992. Colonization by the eastern coyote (*Canis latrans*). *in* A. H. Boer (ed). Ecology and management of the eastern coyote, pages 23–37. Wildlife Research Institute, Fredericton, New Brunswick.

Moran, P. A. P. 1953. The statistical analysis of the Canada lynx cycle. I. Structure and prediction. Australian Journal of Zoology **1**:163–173.

Mowat, G., B. G. Slough, and S. Boutin. 1996. Lynx recruitment during a snowshoe hare population peak and decline in southwest Yukon. Journal of Wildlife Management **60**:441–452.

Mowat, G., B. G. Slough, and R. Rivard. 1994. A comparison of three live capturing devices for lynx: capture efficiency and injuries. Wildlife Society Bulletin **22**:644–650.

Murdoch, W. W. 1969. Switching in generalist predators: experiments on predator specificity and stability of prey populations. Ecological Monographs **39**:335–354.

Murdoch, W. W., and A. Oaten. 1975. Predation and population stability. Advances in Ecological Research **9**:1–131.

Murray, D. L., and S. Boutin. 1991. The influence of snow on lynx and coyote movements: does morphology affect behavior? Oecologia **88**:463–469.

Murray, D. L., S. Boutin, and M. O'Donoghue. 1994. Winter habitat selection by lynx and coyotes in relation to snowshoe hare abundance. Canadian Journal of Zoology **72**:1444–1451.

Murray, D. L., S. Boutin, M. O'Donoghue, and V. O. Nams. 1995. Hunting behaviour of a sympatric felid and canid in relation to vegetative cover. Animal Behaviour **50**:1203–1210.

Nams, V. O. 1990. Locate II user's guide. Pacer Computer Software, Truro, Nova Scotia.

Nellis, C. H., and L. B. Keith. 1968. Hunting activities and success of lynxes in Alberta. Journal of Wildlife Management **32**:718–722.

Nellis, C. H., and L. B. Keith. 1976. Population dynamics of coyotes in central Alberta, 1964–68. Journal of Wildlife Management **40**:389–399.

Nellis, C. H., S. P. Wetmore, and L. B. Keith. 1972. Lynx-prey interactions in central Alberta. Journal of Wildlife Management **36**:320–329.

Nunley, G. L. 1978. Present and historical bobcat population trends in New Mexico and the west. Proceedings of the Vertebrate Pest Conference **8**:177–184.

O'Connor, R. M. 1986. Reproduction and age distribution of female lynx in Alaska, 1961–1971—preliminary results. *in* S. D. Miller and D. D. Everett (eds). Cats of the world: biology, conservation, and management, pages 311–325. National Wildlife Federation, Washington D.C.

O'Donoghue, M., S. Boutin, C. J. Krebs, and E. J. Hofer. 1997. Numerical responses of coyotes and lynx to the snowshoe hare cycle. Oikos **80**:150–162.

O'Donoghue, M., S. Boutin, C. J. Krebs, D. L. Murray, and E. J. Hofer. 1998a. Behavioural responses of coyotes and lynx to the snowshoe hare cycle. Oikos **82**:169–183.

O'Donoghue, M., S. Boutin, C. J. Krebs, G. Zuleta, D. L. Murray, and E. J. Hofer. 1998b. Functional responses of coyotes and lynx to the snowshoe hare cycle. Ecology **79**:1193–1208.

O'Donoghue, M., E. Hofer, and F. I. Doyle. 1995. Predator versus predator. Natural History **104(3)**:6–9.

O'Gara, B. W. 1986. Reliability of scat analysis for determining coyote feeding on large mammals. Murrelet **67**:79–81.

Ozoga, J. J., and E. M. Harger. 1966. Winter activities and feeding habits of northern Michigan coyotes. Journal of Wildlife Management **30**:809–818.

Parker, G. R. 1981. Winter habitat use and hunting activities of lynx (*Lynx canadensis*) on Cape Breton Island, Nova Scotia. *in* J. A. Chapman and D. Pursley (eds). Worldwide Furbearer Conference proceedings, 3–11 August, 1980, pages 221–248. Frostburg, Maryland.

Parker, G. R., J. W. Maxwell, L. D. Morton, and G. E. J. Smith. 1983. The ecology of the lynx (*Lynx canadensis*) on Cape Breton Island. Canadian Journal of Zoology **61**:770–786.

Polis, G. A., C. A. Myers, and R. D. Holt. 1989. The ecology and evolution of intraguild predation: potential competitors that eat each other. Annual Review of Ecology and Systematics **20**:297–330.

Pollock, K. H., S. R. Winterstein, C. M. Bunck, and P. D. Curtis. 1989. Survival analysis in telemetry studies: the staggered entry design. Journal of Wildlife Management **53**:1–15.

Poole, K. G. 1994. Characteristics of an unharvested lynx population during a snowshoe hare decline. Journal of Wildlife Management **58**:608–618.

Poole, K. G. 1995. Spatial organization of a lynx population. Canadian Journal of Zoology **73**:632–641.

Poole, K. G. 1997. Dispersal patterns of lynx in the Northwest Territories. Journal of Wildlife Management **61**:497–505.

Poole, K. G., G. Mowat, and B. G. Slough. 1993. Chemical immobilization of lynx. Wildlife Society Bulletin. **21**:136–140.

Poole, K. G., L. A. Wakelyn, and P. N. Nicklen. 1996. Habitat selection by lynx in the Northwest Territories. Canadian Journal of Zoology **74**:845–850.

Pulliam, H. R. 1988. Sources, sinks, and population regulation. American Naturalist **132**:652–661.

Rosenthal, R. 1978. Combining the results of independent studies. Psychological Bulletin **85**:185–193.

Rosenzweig, M. L. 1966. Community structure in sympatric Carnivora. Journal of Mammalogy **47**:602–612.

Royama, T. 1970. Factors governing the hunting behaviour and selection of food by the great tit (*Parus major* L.). Journal of Animal Ecology **39**:619–668.

Royama, T. 1992. Analytical population dynamics. Chapman and Hall, London.

Saunders, J. K. 1963a. Food habits of lynx in Newfoundland. Journal of Wildlife Management **27**:384–390.

Saunders. J. K. 1963b. Movements and activities of lynx in Newfoundland. Journal of Wildlife Management **27**:390–400.

Schaffer, W. M. 1984. Stretching and folding in lynx fur returns: evidence for a strange attractor in nature? American Naturalist **124**:798–820.

Schaller, G. B. 1972. The Serengeti lion. University of Chicago Press, Chicago.

Schoener, T. W. 1982. The controversy over interspecific competition. American Scientist **70**:586–595.

Shettleworth, S. J., P. J. Reid, and C. M. S. Plowright. 1993. The psychology of diet selection. *in* R. N. Hughes (ed). Diet selection, pages 56–77. Blackwell Scientific Publications, London.

Sinclair, A. R. E., C. J. Krebs, J. N. M. Smith, and S. Boutin. 1988. Population biology of snowshoe hares. III. Nutrition, plant secondary compounds and food limitation. Journal of Animal Ecology **57**:787–806.

Sinclair, A. R. E., and R. P. Pech. 1996. Density dependence, stochasticity, compensation and predator regulation. Oikos **75**:164–173.

Slough, B. G., and G. Mowat. 1996. Population dynamics of lynx in a refuge and interactions between harvested and unharvested populations. Journal of Wildlife Management **60**:946–961.

Smith, J. N. M. 1974. The food searching behaviour of two European thrushes. II. The adaptiveness of search patterns. Behaviour **59**:1–59.

Smith, J. N. M., C. J. Krebs, A. R. E. Sinclair, and R. Boonstra. 1988. Population biology of snowshoe hares. II. Interactions with winter food plants. Journal of Animal Ecology **57**:269–286.

Solomon, M. E. 1949. The natural control of animal populations. Journal of Animal Ecology **18**:1–35.

Staples, W. R. 1995. Lynx and coyote diet and habitat relationships during a low hare population on the Kenai Peninsula, Alaska. MSc thesis. University of Alaska, Fairbanks.

Stenseth, N. C. 1995. Snowshoe hare populations: squeezed from below and above. Science **269**:1061–1062.

Stephenson, R. O., D. V. Grangaard, and J. Burch. 1991. Lynx, *Felis lynx*, predation on red foxes, *Vulpes vulpes*, caribou, *Rangifer tarandus*, and Dall sheep, *Ovis dalli*, in Alaska. Canadian Field-Naturalist **105**:255–262.

Stuart-Smith, A. K., and S. Boutin. 1995. Predation on red squirrels during a snowshoe hare decline. Canadian Journal of Zoology **73**:713–722.

Tinbergen, L. 1960. The natural control of insects in pine woods. I. Factors influencing the intensity of predation by songbirds. Archives Néerlandaises de Zoologie **13**:266–336.

Todd, A. W., and L. B. Keith. 1983. Coyote demography during a snowshoe hare decline in Alberta. Journal of Wildlife Management **47**:394–404.

Todd, A. W., L. B. Keith, and C. A. Fischer. 1981. Population ecology of coyotes during a fluctuation of snowshoe hares. Journal of Wildlife Management **45**:629–640.

Trexler, J. C., C. E. McCulloch, and J. Travis. 1988. How can the functional response best be determined? Oecologia **76**:206–214.

Trostel, K., A. R. E. Sinclair, C. J. Walters, and C. J. Krebs. 1987. Can predation cause the ten-year hare cycle? Oecologia **74**:185–193.

Vander Wall, S. B. 1990. Food hoarding in animals. University of Chicago Press, Chicago.

Van Vuren, D., and S. E. Thompson. 1982. Opportunistic feeding by coyotes. Northwest Science **56**:131–135.

Van Zyll de Jong, C. G. 1966. Food habits of the lynx in Alberta and the Mackenzie District, N. W. T. Canadian Field-Naturalist **80**:18–23.

Ward, R. M. P., and C. J. Krebs. 1985. Behavioural responses of lynx to declining snowshoe hare abundance. Canadian Journal of Zoology **63**:2817–2824.

Wells, M. C., and M. Bekoff. 1982. Predation by wild coyotes: behavioral and ecological analyses. Journal of Mammalogy **63**:118–127.

Wiens, J. A. 1993. Fat times, lean times and competition among predators. Trends in Ecology and Evolution **8**:348–349.

Witmer, G. W., and D. S. DeCalesta. 1986. Resource use by unexploited sympatric bobcats and coyotes in Oregon. Canadian Journal of Zoology **64**:2333–2338.

Wolff, J. O. 1980. The role of habitat patchiness in the population dynamics of snowshoe hares. Ecological Monographs **50**:111–130.

Wood, C. C., and C. M. Hand. 1985. Food-searching behaviour of the common merganser (*Mergus merganser*) I: Functional responses to prey and predator density. Canadian Journal of Zoology **63**:1260–1270.

14

Other Mammalian Predators

MARK O'DONOGHUE, STAN BOUTIN, ELIZABETH J. HOFER,
& RUDY BOONSTRA

There were 12 species of carnivores that were confirmed living within the study area of the Kluane Project. We studied the population dynamics and behavior of only two of these, coyotes and lynx, because these were the two most closely associated with the snowshoe hare cycle. However, we also counted the tracks of all carnivores during winter and recorded direct observations year-round. In this chapter, we briefly summarize the information we gathered on other carnivores at Kluane and discuss interactions of the carnivore group as a whole with each other and their prey.

The cyclical changes in prey abundance that are characteristic of the Holarctic affect not only the main predators of the fluctuating prey, but also a wide variety of other predators that may track cycles and herbivores that are their alternative prey. In North America, both the 10-year cycle of snowshoe hares and the 4-year cycles of lemmings and some species of voles may cause cyclical dynamics in a wide variety of other mammals and birds (Finerty 1980, Keith and Cary 1991). In northern Fennoscandia, trends in numbers of voles are strongly cyclical with a 3- to 4-year period, and this cycle is also observed in abundance of their predators, grouse, hares, and shrews (Hörnfeldt 1978, Angelstam et al. 1985, Henttonen 1985, Lindström et al. 1994).

Boutin et al. (1995) documented changes in the populations of 22 species of vertebrates over 5–19 years at Kluane, including 5 carnivores. Here, we update and expand on their data and discussion.

14.1   Methods

We counted the tracks of all carnivores each winter (October through April) along a 25-km transect that traversed our study area, on days after fresh snowfalls while tracks were distinguishable. We calculated annual means and standard errors of these counts using means for the whole winter for each of eight segments of this transect (1.4–4.9 km in length, separated by topographical features), standardized to tracks per track night per 100 km. These counts gave us indices of the numbers of predators present each winter and also information on the locations of animals that we did not have radio collared (O'Donoghue et al. 1997). During winters when snowshoe hares were abundant (1987–1988 through 1991–1992), their tracks frequently obscured the tracks of other smaller animals, so we only counted the number of weasel tracks on the day after snowfalls for the analyses presented here. From 1986–1987 through 1996–1997, we counted tracks along 12,194 km of transect.

We also recorded the numbers and locations of all tracks of other carnivores that crossed the trails of coyotes and lynx while we were snow tracking (chapter 13). The locations of these trails changed among winters (O'Donoghue et al. 1998), so we used these counts to indicate use of habitats by other predators relative to those used by coyotes and lynx, rather than as indices of abundance among years.

14.2   Red Fox

Red fox, *Vulpes vulpes,* and their tracks were regularly but infrequently observed in the Kluane study area from 1986 through 1997. The number of tracks counted along our transect remained fairly low in all winters except 1991–1992 (figure 14.1). During that

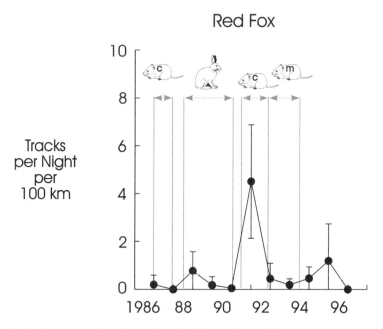

Red Fox

Tracks per Night per 100 km

Figure 14.1 Abundance of tracks (± SE) of red foxes along a 25-km transect run with snow-mobiles during winter from 1986–1987 through 1996–1997. Dashed lines demarcate periods of peak densities of *Clethrionomys* (>500 voles/100 ha; vole silhouettes marked with "c"), snowshoe hares (>125 hares/100 ha), and *Microtus* (>1000 voles/100 ha; vole silhouette marked with "m"). Captures of *Microtus* on forested grids indicate that they may have also reached high numbers in 1988, but their preferred meadow habitats were not trapped in that year.

winter, we observed a large influx of fox tracks in March on our transect (36 tracks in March, relative to an average of 6.3 fox tracks per whole winter in other years). Fox tracks were concentrated (87% of all fox tracks) in the first 9 km of the transect, mostly in two large, shrubby, open areas, in all winters except 1991–1992. During March of 1992, however, fox tracks were found along the entire transect, and only 39% of tracks were in this first section.

Based on incidental observations of their tracks, foxes were abundant in the alpine areas surrounding the forested part of our study area. In the valley, coyotes were much more abundant: we saw 10 times as many coyotes as foxes while in the field, and their tracks were 20 times more abundant. We suspected that only 1 radio-collared hare was killed by a fox in our study area, relative to 217 confirmed coyote kills. When hunting in the valley, foxes used habitats with less dense overstories than either coyotes or lynx (91% of their crossings of coyote and lynx trails were in habitats with less than 50% cover, relative to 60% use of these habitats by coyotes and lynx; $n = 32$, $\chi^2 = 8.33$, $p = 0.004$). While live trapping for coyotes and lynx, we trapped four red foxes—one a juvenile male, and three adult males. The mean weight of the adults was 6.5 ± 0.9 (SD) kg.

In summary, we saw little evidence that red foxes responded numerically to either changes in the numbers of hares or voles in our study area. However, as foxes were likely more abundant in alpine areas surrounding our study area, we may have missed any changes in fox numbers that did occur. The large increase in the number of fox tracks that we observed in 1991–1992 may have been the result of increased movements of foxes, resulting from the cyclic decline in numbers of hares or ptarmigan at higher elevations. We did not measure the numbers of hares in the zones of dense subalpine willows, but they were abundant there at the peak in hare abundance. In a study area approximately 60 km from ours, foxes hunted hares heavily in areas of dense shrubs at a cyclic peak and early decline (Theberge and Wedeles 1989), but relied more heavily on alternative prey, mostly voles and ground squirrels, when hares were less abundant (Jones and Theberge 1983).

Our results contrast with the conclusions from fur records that suggest populations of red foxes may show 4-year or 10-year cycles in North America, depending on their main prey (Finerty 1980). Likewise, in northern and central Fennoscandia, numbers of foxes fluctuate in 3- to 4-year cycles with those of voles (Hörnfeldt 1978, Angelstam et al. 1985, Lindström 1989), and both observational and experimental data suggest that the resulting cyclical predation by foxes may contribute to cycles in numbers of grouse and hares, alternative prey of the foxes (Marcström et al. 1988, 1989, Small et al. 1993, Lindström et al. 1994).

In our study area, foxes may have avoided using the valley because of the presence of coyotes. There is ample evidence from elsewhere in their range that the presence of coyotes generally leads to lower numbers of foxes (Peterson 1995), and this is supported by local accounts of much higher fox numbers in our study area before coyotes colonized the region. Foxes are sometimes directly killed by coyotes (Peterson 1995), and they frequently establish their home ranges (Major and Sherburne 1987, Harrison et al. 1989) and dens (Voigt and Earle 1983, Sargeant et al. 1987) outside or on the periphery of coyote territories.

On a worldwide basis, red foxes are most abundant in areas with heterogeneous habitats and, unlike coyotes, their range extends into the tundra (Larivière and Pasitschniak-Arts 1996). Foxes regularly use open habitats, especially when snow depths are lower and voles are more vulnerable to predation (Halpin and Bissonette 1988, Lindström and Hörnfeldt 1994). Local differences in habitat use may allow coyotes and red foxes to persist where they are sympatric (Major and Sherburne 1987, Theberge and Wedeles 1989). The subalpine shrubby areas around our study area were seldom used by coyotes or lynx until late in the cyclic decline of hares. When lynx moved into these areas in 1992–1993, we observed one case of predation by lynx on a fox (O'Donoghue et al. 1995). In Alaska, there are a number of records of lynx preying on red foxes, especially during cyclic lows of hares (Stephenson et al. 1991). Changes in predation pressure on foxes may therefore be an indirect effect of the hare cycle on their abundance.

14.3   Wolf

Wolves, *Canis lupus*, were present each year in our study area, and the abundance of their tracks fluctuated irregularly along the track transect during this study (figure 14.2). One pack of wolves passed through the area periodically, and signs of single animals were seen more frequently; group sizes ranged from one to nine. We observed the most wolf tracks during the winter of 1989–1990, at the peak of the hare cycle (figure 14.2), but

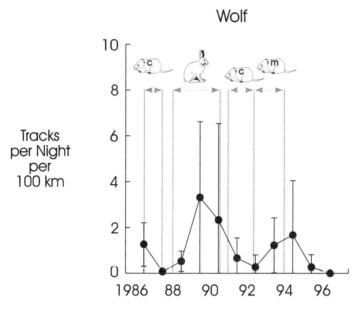

Figure 14.2 Abundance of tracks (± SE) of wolves along a 25-km transect run with snow-mobiles during winter from 1986–1987 through 1996–1997. Dashed lines demarcate periods of peak densities of *Clethrionomys* (>500 voles/100 ha; vole silhouettes marked with "c"), snowshoe hares (>125 hares/100 ha), and *Microtus* (>1000 voles/100 ha; vole silhouette marked with "m").

numbers were highly variable. The percentage of observations that were of lone wolves changed from 40% during the cyclical increase (6 of 15 observations), to 84% at the peak (16 of 19), to 57% during the decline (4 of 7), to 20% at the low (1 of 5).

There is no clear evidence of a 10-year cycle in numbers of wolves in North America (Finerty 1980). Although wolves regularly prey on snowshoe hares, ungulates are their main prey throughout their range in North America (Mech 1970, 1974, Banfield 1974). Hares may represent 1–2% of their diet in winter (Carbyn et al. 1993), and up to 12% in summer (Thurber et al. 1992, Carbyn et al. 1993). In our study area, only four radio-collared hares were killed by wolves.

Our data suggest that the percentage of lone wolves was highest when the abundance of hares was at its peak. Wolves have been observed actively hunting hares, and a super-abundant supply of hares could increase the survival of lone animals. However, despite many studies of wolves, there are no clear data that indicate a direct link between the hare cycle and wolf survival. The abundance of moose was low (<0.1/km^2), and there were no caribou in our study area, which may increase the frequency of solitary living by wolves (Messier 1985). Our data on numbers and group sizes of wolves were likely affected in later years by a wolf-control program started by the territorial government in 1993 north of our study area.

Wolves may exclude coyotes over large areas (Peterson 1995), and, locally, coyotes avoid areas frequented by wolves in some places (Thurber et al. 1992). Exclusion of coyotes benefits red foxes, so if densities of wolves in our study area had been higher, red foxes may have been more abundant in the valley, as proposed by Peterson (1995).

14.4   Weasel

There were two species of weasels present in our study area, the short-tailed weasel or ermine, *Mustela erminea,* and the least weasel, *Mustela nivalis.* We did not distinguish between their tracks, but based on our observations, short-tailed weasels were more abundant. Numbers of weasels remained low from 1986–1987 through 1990–1991, based on our track transect, and then increased steadily to a high in 1994–1995 (figure 14.3). The period of increase corresponded with successive highs in the abundance of red-backed voles and *Microtus,* with peak numbers of weasels occurring in the winter after the decline in *Microtus* abundance (see chapter 10). The number of weasel tracks declined again by the winter of 1995–1996.

Weasels were mostly found in more open habitats along the trails of coyotes and lynx. Approximately 39% of their crossings were in habitats with less than 25% overstory cover, relative to 22% of this habitat available ($\chi^2 = 154.16$, $p < .0001$). This was especially

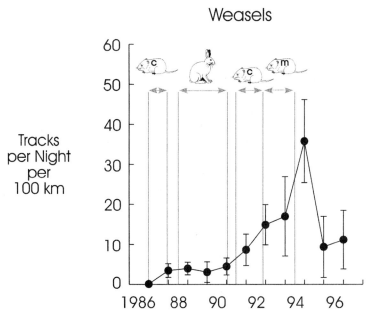

Figure 14.3 Abundance of tracks (± SE) of short-tailed and least weasels along a 25-km transect run with snowmobiles during winter from 1986–1987 through 1996–1997. Dashed lines demarcate periods of peak densities of *Clethrionomys* (>500 voles/100 ha; vole silhouettes marked with "c"), snowshoe hares (>125 hares/100 ha), and *Microtus* (>1000 voles/100 ha; vole silhouette marked with "m").

true in years when *Microtus* were abundant; before 1992–1993, weasels used these open habitats as expected.

Our data show that weasels did not respond numerically to the snowshoe hare cycle in our study area. This is contrary to evidence from incidental trapping in studies in Alberta (Keith and Cary 1991), where weasels (*M. erminea* and *M. frenata*) did show a 10-year cycle in numbers. Based on fur records, however, there is no evidence that the hare cycle affected the abundance of short-tailed weasels (Finerty 1980). Weasels killed only two radio-collared hares in our study area.

Weasels clearly responded to the 1991–1994 increase in numbers of voles at Kluane, and their population trend was correlated with that of *Microtus* in meadow habitats during our study with a 1-year lag ($r = .87$, $p = .027$). Voles are the main prey of weasels throughout their circumboreal ranges (King 1983, Sheffield and King 1994). There is ample evidence that weasels show strong numerical responses to changes in vole abundance in many areas (Fitzgerald 1977, Tapper 1979, Korpimäki et al. 1991). Short-tailed weasels have delayed implantation (King 1983), and hence their numerical responses to changes in prey abundance are typically delayed. Empirical studies in northern Fennoscandia show that the abundance of weasels (*Mustela nivalis*) fluctuates cyclically, following the 3- to 4-year cycles in numbers of voles, mostly *Microtus agrestis* (Korpimäki et al. 1991, Korpimäki 1993). These data have been used in models to suggest that predation by the specialist weasels causes the cyclic dynamics of voles in northern Fennoscandia (Hanski et al. 1991, 1993, Hanski and Korpimäki 1995).

Our index of weasel abundance was not correlated with density estimates of *Clethrionomys* (chapter 10). Although 3- to 4-year cycles are characteristic of some populations of voles in North America (Krebs and Myers 1974), abundance of *Clethrionomys* generally does not cycle regularly (Gilbert and Krebs 1991). In our study area, numbers of red-backed voles were high in 1984, 1987, and 1991–1992 (Gilbert and Krebs 1991, Boutin et al. 1995). We did not see a numerical response of weasels to high numbers of red-backed voles in 1987. Although the abundance of *Microtus* was not monitored in meadow habitats before 1989, captures on forested grids suggest that they reached high numbers in 1988 (chapter 10). The peak in numbers of voles in 1987 and 1988 was of shorter duration and, at least for *Clethrionomys,* of lower densities than the peak in the 1990s; this may be why weasels did not respond to the 1987–1988 vole peak.

We suggest that the abundance of weasels in our study area was limited by low densities of small mammals, particularly in spring. Erlinge (1974) and Tapper (1979) found that least weasels required minimum spring densities of field voles (*Microtus agrestis*) of 10/ha and 14/ha in southern Sweden and England, respectively, before they would breed. The densities of voles at Kluane seldom approached those levels (chapter 10). In addition, only about 7% of our study area consisted of the meadow habitats preferred by *Microtus,* which are apparently more susceptible to predation than *Clethrionomys* (Henttonen et al. 1987).

The use of more open habitats by weasels at Kluane likely reflects their preference for hunting *Microtus.* Throughout their range, weasels use a wide range of habitat types, but generally avoid dense forest and open areas with no shelter (King 1983, Sheffield and King 1994). Risk of predation is thought to cause the avoidance of open areas, but during winter, weasels have subnivean refuges from many predators, and they can hunt in meadows with more protection.

Wolverine

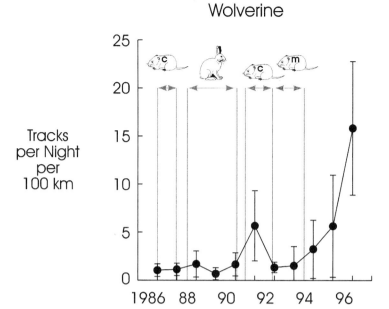

Tracks per Night per 100 km

Figure 14.4 Abundance of tracks ($\pm$ SE) of wolverines along a 25-km transect run with snow-mobiles during winter from 1986–1987 through 1996–1997. Dashed lines demarcate periods of peak densities of *Clethrionomys* (>500 voles/100 ha; vole silhouettes marked with "c"), snowshoe hares (>125 hares/100 ha), and *Microtus* (>1000 voles/100 ha; vole silhouette marked with "m").

14.5 Wolverine

Wolverines, *Gulo gulo,* were regular residents in the Kluane study area. They typically travel over large home ranges in the north (up to over 900 km²; Banci 1994), so it is likely that the home ranges of wolverines using our study area also extended into the alpine areas around it. The abundance of wolverine tracks along our track transect stayed fairly constant and low in all winters but 1991–1992 until 1994–1995, when numbers of tracks began to increase until the last winter of our study, 1996–1997 (figure 14.4).

Our data show no clear evidence that numbers of wolverines were related to the abundance of hares or voles in our study area. Population trends from fur returns show some evidence of a 10-year cycle in North America (Finerty 1980). We do not know the cause of the increase in numbers of wolverine tracks during the last 3 years of this study, but our subsequent observations in the study area in 1997–1998 and 1998–1999 suggest that this increasing trend did not continue.

Throughout their range, wolverines are opportunistic feeders, and they eat a wide variety of prey, hares and voles included, and carrion (Pasitschniak-Arts and Larivière 1995, Banci 1994). Wolverines killed only two radio-collared hares in our study area, but studies in the Yukon have shown that hares are an important component of wolverine diets at both high and low phases of the hare cycle (Banci 1994).

14.6   Marten

Martens, *Martes americana,* were uncommon in our study area, despite the fact that most of the forest was mature spruce, their preferred habitat (Clark et al. 1987, Buskirk and Ruggiero 1994). Only three observations of martens were made during this study, and their tracks were recorded regularly but infrequently along the trails of coyotes and lynx. Martens are uncommon throughout a large area of the southwest Yukon (Slough 1989).

Voles, particularly *Microtus* when they are available, and *Clethrionomys* when they are not, are the staple food of martens throughout their range, but they also feed on a wide range of prey up to the size of hares (Clark et al. 1987, Buskirk and Ruggiero 1994). Martens killed only two radio-collared hares in our study area. Based on fur records, martens may show 4-year or 10-year cycles in abundance, depending on their main prey (Finerty 1980). We suggest that the abundance of martens in our study area, as with weasels, was limited by the low densities of small mammals.

14.7   Mink and Otter

Mink, *Mustela vison,* and river otters, *Lutra canadensis,* were both present along creeks and lakes in the Kluane study area. We seldom counted their tracks or made direct observations of either species, because of their restricted distribution.

Both species prefer streams and lakes with banks where they can use burrows created by beavers, or they use the lodges of beavers (Banfield 1974, Larivière and Walton 1998). In Alberta, incidental trapping records of mink suggested that their numbers cycle with a 10-year periodicity (Keith and Cary 1991).

14.8   Synthesis and Conclusions

Our data on the abundance of carnivores besides coyotes and lynx, although limited, suggest that their population trends were not tightly tied to those of hares at Kluane. Tracks of red foxes, wolves, weasels, and wolverines were encountered frequently enough in our study area that we would have detected large changes in abundance. Martens were uncommon, and mink and otters were restricted to riparian habitat, so our conclusions regarding these species are less robust.

Numbers of weasels were related to the abundance of voles, especially *Microtus.* Weasels did not respond numerically to high numbers of voles in 1987 and 1988, whereas they did increase from 1992 to 1995, when first *Clethrionomys* and then *Microtus* reached high numbers. Therefore, vole populations may need to be high for a number of years in order for numbers of weasels to increase in our study area. Population trends of other carnivores, notably red foxes, were not related to vole abundance in the Kluane study area.

The responses of boreal predators to cyclic fluctuations in numbers of their prey differ between North America and Fennoscandia. In central and northern Fennoscandia, 3- to 4-year population cycles of voles and their predators are the dominant influence in generating synchronous cycles for a host of alternative prey, including hares, grouse, and shrews. In the North American boreal forest, though, the 10-year cycle of snowshoe hares appears to be the main dynamic driving the population trends of larger vertebrate predators only, and these cycles may then be imposed on some alternative prey such as grouse and ground squirrels, but not on others such as voles and red squirrels. Boutin et al. (1995) speculated that the differences between the vertebrate communities in North America and Fennoscandia may be because of the numerical dominance of hares and the resulting evolution of larger-bodied predators that are inefficient at hunting voles, and the greater abundance of *Clethrionomys* than the more vulnerable *Microtus.*

There are no clear data showing that the abundance of wolves and their main prey, moose and caribou, are related to the hare cycle anywhere they coexist. The communities of large and small mammals may therefore function independently in the boreal forest. We did, however, observe changes in the frequency of lone wolves over the hare cycle at Kluane, and this relationship bears further study.

Relations among predatory species in the North American boreal forest are characterized by behavioral avoidance and intraguild predation (Polis et al. 1989) among many species (O'Donoghue et al. 1995). Sympatric carnivores typically use different habitats or prey species, which may reduce competition for resources (Rosenzweig 1966), as was the case for red foxes and coyotes in our study area. We do not know whether the predation that we observed among predators affected their abundance, or was mostly compensatory.

There is still a great deal to learn about how boreal carnivores are affected by cyclical fluctuations of hares and voles and how they interact with each other. Few studies have been devoted to examining the community relationships of the less abundant carnivores in boreal North America, and none, to our knowledge, has looked in detail at community structure and functional relationships among the whole group of carnivores. Long-term studies such as the Kluane Project and Keith's work in Alberta (Keith and Cary 1991) have gathered data indexing population trends of these species, but research specifically aimed at them will be necessary to further clarify their roles in the boreal food web.

Literature Cited

Angelstam, P., E. Lindström, and P. Widén. 1985. Synchronous short-term fluctuations of some birds and mammals in Fennoscandia—occurrence and distribution. Holarctic Ecology **8**:285–298.

Banci, V. 1994. Wolverine. *in* L. F. Ruggiero, K. B. Aubry, S. W. Buskirk, L. J. Lyon, and W. J. Zielinski (eds). The scientific basis for conserving forest carnivores: American marten, fisher, lynx, and wolverine in the western United States, pages 99–127. General Technical Report RM-254, Fort Collins, Colorado. U.S. Department of Agriculture, Forest Service, Rocky Mountain Forest and Range Experiment Station.

Banfield, A. W. F. 1974. The mammals of Canada. University of Toronto Press, Toronto, Ontario.

Boutin, S., C. J. Krebs, R. Boonstra, M. R. T. Dale, S. J. Hannon, K. Martin, A. R. E. Sinclair, J. N. M. Smith, R. Turkington, M. Blower, A. Byrom, F. I. Doyle, C. Doyle, D. Hik, L. Hofer, A. Hubbs, T. Karels, D. L. Murray, V. Nams, M. O'Donoghue, C. Rohner, and S. Schweiger. 1995. Population changes of the vertebrate community during a snowshoe hare cycle in Canada's boreal forest. Oikos **74**:69–80.

Buskirk, S. W., and L. F. Ruggiero. 1994. American marten. *in* L. F. Ruggiero, K. B. Aubry, S. W. Buskirk, L. J. Lyon, and W. J. Zielinski (eds). The scientific basis for conserving forest carnivores: American marten, fisher, lynx, and wolverine in the western United States, pages 7–37. U.S. Forest Service General Technical Report RM-254, Washington, DC.

Carbyn. L. N., S. M. Oosenbrug, and D. W. Anions. 1993. Wolves, bison and the dynamics related to the Peace-Athabasca Delta in Canada's Wood Buffalo National Park. Circumpolar Research Series No. 4. Canadian Circumpolar Institute, Edmonton, Alberta.

Clark, T. W., E. Anderson, C. Douglas, and M. Strickland. 1987. *Martes americana.* Mammalian Species 289. American Society of Mammalogists.

Erlinge, S. 1974. Distribution, territoriality and numbers of the weasel *Mustela nivalis* in relation to prey abundance. Oikos **25**:308–314.

Finerty, J. P. 1980. The population ecology of cycles in small mammals. Yale University Press, New Haven, Connecticut.

Fitzgerald, B. M. 1977. Weasel predation on a cyclic population of the montane vole (*Microtus montanus*) in California. Journal of Animal Ecology **46**:367–397.

Gilbert, B. S., and C. J. Krebs. 1991. Population dynamics of *Clethrionomys* and *Peromyscus* in southwestern Yukon 1973–1989. Holarctic Ecology **14**:250–259.

Halpin, M. A., and J. A. Bissonette. 1988. Influence of snow depth on prey availability and habitat use by red fox. Canadian Journal of Zoology **66**:587–592.

Hanski, I., L. Hansson, and H. Henttonen. 1991. Specialist predators, generalist predators, and the microtine rodent cycle. Journal of Animal Ecology **60**:353–367.

Hanski, I., and E. Korpimäki. 1995. Microtine rodent dynamics in northern Europe: parameterized models for the predator-prey interaction. Ecology **76**:840–850.

Hanski, I., P. Turchin, E. Korpimäki, and H. Henttonen. 1993. Population oscillations of boreal rodents: regulation by mustelid predators leads to chaos. Nature **364**:232–235.

Harrison, D. J., J. A. Bissonette, and J. A. Sherburne. 1989. Spatial relationships between coyotes and red foxes in eastern Maine. Journal of Wildlife Management **53**:181–185.

Henttonen, H. 1985. Predation causing extended low densities in microtine cycles: further evidence from shrew dynamics. Oikos **45**:156–157.

Henttonen, H., T. Oksanen, A. Jortikka., and V. Haukisalmi. 1987. How much do weasels shape microtine cycles in the northern Fennoscandian taiga? Oikos **50**:353–365.

Hörnfeldt, B. 1978. Synchronous population fluctuations in voles, small game, owls, and tularemia in northern Sweden. Oecologia **32**:141–152.

Jones, D, M., and J. B. Theberge. 1983. Variation in red fox, *Vulpes vulpes,* summer diets in northwest British Columbia and southwest Yukon. Canadian Field-Naturalist **97**:311–314.

Keith, L. B., and J. R. Cary. 1991. Mustelid, squirrel, and porcupine population trends during a snowshoe hare cycle. Journal of Mammalogy **72**:373–378.

King, C. M. 1983. *Mustela erminea.* Mammalian Species 195. American Society of Mammalogists.

Korpimäki, E. 1993. Regulation of multiannual vole cycles by density-dependent avian and mammalian predation? Oikos **66**:359–363.

Korpimäki, E., K. Norrdahl, and T. Rinta-Jaskari. 1991. Responses of stoats and least weasels to fluctuating food abundances: is the low phase of the vole cycle due to mustelid predation? Oecologia **88**:552–561.

Krebs, C. J., and J. H. Myers. 1974. Population cycles in small mammals. Advances in Ecological Research **8**:267–399.

Larivière, S., and M. Pasitschniak-Arts. 1996. *Vulpes vulpes.* Mammalian Species 537. American Society of Mammalogists.

Larivière, S., and L. R. Walton. 1998. *Lontra canadensis.* Mammalian Species 587. American Society of Mammalogists.

Lindström, E. 1989. Food limitation and social regulation in a red fox population. Holarctic Ecology **12**:70–79.

Lindström, E. R., H. Andrén, P. Angelstam, G. Cederlund, B. Hörnfeldt, L. Jäderberg, P.-A.

Lemnell, B. Martinsson, K. Sköld, and J. E. Swenson. 1994. Disease reveals the predator: sarcoptic mange, red fox predation, and prey populations. Ecology **75**:1042–1049.

Lindström, E. R., and B. Hörnfeldt. 1994. Vole cycles, snow depth and fox predation. Oikos **70**:156–160.

Major, J. T., and J. A. Sherburne. 1987. Interspecific relationships of coyotes, bobcats, and red foxes in western Maine. Journal of Wildlife Management **51**:606–616.

Marcström, V., L. B. Keith, E. Engren, and J. R. Cary. 1989. Demographic responses of arctic hares (*Lepus timidus*) to experimental reductions of red foxes (*Vulpes vulpes*) and martens (*Martes martes*). Canadian Journal of Zoology **67**:658–668.

Marcström, V., R. E. Kenward, and E. Engren. 1988. The impact of predation on boreal tetraonids during vole cycles: an experimental study. Journal of Animal Ecology **57**:859–872.

Mech, L. D. 1970. The wolf. Ecology and behavior of an endangered species. University of Minnesota Press, Minneapolis.

Mech, L. D. 1974. *Canis lupus*. Mammalian Species 37. American Society of Mammalogists.

Messier, F. 1985. Solitary living and extraterritorial movements of wolves in relation to social status and prey abundance. Canadian Journal of Zoology **63**:239–245.

O'Donoghue, M., S. Boutin, C. J. Krebs, and E. J. Hofer. 1997. Numerical responses of coyotes and lynx to the snowshoe hare cycle. Oikos **80**:150–162.

O'Donoghue, M., S. Boutin, C. J. Krebs, D. L. Murray, and E. J. Hofer. 1998. Behavioural responses of coyotes and lynx to the snowshoe hare cycle. Oikos **82**:169–183.

O'Donoghue, M., E. Hofer, and F. I. Doyle. 1995. Predator versus predator. Natural History **104(3)**:6–9.

Pasitschniak-Arts, M., and S. Larivière. 1995. *Gulo gulo*. Mammalian Species 499. American Society of Mammalogists.

Peterson, R. O. 1995. Wolves as interspecific competitors in canid ecology. *in* L. N. Carbyn, S. H. Fritts, and D. R. Seip (eds). Ecology and conservation of wolves in a changing world, pages 315–324. Canadian Circumpolar Institute, Edmonton, Alberta.

Polis, G. A., C. A. Myers, and R. D. Holt. 1989. The ecology and evolution of intraguild predation: potential competitors that eat each other. Annual Review of Ecology and Systematics **20**:297–330.

Rosenzweig, M. L. 1966. Community structure in sympatric Carnivora. Journal of Mammalogy **47**:602–612.

Sargeant, A. B., S. H. Allen, and J. O. Hastings. 1987. Spatial relations between sympatric coyotes and red foxes in North Dakota. Journal of Wildlife Management **51**:285–293.

Sheffield, S. R., and C. M. King. 1994. *Mustela nivalis*. Mammalian Species 454. American Society of Mammalogists.

Slough, B. G. 1989. Movements and habitat use by transplanted marten in the Yukon Territory. Journal of Wildlife Management **53**:991–997.

Small, R. J., V. Marcström, and T. Willebrand. 1993. Synchronous and nonsynchronous population fluctuations of some predators and their prey in central Sweden. Ecography **16**:360–364.

Stephenson, R. O., D. V. Grangaard, and J. Burch. 1991. Lynx, *Felis lynx*, predation on red foxes, *Vulpes vulpes*, caribou, *Rangifer tarandus*, and Dall sheep, *Ovis dalli*, in Alaska. Canadian Field-Naturalist **105**:255–262.

Tapper, S. 1979. The effect of fluctuating vole numbers (*Microtus agrestis*) on a population of weasels (*Mustela nivalis*) on farmland. Journal of Animal Ecology **48**:603–617.

Theberge, J. B., and C. H. R. Wedeles. 1989. Prey selection and habitat partitioning in sympatric coyote and red fox populations, southwest Yukon. Canadian Journal of Zoology **67**:1285–1290.

Thurber, J. M., R. O. Peterson, J. D. Woolington, and J. A. Vucetrich. 1992. Coyote coexistence with wolves on the Kenai Peninsula, Alaska. Canadian Journal of Zoology **70**:2494–2498.

Voigt, D. R., and B. D. Earle. 1983. Avoidance of coyotes by red fox families. Journal of Wildlife Management **47**:852–857.

PART V

AVIAN PREDATORS

15

Great Horned Owls

CHRISTOPH ROHNER, FRANK I. DOYLE, & JAMES N. M. SMITH

Great horned owls (*Bubo virginianus* Gmelin) are large, long-lived, generalist preda-tors. Diets of great horned owls are extremely variable and include a wide variety of species (ranging in size from insects to lagomorphs; Donazar et al. 1989), although the average prey size of great horned owls is larger than for smaller owl species (Marti 1974). Great horned owls typically ambush their prey from elevated perches and may be most successful in a mix of open and forested habitat (Johnson 1993, Rohner and Krebs 1996). Nevertheless, they are widely distributed in most landscapes across North and South America, including the whole range of boreal forest (Voous 1988, Houston et al. 1998; See figure 2.7). Great horned owls form lasting pair bonds and defend territories year-round (Petersen 1979, Rohner 1996). Great horned owls are the largest avian predators occurring widely at high densities in the boreal forest (body mass in June: 18 female adults, 1.61 ± 0.26 kg, 4 male adults, 1.23 ± 0.07 kg; C. Rohner, unpublished data).

Several links to the 10-year population cycle of snowshoe hares (*Lepus americanus*) have been recognized. Great horned owls include a higher proportion of hares in their diet and increase reproduction during the peak of the cycle, and irruptions of great horned owls into southern Canada and the northern United States occur during the decline phase (Rusch et al. 1972, McInvaille and Keith 1974, Adamcik et al. 1978, Houston 1987, Keith and Rusch 1988, Houston and Francis 1995). Maximum life span can exceed 20 years (Houston and Francis 1995); therefore, these predators may survive more than one snow-shoe hare cycle. So far, radio-telemetry and experimental approaches have not been used to address the population ecology and social structure of great horned owls in the boreal forest.

In this chapter, we complement the main experiments of the project with more specific results on the mechanism of population processes that are relevant in the context of trophic interactions in the boreal forest, and we provide data that are essential for modeling this predator as a component in the food web of the vertebrate community (see chapter 18).

First, we address the question of how great horned owls respond in their demography to changing hare densities. In particular, we examine the "numerical response" (Holling 1959) in relation to social status, and we directly estimate the size of the pool of secretive, nonterritorial floaters.

Second, we focus on the question of how the diet of great horned owls varied during the course of the snowshoe hare cycle. More specifically, we used estimates of prey den-sity to calculate the preferences for different prey, and we integrated the available infor-mation to construct functional responses (Holling 1965, Fujii et al. 1986) by great horned owls to varying hare densities. Much is known about how predators optimize their forag-ing decisions (e.g., review in Stephens and Krebs 1986), but most implications on popu-lation dynamics were derived from functional responses of predators to prey density (Holling 1959, Fujii et al. 1986). It is particularly the distinction between opportunistic foraging (type-2 response) and prey switching (type-3 response) that changes the dynam-ics of predator–prey interactions (e.g., Rosenzweig and MacArthur 1963, Holling 1965, Murdoch and Oaten 1975; examples in Messier 1994, Caughley and Sinclair 1994).

Third, we investigated the social structure of great horned owls and examined the ques-tion of how individual behavior related to population processes. In particular, we ad-dressed the question of whether territorial behavior (social exclusion) limited population growth and breeding density in great horned owls (see also Newton 1992) or whether food was sufficient to explain the dynamics of demographic change. We further demonstrate that individual behavior is necessary to explain the time lag of this predator to a prey cycle.

Finally, we asked at what spatial scale great horned owls responded to the large-scale manipulations of prey densities (chapter 8). Predators can concentrate their foraging effort or even aggregate in hot spots of high prey density, and spatially shifting predators have been hypothesized to synchronize population cycles in small mammals (e.g., Pitelka et al. 1955, Angelstam et al. 1984, Lindström et al. 1987, Ydenberg 1987, Ims and Steen 1990, Korpimäki and Norrdahl 1991, Korpimäki 1994). Few empirical data are available on such aggregative responses, and we discuss our findings in the context of whether territoriality can act as a social fence (see also Hestbeck 1982), which limits access and spatial aggregations of predators where prey is locally abundant.

15.1   Methods

15.1.1   Population Census

The population data for great horned owls span the years 1988–1996, with the most intensive monitoring and additional studies from 1989–1992. Great horned owls were censused in late winter and early spring on a 100-km^2 plot within the main study area (see CD-ROM frame 68). Individual pairs were identified when hooting simultaneously with neighbors at dawn and dusk, and obvious disputes between hooting males or pairs were used to map territorial boundaries. When necessary, we used playbacks of calls to elicit territorial responses of owners and their neighbors. Most males were individually known, not only because of radio tagging but also because of their distinctly different hoots. These differences were later verified with sonograms from recordings at the nest (C. Rohner unpublished data; method as used for *Strix aluco,* Galeotti 1990). Observations of territorial activity were made almost daily from early February until late April (at least 300 h in each year). Details on owl territories are given in Rohner (1997).

We intensively monitored a subsample of territories to assess reproductive parameters. The proportion of owl pairs that did not breed or failed early before producing nestlings was established from two methods: (1) monitoring breeding activities by females with radio transmitters (checked at least twice per week), and (2) systematic search for nests by triangulation of hooting owls according to Rohner and Doyle (1992). A search for breeding activity was recorded negative if no nest was found within 5 h of systematic search and if searches for begging calls of fledglings at a later stage were also negative.

To avoid possible effects of disturbance, we only checked nests when chicks were estimated to be at least 1 week old. A total of 116 nestlings were measured during 1989–1991, and their ages were determined from feather measurements (Rohner and Hunter 1996). Clutch initiation was back-calculated using the oldest chick, assuming that incubation commenced with the first egg and required 33 days until hatching (review in Houston et al. 1998). We determined postfledging survival according to Rohner and Hunter (1996).

15.1.2 Monitoring Diets

In May and June, nestlings were transferred shortly before fledging to elevated tethering platforms (Petersen and Keir 1976), where the parents kept feeding them, and they were systematically monitored for about 5 weeks (CD-ROM frame 33). This method allowed short-term brood size manipulations and collection of data on diets (details in Rohner 1995, Rohner and Hunter 1996, Rohner and Smith 1996).

We sampled summer diets of territorial owls during May–July, when pellets were collected from breeding birds at nests and at roost sites of owls located by telemetry. The results of pellet analysis were expressed as the percentage of a prey species of the total biomass, calculated by adding up diagnostic bones of each prey species for a minimum estimate of prey items (details in Rohner 1995). Because sample sizes were small for winter diets and for the summer diets in 1993–1995, the number of diagnostic bones may not have been large enough to avoid rounding errors. Therefore, we directly estimated for each pellet the proportion of different prey species based on all bones and fur encountered, and then calculated the proportion of prey species as an average across pellets. A test on subsamples confirmed that results of the two methods can be directly compared (details in Rohner 1995).

15.1.3 Radio Telemetry

Survival estimates and information on movements were based on individual great horned owls monitored by radio telemetry. We captured 21 territorial adult owls with mist nets and cage traps (CD-ROM frame 32) and equipped 55 owlets with radio transmitters before they fledged (breakdown of sample sizes in table 15.1, figure 15.2b). Successful dispersers were later monitored intensively (3 born in 1988, 11 in 1989, and 16 in 1990), and 9 remained as nonterritorial floaters in the study area (details in Rohner 1996, 1997). The radios weighed 50 g including a shoulder harness of teflon ribbon for attachment as a backpack ($<5\%$ of body weight, Kenward 1985; See CD-ROM frame 39). Battery life was 2–2.5 years. The radios were equipped with a two-phase activity switch (sensitive to movement and change of angle).

All floaters and territory holders with transmitters were normally monitored once per week (for the presentation of weekly data, locations in addition to the weekly sampling intervals were excluded). Most checks were conducted with handheld equipment from the Alaska Highway, which follows the valley bottom for the whole length of the study area (see CD-ROM frame 40). In addition, the entire area and its surroundings were searched for radio signals from helicopter or fixed-wing aircraft at least twice per year (in autumn after dispersal and in spring after the onset of breeding).

Telemetry work on space use of great horned owls concentrated on periods of 3 weeks (20–21 days) for each year, with locations obtained on consecutive nights for each bird if possible (this was not achieved in the first year of data collection, thus the total monitoring periods for each bird in 1989 were 27, 28, and 41 days). These periods of intensive monitoring were conducted 24 July–8 September 1989, 7–26 September 1990, 5–26 September 1991, and 12 June–3 July 1992. A detailed breakdown of sample sizes is presented in table 15.1. All of the territorial owls were females, except in 1992, when female no. 503 emigrated from the study area and her mate, no. 564, was monitored instead (details in Rohner and Krebs 1998).

Table 15.1 Summary of sample sizes and precision of telemetry locations used for specific comparisons (see text for details).

	Floaters		Territory Holders			
	1990	1991	1989	1990	1991	1992
All Locations						
N owls	6	8	—	15	20	—
N locations/owl (mean ± SE)	80.3 ± 12.2	61.3 ± 7.8	—	17.3 ± 3.1	35.1 ± 5.4	—
95% Error area (median, km²)	0.46	0.21	—	0.08	0.07	—
95% Error area (quartiles, km²)	0.14–1.79	0.08–0.84	—	0.02–0.28	0.02–0.19	—
Weekly Locations						
N owls	6	8	—	—	—	—
N locations/owl (mean ± SE)	42.7 ± 2.6	36.9 ± 2.0	—	—	—	—
95% Error area (median, km²)	0.58	0.25	—	—	—	—
95% Error area (quartiles, km²)	0.19–1.79	0.09–1.45	—	—	—	—
September (3 weeks)						
N owls	4	3	3	5	5	6
N locations/owl (mean ± SE)	19.8 ± 0.25	17.3 ± 2.2	18.7 ± 1.2	20.0 ± 0.0	20.0 ± 0.0	21.0 ± 0.0
95% Error area (median, km²)	0.15	0.13	0.19	0.10	0.09	0.05
95% Error area (quartiles, km²)	0.05–0.34	0.13–0.29	0.05–0.37	0.02–0.26	0.03–0.20	0.02–0.15

15.1.4   Analysis of Telemetry Data

Telemetry locations were obtained by triangulating owls with handheld equipment. We used topographical maps in the field to plot the locations and assess the number of bearings needed for reliable estimates. The triangulations were then analyzed with the program Locate II (Nams 1990) for calculating exact locations and distances. Median 95% error ellipses (Lenth estimator; Saltz and White 1990) are given in table 15.1 to allow an assessment of precision for telemetry locations. The accuracy of telemetry locations was assessed by triangulating five transmitters that were placed in trees at a height of 4.5–5.5 m. The deviation of these telemetry locations (error area of 0.052 ± 0.018 km) from the site coordinates obtained by GPS (Global Positioning System) was 0.101 ± 0.027 km.

Space use was measured by utilization distributions based on clustering methods, and all calculations were performed using the program Ranges IV (Kenward 1990). From a center of closest locations, an increasing percentage of nearest-neighbor locations were added, resulting in a cumulative increase of core area used. Mononuclear clustering was centered around the harmonic mean location only, whereas multinuclear clustering allowed for separate clusters of closest locations. Home range sizes were then derived for different levels of core percentages (Kenward 1987). Patchiness was calculated as "part areas," which are the areas used at a specific core percentage and expressed as a portion of the total area (details in Kenward 1987). These procedures allowed a more sensitive approach to recognizing biases due to outliers and different patterns of space use. For the monitoring period in September 1991, we excluded three territorial owls from analysis because of extreme long-distance movements during several days (these extraterritorial movements are described in Rohner 1996).

15.1.5 Statistical Analyses

All arithmetic means are reported with standard errors and all probabilities are two-tailed unless otherwise specified. We calculated correlation coefficients as Spearman rank correlations. For statistical testing, nonparametric tests were used wherever possible (all analyses of variance were calculated with log-transformed data, or with arcsine-transformed data for percentages). The testing of bootstrap hypotheses followed the guidelines of Hall and Wilson (1991), and two-sided probabilities were derived from 500 simulations (see also Rohner 1997).

15.2 Demography

15.2.1 Reproduction and Population Productivity

Great horned owls showed a strong reproductive response to the snowshoe hare cycle (figure 15.1, table 15.2). During peak densities of hares in 1989 and 1990, 1.7 offspring per resident pair were estimated to reach independence in autumn. As hare density started to decline in 1991, productivity fell by 82% to 0.3 offspring per pair, and during the lowest phase from 1992 to 1994 reproduction of great horned owls ceased altogether. From 1995 onward, owl productivity recovered along with increasing hare densities. The calculations of population productivity are provided in table 15.2, and more details on the reproductive parameters involved are described in the following paragraphs (see also Rohner 1996, Rohner and Hunter 1996).

The proportion of pairs breeding successfully not only had the greatest effect on population productivity but was also the parameter most closely related to snowshoe hare densities (figure 15.1B). During 1989–1991, 14–22% of territorial owl pairs did not breed or failed early, with results from radio-tagged owls and systematic searches for breeding activity in monitored territories being similar (Rohner 1996). Of 17 females that did not produce young and were monitored by radio telemetry in 1989–1992, only 2 (11%) laid eggs, indicating that inhibition of breeding activity occurred at a very early stage. In 1992, there were no signs of nesting attempts, and mates of three monitored pairs did not even roost together, as typically found in reproductive years (Petersen 1979, Rohner and Doyle 1992). This drop in the proportion of owl pairs producing young from 1989–1991 to 1992 was statistically significant ($\chi^2 = 17.5$, df $= 1, p < .001, n = 123$). The situation remained unchanged during the lowest phase of the cycle during 1993 and 1994, with only one

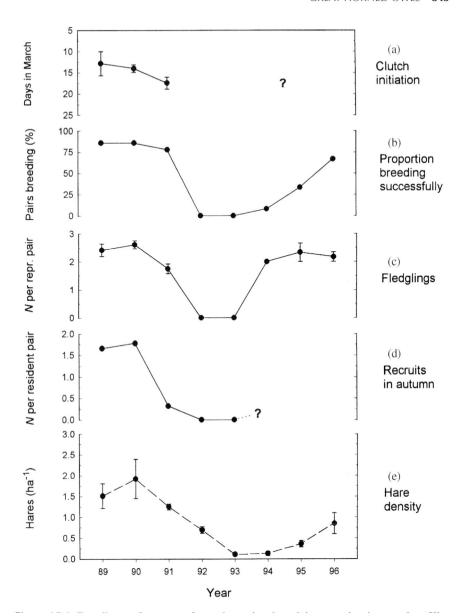

Figure 15.1 Breeding performance of great horned owls and the snowshoe hare cycle at Kluane Lake, southwestern Yukon (sample sizes in table 15.2; time periods with missing data are indicated by question marks). (a) Laying day in March; (b) breeding rate (proportion of territorial pairs producing nestlings), measured 1989–1992 only; (c) number of fledglings per successful nest; (d) number of juveniles (per resident pair) reaching independence and dispersing in autumn, measured 1989–1992 only; (e) snowshoe hare densities (winter estimates ± SE). Modified from Rohner (1996).

Table 15.2 Reproductive parameters and estimation of the autumn recruitment rate of great horned owls at Kluane Lake, Yukon, during 1989–1996.

Reproductive Parameter	1989	1990	1991	1992	1993	1994	1995	1996
Number of owl pairs monitored	14	21	27	25	17	13	12	9
Proportion of pairs with successful nests (b_i)	0.86	0.86	0.74	0.00	0.00	0.008	0.33	0.67
Fledglings per successful nest (FL_i) ±SE	0.242 ± 0.21	2.61 ± 0.13	1.75 ± 0.15	—	—	2.00	2.33 ± 0.33	2.17 ± 0.17
Number of fledglings radio tagged	15	22	18	—	—	0	0	0
Postfledgling survival (s_{FL_i}) (20 weeks) ± SE	0.800 ± 0.208	0.795 ± 0.090	0.232 ± 0.077	—	—	NA	NA	NA
Autumn recruitment rate (offspring per resident pair; R_i)	1.66	1.78	0.32	0.00	0.00	NA	NA	NA

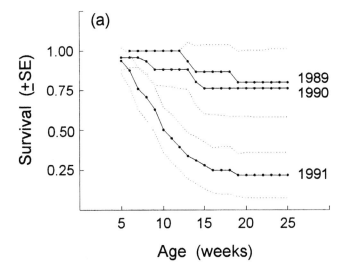

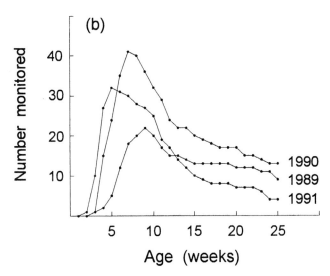

Figure 15.2 Postfledging survival rates of juvenile great horned owls, 1989–1991. (a) Survival curves with 95% confidence limits (combined for 1989 and 1990); (b) sample sizes for each time interval. From Rohner and Hunter (1996).

Table 15.3 Causes of mortality of juvenile great horned owls older than 35 days (from Rohner and Hunter 1996).

Cause of Mortality	1989	1990	1991
High levels of parasitism	0	3	10
Predation or signs of scavenging	0	0	11
Found intact (disease, starvation, unknown)	3	3	3
Traffic mortalities (road kills)	1	3	2
Total mortalities	4	9	26
Total no. juveniles monitored	24	36	34

breeding pair (3% of population). Breeding attempts rose again to 42% in 1995 and 89% in 1996 (some of these attempts failed early; table 15.2).

There was a trend toward later clutch initiation from 1989 to 1991 (figure 15.1A), but this result was not significant statistically. In 1991, brood size was lower, and nestling mortalities increased (Rohner and Hunter 1996). This resulted in fewer fledglings produced per successful nest in 1991 compared to 1989–1990 (1.75 ± 0.79 SD, $n = 27$, $p < .005$, df $= 2$; Kruskal-Wallis test; figure 15.1C). But note that the number of fledglings per succesful nest was relatively constant at whatever hare densities owls attempted to breed. The most dramatic change of reproductive parameters in 1991 occurred between fledging and dispersal, when survival fell to 29% of previous levels (figure 15.2). The causes of mortality during this period are presented in table 15.3. It is interesting that only few owlets died of starvation, but extreme damage by blood-sucking flies (Simuliidae) and a blood parasite (*Leucocytozoon ziemanni*) were common (details in Hunter et al. 1997). The effect on birds by black flies and diseases transmitted by them in the boreal forest may have been underestimated. Great horned owls tried to escape black flies by roosting closer to the ground and more in the open (Rohner et al. 2000), although survival rates dropped despite this change in behavior. The main cause of proximate mortality was likely an interaction between food shortage and parasitism (Rohner and Hunter 1996). For information on dispersal and age at first breeding, see Rohner (1996).

To test whether great horned owls typically fail to reproduce during the cyclic low of snowshoe hares in the boreal forest, an external set of data was used. By 1978, D. Mossop (Renewable Resources, Yukon territorial government) had established routine inspections of raptors and owls that were found injured or dead and reported to conservation officers. This information is based on the area of the entire Yukon and covers two snowshoe hare peaks (replication over space and time). The same pattern as in our study area was apparent (figure 15.3). Although injured or dead adult owls were reported throughout the entire length of the cycle, there were no juveniles during the years of lowest hare densities during 1984–1986.

15.2.2 Survival and Emigration

Survival of both territorial owls and floaters was high during the peak phase of the cycle (table 15.4). Territorial adults survived at a yearly rate of 95.1% (± 3.4% SE) in 1989–1990 and 1990–1991. Floaters in their first or second year of life had an equally high sur-

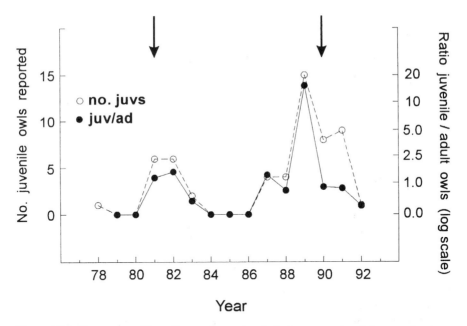

Figure 15.3 Occurrence of juvenile great horned owls during two peaks of the snowshoe hare cycle in the Yukon (56 juveniles, 70 adults, arrows indicate years of highest hare density in Kluane); based on numbers of injured and dead great horned owls that were reported to Renewable Resources, Yukon territorial government (data from D. Mossop, personal communication).

Table 15.4 Survival and emigration of great horned owls at Kluane Lake, Yukon, as determined by radio telemetry from autumn 1989 to autumn 1992.

Time Period	Hare Density	Social Class	Survival ±SE	Residency ±SE	Total No. Monitored (weekly avg.)
1989–90	Peak	Territorial	0.947 ± 0.051	1.000	19 (14)
		Floater	1.000	1.000	8 (8)
1990–91	First year decline	Territorial	0.955 ± 0.047	0.950 ± 0.049[a]	22 (19)
		Floater	1.000	0.696 ± 0.136[a,b]	19 (13)
1991–92	Second year decline	Territorial	0.819 ± 0.132[a]	0.668 ± 0.136[a,b]	18 (13)
		Floater	0.400 ± 0.219[a,b]	0.600 ± 0.268	10 (4)
1989–92	Overall	Territorial	0.905 ± 0.073	0.860 ± 0.136	22 (16)
		Floater	0.701 ±0.174	0.748 ± 0.225	19 (8)

Given are yearly survival rates (s_{Ti} and s_{Fi}) and yearly residency rates (e_{Ti} and e_{Fi}) for territorial owls and floaters. Survival rates are (1−mortality); residency rates are (1−emigration). All rates (including overall calculations) are annual rates. From Rohner (1996).

[a]$p < .05$ for difference between social classes (within individual years).

[b]$p < .05$ for difference to previous year (within social classes).

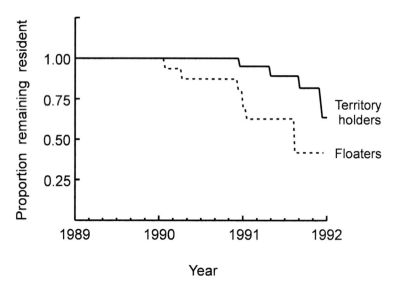

Figure 15.4 Emigration of adult owls (territorial) and young owls (first and second year, floaters) based on radio telemetry. Presented is the residency rate (1 = all owls remain resident; 0 = all owls emigrate). Sample sizes are given in figure 15.2b. From Rohner (1996).

vival rate during peak hare densities. The values in table 15.4 even exceed those for adult owls, but this is likely the result of a slight rounding error because the sample size for young birds was too small to detect survival differences of <5%.

Hare densities did not recover during summer 1991 and continued to decline in the following years. The survival of territory owners decreased by 13.2% in 1991–1992 compared to previous levels (table 15.4; difference 1990–1991 versus 1991–1992, $p = .12$). During the same time, the survival of floaters dropped by >60% (table 15.4, difference 1990–1991 versus 1991–1992, $p < .01$). This sharper decline in survival of floaters was also reflected in a significant difference in survival between floaters and territorial owls in 1991 ($p = .02$).

Territory holders showed extreme site fidelity during highest hare densities, but an increase in extraterritorial movements was observed as the hare population declined in 1991–1992 (Rohner 1996). None of 18 territorial owls was recorded outside its territory during 1989 and 1990. Extraterritorial movements first occurred in September 1991, when three females moved 15 km, 28 km, and >30 km away. These birds returned within 2–14 days to their territories. By October 1992, 37% of monitored territory owners (7 of 19) had shown extraterritorial movements (difference between years: Fisher's Exact test, df = 1, $p = .03$).

A similar trend for increased movements when hare densities declined was observed for emigration rates (proportion of owls leaving the study area permanently, figure 15.4). None of the territory holders and floaters monitored in the study area left during the peak of the hare cycle before autumn 1990. Consistent with the trend in the survival data (table 15.4), floaters were affected before territorial owls. In 1990, emigration of floaters was significantly higher than in territorial owls ($p = .02$), and the first territory owners started to emigrate with a 1-year time lag relative to floaters (figure 15.4).

15.2.3   Estimating Numerical Responses

The density of nonterritorial floaters was estimated based on productivity, survival, and emigration (tables 15.2, 15.4; population model in Rohner 1996). The results are shown in figure 15.5A. Even when assuming that no floaters were present in spring 1988 for a minimum estimate, the numbers rose quickly from zero to densities similar to territorial owls (figure 15.5B). The beginning of the hare decline in the winter of 1990–1991 resulted in an immediate reduction in population growth due to emigration and lowered production of recruits by territorial pairs. Floater densities reached a peak with a time lag of 1 year relative to the hare cycle, and then dropped sharply from 1991 onward because of increased emigration and mortality and because no additional juveniles were produced locally that could have compensated for losses in the nonterritorial segment of the population.

The number of territorial owls in the study area increased almost linearly from 1988 to 1992 (figure 15.5B). Even when the hare population started to decline in 1990–1991, the number of owl territories kept rising until spring 1992. Then, with a time lag of 2 years relative to the hare cycle, the number of territories dropped from 22–25 pairs/100 km^2 to 7–14 pairs/100 km^2 in 1993. The decline of territorial owls appears to have leveled off at 10–11 pairs/100 km^2 during 1994–1995, and the density was still at 10 pairs/100 km^2 in 1996.

The numerical response of the total population of great horned owls during 1988–1993 is given in figure 15.5B. Because the territorial segment represented a nearly linear component, the sum of densities or overall pattern more closely resembled the floater response with (1) an immediate reduction in population growth as hare densities declined, and (2) with a decline that was delayed by 1 year relative to the hare cycle.

15.3   Foraging Behavior

15.3.1   Diet

Great horned owls foraged on a wide variety of prey species from the size of a beetle, dragonfly, or warbler (<15 g) to prey of the size of a snowshoe hare, muskrat (*Ondatra zibethica*), or mallard (*Anas platyrhynchos*) (all >1.2 kg). Snowshoe hares were clearly the predominant prey in summer diets during the peak of the cycle (83.2–90.1% during 1989–1991) but declined to 18.8–27.4% during the low phase in the summers of 1992–1994 (figure 15.6A; a more detailed account is given for 1989–1992 in table 15.5). During this time, hares were mostly substituted by voles (11.7–59.1%), red squirrels (*Tamasciurus hudsonicus;* 1.8–33.4%), ground squirrels (*Spermophilus parryii;* 1.7–22.5%), birds (9.3–19.6%), and other mammals such as muskrats, woodrats (*Neotoma* sp.), northern flying squirrels (*Glaucomys sabrinus*), and least weasels (*Mustela nivalis;* 0–10.4%). Among birds, all major groups were represented including ducks, grouse, shorebirds, corvids, woodpeckers, and other owls and raptors (see also Rohner and Doyle 1992, Rohner et al. 1995). It appears that waterfowl migration in spring made up a substantial portion in the diet of breeding great horned owls.

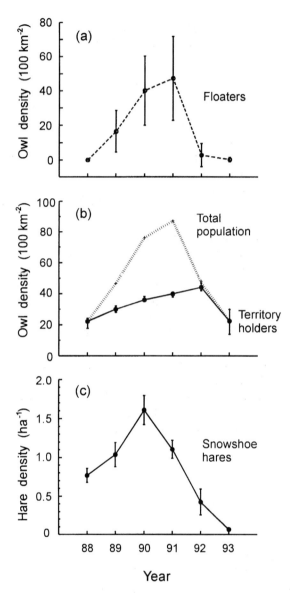

Figure 15.5 Numerical response of great horned owls (spring densities) to the snowshoe hare cycle. (a) Density of nonterritorial owls (floaters) estimated by the population model (with standard deviation); (b) census of the territorial owl population (with minimum and maximum estimates) and estimate of the total population (sum of territorial and nonterritorial owls); (c) spring densities of snowshoe hares (with SEs) from live trapping. From Rohner (1996).

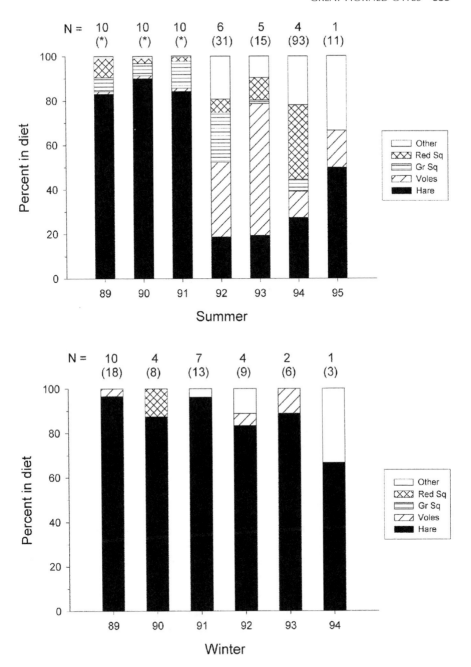

Figure 15.6 Proportions of main prey species in the diet of great horned owls in the boreal forest at Kluane Lake Yukon, during 1989–1995, based on pellet analysis. Sample sizes (number of territories sampled, with total number of pellets collected in parentheses) are given above the bars. Asterisks refer to years with >500 pellets and the total prey biomass accounted for in table 15.5. (a) Summer diet (mid-April–mid-October); (b) winter diet (mid-October–mid-April).

Table 15.5 Great horned owl diet, 1989–1992 at Kluane Lake, Yukon, according to minimum counts of prey species from pellets collected during the breeding season (10 territories 1989–1991, 6 territories 1992).

	% Mass (n)				
Prey Species	1989	1990	1991	1992	1989–92
Adult hares	53.8 (47)	42.8 (42)	50.5 (33)	0 (0)	48.0 (122)
Juvenile hares	29.3 (163)	47.3 (295)	33.7 (140)	18.8 (4)	37.2 (602)
Ground squirrels	6.3 (22)	5.6 (22)	12.2 (32)	22.5 (3)	7.8 (79)
Red squirrels	8.5 (37)	2.2 (11)	1.8 (6)	6.0 (1)	4.3 (55)
Voles	1.1 (44)	1.2 (54)	1.4 (42)	33.4 (52)	1.6 (192)
Birds	1.0 (4)	0.9 (4)	0.3 (1)	19.3 (3)	1.0 (12)
Total	100 (317)	100 (428)	100 (254)	100 (63)	100 (1062)

Proportions of prey categories are expressed as the percentage of total prey mass in the diet (number of prey items in parentheses).

Winter diets of great horned owls were remarkably different from summer diets. The proportion of snowshoe hares in winter diets was similar to that in sumer diets as long as hare densities were high, but an important difference was observed during the hare decline (figure 15.6B). In winter, the proportion of hares remained high (66.7–96.2%) despite the low hare numbers. Apparently, in the boreal forest at this latitude, alternative prey is difficult to find in winter. Most birds have migrated to southern climates, and most small mammals are either hibernating or active under the snow (see also Pruitt 1978). We suspect that we may have somewhat underestimated the proportion of less frequent prey items such as grouse because our small sample sizes were likely to result in rounding errors of small fractions. Most pellets were collected in the last part of winter, and more information on early winter diets is desirable.

15.3.2 Prey Preferences

Although great horned owls are known as generalist predators, there were pronounced differences in the selectivity of certain prey species by great horned owls (table 15.6). Snowshoe hares were 1.8 times more prevalent in the diet than expected from the relative availability in the environment. Voles were taken approximately in proportion to measured densities, and ground squirrels were slightly avoided. Red squirrels were the least preferred species and were 9.2 times less prevalent than snowshoe hares (table 15.6). There were strong differences between seasons.

During summer, the values of Manly's alpha (preference index) were 0.44 ± 0.11 and 0.32 ± 0.06 for voles and hares, respectively (table 15.6). When plotting use versus availability, this resulted in a convex curve above the line of equal representation (figure 15.7). For hares, the data points at low densities are close to the line of equal representation, and prey switching (sensu Murdoch and Daten 1975) may occur. To establish the vulnerability of juvenile and adult hares to predation by great horned owls, we calculated Manly's alpha for diet samples from 30 territories during the summers of 1989–1991 (14 different territories) and compared these data to a more detailed estimation of age-specific hare

Table 15.6 Prey selection by great horned owls in the boreal forest at Kluane Lake in southwestern Yukon.

Prey Category	Summer (n = 7)		Winter (n = 6)		All Seasons
	Factor[a]	Manly's α	Factor[a]	Manly's α	Factor[a]
Snowshoe hares	1.27	0.32 ± 0.06	2.51	0.83 ± 0.09	1.84 ± 0.25
Voles	1.78	0.44 ± 0.11	1.13	0.12 ± 0.09	1.13 ± 0.32
Red squirrels	0.24	0.06 ± 0.02	0.20	0.05 ± 0.05	0.20 ± 0.08
Ground squirrels	0.71	0.18 ± 0.06	0.71	–	0.71 ± 0.22

Given are means and standard errors for summers (1989–1995), winters (1989–1994), and both seasons pooled (numbers of territories and pellets as in figure 15.7). Manly's α is an index describing the relative preference for prey categories in each season (sum = 1). The preference factor describes deviations from equal representation in diet and environment and is comparable across seasons (preference >1, avoidance <1).

[a]Factor = $(\alpha_i / \Sigma \alpha_i / m_i)$, where m_i = number of prey categories.

densities in summer (Rohner and Krebs 1996). The preference of great horned owls for juvenile hares was much higher than the preference for adult hares (juvenile, α = 0.68 vs. adult, α = 0.32, SE = 0.04, $p < .001$; Wilcoxon rank test for paired samples).

Ground squirrels were taken at similar proportions as available during summer (Manly's α = 0.18 ± 0.06, slightly below the neutral α of 0.25). Red squirrels were the least preferred of these prey species (table 15.6), resulting in a clearly concave curve below the line of equal proportions (figure 15.7).

Winter results were remarkably different (table 15.6). As illustrated in figure 15.8, there was a strong preference for snowshoe hares (α = 0.83 ± 0.09, factor 2.5), whereas voles were underrepresented in winter diets (α = 0.12 ± 0.09, factor 0.4), presumably because they were less accessible due to snow cover. Red squirrels were selected for 1.5 times less than in summer, possibly due to the reduced activity of red squirrels in winter.

15.3.3 Estimating Functional Responses and Predation Impact

Functional responses predict specific kill rates of predators in relation to changing prey densities (e.g., Holling 1959). Such measurements of the number of prey killed by a predator per unit time are extremely difficult to obtain for elusive species such as great horned owls. We estimated the kill rate, KR, of a prey type i as

$$KR_i = D_i \, M_c \, A \, R \, W_i \, t, \tag{1}$$

where D_i = proportion of prey type i in diet, M_c = average biomass consumed per day, A = activity level, R = increase needed for reproduction (e.g., mass delivered to dependent young), W_i = wastage of prey type i (parts of kill that are not consumed), and t = number of days over which kill rate is calculated. A minimum estimate of M_c can be obtained from studies of existence metabolism in captive owls. Data from allometric equations, specific metabolic measurements, and feeding trials indicate a daily consumption of approximately 0.15 kg live prey mass per owl (Craighead and Craighead 1956, Kendeigh et

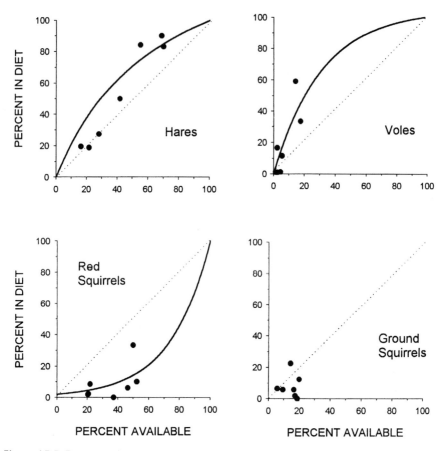

Figure 15.7 Responses in summer diets of great horned owls to changes in relative availability of the major prey species. Preference for a prey type is indicated when points lie above the line of equal representation; avoidance is indicated when points are below the line. A change from underrepresentation at low prey densities to overrepresentation at high densities is evidence for prey switching.

al. 1977, Kasparie 1983). This consumption is further modified by the costs of reproduction, R, and other increases of activity levels in the field (e.g., cost of flight; A has been estimated from 1.01 to 2.5; Kasparie 1983, Wijnandts 1984). We approximated R by calculating the number of fledglings per territorial owl (table 15.2). Because data are not available on A and W for great horned owls, we did not use these parameters in our calculations. Therefore, the functional responses presented are minimum estimates.

Overall, great horned owls showed a strong functional response to the snowshoe hare cycle (figure 15.9c) and closely followed the theoretical prediction of a type-2 functional response (Holling 1959, Fujii et al. 1986):

$$KR = a N_0/(1 + a T_H N_0) \qquad \text{with } T_H = 1/A_{max}, \tag{2}$$

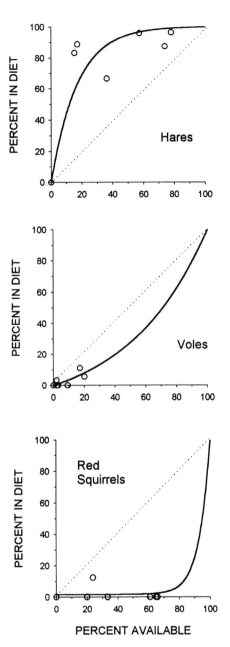

Figure 15.8 Responses in winter diets of great horned owls to changes in relative availability of the major prey species. Preference for a prey type is indicated when points lie above the line of equal representation; avoidance is indicated when points are below the line. A change from underrepresentation at low prey densities to overrepresentation at high densities is evidence for prey switching.

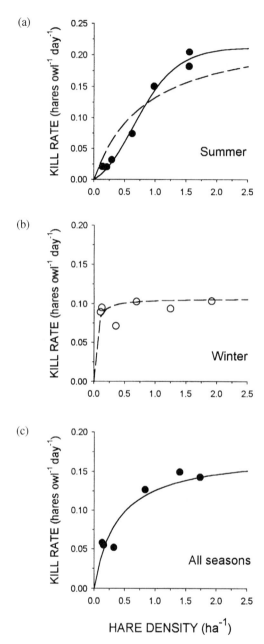

Figure 15.9 Minimum estimates of functional responses by great horned owls (kills per owl per day) to changing snowshoe densities, with curves from Holling's disc equation fitted to the data (Fujii et al. 1986). (a) Summer (mid-April–mid-October, average of hare densities in spring and autumn); (b) winter (mid-October–mid-April, average of hare densities in autumn and spring); (c) all seasons (average of summer and winter).

whereas N_0 = hare density/(ha^{-1}), a = 0.456 (coefficient indicating prey preference), A_{max} = 0.174 hares(owl^{-1}day^{-1}), asymptotic maximum of kill rate, KR. In other words, the handling time, T_H (or time interval between kills), was estimated at 5.7 days on average throughout the year. But figure 15.9 also shows a strong difference in the functional responses during summer and winter.

During summer, the kill rate of great horned owls responded strongly to changing snowshoe hare densities. We estimated A_{max} at 0.24 hares(owl^{-1}day^{-1}) from a maximum increase, M_c, of 2.5 fledglings per owl pair, and using a type-2 functional response, curve fitting yielded a = 0.297 (n = 7, r^2 = 0.84). However, a sigmoid type-3 (Holling 1959, Fujii et al. 1986) provided a better fit to the data:

$$KR = a\,N_0/[a\,N_0/a\,A_{max} + \exp(-2\,N_0)], \tag{3}$$

with a parameter estimation for both a = 0.063 and A_{max} = 0.213 hares(owl^{-1}day^{-1}) (n = 7, r^2 = 0.98). A different approach to estimate the kill rate during breeding consisted of calculating the prey biomass from remains found in pellets (table 15.5). The daily biomass brought to owlets during 1989–1991 was 353 ± 50 g, 375 ± 58 g, and 192 ± 43 g, respectively. This translates into a kill rate of 0.21, 0.24, and 0.12 hares per day, and these results were similar to the estimates based on energetic demands (figure 15.9a) of 0.18, 0.20, and 0.15 hares per day. A third method based on the dry weight of pellets yielded somewhat higher estimates. In 1989, 26.2 ± 4.2 g of dry pellet mass were delivered daily to tethered owlets. Assuming the assimilation efficiency measured by Kasparie (1983), this translates into a total of 502 g of live prey, or an estimated kill rate of 0.30 hares per day.

During winter, great horned owls were almost entirely dependent on snowshoe hares as their main diet, resulting in a flat curve with estimated kill rates close to the asymptote throughout the hare cycle (a = 3.678, A_{max} = 0.11, n = 6, r^2 = .98).

These kill rates from functional responses were then combined with the numerical response (figure 15.5) to produce a minimum estimate of the predation impact on the snowshoe hare population (table 15.7). The results indicate that great horned owls killed at least 5–10% of the hare population in peak winters, and possibly had an even larger impact during declining and low densities.

Table 15.7 Estimates of predation impact by great horned owls on snowshoe hares during winter at Kluane Lake, Yukon (see text for details).

Year	Hares/km^2	Kill Rate (hares/owl/day)	Owls/km^2	Days Exposed	Hares Killed/ km^2	Predation Impact (%)
1988–89	228	0.10	0.53	180	9.85	4.3
1989–90	235	0.09	0.78	180	13.15	5.6
1990–91	162	0.10	0.94	180	17.35	10.7
1991–92	109	0.09	0.49	180	7.92	7.3
1992–93	13	0.10	0.29	180	4.97	38.2
1993–94	20	0.07	0.21	180	2.72	13.6

15.4   Social Organization

15.4.1   Social Status and Vocal Activity

The long life spans of radio transmitters allowed us to examine the integration of fledglings into the breeding population. Only 15% (3 of 20) settled in territories before the end of their first year of life (for details, see Rohner 1996). None of the 9 owls that were further monitored in the study area to the end of their second year of life settled during that time, or showed any sign of hooting or other territorial defense. To test whether these nonterritorial floaters would normally be included in a census, we monitored a number of radio-tagged owls within hearing range to record their hooting activity from 3 March to 27 April 1990. Hooting activity was measured as the duration of bouts, each of them considered to be finished when more than 5 min elapsed between hoots.

Almost all territorial males, and often also females, gave territorial challenges at least for a short time, particularly at dusk and dawn (see also Rohner and Doyle 1992). In 11 territories that were monitored for a total of 32.0 h between dusk and midnight, all males were recorded giving territorial challenges. Their hooting bouts lasted 26.7% of the total time. Of six individual floaters that were monitored for a total of 16.8 h between dusk and midnight, none of them gave a territorial challenge or any other call.

During the same time period, known territorial and nonterritorial owls were tested for their responsiveness to playback. Territorial challenges were broadcast at irregular intervals for a total duration of 20 minutes from a tape-recorder, and each individual was tested in one trial. Seventeen out of 24 territorial males (70.8%) responded vocally. Two out of six floaters approached the speaker as concluded from telemetry readings, but none of them responded with a vocal signal that would have allowed their detection during a standard census (Fisher's Exact test, $p < .01$, df = 1, $n = 30$).

15.4.2   Stability and Size of Home Ranges

In contrast to territorial owls, which formed stable pairs on distinct territories, floaters showed a variety of movement patterns (Rohner 1997). Some individuals concentrated their space use more than others, brief long-distance movements occurred, and one floater shifted its home range >20 km. Despite this variation, there was a consistent pattern of stable home ranges with occasional movements beyond the area normally used. Median shifts in the centers of activity between subsequent 4-month periods ranged from 2.1 to 4.1 km for floaters, but were only 0.2–0.9 km for territorial owls (Mann-Whitney $p < .05$ for all 4-month periods, $n > 5$). As a comparison, we also calculated median shifts in home range centers for the entire monitoring periods of individual birds (average 12.6 ± 0.6 months per owl). These shifts ranged from 0.6 to 28.3 km (median 5.8 km) for floaters and from 0.2 to 3.11 km (median 1.2 km) for territorial birds ($p = .006$, Mann-Whitney $U = 128$, $n_1 = 9$, $n_2 = 17$).

Territory sizes were much smaller than floater home ranges. Based on weekly locations, floaters covered a 90% area of 12.0–48.3 km^2 in 1990 and 4.75–69.4 km^2 in 1991. On average, these values were 26.1 ± 5.7 km^2 and 24.8 ± 8.1 km^2 (details in Rohner 1997). In contrast, there were 18–19 territorial pairs per 100 km^2 in 1990 (Rohner 1996; i.e., average territory size was 5.26–5.56 km^2). In 1991, the boundaries of 16 territories were mapped by observing encounters of hooting males. Territory sizes ranged from 2.30 to 8.83 km², with an average of 4.83 ± 0.40 km².

A more direct comparison of space use between territorial and nonterritorial owls consisted of a 3-week period in September 1990 and 1991 with locations for each night (Rohner 1997). There was a significant effect of social class not only for 90% areas but throughout different core percentages, while patchiness in space use and year effects were not significant (Rohner 1997).

Home range sizes of territorial owls decreased from 1989 to 1992 ($r = -.64$, $n = 19$, $p < .001$, details in Rohner and Krebs 1998). Because hare densities declined during the same time, it is an obvious hypothesis that the observed changes in space use are a direct function of varying prey density. Hare density, however, explained only a small portion of the variance in home range size ($r^2 = .163$, $F_{1,2} = 0.39$, $p = .59$). At the same time, the owl population increased, leading to a more densely packed array of territories in the study area. Thus, declining territory size explained considerably more of the variation in space use of these owls than prey density ($r^2 = .846$, $F_{1,2} = 10.95$, $p = .08$, details in Rohner and Krebs 1998).

15.4.3   Effect of Territoriality on Spacing of Owls

Nonterritorial owls overlapped broadly with other owls of the same social class (figure 15.10). On average, mononuclear 90% areas overlapped by 23.3 ± 4.8% and multinuclear 90% areas overlapped by 28.8 ± 6.4% ($n = 23$ overlappers and $n = 18$ overlappers, respectively, only for combinations of floaters that were monitored simultaneously and had >1% overlap). There were no consistent differences between 1990 and 1991. Some overlapped with up to four other monitored floaters (figure 15.10), and the highest overlap observed with one other floater was 87.8% (mononuclear 90% area in 1991).

Floaters were not restricted to areas outside of established territories and intruded widely into several territories (all mononuclear 90% ranges of figure 15.10 overlapped with at least five territories in the area of figure 15.11 where territorial boundaries were known). On a finer scale, however, some spatial segregation became apparent (figure 15.11). Four of five floaters were located significantly closer to territorial boundaries than expected from a random pattern (Rohner 1997). The median distance of random points to territorial boundaries was 0.343 km, the overall median of the results for individual floaters (not the median of the pooled data) was 0.229 km. This deviation of 33% was significantly different from random (bootstrap $p < .001$).

15.4.4   Territorial Behavior and Limitation
           of Population Increase

Predators are usually larger, live longer, and have fewer offspring than their prey (Taylor 1984, Stearns 1992). This translates into a lower rate of increase for populations of long-lived species, as illustrated in figure 15.12. The yearly finite rate of increase at Kluane was $\lambda = 1.22$ for great horned owls in comparison to $\lambda = 1.5-1.8$ for snowshoe hares (details of calculation in Rohner 1995). An almost identical pattern was found in Rochester, Alberta, during an earlier cycle with peak hare densities in 1971 (analysis of data from Adamcik et al. 1978; figure 15.12B). Although the values for absolute densities

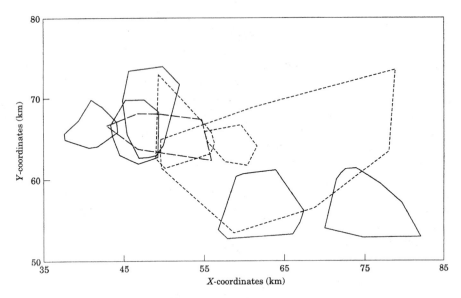

Figure 15.10 Spatial overlap among nonterritorial owls in the study area. The home ranges presented are based on 90% areas calculated by mononuclear clustering. Five owls monitored both in 1990 and in 1991 are identified by solid lines, one owl monitored only in 1990 by a broken line, and three owls monitored only in 1991 by dotted lines. From Rohner (1997).

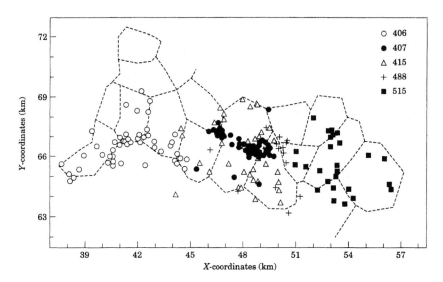

Figure 15.11 Locations of floaters relative to territorial boundaries during the period of September 1990–June 1991. Five individual floaters (see table 15.8) are represented with different symbols and a total of 198 locations. All locations are shown within the minimum convex polygon that connects the outermost corners of these known territories. Less precise locations with 95% error areas >0.5 km^2 were excluded. From Rohner (1997).

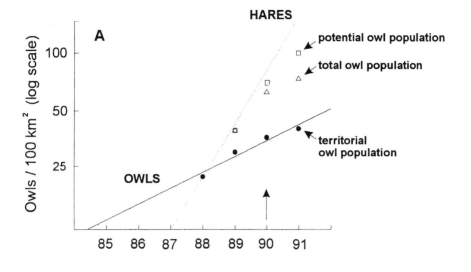

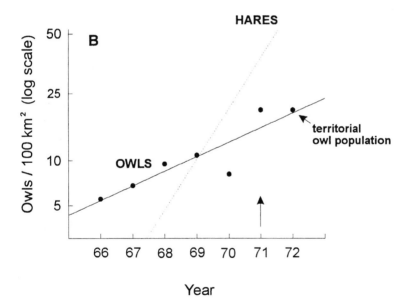

Figure 15.12 Annual rates of increase (slopes) in populations of great horned owls (solid line, symbols) and snowshoe hares (broken line, slope overlaid onto owl scale). Years of peak hare densities are indicated by arrows. (a) Kluane, Yukon; (b) Rochester, Alberta (data from Adamcik et al. 1978). From Rohner (1995).

were somewhat different, the rates of increase were similar, with $\lambda = 1.26$ for owls and $\lambda = 1.8$ for hares.

This result may not be surprising, but it is interesting that the differences in population growth almost disappeared when nonterritorial floaters were taken into account ($\lambda = 1.5$; figure 15.12A). This phenomenon is even more pronounced when a population model is used to simulate the potential rate of increase under the assumption that young owls (which in reality became floaters) could establish territories and breed as yearlings ($\lambda = 1.66$).

These findings suggest that territoriality, or, more precisely, the social exclusion from breeding, limited the increase of the predator population. This can also be illustrated by testing demographic parameters for density dependence. In agreement with the hypothesis of social exclusion, the population growth of territorial owls declined with increasing density of territories already established (figure 15.13A). Meanwhile, the floater pool of nonterritorial owls increased as territories became packed more densely (figure 15.13B).

The best evidence for the limitation of breeding densities by social exclusion is drawn from removal experiments (reviews in Watson and Moss 1970, Newton 1992). While monitoring radio-marked birds, we observed six vacancies in territories, which served as natural removal experiments. Territory holders either died or emigrated, and we recorded whether these vacancies were filled with new birds. In at least five of six vacancies, such replacements occurred (details in Rohner 1995).

All of the above results indicate that competition among owls was severe, and that this predator population was food limited even at extremely high prey densities. Usually, predators have been assumed to experience superabundant food in peak years of cyclic prey (e.g., Lack 1946). A combination of brood-size manipulations and food additions, however, suggested that this was not the case (Rohner and Smith 1996). Owlets in temporarily enlarged broods lost weight, apparently because their parents could not easily increase provision rates (Rohner and Smith 1996). In addition, mothers of enlarged broods ranged farther away from nests, probably because they were forced to hunt under food stress (Rohner and Smith 1996). In agreement with this hypothesis, food additions to enlarged broods rapidly returned the parental behavior to normal (Rohner and Smith 1996).

15.4.5 Social Behavior and the Time Lag in the Numerical Response

Predator populations often decline with a time lag after the peak of prey cycles. This delay can be explained by two hypotheses (Rohner 1995): densities of the main prey species are extremely high, leave predators with superabundant food, and buffer their decline (single prey hypothesis, SPH). Alternatively, the availability of other prey is high and thus delays the decline of generalist predators (multiple prey hypothesis, MPH).

Social exclusion among owls seemed to set a ceiling to their peak densities (figure 15.5B). This situation is in agreement with the SPH, but the presence of a large floater pool, the complete failure to breed during the prey decline (figure 15.1, table 15.2), and indications of food stress during the brood-size manipulations suggest that food was not superabundant when great horned owls showed a time lag in their numerical response to the decline of snowshoe hares.

The predictions of the MPH were met for summer diets, when great horned owls took

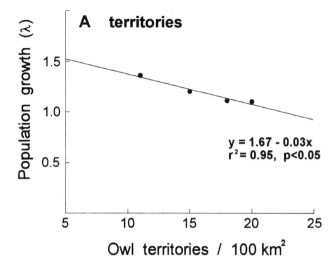

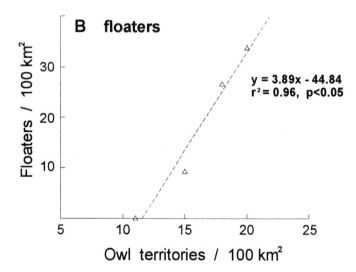

Figure 15.13 Social behavior and the limitation of population growth in great horned owls at Kluane, Yukon. (A) Growth rates of the territorial population decline as numbers of owl territories increase in the area (inverse density-dependent growth rate); (B) numbers of nonterritorial floaters increase as territories are packed more densely (density-dependent increase). From Rohner (1995).

more alternative prey and fewer hares during the low phase of the cycle (figure 15.6A). But during winter, the proportion of hares as the main prey remained high throughout the cycle (figure 15.6B), probably due to a lack of other available prey at these northern latitudes. This result emphasizes the dependence of great horned owls on snowshoe hares.

Therefore, neither the SPH nor the MPH are satisfactory explanations for the delayed decline of great horned owls relative to the snowshoe hare cycle. The key to understanding the processes involved seems to lie in the behavior of individuals (see 15.6.4).

15.5 Responses to Large-Scale Experiments

15.5.1 Space Use in Territories with Experimental Hot Spots of Prey

The contrast between prey density on artificially created hot spots and on the study area in general increased drastically from 3.8- to 10.3-fold as the decline of the snowshoe hare cycle progressed (Rohner and Krebs 1998; see chapter 8). Therefore, we investigated owls on two experimental territories that were accessible for telemetry work in more detail in 1992, when the response of predators was expected to be greatest. Food shortages for great horned owls during this year were indicated by a complete lack of breeding attempts and elevated rates of emigration and mortality by resident owls (details in Rohner 1996).

Results from experimental owls did not confirm a direct effect of prey density on home range use. Owls on territories with hot spots did not differ in home range size from owls on control territories (ANCOVA, $F_{1,16} < 0.1$, $p = 0.99$; details in Rohner and Krebs 1998). However, a more powerful randomization test revealed clear concentrations in space use of owls on territories with experimental hot spots. We modeled the null hypothesis of uniform space use by randomly generating 5000 points within the known boundaries of these territories. The distances of these points to the center of the territory (geometric mean) were then calculated and grouped into classes of 200 m, thus representing an expected frequency distribution of uniform space use within a territory. The actual telemetry locations of experimental owls were then compared to this expected distribution. If the owls behaved according to the null hypothesis, no systematic and significant deviation from the expected distribution should occur. This, however, was not the case. The results showed regions that were used more frequently than expected for both experimental owls. For owl no. 544, the mean of positive deviation from the expected values was 5.35 ($p < .01$); for owl no. 503 this mean was 9.20 ($p = .02$).

Were these concentrations in space use related to hot spots in hare abundance? Telemetry locations were compared to the experimental blocks of manipulated prey densities, and the distances to the center of these 1-km^2 blocks were calculated (figure 15.14). Both experimental owls showed a positive response when compared to the expected distributions as calculated from points randomly generated within the boundaries of their territories. For owl no. 544, the distance was 0.846 ± 0.050 km ($n = 152$) from the center of the hot spot (mean expected distance 1.534 km, bootstrap $p < .001$). For owl no. 503, this distance was 0.741 ± 0.087 km ($n = 41$, mean expected distance 1.179 km, bootstrap $p < .001$).

Data from pellet analysis of territorial owls support that these concentrations in space use were related to foraging activity. Summer diets of great horned owls consisted of

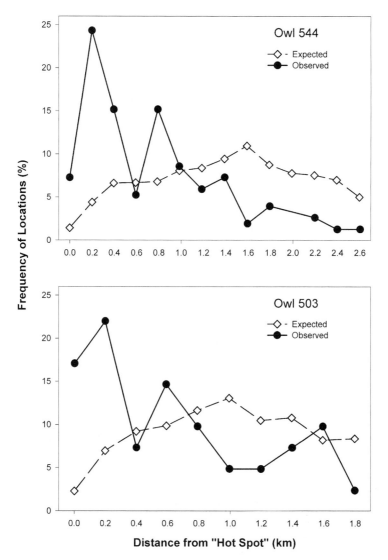

Figure 15.14 Concentrations in space use of two experimental owls relative to the center of 1-km² blocks with increased prey density. Locations were grouped into distance classes and shown as frequency distribution of observed locations (filled symbols) and expected frequencies from random locations in the specific territories (open symbols). From Rohner and Krebs (1998).

82.6–86.0% snowshoe hares in 1989–1991 when hare abundance was high ($n = 13$ different territories; see also pooled data in figure 15.6). In 1992, this portion dropped to 12.7 ± 8.5% (range 0–42.3%, $n = 5$ territories). We were able to collect pellets from one experimental owl in summer 1992. In contrast to controls, the diet of experimental owl no. 544 consisted of a high portion (64.7%) of snowshoe hares (see also figure 15.6).

Table 15.8 Shifts in space use by nonterritorial great horned owls in relation to experimental hot spots of high prey density.

Individual	N_1	N_2	D_1 (km)	D_2 (km)	D_2-D_1 (km)	p
406	71	20	10.858	11.070	0.211	.153
407	61	5	5.221	7.091	1.876	.092
415	72	14	4.454	4.967	0.513	.674
417	25	22	4.564	3.545	−1.019	.258
425	52	15	8.723	11.380	2.657	.003
433	60	13	5.178	6.115	0.937	.076
505	25	31	4.890	12.700	7.810	.001
515	38	19	1.800	3.667	1.868	.021

For each monitored owl, the median distances of weekly telemetry locations to the center of the closest experimental block are shown for the peak (D_1, January 1990–May 1991) and decline phase of the snowshoe hare cycle (D_2, June 1991–September 1992), with sample sizes of telemetry locations (N_1, N_2) and levels of significance for the difference D_1-D_2 (U-test, two-tailed). From Rohner and Krebs (1998).

15.5.2   Predator Movements from Poor Patches to Rich Patches

During the decline phase of the hare cycle, territorial owls were predicted to abandon poor territories and intrude at least temporarily into territories with hot spots of high prey abundance. Despite intensive monitoring, we did not observe any owners leaving their territories to use food-addition blocks or changing to a nomadic strategy on the >350 km² study area (for details, see Rohner 1996).

Although nonterritorial floaters were present in the study area, these owls did not aggregate in areas with food enrichment, and despite intensive monitoring, we were unable to detect an association of floaters with hot spots of high prey abundance. Table 15.8 summarizes the results on potential shifts of 8 nonterritorial owls toward hot spots during a time when the abundance of hares in these patches increased from 3.8- to 10.3-fold compared to control levels in the study area (see Rohner and Krebs 1998). There was no distinct pattern, and the only significant shifts in space use by three owls were opposite to the predicted direction.

15.6   Discussion

15.6.1   Large Floating Population When Resources Are Abundant

This study demonstrates a profound effect of resource availability on reproduction, juvenile survival, and emigration in a predator of cyclic prey. The fraction of nonterritorial owls rose to 40–50% of the population during years of high food abundance. The numerical responses of territorial and nonterritorial birds were qualitatively different, and the notion that such a high proportion of secretive floaters can occur may require a reassessment of results that are entirely based on the territorial and therefore visible fraction of a population (Rohner 1996).

In considering both academic research and applied management questions, many studies have attempted to examine the impact of predators on prey populations (review in Krebs 1994). A common method is to census predator and prey populations, identify the diet of predators, and calculate the predation mortality among prey (e.g., Craighead and Craighead 1956, Keith et al. 1977, Petersen 1979, Angelstam et al. 1984, Trostel et al. 1987, Korpimäki and Norrdahl 1991, Korpimäki 1993). In this study, the proportion of secretive floaters was high, and predation pressure on prey would have been severely underestimated when taking only territorial birds into account. This problem may also arise in other systems where floaters are difficult to detect.

Our results also demonstrate how a large proportion of secretive floaters can delay the detection of population declines in conservation studies. When, as here, floaters are more affected by decreasing habitat quality than are territorial birds, traditional monitoring programs based on censusing territories will not reveal these declines at an early stage (Wilcove and Terborgh 1984). For example, in a scenario for spotted owl populations, Franklin (1992) estimated that declines in territorial owls could not be detected for 15 or more years when floaters were present even at low densities.

15.6.2 Factors Affecting Functional Responses and Predation Impact

Great horned owls showed strong preferences for specific prey species, mostly for snowshoe hares and voles and, among vole species, *Microtus* over *Clethrionomys* (see also Rohner et al. 1995). A major behavioral mechanism accounting for these differences lies in the habitat selection of both predator and prey. The hunting success of great horned owls on snowshoe hares was highest in open habitat with shrub and tree cover below average (Rohner and Krebs 1996), and snowshoe hares appear to select for dense cover as an antipredator behavior (Hik 1995). Voles are less accessible to great horned owls during winter because they are active below snow cover, and, at the same time, the loss of leaves reduces shrub cover for snowshoe hares outside the growing period and exposes them to higher predation by owls (see also Sonerud 1986). Several lines of evidence suggest that great horned owls prefer to hunt in open habitat (Rohner and Krebs 1996), and since field voles (in particular *Microtus pennsylvanicus*) prefer open grassy habitat, this may also account for the surprisingly low representation of the forest-dwelling red-backed voles (*Clethryonomys rutilus*) in the diet of great horned owls. We assume that this basic pattern based on habitat is further modified by other antipredator behavior of specific prey species. For example, ground squirrels use social alert calls and burrows to escape predation, and red squirrels are more agile than a pursuing owl in spiraling branches and tree trunks (K. McKeever, personal observation).

The shape of the functional response of great horned owls to changing hare densities deviated from a typical type-2 curve in summer and suggested that there may be an element of active prey switching (type-3 response, sensu Holling 1959, Fujii et al. 1986). In figure 15.7, the data points at low hare densities do not clearly cross the line of equal representation as predicted by prey switching (Murdoch and Oaten 1975), but they are nevertheless close. However, a minor bias in our calculations may account for this result. Because snowshoe hares are the largest portion of vertebrate biomass in the boreal forest and we only included the four major prey groups to calculate relative availabilities, a complete array of prey species would reduce the relative portion of snowshoe hares at low densities and shift the data points closer to the convex curve that is expected with no active change in behavior.

More significantly, prey switching in summer may be caused by the active change in breeding behavior. At low hare densities, great horned owls do not attempt to breed, whereas they switch to producing large broods at high hare abundance. This increases the energy requirements dramatically, and hunting males providing for families will show higher kill rates than predicted by a simple type-2 curve (figure 15.9). Furthermore, breeding central-place foragers may ignore small prey such as voles and actively select for larger prey (i.e., snowshoe hares) when provisioning young, but not when they are refraining from reproduction (e.g., Stephens and Krebs 1986, Sonerud 1992). We find it interesting that others have also reported almost linear (and not typical type-2) functional responses of avian predators in summer (e.g., McInvaille and Keith 1974, Korpimäki and Norrdahl 1991), and we add that such elongated or quasi–type-3 responses should be expected on theoretical grounds in avian predators that adjust their breeding effort in relation to prey density.

The flat shape of our type-2 functional response in winter is probably a result of sampling limitations. We calculated the asymptote A_{max} based on energetic considerations, and, lacking actual data, we assumed that the energetic requirements are constant through the whole range of hare densities. However, this is not very likely. At high snowshoe hare densities, we observed in the field that considerable wastage or surplus killing of snowshoe hares occurs and that territorial and courtship activity was considerably elevated compared to the low phase of the cycle, therefore indicating that A_{max} in figure 15.9B is an underestimate. At low hare densities, great horned owls may reduce their energy demands drastically (Kasparie 1983), and the functional response in great horned owls may increase more gradually than we assumed. In addition, great horned owls may sample hares more efficiently than we did and hunt at patches of above average density, and the observed high values at extremely low hare densities may correspond to somewhat higher densities than we measured by live-trapping grids that are relatively small compared to an owl territory.

Overall, our functional responses were similar to those measured by Keith's group (McInvaille and Keith 1974, Keith et al. 1977), with an asymptotic kill rate of 0.2–0.3 hares/day during April–May and similar limitations in estimating procedures. Our cross-comparison with biomass brought to nests indicated that we also measured minimum estimates in summer (calculations based on diagnostic bones underestimate the actual number of prey, and our pellet samples at nests did not include the consumption by the parents). Although data are not currently available, we conservatively speculate that our functional responses underestimate the real kill rates by a factor of 1.5–3.

Minimum predation rates of 5–10% of the hare population at high densities are considerable and may add substantially to the effect of other predators (see chapter 13). In contrast to other predators, the numerical response of great horned owls is only three- to fourfold. This means that a generalist predator will remain at relatively high densities during low hare densities, with potentially high predation impacts on snowshoe hares and alternative prey species during the low phase of the snowshoe hare cycle (see chapter 8).

15.6.3   Limitation of Population Growth at Peaks of Cyclic Prey

This study provides several lines of evidence that territorial behavior excluded subordinate owls from breeding even at peak abundance of prey. Floaters were located more often than expected at the periphery of established territories, indicating that territorial behavior restricted their movements (Rohner 1996). Furthermore, replacements of territorial vacancies were in agreement with the hypothesis that social behavior limited the number of owl territories; there were inverse density-dependent growth rates in the territorial population, and the accumulation of floaters was density dependent (Rohner 1995). A combination of brood-size manipulations and food additions suggested that there was competition over limited resources even at peak densities of the snowshoe hare cycle (Rohner and Smith 1996).

It is now well established that territorial behavior limits the breeding density of many raptors and owls (Southern 1970, Newton 1976, Village 1983, Hirons 1985, Newton 1992). It is interesting that not only stable populations, but also fluctuating populations linked to prey cycles of extreme amplitude may reach population ceilings without superabundant food. A similar situation was suggested for Tengmalm's owls (*Aegolius funereus*) responding to vole cycles in Fennoscandia (Korpimäki 1988, 1989). Such ceilings in the numerical response of predators self-regulation of predator populations, or social interference among predators have rarely been demonstrated empirically, although they have been assumed for models of predator–prey dynamics, and their consequences on stability in population interactions and on cascading effects in trophic levels of communities have been discussed (e.g., Rosenzweig and MacArthur 1963, Arditi and Ginzburg 1989, Hanski 1991, Caughley and Sinclair 1994).

15.6.4   Time Lag in the Numerical Response of a Predator of Cyclic Prey

Time lags are an important feature of theoretical models addressing the causes of population cycles (e.g., Ginzburg and Taneyhill 1994, review in Krebs 1994). Several hypotheses can explain why predators of cyclic prey should decline with a time lag relative to prey populations. This study suggests that time lags in some vertebrate predators may not be caused by an excess of food but a change in individual strategy. Territorial great horned owls monopolized a disproportionate amount of resources, while others were excluded. These "family territories" allowed not only a high reproductive success during optimal conditions, but also a sufficient overhead of prey when owls switched off reproduction and adopted a more conservative strategy during the decline phase of the cycle (Rohner 1995, see also Kasparie 1983). In agreement with this hypothesis, the prey decline affected survival and emigration rates of subordinate owls before those of territory holders (Rohner 1996).

The importance of behavioral changes in causing the time lag in the numerical response may also apply to other vertebrate predators. For example, avian predators with less pronounced territorial behavior showed no delay in their numerical response to vole cycles in Finland (Korpimäki and Norrdahl 1991, Korpimäki 1994). Goshawks (*Accipiter gentilis*), which may not have the option of conserving much energy because fast flights are essential to their search and hunting success (Kenward 1982, Widén 1984), also responded immediately to the decline of snowshoe hares both at Rochester and Kluane (Keith et al. 1977, Doyle and Smith 1994; see chapter 16).

15.6.5 Factors Limiting Spatial Aggregation of Predators

Experimental owls on food-enriched territories concentrated their foraging effort on experimental hot spots. Although conventional methods of estimating home range size and patchiness in space use failed to detect a difference between experiments and controls, more specific measurements showed a clear response. Both the distribution of owl locations within known territory boundaries and the distance of locations in relation to the food-addition blocks were significantly different from a random prediction in two replicates. Differences in the use of snowshoe hares in the diet by experimental versus control owls confirmed the concentration of foraging effort on food-enriched patches within a territory.

At a larger scale, however, neither territorial owls nor floaters showed any tendency to leave poorer patches and move towards hot spots, and the territorial system of great horned owls was robust to extreme variations in prey density. The spatial scale of the experimental hot spots may not have been large enough to present a detectable patch (or a patch of sufficient rewards) to attract great horned owls other than the specific territory owners, but nonterritorial floaters ranged widely in the study area and were temporarily located at many different sites.

More likely, great horned owls did not follow an ideal free distribution (sensu Fretwell 1972, review in Milinski and Parker 1991), and territory holders prevented aggregations of other owls on hot spots. According to territory economics and the threshold model of territoriality (Brown 1964, Davies 1980, Carpenter 1987), short-term feeding territories should be abandoned when food and intruder pressure increase beyond an upper threshold where the cost of defense exceeds the benefits of exclusive access. Despite extreme variations in prey density, this was not the case for great horned owls. Although intrusions by nonterritorial owls were frequent, this robust territorial system did not only impose self-regulation (sensu Caughley and Sinclair 1994) on the growth of a predator population, but also represented a ceiling to spatial aggregations of predators where prey was abundant.

Literature Cited

Adamcik, R. S., A. W. Todd, and L. B. Keith. 1978. Demographic and dietary responses of great horned owls during a snowshoe hare cycle. Canadian Field-Naturalist **92**:156–166.
Angelstam, P., E. Lindström, and P. Widén. 1984. Role of predation in short-term population fluctuations of some birds and mammals in Fennoscandia. Oecologia **62**:199–208.
Arditi, R. and L. R. Ginzburg. 1989. Coupling in predator-prey dynamics: ratio-dependence. Journal of Theoretical Biology **139**:311–326.

Brown, J. L. 1964. The evolution of avian diversity in avian territorial systems. Wilson Bulletin **76**:160–169.

Carpenter, F. L. 1987. Food abundance and territoriality: to defend or not to defend? American Zoologist **27**:387–399.

Caughley, G., and A. R. E. Sinclair. 1994. Wildlife Ecology and Management. Blackwell Scientific, Oxford.

Craighead, J. J., and F. C. Craighead. 1956. Hawks, owls and wildlife. Stackpole, Harrisburg, Pennsylvania.

Davies, N. B. 1980. The economics of territorial behaviour in birds. Ardea **68**:63–74.

Donazar, J. A., F. Hiraldo, M. Delibes, and R. R. Estrella. 1989. Comparative food habits of the eagle owl *Bubo bubo* and the great horned owl *Bubo virginianus* in six palearctic and nearctic biomes. Ornis Scandinavica **20**:298–306.

Doyle, F. I., and J. N. M. Smith. 1994. Population responses of northern goshawks to the 10-year cycle in numbers of snowshoe hares. Studies in Avian Biology **16**:122–129.

Franklin, A. B. 1992. Population regulation in northern spotted owls: theoretical implications for management. *in* D. McCullough and R. Barrett (eds). Wildlife 2001: populations, pages 815–827. Elsevier, London.

Fretwell, S. D. 1972. Populations in a seasonal environment. Princeton University Press, Princeton, New Jersey.

Fujii, K., C. S. Holling, and P. M. Mace. 1986. A simple generalized model of attack by predators and parasites. Ecological Research **1**:141–156.

Galeotti, P. 1990. Territorial behaviour and habitat selection in an urban population of the tawny owl *Strix aluco L.* Bolletino di Zoologia **57**:59–66.

Ginzburg, L. R., and D. E. Taneyhill. 1994. Population cycles of forest Lepidoptera: a maternal effect hypothesis. Journal of Animal Ecology **63**:79–92.

Hall, P., and S. R. Wilson. 1991. Two guidelines for bootstrap hypothesis testing. Biometrics **47**:757–762.

Hanski, I. 1991. The functional response of predators: worries about scale. Trends in Ecology and Evolution **6**:141–142.

Hestbeck, J. B. 1982. Population regulation of cyclic mammals: the social fence hypothesis. Oikos **39**:157–163.

Hik, D. S. 1995. Does risk of predation influence population dynamics? Evidence from the cyclic decline of snowshoe hares. Wildlife Research **22**:115–129.

Hirons, G. J. M. 1985. The effects of territorial behaviour on the stability and dispersion of tawny owl (*Strix aluco*) populations. Journal of Zoology (London) B **1**:21–48.

Holling, C. S. 1959. Some characteristics of simple types of predation and parasitism. Canadian Entomologist **91**:385–398.

Holling, C. S. 1965. The functional response of predators to prey density and its role in mimicry and population regulation. Memoirs of Entomological Society of Canada **45**:1–60.

Houston, C. S. 1987. Nearly synchronous cycles of the great horned owl and snowshoe hare in Saskatchewan. *in* R. W. Nero, R. J. Clark, R. J. Knapton, and R. H. Hamre (eds). Biology and conservation of northern forest owls: symposium proceedings, pages 55–58. General Technical Report Rm-142, USDA Forest Service, Washington, DC.

Houston, C. S., and C. M. Francis. 1995. Survival of great horned owls in relation to the snowshoe hare cycle. Auk **112**:44–59.

Houston, C. S., D. W. Smith, and C. Rohner. 1998. Great horned owl (*Bubo virginianus*). *In* A. Poole and F. Gill (eds). The birds of North America, no. 372. American Ornithologists' Union, Washington, DC.

Hunter, D. B., C. Rohner, and D. C. Currie. 1997. Mortality in fledgling great horned owls from black fly hematophaga and leucocytozoonosis. Journal of Wildlife Diseases **33**:486–491.

Ims, R. A., and H. Steen. 1990. Geographical synchrony in microtine population cycles: a theoretical evaluation of the role of nomadic avian predators. Oikos **57**:381–387.

Johnson, D. H. 1993. Spotted owls, great horned owls, and forest fragmentation in the central Oregon Cascades. MSc thesis. Oregon State University, Corvallis.

Kasparie, J. A. 1983. Some physiological and behavioral responses of the great horned owl (Bubo virginianus) to winter conditions. MSc thesis. University of Missouri-Columbia, Columbia.

Keith, L. B., and D. H. Rusch. 1988. Predation's role in the cyclic fluctuations of ruffed grouse. *In* H. Ouellet (ed). Acta XIX Congressus Internationalis Ornithologici 1986, pages 699–762. National Museum of Natural Sciences, Ottawa.

Keith, L. B., A. W. Todd, C. J. Brand, R. S. Adamcik, and D. H. Rusch. 1977. An analysis of predation during a cyclic fluctuation of snowshoe hares. Proceedings of the International Congress of Game Biology **13**:151–175.

Kendeigh, S. C., V. R. Dol'nik, and V. M. Gavrilov. 1977. Avian energetics. *In* J. Pinowski and S. C. Kendeigh (eds). Granivorous birds in ecosystems, pages 127–204. Cambridge University Press, New York.

Kenward, R. E. 1982. Goshawk hunting behaviour, and range size as a function of food and habitat availability. Journal of Animal Ecology **51**:69–80.

Kenward, R. E. 1985. Raptor radio-tracking and telemetry. ICBP Technical Publication **5**:409–420.

Kenward, R. E. 1987. Wildlife radio-tagging: equipment, field techniques, and data analysis. Academic Press, London.

Kenward, R. E. 1990. Ranges IV. Institute of Terrestrial Ecology, Furzebrook, Wareham, UK.

Korpimäki, E. 1988. Costs of reproduction and success of manipulated broods under varying food conditions in Tengmalm's owl. Journal of Animal Ecology **57**:1027–1039.

Korpimäki, E. 1989. Breeding performance of Tengmalm's owl *Aegolius funereus:* effects of supplementary feeding in a peak vole year. Ibis **131**:51–56.

Korpimäki, E. 1993. Regulation of multiannual vole cycles by density-dependent avian and mammalian predation? Oikos **66**:359–363.

Korpimäki, E. 1994. Rapid or delayed tracking of multi-annual vole cycles by avian predators? Journal of Animal Ecology **63**:619–628.

Korpimäki, E., and K. Norrdahl. 1991. Numerical and functional response of kestrels, short-eared owls, and long-eared owls to vole densities. Ecology **72**:814–826.

Krebs, C. J. 1994. Ecology: the experimental analysis of distribution and abundance, 4th ed. Harper Collins, New York.

Lack, D. 1946. Competition for food by birds of prey. Journal of Animal Ecology **15**:123–129.

Lindström, E., P. Angelstam, P. Widén, and H. Andrén. 1987. Do predators synchronize vole and grouse fluctuations? An experiment. Oikos **48**:121–124.

Marti, C. D. 1974. Feeding ecology of four sympatric owls. Condor **76**:45–61.

McInvaille, W. B. and L. B. Keith. 1974. Predator-prey relations and breeding biology of the great horned owl and red-tailed hawk in central Alberta. Canadian Field-Naturalist **88**:1–20.

Messier, F. 1994. Ungulate population models with predation: a case study with the North American moose. Ecology **75**:478–488.

Milinski, M., and G. A. Parker. 1991. Competition for resources. *in* J. R. Krebs and N. B. Davies (eds). Behavioural ecology: an evolutionary approach, pages 137–168. Blackwell Scientific, Oxford.

Murdoch, M. W. and A. Oaten. 1975. Predation and population stability. Advances in Ecological Research **9**:2–131.

Nams, V. O. 1990. Locate II. Pacer, Truro, Nova Scotia.

Newton, I. 1976. Population limitation in diurnal raptors. Canadian Field-Naturalist **90**:274–300.

Newton, I. 1992. Experiments on the limitation of bird numbers by territorial behaviour. Biological Reviews **67**:129–173.

Petersen, L. R. 1979. Ecology of great horned owls and red-tailed hawks in southeastern Wisconsin. Technical Bulletin 111: Department of Natural Resources, Madison, Wisconsin.

Petersen, L. R., and J. R. Keir. 1976. Tether platforms—an improved technique for raptor food habits study. Raptor Research **10**:21–28.

Pitelka, F. A., P. Q. Tomich, and G. W. Treichel. 1955. Ecological relations of jaegers and owls as lemming predators near Barrow, Alaska. Ecological Monographs **25**:85–117.

Pruitt, W. O. 1978. Boreal ecology. Studies in Biology **91**:1–73.

Rohner, C. 1995. Great horned owls and snowshoe hares: what causes the time lag in the numerical response of predators to cyclic prey? Oikos **74**:61–68.

Rohner, C. 1996. The numerical response of great horned owls to the snowshoe hare cycle: consequences of non-territorial 'floaters' on demography. Journal of Animal Ecology **65**:359–370.

Rohner, C. 1997. Non-territorial 'floaters' in great horned owls: space use during a cyclic peak of snowshoe hares. Animal Behavior **53**:901–912.

Rohner, C., and F. I. Doyle. 1992. Methods of locating great horned owl nests in the boreal forest. Journal of Raptor Research **26**:33–35.

Rohner, C., and D. B. Hunter. 1996. First-year survival of great horned owls during a peak and decline of the snowshoe hare cycle. Canadian Journal of Zoology **74**:1092–1097.

Rohner, C., C. J. Krebs, D. B. Hunter, and D. C. Currie. 2000. Roost site selection of great horned owls in relation to black fly activity: an anti-parasite behavior? Condor **102**:950–955.

Rohner, C., and C. J. Krebs. 1996. Owl predation on snowshoe hares: consequences of antipredator behaviour. Oecologia **108**:303–310.

Rohner, C., and C. J. Krebs. 1998. Responses of great horned owls to experimental 'hot spots' of snowshoe hare density. Auk **115**:694–765.

Rohner, C., and J. N. M. Smith. 1996. Brood size manipulations in great horned owls *Bubo virginianus:* are predators food limited at the peak of prey cycles? Ibis **138**:236–242.

Rohner, C., J. N. M. Smith, J. Stroman, F. I. Doyle, M. Joyce, and R. Boonstra. 1995. Northern hawk owls (*Surnia ulula caparoch*) in the nearctic boreal forest: prey selection and population consequences of multiple prey cycles. Condor **97**:208–220.

Rosenzweig, M. L., and R. H. MacArthur. 1963. Graphical representation and stability conditions of predator-prey interactions. American Naturalist **97**:209–223.

Rusch, D. H., E. C. Meslow, L. B. Keith, and P. D. Doerr. 1972. Response of great horned owl populations to changing prey densities. Journal of Wildlife Management **36**:282–296.

Saltz, D., and G. C. White. 1990. Comparison of different measures of telemetry error in simulated radio-telemetry locations. Journal of Wildlife Management **54**:169–174.

Sonerud, G. A. 1986. Effect of snow cover on seasonal changes in diet, habitat, and regional distribution of raptors that prey on small mammals in boreal zones of Fennoscandia. Holarctic Ecology **9**:33–47.

Sonerud, G. 1992. Functional responses of birds of prey: biases due to the load-size effect in central place foragers. Oikos **63**:223–232.

Southern, H. N. 1970. The natural control of a population of tawny owls (*Strix aluco*). Journal of Zoology (London) **162**:197–285.

Stearns, S. C. 1992. The evolution of life histories. Oxford University Press, Oxford.

Stephens, D. W., and J. R. Krebs. 1986. Foraging theory. Princeton University Press, Princeton, New Jersey.

Taylor, R. T. 1984. Predation. Chapman and Hall, New York.

Trostel, K., A. R. E. Sinclair, C. J. Walters, and C. J. Krebs. 1987. Can predation cause the 10-year hare cycle? Oecologia **74**:185–192.

Village, A. 1983. The role of nest-site and territorial behaviour in limiting the breeding density of kestrels. Journal of Animal Ecology **52**:635–634.

Voous, K. H. 1988. Owls of the Northern Hemisphere. Collins Sons, London.

Watson, A., and R. Moss. 1970. Dominance, spacing behaviour and aggression in relation to population limitation in vertebrates. *in* A. Watson and R. Moss (eds). Animal populations in relation to their food resources, pages 167–218. Blackwell Scientific, Oxford.

Widén, P. 1984. Activity patterns and time-budget in the goshawk *Accipiter gentilis* in a boreal forest area in Sweden. Ornis Fennica **61**:109–112.

Wijnandts, H. 1984. Ecological energetics of the long-eared owl *(Asio otus)*. Ardea **72**:1–92.

Wilcove, D. S., and J. W. Terborgh. 1984. Patterns of population decline in birds. American Birds **38**:10–13.

Ydenberg, R. C. 1987. Nomadic predators and geographical synchrony in microtine population cycles. Oikos **50**:270–272.

16

Raptors and Scavengers

FRANK I. DOYLE & JAMES N. M. SMITH

The principal study to date of the responses of raptorial birds to the snowshoe hare cycle was conducted by Lloyd Keith and colleagues in a mixed farmland/aspen parkland landscape at Rochester, Alberta. They found that great horned owls (*Bubo virginianus*) showed strong reproductive and numerical responses to hare abundance (McInvaille and Keith 1974, Adamcik et al. 1978). In contrast, red-tailed hawk (*Buteo jamaicensis*) numbers at Rochester remained stable, even though their reproductive success declined as hares disappeared from the prey base (Adamcik et al. 1979). In Alaska, numbers and breeding success of northern goshawks responded positively to increasing hare abundance (McGowan 1975). Also in Alaska, golden eagles (*Aquila chrysaetos*) reproduced more successfully during a period of high hare and ptarmigan abundance (McIntyre and Adams 1999). The community and population level responses of other raptors to vole cycles have been studied intensively in Europe (e.g., Korpimäki and Norrdahl 1991) and less intensively in temperate North America (e.g., Phelan and Robertson 1978).

Raptorial birds vary greatly in their migratory behavior (see below), but nearly all breed in territorial pairs that rear only a single brood of young per year (Newton 1979). Most raptors exhibit size and behavioral dimorphisms. Females are larger and are solely responsible for building and guarding the nest, incubation, and brooding. They feed small young with food brought to the nest by the male until the chicks are half grown, when the female also starts to hunt to help meet the increased food demands of the growing brood. In corvids, the sexes are more similar in size and share parental duties more evenly.

Raptorial and scavenging birds (hawks, owls, and corvids) composed a substantial fraction (21 of 87 species; table 16.1) of the vertebrates that we studied at Kluane. They varied in size from the 73-g gray jay (*Perisoreus canadensis*), which rarely eats prey larger than a nestling bird, to the 5-kg bald eagle (*Haliaeetus leucocephalus*) that consumes birds, fish, and mammals that often exceed its own mass.

Our aim in this chapter is to describe how these 21 avian predators and scavengers responded to the changes in prey through the hare cycle. Boutin et al. (1995) provided a preliminary account of these responses. For each of 12 breeding species, we describe diets and functional and numerical and reproductive responses to the hare cycle. We also provide some information on mortality factors affecting raptors.

16.1   Methods

16.1.1   General Approach

The methods used in studies of raptors have a long and successful history worldwide (Craighead and Craighead 1956, Newton 1979). Most previous research on avian predators and scavengers has been conducted in developed areas of North America (e.g., McInvaille and Keith 1974, Grant et al. 1991) and in the heavily managed forests of boreal Europe (Hörnfeldt 1978, Newton 1986, Korpimäki 1988, Hörnfeldt et al. 1990). In such areas, roads and tracks allow ready access by vehicle, and breeding pairs and nests can be found readily from the air or ground in open country and deciduous forests (e.g., Craighead and Craighead 1956, Luttich et al. 1971, McInvaille and Keith 1974, Grant et al.

Table 16.1 Status of the 21 species of raptors present at Kluane.

Species	Status	Present Each Year	Breeds in Study Area	Breeds Every Year	Found throughout study area	Habitat
Great horned owl	C	Y	Y	N	Y	C, S, O
Northern hawk owl	C/R	N	Y	N	N	C, S
Short-eared owl	R	N	N	N	U	O
Boreal owl	C/R	U	Y	N	N	C
Saw-whet owl	R	N	N	N	U	U
Great gray owl	R	N	N	N	U	U
Red-tailed hawk	C	Y	Y	Y	Y	C, S, O
Rough-legged hawk	F	Y	N	N	N	C, S, O
Northern harrier	C	Y	Y	Y	Y	O
Gyrfalcon	R	N	N	N	U	O
Peregrine falcon	R	N	N	N	U	O
Merlin	F	Y	N	N	U	O
American kestrel	C	Y	Y	Y	N	O
Northern goshawk	C	Y	Y	Y	Y	C, S, O
Sharp-shinned hawk	R	Y	Y	U	U	S, O
Osprey	R	Y	N	N	U	L
Golden eagle	C	Y	Y	N	N	S, O
Bald eagle	C	Y	Y	N	N	C, S, O, L
Common raven	C	Y	Y	N	Y	C, S, O
Black-billed magpie	C	Y	Y	Y	N	S, O
Gray jay	C	Y	Y	Y	Y	C, S

F = frequently seen or heard (at least once a month); R = rare (<3 records per year) C/R = switched from common to rare; Y = yes; N = no. Habitat key: C = closed forest, O = open forest, S = shrubs/meadows, L = lakes, U = unknown.

1991). Also, nest boxes have often been used successfully for studying cavity-nesting species (e.g., Hörnfeldt and Eklund 1989, Korpimäki and Norrdahl 1991).

Access to most of the Kluane study area was only possible on foot, and most raptor nests were in dense white spruce (*Picea glauca*) foliage and were not visible from the ground or air. We therefore needed to use new methods to assess population densities (e.g., Rohner and Doyle 1992a), and we had to conduct most surveys on foot. We chose not to use nest boxes extensively to avoid confounding our experimental design by increasing the densities of cavity-nesting species. A few nest boxes, however, were erected away from our study plots, and these proved useful for studying diets of boreal owls (*Aegloius funereus*) and American kestrels (*Falco sparverius*).

For all raptor species and the two larger corvids, common ravens (*Corvus corax*) and black-billed magpies (*Pica pica*), we estimated the population density by mapping territories and nest sites within the "intensive search area" (see below) from March to July 1989–1995. Additional information was available for some species in 1988 and 1996. Three secretive species (northern goshawk, *Accipiter gentilis;* sharp-shinned hawk, *A. striatus;* and boreal owl) could not be enumerated as reliably as other breeding species (see below).

16.1.2 Population Surveys: The Intensive Search Area

We surveyed all species in a 100-km^2 area (see CD-ROM frame 68), that we called the "intensive search area." The intensive search area had habitats and prey species characteristic of the study area as a whole and included most of our experimental treatment plots. The entire intensive area was surveyed on transects 250 m apart each May and June. Transects were mostly walked at a steady pace from dawn to early afternoon (0430–1400 h), but some areas were covered at other times. Night surveys were mainly conducted from roads and tracks. The locations of individuals heard and seen by day and by night were mapped, and we searched for nests at the time or on later visits. In open areas, we looked for perching or flying raptors. In dense forest, we listened for calling raptors or alarm calls from prey and looked for plucking posts and whitewash.

These general methods provided estimates of numbers of breeding pairs in the intensive search area and samples of nests for assessing breeding success. Additional methods for particular species are described in the species accounts below. We also used data from additional nests outside the intensive search area to supplement our studies of diets and reproductive success.

To assist the small team of raptor specialists, we trained other project workers to identify raptors by sight and vocalizations. These reports provided indices of change in numbers for all raptor species and clues about the locations of pairs and nests for the specialists to follow up.

16.1.3 Reproductive Success

We obtained information on reproductive success by visiting nests after hatching and by counting large young in the nest shortly before fledging or newly fledged young near the nest. Because we generally did not check nests during incubation, we obtained few data on clutch sizes.

16.1.4 Diets

Traditional methods for studying raptor diets can be intrusive (e.g., the use of nest platforms and nest boxes; Newton 1979). Diets of most species at Kluane were only studied at natural nest sites, although we used nestling platforms for horned owls in some years (see chapter 15). We minimized disturbance at nests early in the breeding season (when adults and chicks are most vulnerable to food shortages and poor weather) by only beginning to visit nests when young were about 1 week old. Regurgitated pellets and prey remains left near nests provided most information on diets. In a few species (northern hawk owl, American kestrel, northern goshawk, northern harrier), information from pellets and remains was supported by direct observations from blinds or natural vantage points.

Pellets and Prey Remains   Pellets were collected every 2–5 days at nest sites and oven dried at 60°C for 48 h. Pellets were then teased apart (Marti 1987), and prey species were identified using skeletal, nail, beak, dental, hair (color, banding patterns, scale and cell pattern), and feather (color, shape, and size) traits. We maintained a reference collection of local prey specimens and also used the collections of Cowan Vertebrate Museum at the University of British Columbia and published keys (Moore et al. 1974, Kennedy and Carbyn 1981). All prey individuals in pellets were counted, using the minimum number of individuals possible from the diagnostic fragments found (e.g., three left jaws of a species = three individuals; Rohner et al. 1995). Prey remains were collected at the same time as pellets from in and around raptor nests. We also sexed and aged individual prey when the remains allowed us to do so (see Doyle and Smith 1994).

We obtained other information on raptor diets from chance encounters with birds consuming prey. Although such information is probably biased toward larger prey, which take longer to consume, these encounters provided data on diets outside the breeding season and revealed some surprising linkages between predators at Kluane.

Radio Tagged Prey   Part of the population of several larger prey species (spruce grouse, red squirrel, ground squirrel, and snowshoe hare) was radio tagged throughout the study. These collared prey were monitored daily, and we could often determine the identity of their killers.

Scavenging Experiment   We set up a scavenging experiment to estimate rates of removal of dead radio-collared hares. Twenty road-killed hare carcasses were fitted with radio collars and laid out on the snow surface in March 1991 and checked daily. Tracks or wing marks in the snow or location of the radio collar often revealed the identity of the scavengers of these carcasses.

16.1.5   Data Analysis

We related raptor and corvid population densities and breeding success to fluctuations in prey population density and the timing of breeding in spring using linear and logarithmic regression and Pearson and Spearman correlation coefficients. Because the number of study years was limited and many species were studied, we used correlation coefficients primarily as descriptions of data. Hare densities in spring were mean values from all control grids (chapter 8). Densities of squirrels and voles were taken from chapters 9 and 10, respectively.

We used Levins's (1968) niche width index, $B = 1/p_j^2$ where p_j is the proportion of prey taxon j in the diet, to quantify the diet breadths of raptors. This index ranges from 1 to n, where 1 is the narrowest diet width possible.

16.2   Predicted Responses of Raptors and Corvids

We expected the large resident species that feed on hares in winter (horned owl, goshawk) to show strong numerical and reproductive responses to varying hare densities (Adamcik et al. 1978, chapter 15) and to broaden their diets to include alternative prey as hare populations declined. Based on previous work (Adamcik et al. 1979), we did not expect a strong population response from the migratory red-tailed hawk, but we expected all larger raptors to show functional responses to hares near the peak of the cycle. Because of the large spatial scales at which most of these birds operate, we did not expect to see strong responses to our experimental treatments (but see chapter 15 for an exception). As a result, we did not routinely compare data from different experimental treatments.

We expected the small raptors to depend on small prey species and not to respond directly to changes in hare numbers. We looked, however, for indirect effects in the food web (Bonsall and Hassell 1998)—for example, release from predation (Crooks and Soulé 1999)—that could influence these small species. We expected corvids to respond positively to the hare peak because of increased scavenging opportunities at high hare numbers.

16.3   Responses by Groups and Individual Species

We now treat each group of raptors in detail in light of these predictions. For each group, we first summarize patterns of diet and the functional, numerical, and reproductive responses to the hare cycle. We then present detailed information for each well-studied species, except for the horned owl, which was treated in detail in the previous chapter. We do, however, present some data for horned owls in the figures in this chapter to facilitate comparisons with other raptors. In the species accounts, we describe our methods and present data on diets and on functional responses to hares and other prey. We then consider numerical and reproductive responses. Information on body mass is taken from Dunning (1993) and from accounts in the *Birds of North America* series.

16.3.1   Large Resident Raptors

The large resident raptors group includes two species, the great horned owl and the northern goshawk, both of which varied strongly in abundance through the hare cycle (figure 16.1a,b). Both horned owls and goshawks fed heavily on hares when they were available, but the horned owl relied on hares to a greater extent than did the goshawk (tables 15.5, 16.2), and was unable to breed when hares were scarce (figure 16.2a). As a result, the horned owl showed a stronger numerical response to the hare cycle than the goshawk, and hare densities affected its dynamics with a 2-year lag (figure 16.1a).

Northern Goshawk   Northern goshawks are large but inconspicuous forest hawks (mean female mass [MFM] = 1.14 kg, mean male mass [MMM] = 0.91 kg). Goshawks from the boreal forest are generally considered to be migratory (e.g., Mueller et al. 1977), but they were resident at Kluane. During the hare decline in 1990–1991, some goshawks may have migrated south in winter, but others remained resident and several birds died *in situ* that winter (Doyle and Smith 1994).

We trapped goshawks in late winter with falling-lid traps and fitted some individuals with tail-mounted radios (Doyle and Smith 1994; CD frames 23, 31), which allowed us to locate a sample of nests. Most nests were in dense spruce stands, often near water. Once nest sites were known, we checked these each spring and looked for additional nesting pairs by broadcasting female alarm calls (Squires and Reynolds 1997) and listening for responses.

As others have found (Squires and Reynolds 1997), goshawks had broad diets. They ate arctic ground squirrels, red squirrels, spruce grouse, ptarmigan, and other birds in sum-

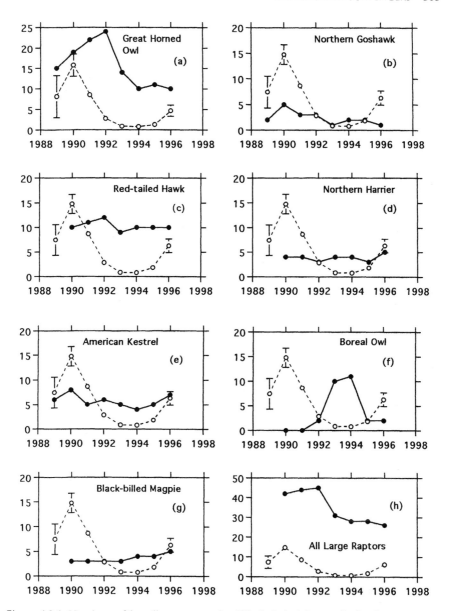

Figure 16.1 Numbers of breeding raptor pairs (filled circles) in standardized survey areas (methods vary with species; for details, see text) in relation to snowshoe hare densities in spring (hares/10 ha, open circles) at Kluane from 1989 to 1996.

mer, and they were the only raptors recorded eating northern flying squirrels (*Glaucomys sabrinus*; table 16.2). They did not, however, exhibit strong functional responses to any of these smaller prey species. During the hare peak, hares made 50% of the biomass of prey brought to nests, but a wide range of prey species continued to be eaten (table 16.2). When adult hares were dropped from the diet entirely in 1993, the biomass of grouse

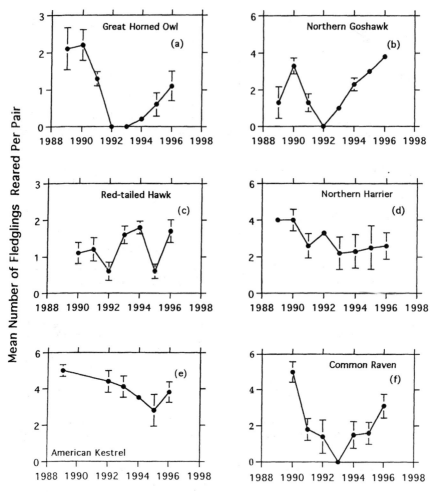

Figure 16.2 Annual variation in fledgling production by six species of predatory birds followed consistently through the hare cycle.

and ptarmigan eaten increased sharply and remained high through 1995 (table 16.2). Goshawks exhibited a type-1 or type-2 functional response to hare densities (figure 16.3a). They usually took live prey, but also fed on 2 of the 20 hares used in the scavenging experiment (C. Doyle personal communication, see also Squires and Reynolds 1997).

The number of territories in the intensive area (figure 16.1b), peaked at five in 1990, the year of peak hare numbers, and then declined to one by the summer of 1993. Goshawks showed a significant numerical response to spring hare densities ($r^2 = .537, p = .039$). Goshawks laid their eggs in late March and early April, with chicks fledging in the first and second week of July (figure 16.4). Successful nests were found in every year except 1992, and the number of young reared per pair was unrelated to hare density in spring (figure 16.5b).

Table 16.2 Diets and diet widths of northern goshawks during the breeding season each year.

Prey Species or Group	1989	1990	1991	1992	1993	1994	1995	1996
Lepus americanus (adult)	14	14	20	11	0	3	9	19
Lepus americanus (juvenile)	10	25	17	0	0	5	5	6
Spermophilus parryii	18	33	21	11	17	15	12	13
Tamiasciurus hudsonicus	34	4	21	0	8	19	9	19
Ptarmigan and grouse	13	14	11	22	33	37	38	21
Other birds	9	10	10	56	33	21	28	19
Glaucomys sabrinus	1	0	0	0	0	0	0	2
Mice and voles	1	0	0	0	8	0	0	0
Prey animals	91	229	81	9	14	73	66	47
No. of nests	3	8	4	1	1	3	3	3
Diet width (prey remains)	4.3	3.4	4	2.6	3.9	4.1	3.8	5.1

Values for each prey species are percentages of all individuals identified in pellets.

16.3.2 Large Migratory Raptors

The two species in the large migratory raptors group were the red-tailed hawk and the bald eagle. Red-tails were specialized predators of squirrels at Kluane. They showed a functional response to hare densities, but no numerical response. Bald eagles (MFM = 5.35 kg, MMM = 4.13 kg) were not studied intensively, and we thus do not consider them in detail. We did, however, find that some bald eagles shifted from being migrants to being resident near the peak of the hare cycle and that their reproductive success was low. Only 3 of 16 nesting attempts by the 2 resident pairs that bred near two large lakes in the study area (CD frame 68) were successful, with four chicks fledged in total.

Red-tailed Hawk   Red-tailed (Harlan's) hawks of the northwestern boreal forest (*Buteo j. harlani*) are usually dark and often lack the red tail (Mindell 1983). Of the 21 species discussed here, this large migratory hawk (MFM = 1.22 kg, MMM = 1.03 kg) was 1 of only 4 raptors that bred at Kluane every year (table 16.1).

We studied red-tails from 1990 onward by noting the presence of soaring birds and following them to their traditional nest sites, which were then monitored annually. Variable plumage helped us to identify many individuals, which stood out due to the fortunate mix of dark- and light-phase morphs. Each year from 1990, 56–80% of birds had unique markings. Nests were usually near the top of an open spruce tree on the edge of a clearing or along a ridge. Pairs were regularly spaced throughout the study area (table 16.1; CD-ROM frame 68). Diets were estimated by climbing to accessible nests at the end of the breeding season to collect prey remains. Some nest trees, however, were too exposed to climb safely.

Red-tailed hawks elsewhere prey mainly on medium-sized mammals (e.g., Adamcik et al. 1979, Restani 1991). This was also true at Kluane. Arctic ground squirrels and red squirrels dominated the diet in 6 of 7 years, and these two squirrels made up >47% of diets at nests in all years (table 16.3). The proportion of pairs that successfully fledged young

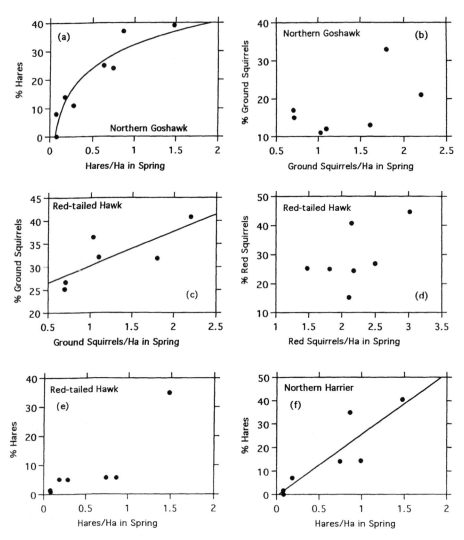

Figure 16.3 Functional responses by three species of predatory birds to changing prey densities. Best-fit lines or curves are shown when regressions were statistically significant ($p < .05$) and not dependent on a single outlying point.

was higher in years when young red squirrels ($r_s = .81, n = 8$) and arctic ground squirrels ($r_s = 1.0, n = 6$) emerged from the nest early (F. I. Doyle in preparation). The proportions of both squirrels brought to nests increased with their densities, but the relationship was tighter for the arctic ground squirrel (figures 16.3c,d). Hares were taken more frequently than any other prey only at the peak of the hare cycle in 1990, but otherwise were taken at similar frequencies regardless of hare density (figure 16.3e). Voles were taken commonly, particularly during the high vole years of 1992 and 1993 (table 16.3, chapter 10), but they contributed only a small portion of the total biomass of food brought

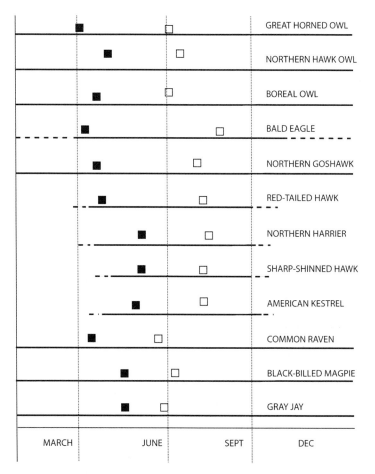

Figure 16.4 Timing of breeding by raptors and corvids at Kluane. Solid line = species present; dashed line = species' presence/absence alters annually; solid squares = timing of egg laying; open squares = timing of fledging.

to nests. Birds also contributed regularly to diets, but again the biomass consumed was small relative to that of squirrels (table 16.3). Weasels (*Mustela erminea* spp.) were taken on two occasions.

Unlike the goshawk and horned owl, neither the density nor the breeding success of red-tails was related to snowshoe hare densities (figures 16.1c, 16.2c). The density of occupied territories remained almost constant throughout the study (figure 16.1c), with only a slight increase in the numbers of females because of polygynous trios of a male and two females on three territories in 1991 and 1992 (Doyle 1995b). Breeding success was variable (figure 16.2c), but all known females attempted to breed each year. In 1992, the second year of the hare decline, 9 of 12 nests (75%) failed. Four of these failures were due to predation by horned owls; the cause of the other five failures was unknown. After this poor year, however, breeding success was high for the next 2 years (96% of 23 nests) and

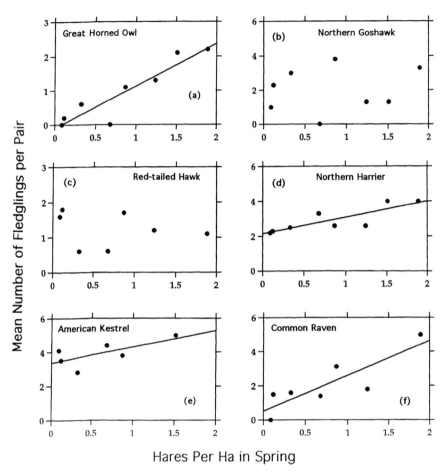

Figure 16.5 Reproductive responses of six predatory birds to variation in snowshoe hare densities in spring. Lines fitted by linear regression, where relationships where statistically significant.

in 1996 (figure 16.2c), but success was poor (44% of 16 nests) in 1995, a time of low hare and declining vole numbers.

16.3.3 Small Migratory Raptors

The small migratory raptors group included the medium-sized northern harrier and the small American kestrel and sharp-shinned hawk. All three species fed on juvenile hares, and the harrier did so extensively at the hare peak. The harrier exhibited both functional and reproductive responses to hare densities, and, to our surprise, the kestrel exhibited a numerical response.

Northern Harrier The northern harrier is a medium-sized (MFM = 513 g, MMM = 358 g), long-winged raptor, adapted to hunt near the ground and specialized to detect prey

Table 16.3 Diets and diet widths of red-tailed hawks during the breeding season each year.

Prey Species or Group	1989	1990	1991	1992	1993	1994	1995
Lepus americanus	5.6	34.8	5.6	5	0.8	1.4	5.1
Spermophilus parryii	31.5	31.8	40.8	36.4	25.2	26.7	32.1
Tamiasciurus hudsonicus	40.7	15.2	25.3	25	24.4	44.6	26.9
Birds	5.6	6.8	9.9	12.1	13.8	9.1	19.2
Voles	16.7	11.4	16.9	20.7	35	17.2	16.7
Mustelids	0	0	1.4	0	0	0.7	0
Eutamias minimus	0	0.8	0	0.7	0.8	0.4	0
Prey animals	54	132	143	139	244	284	78
No. of nests	2	7	6	4	8	14	7
Diet width	3.34	3.8	3.67	3.92	3.78	3.24	4.12

Values for each prey species are percentages of all individuals identified in pellets.

by sound as well as by sight (MacWhirter and Bildstein 1996). We studied these birds by watching open areas for courtship from a blind or from dense cover. Once nest building and incubation had begun, close observations of the female revealed the ground nest. Aggressive interactions with conspecifics and other species (particularly ravens) also helped us locate breeding sites.

Harriers bred every year at Kluane in meadows and marshes (table 16.1), and their diets were studied every year except 1992. Harriers arrived in mid- to late April, and most suitable habitat was occupied by early May (table 16.4). Eggs were laid in late May to

Table 16.4 Diets and diet widths of northern harriers during the breeding season each year.

Prey Species or Group	1988	1989	1990	1991	1993	1994	1995
Lepus americanus	14.3	14	40.5	34.8	0	1.6	7
Spermophilus parryii	14.3	5.3	2.4	1.1	3.6	14.1	4.7
Tamiasciurus hudsonicus	10.7	1.8	0	4.3	3	2.1	4.7
Birds	21.4	14	26.2	42.4	26.2	26.7	40.7
Microtus pennsylvanicus	14.3	47.4	19	7.6	38.1	30.9	27.9
M. oeconomus	17.9	12.3	9.5	7.6	17.3	17.3	8.1
M. miurus	0	0	0	0	0.6	0	0
Phenacomys intermedius	0	1.8	0	0	2.4	2.1	2.3
Clethrionomys rutilus	3.4	0	0	1.1	7.1	2.6	0
Peromyscus maniculatus	0	0	0	1.1	0	0	1.2
Eutamias minimus	3.4	1.8	2.4	1.1	0	0.5	0
Shrews	0	1.8	0	1.1	0	0	2.3
Synaptomys borealis	0	0	0	0	1.8	1.6	1.2
Prey animals	28	57	42	92	168	191	86
No. of nests	1	1	2	5	5	6	3
Number of pellets	19	29	32	76	99	124	73
Diet width	6.5	3.5	3.6	3.2	4	4.6	3.8
Diet width (mice and voles summed)	4.4	2.4	3.2	3	2	2.6	2.9

Data from pellets at nests.

early June, and most young fledged by late July. Polygyny is common in southern populations of harriers (MacWhirter and Bildstein 1996), but we only noted monogamous pairs at Kluane.

MacWhirter and Bildstein (1996) reported that harriers in the northern part of their range feed almost entirely on small mammals. Harrier diets at Kluane were dominated numerically by voles and by smaller birds, and juvenile and subadult hares were taken commonly in 1990 and 1991 (table 16.4). Diet width did not change noticeably with changing hare densities, but snowshoe hares were dominant numerically in 1990 and in biomass in 1991. A few arctic ground squirrels, chipmunks, red squirrels, and shrews were also taken. Harriers occasionally fed on road kills along the Alaska Highway. A female harrier was also seen killing a wading bird, which was then stolen by an adult bald eagle (F. Doyle personal observation).

Harrier numbers were fairly stable through the hare cycle (figure 16.1d). The density of pairs fluctuated between 3 and 5 per 100 km ^{2}. Numbers and breeding success were unrelated to densities of either *Microtus* or *Clethrionomys* voles in either the same or in the previous year. During 3 years of high hare numbers (1989–1991) all eight nests fledged chicks, but only 12 of 19 nests (63%) were successful thereafter. Chick production was positively related to hare density in spring ($r^2 = .499, p = .05$; figure 16.5d).

American Kestrel The American kestrel was the smallest (MFM = 120 g, MMM = 110 g) migratory raptor at Kluane. Pairs of this conspicuous species were often found perched around open meadows and marshland (table 16.1) near the Alaska Highway. By visiting these habitats, we located all breeding pairs within 500 m of the highway and located their nest cavities by subsequent intensive monitoring. Kestrels arrived in late March and early April, but egg laying did not begin until late May (table 16.4).

Prey brought by American kestrels to eight nests were identified using super 8 cameras in 1995 and 1996. At hatching, cameras were placed 10–20 m from the nest and were gradually moved closer until they were 2 m from the nest hole 2 weeks after hatching. The parent and its prey were photographed as the parent perched on a mechanical trigger at the edge of the nest hole. Voles, small birds, and least chipmunks (*Tamias minimus*) were the main prey (table 16.5). Insects, particularly dragonflies and grasshoppers, were taken more often than any one vertebrate prey, but they contributed little to the biomass delivered to nests. It is likely that hunting males fed mainly on insects while bringing mainly vertebrate prey to nests. The largest prey taken was a small juvenile snowshoe hare. A radio collar was found beneath a kestrel's nest cavity, with juvenile hare fur protruding from the nest hole. The kestrel was the only species at Kluane that ate an amphibian (a wood frog, *Rana sylvatica,* the only amphibian found at such northern latitudes in North America).

Relative densities (figure 16.1e) ranged from four to eight pairs per 12 km of highway over 9 years and were significantly correlated with hare densities in spring ($r^2 = .603, p = .023$), but not with vole densities in spring or in the previous summer. The proportion of successful nests varied from 70% in 1995 to 100% in 1989 and 1990, and the number of chicks fledged per nest (figure 16.2e) was unrelated to vole densities in either spring or summer. Two of six nest failures were due to predation; the other four failed from unknown causes.

Table 16.5 Diets and diet widths of American kestrels in the summers of 1995 and 1996.

Prey Species or Group	1995	1996
Voles	16.7	31
Tamias minimus	13.9	11.9
Small birds	30.6	20.8
Rana sylvatica	1.4	0.6
Shrews	0	0.6
Peromyscus maniculatus	0	0.6
Unidentifiable small mammals	0	10.1
Arthropods	37.5	24.4
Prey animals	72	168
No. of nests	3	5
Diet width	3.6	4.6

Data are from photographs of adults carrying prey at the nest entrance.

Sharp-shinned Hawk   The small sharp-shinned hawk (MFM = 174 g, MMM = 103 g) is a specialized predator of small songbirds (Ehrlich et al. 1988). It was a regular, but rarely seen, summer migrant at Kluane (table 16.1). The sharp-shin was the breeding bird that we were least able to study successfully. To locate sharp-shinned hawks, we listened for vocalizations and broadcast their calls by tape recorder (Rohner and Doyle 1992b, Doyle and Smith 1994). The few breeding birds we located behaved inconspicuously, and few data were obtained on numbers or annual changes in diet. The sharp-shin was the last migrant raptor to arrive in the study area (in late April–early May). Egg laying began in late May and early June (figure 16.4), with the first young fledging in late July.

In 1992, prey remains were collected at two nest sites (table 16.6). As expected (Joy et al. 1994, Newton 1979), passerines dominated the diet. The Lapland longspur (*Calcarius lapponicus*), an early season migrant that does not breed at Kluane, was the most commonly taken species. The American robin (*Turdus migratorius*) and Swainson's thrush (*Catharus ustulatus*), two common and large passerines at Kluane (figure 12.2) were also important prey species. A juvenile hare was taken in 1992, when hares were scarce. One to three pairs per 100 km^2 were located annually (CD-ROM frame 68), and three nests found before fledging each fledged five chicks.

No sharp-shinned hawks were brought as prey to the nests by other raptors nor were any taken by mammalian predators in our study area. Thus, it appears that this species really was uncommon and not simply overlooked.

16.3.4 Small Owls

Two species of owls, the hawk owl and the boreal owl, made up the small owl group. They shared a primary diet of voles, and numbers of both species fluctuated, as they are known to do in Europe (Hörnfelt 1978, Sonerud 1997). Both species took juvenile hares, and hawk owls also preyed on adult hares. Boreal owls responded numerically to fluctuating vole numbers, and hawk owls responded numerically to combined hare and vole numbers.

Table 16.6 Diets of sharp-shinned hawks
in summer 1992.

Prey Species	Percent
Calcarius lapponicus	21.9
Turdus migratorius	14.6
Catharus ustulatus	12.2
Dendroica coronata	12.2
Wilsonia citrina	7.3
Junco hyemalis	7.3
Bombycilla garrulus	4.9
Zonotrichia leucophrys	4.9
Sialia currucoides	2.4
Dendroica petechia	2.4
Dendroica striata	2.4
Hirundo pyrrhonota	2.4
Perisoreus canadensis	2.4
Lepus americanus (juvenile)	2.4
Prey animals	41
No. of nests	2

Data are from prey remains identified near nests.

Northern Hawk Owl   The northern hawk owl (*Surnia ulula;* MFM = 345 g, MMM = 299 g) is noted for its unpredictable population dynamics in Europe (Sonerud 1997), and it sometimes irrupts south from the boreal forests of North America (Duncan and Duncan 1998). Hawk owls at Kluane often perched at the tops of snags close to the nest cavity, so their nests were readily located. They were resident in some years (figure 16.1). In other years, a few birds of unknown status were seen in spring and fall. Egg laying occurred in mid- to late April, and the chicks fledged by mid-June (figure 16.4).

The diets of hawk owls were dominated numerically by voles (table 16.7). However, during the 2 years of peak hare density (1989–1990), juvenile hares up to 240 g in size contributed considerably to biomass in nestling diets (Rohner et al. 1995). Diet width decreased in years with vole peaks and broadened in the 2 years of peak hare densities (table 16.7). Diets were dominated by *Microtus pennsylvannicus* and *M. oeconomus* in 1988 and 1993 when vole numbers were high (chapter 10). We were not surprised to find hawk owls preying on juvenile hares, but we were surprised that a supposed vole specialist ate both adult hares and spruce grouse (Rohner et al. 1995). A hawk owl ate 1 of 20 hares put out in the scavenging experiment in 1991 (C. Doyle personal communication).

Hawk owl numbers were highest (two pairs per 12 km^2 of highway transect) during the hare peak, and numbers were positively related to densities of snowshoe hares in spring (r^2 = .502, p = .049). Hawk owls were absent in 1992, bred again in 1993, but disappeared during the last 3 years. Only one nest failed, in 1991. Four or five chicks fledged from each of the seven successful nesting attempts. Hawk owl density increased in years of high combined hare and vole numbers (r_s = .77, p = .07, n = 6; details in Rohner et al. 1995).

Table 16.7 Summer diets (from Rohner et al. 1995) and diet widths of northern hawk owls for 4 years.

Prey Species or Group	1988	1989	1990	1993
Microtus pennsylvanicus	31.6	15	3.1	51.8
Microtus oeconomus	25.4	22.5	27.6	18.1
Microtus spp.	20.6	20	23.5	0
Synaptomys borealis	6.1	0	7.1	7.2
Peromyscus maniculatus	0.4	0	1	0
Phenacomys intermedius	7	5	9.2	7.2
Clethrionomys rutilus	0.9	10	7.1	7.2
Lepus americanus (juveniles)	3.5	15	14.3	0
Lepus americanus (adults)	0.4	2.5	0	0
Eutamias minimus	0.4	0	0	0
Tamasciurus hudsonicus	2.2	5	2	0
Spermophilus parryii	0	5	4.1	1.2
Mustela nivalis	0.4	0	1	1.2
Shrews	0	0	0	4.8
Dendragapus canadensis	0.4	0	0	0
Perisoreus canadensis	0.4	0	0	1.2
Prey animals	228	40	98	83
No. of nests	3	1	3	2
Diet width	4.6	6.5	5.8	3.1
Diet width (all mice and voles summed)	1.2	1.8	1.6	1.2

Data are percentages of individuals in pellets collected near nests.

Boreal Owl   The small boreal (Tengmalm's) owl (MFM = 167 g, MMM = 101 g) is known as a nomad in Fennoscandia and, like the hawk owl, is known for its ability to track shifting patches of small mammal numbers (Korpimäki 1985). Like goshawks and hawk owls, boreal owls sometimes irrupt southward from North American boreal forests (Campbell et al. 1990, Hayward and Hayward 1993). We surveyed for boreal owls by listening during nocturnal transects and by broadcasting territorial hoots and listening for responses. Eggs were laid from late March to mid-April (figure 16.4).

Insufficient data were collected to examine diets of boreal owls across the hare cycle. Pellets from seven nests in 1994 (table 16.8) contained mostly voles (95%), supplemented by a few small birds and shrews. Heather voles, *Phenacomys intermedius,* and deer mice, *Peromyscus maniculatus,* two forest-dwelling small mammals, were taken much more by boreal owls than by any other raptor, suggesting that boreal owls foraged in closed forest more than other predatory birds (table 16.1). At one nest in 1995, shrews accounted for 15 of the 22 prey items; 2 juvenile hares were also taken. In a separate incident, a perched boreal owl was located eating a freshly killed, radio-tagged juvenile hare.

All nine nests studied produced young, with five chicks fledging from each of six nests in 1994, two from one nest in 1995, and three from one nest in 1996. Fledging success of boreal owls at Kluane was higher than is typical for other populations (Hayward and Hay-

Table 16.8 Summer diets of boreal owls in 1994 and 1995.

Prey Species or Group	1994	1995
Lepus americanus (juveniles)	0.5	9.1
Spermophilus parryii	0.9	0
Birds	5.1	9.1
Microtus pennsylvanicus	15.7	4.5
Microtus oeconomus	31.8	0
Mictotus miurus	9.2	0
Phenacomys intermedius	12.9	0
Clethrionomys rutilus	11.5	4.5
Peromyscus maniculatus	10.1	0
Soricidae	2.3	72.7
Prey animals	217	22
No. of nests	7	1
Diet width	5.6	1.8
Diet width (all voles and mice)	1.2	1.8

Data are percentages of individuals in pellets collected near nests.

ward 1993), perhaps because the principal predator of boreal owl nests, the marten (*Martes americana,* Hayward and Hayward 1993), was absent at Kluane.

Boreal owl numbers fluctuated sharply (figure 16.1f). However, breeding males did not always give territorial hoots in spring (see below), which adds uncertainty to our estimates of numbers. Boreal owls bred in the study area in 5 of 7 years from 1990 (table 16.1). Birds were concentrated in the northwest corner of the study area (CD-ROM frame 68). Numbers of calling males peaked at 11 per 24 km^2 of road transect in 1994 (figure 16.1f), and numbers were positively correlated with combined densities of *Microtus* and *Clethrionomys* voles the previous summer ($r^2 = .735, p = .014$).

16.3.5 Corvids

The three species of corvids at Kluane, common ravens, black-billed magpies, and gray jays, were studied in only moderate detail. Corvids are typically omnivores and scavengers, and this was also the case at Kluane. The large raven showed a strong reproductive response to the hare cycle, and the small gray jay responded to the supplemental feeding experiment by decreasing the sizes of its territories.

Common Raven   The common raven is one of the world's largest passerines (MFM = 1.16 kg, MMM = 1.24 kg). Ravens were resident at Kluane, and adults were territorial (see also Ratcliff 1997). Young birds and nonbreeders were not studied at Kluane but are known to live in loose flocks elsewhere and to be subordinate to territorial adults (Heinrich 1994). Territorial ravens were highly visible, and their density was readily monitored from the highway because birds flying along the highway always turned around or flew away from the road at predictable places. Repeated observations of these points allowed us to map territories. Breeding began in mid-March, and eggs were laid in late March to early April (figure 16.4). Chicks fledged in early June.

The limited information available on raven diets corresponded with our expectations (Heinrich 1989, Ratcliff 1997) that they would be habitual scavengers and occasional predators. Ravens were frequently seen taking road-killed squirrels and hares and were regularly found at the remains of adult snowshoe hares killed in the winter. In the scavenging experiment in 1991, 16 of the 20 hares set out were eaten by ravens (C. Doyle personal communication). On another occasion, F. D. saw a pair of ravens attack and kill a juvenile male American kestrel as it left its nest cavity.

The density of breeding ravens was stable at $2/100$ km^2, except from 1990 to 1992, at the peak in the snowshoe hare cycle, when a third pair established a territory for 2 years. The number of chicks fledged per pair was greater in springs with higher hare densities ($r^2 = .767, p = .01$; figure 16.2f), and no chicks were reared by four nesting pairs in 1993, at the nadir of the hare cycle (figure 16.2f).

Black-billed Magpie   The black-billed magpie is a small (MFM = 166 g, MMM = 189 g) and conspicuous corvid that prefers open habitats and breeds in loose colonies (Birkhead 1991). Magpies at Kluane nested at the base of steep, open hillsides adjacent to a lakeshore. Nests were easily located from a high point or blind above the nest area, as birds called frequently while they carried nest material. Active nests were found every year, with egg laying in early May, and chicks fledging in mid-June (figure 16.4).

No quantitative information was collected on the diet of the magpie. We saw them foraging on open hillsides (probably for grasshoppers and other arthropods), on road-killed insects in the summer, and on road-killed mammals throughout the year. F. D. saw a magpie kill a red-backed vole that had just been released from a trap. Elsewhere, magpies primarily eat insects in summer (Linsdale 1937), but also carrion, birds' eggs, and small mammals. In winter, fruit and seeds form a large part of the diet (Birkhead 1991).

The density of pairs was stable at $3/100$ km^2 from 1990 to 1993, and it increased to $5/100$ km^2 by the end of the study in 1996 (figure 16.1g). The number of chicks fledging per nest ranged from 3 to 5 from 12 nests, with means ranging from 1.29 per pair in 1994 to 3.14 in 1995. The proportion of pairs fledging young ranged from 3 of 7 in 1994 to 4 of 5 in 1992, but information was not collected every year.

Gray Jay   The gray jay, our smallest corvid (mean mass = 73 g), was resident throughout the study area in forested habitats (Delehanty 1995). Some gray jays were trapped in hare live traps and color banded in 1993 and 1994. They lived on territories of 15–25 ha (Delehanty 1995) and were the only predators or scavengers to respond numerically to the food addition treatment (see Chapter 12). Breeding pairs were located every year (table 16.1) when independent young jays joined their parents in mid-May, suggesting that most eggs were laid in late March or early April. In the two nests that we found, eggs were laid in early May and the chicks fledged in early June, unusually late dates for the species (Strickland and Ouellet 1993).

We lacked quantitative information on jay diets, reproduction, and densities through the cycle. Gray jays are omnivores, relying heavily on stored food in winter (Strickland and Ouellet 1993), and are probably major predators on the eggs and nestlings of shrub-nesting passerines at Kluane (Pelech 1999). A gray jay was seen killing a newborn snowshoe hare on one occasion (Sovell 1993).

16.3.6 Other Raptors

Nine other raptor species were seen in the study area (table 16.1). Some, like the small merlin (*Falco columbarius;* MFM = 218 g, MMM = 163 g) and large golden eagle (*Aquila chrysaetos;* MFM = 4.91 kg, MMM = 3.48 kg), were seen throughout summer every year and bred on the mountain slopes around the study area. Golden eagles are migrants at Kluane and in interior Alaska and arrive in late March or early April (McIntyre and Adams 1999). We have no information on their diets or breeding success, but they undoubtedly took snowshoe hares and arctic ground squirrels from our study area, and possibly also nestling red-tailed hawks and great horned owls. We kept an injured adult horned owl briefly, and it was extremely attentive to golden eagles soaring 1000 m above its perch. Merlins probably ate smaller birds in open habitats. Peregrine falcons (*Falco peregrinus*) were seen less frequently, but we believe that they also bred near Kluane.

The remaining species were all uncommon to rare. Two were annual migrants through the area: rough-legged hawks, *Buteo lagopus,* and ospreys, *Pandion haliaetus,* in spring and fall. Two of the four remaining species, gyrfalcons (*Falco rusticolus;* MFM = 1.76 kg, MMM = 1.17 kg) and short-eared owls (*Asio flammeus;* MFM = 378 g, MMM = 315 g), were only seen occasionally. A third species, the great gray owl (*Strix nebulosa;* MFM = 0.90 kg, MMM = 0.80 kg) is a boreal forest resident (Bull and Duncan 1993), but prefers moist forest habitat, which was scarce at Kluane. It was only seen once. Finally, the northern saw-whet owl (*Aegolius acadicus;* MFM = 91 g, MMM = 81 g) is at the very northern limit of its range at Kluane. A hooting male was heard in 1992.

16.3.7 Intraguild Predation

During our studies at Kluane, raptors sometimes killed each other, or mammalian predators took them (O'Donoghue et al. 1995). For example, a horned owl killed an adult female goshawk and her nestlings in 1991 (Rohner and Doyle 1992b), and a wolverine (*Gulo gulo*) killed and ate a second brood of goshawk nestlings in the same year (Doyle 1995a).

Most of our evidence for intraguild predation (O'Donoghue et al. 1995) came from raptor parts identified in the pellets or prey remains of other raptors. Raptor–raptor predation was surprisingly common. Fifty-two cases of intraguild predation by five avian and one mammalian predator species were accumulated over the hare cycle (table 16.9). Intraguild predation affected both large and small species. The abundant great horned owl was the most frequent predator on other raptors, including other horned owls (table 16.9). There were many more cases of predation on raptors by birds (42) than by mammals (6). Even young of the large bald eagle were taken (by a wolverine).

Intraguild predation rose sharply, particularly on broods of nestlings, in 1991, the first year of the snowshoe hare decline (figure 16.6). It remained common for the next 3 years, but declined sharply in 1995 (figure 16.6).

We suspected that predation risk had strong behavioral effects on the predatory birds at Kluane. Young "floater" horned owls behaved inconspicuously and did not hoot (chapter 15). Breeding boreal owls, which are often eaten by larger predators (Hayward and Hayward 1993; table 16.9), seldom gave territorial hoots in 1992 and 1993, during the peak period of intraguild predation (figure 16.6). On one occasion when we broadcast boreal owl hoots, an adult horned owl flew in silently to the speaker.

Table 16.9 Cases of intraguild predation involving raptors over the study.

Predator Species	Prey Species										
	HO	NG	RTH	NH	RV	NHO	BO	AK	BE	UNK	Total
Horned owl	2	5	4	1	1	4	2	0	0	0	19
Northern goshawk	0	0	0	0	0	1	0	2	0	2	5
Red-tailed hawk	0	0	0	0	0	0	4	1	0	0	5
Northern harrier	0	0	0	0	0	1	0	1	0	2	4
Raven	0	0	0	0	0	0	0	1	0	0	1
Unknown bird	2	0	3	0	0	2	0	1	0	1	9
Wolverine	1	1	0	0	0	0	0	0	1	0	3
Unknown mammal	0	0	0	4	0	0	0	0	0	0	4
Unknown	0	1	1	0	0	0	0	0	0	0	2
Total	5	7	8	5	1	8	6	6	1	5	52

Cases include depredation of young at nests and killed fledglings and adult birds in summer and winter. HO = horned owl, NG = northern goshawk, RTH = red-tailed hawk, RV = common raven, NHO = northern hawk owl, BO = boreal owl, AK = American kestrel, BE = bald eagle. UNK = individual raptors of unknown species identified from parts (e.g., beaks, legs) found in pellets at nests of known predators.

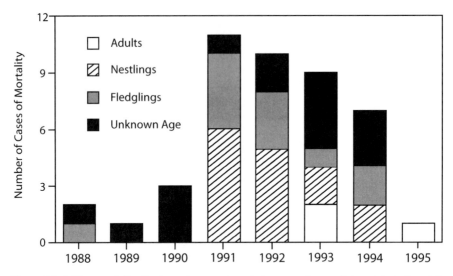

Figure 16.6 Temporal distribution of cases of predation on avian predators detected mainly during analysis of pellets and prey remains. The different shading patterns represent different age categories of the prey species.

16.4 Discussion

16.4.1 Functional, Numerical, and Reproductive Responses to Hare Densities

Functional Responses   We were able to calculate functional responses for four species: the great horned owl (see chapter 15), the northern goshawk, the red-tailed hawk, and the northern harrier. All four species responded by increasing the proportion of hares in their diets at high hare densities (figures 15.7, 16.3). The horned owl and the northern harrier showed particularly strong responses by rarely bringing hares to their nestlings when hares were rare, but by bringing little else when hares were common (figure 15.5, table 16.4). In contrast, the northern goshawk commonly brought hares to nests even when hares were scarce (table 16.2), whereas the red-tailed hawk only fed hares to its young often in one year at the peak of the hare cycle (table 16.3).

Numerical Responses   The horned owl showed a strong numerical response to hare densities, but it lagged 2 years behind the hare cycle, as it did at Rochester, Alberta (Adamcik et al. 1978). The mechanisms underlying this relationship are clear. Young owls from a pool of nonterritorial floaters continued to recruit to the territorial population after the hare peak in 1990 (figure 15.5), but recruitment eventually stopped in 1993 (figure 16.1a). At the same time, the annual survival of adults fell from 95% in 1989–1991 to 82% in 1991–1992 (chapter 15) and to approximately 60% in 1992–1993 (figure 16.1a). The total numerical response of all large raptors to the hare cycle (figure 16.1h) was dominated by the response of the great horned owl, as the other large species were either uncommon and/or did not respond strongly to changing hare densities.

The American kestrel, the northern hawk owl, and the northern goshawk showed more immediate numerical responses to spring hare density. Numbers of a fourth species, the common raven, increased from two to three breeding pairs at the peak of the cycle in the intensive search area. We speculate that breeding goshawks were in better condition in springs with high hare numbers, and we know that they survived poorly in 1991–92 during the hare decline, when five adults were found dead (Doyle and Smith 1994). We are, however, uncertain of the magnitude of the numerical response because our estimates of goshawk numbers were probably low. The hawk owl and kestrel rarely or never ate adult hares in summer. We speculate that their numerical responses stemmed from good breeding success the previous year (perhaps as a result of feeding juvenile hares to their nestlings) and from poor survival or emigration in winter after hares declined (hawk owl only).

The absence of a numerical response to hare densities by the migratory red-tailed hawk was expected from previous work in Alberta (Adamcik et al. 1979), where red-tailed hawks also showed extremely stable breeding populations through a hare cycle. Red-tails at Kluane only ate hares frequently in the peak year, 1990 (table 16.3).

Because horned owl and goshawk numbers increased at high hare densities, and because hares were a key winter food for them, these resident species added to the high predator pressure exerted on hares by lynx and coyotes (chapter 13) in the early winters of the decline.

Despite approximately synchronous population cycles by *Clethrionomys* and *Microtus* voles (chapter 10), only one species showed a delayed numerical response to fluctuating vole densities, the boreal owl. Boreal owl numbers were high one year after high vole numbers, presumably because young male owls recruited and began to hoot in spring after good chick production the previous year. The boreal owl also responds rapidly to vole cycles in Fennoscandia (Sonerud 1997).

Reproductive Responses   Three species showed clear reproductive responses, producing more fledglings in years of high hare density: the great horned owl, the northern harrier, and the common raven (figures 16.2a,d,f). The strongest response was by horned owls, which did not attempt to breed when hares were scarce. Northern harriers had broad diets throughout the hare cycle, but hares dominated numerically in 1990 and 1991 and contributed most of the prey biomass brought to nestlings from 1988 to 1991 (table 16.4). Finally, ravens bred much more successfully in years with high hare numbers, and all nests in 1993 failed when hares were extremely scarce (figure 16.2f). Ravens undoubtedly benefited from increased scavenging opportunities provided by road kills on the Alaska Highway at the peak of the cycle and also may have taken more juvenile hares in peak summers.

We expected the bonanza of hares at the peak of the cycle to generate high reproductive success in species that prey regularly on hares. Indeed, during the hare peak, reproductive success of several raptors and corvids approached the maximum level reported in any study of the species (horned owls, Houston et al. 1999; magpies, Birkhead 1991;

goshawks, Squires and Reynolds 1997, Ethier 1999; ravens, Ratcliff 1997; harriers, MacWhirter and Bildstein 1996; hawk owls, Sonerud 1997). Only the red-tailed hawk, a less enthusiastic hare-eater than the other species, was an exception to this pattern.

16.4.2   Resource Partitioning and Diet Width

Most raptors at Kluane had generalized diets and showed opportunistic shifts in diet as hare and vole numbers fluctuated. Three of the five species for which we had diet data over much of the hare cycle (goshawks, red-tailed hawks, and harriers) took a wide range of prey at all times, but nevertheless showed preferences (tables 15.6, 16.2, 16.3, 16.4, 16.7). Goshawks depended more on hares and adult arctic ground squirrels, whereas red-tailed hawks depended more on red squirrels and juvenile arctic ground squirrels. Goshawks in Europe mainly eat birds (Widén 1987), and this was also true at Kluane in 1993–1995, after hare abundance had declined. Harriers took mainly small mammals and birds, but included many other items in their diets (table 16.4). Data for a small species, the kestrel, also revealed that a variety of prey (voles, small birds, and insects) were taken.

Two species were more specialized in their diet. One, the northern hawk owl, was always a narrow specialist on small mammals, although it ate a variety of vole species (table 16.7) and some hares at the peak of the cycle. The great horned owl switched sharply from specializing strongly on hares at the peak of the cycle (table 15.5) to eating a range of alternative prey in 1992–1995 when hares were relatively rare (figure 15.7). Limited information on the boreal owl also suggested a specialized diet (forest-dwelling voles and shrews) that was similar to the diet of the hawk owl (see also Kertell 1986, Korpimäki 1992).

All of the species that we studied ate juvenile hares at the peak of the cycle. Horned owls may even have timed their breeding to exploit the sharp peak in prey biomass that accompanies the birth of the first litter of hares (F. Doyle unpublished data).

Few studies of northern raptor assemblages have been conducted elsewhere. In the boreal forests of northern Sweden (Hörnfeldt and Eklund 1989) and western Finland (Korpimäki and Norrdahl 1991, Korpimäki 1992) a different suite of avian predators uses a different prey base, mainly composed of regularly cycling voles. Raptors specializing on voles at Kluane took a wider range of prey than the same or similar species did in Sweden and Finland.

16.4.3   Intraguild Predation

Studies of predators in food webs often focus on vertical linkages, but differences in size and strength among predators allow them to prey on each other (e.g., Crooks and Soulé 1999), particularly when herbivorous prey are in short supply. The first substantial evidence that avian predators kill each other regularly came when large numbers of predators were fitted with mortality-sensing radios. In studies reviewed by Newton (1998), other predators killed 38.2% of 178 radio-collared birds of prey. In contrast, in 3,808 cases of mortality of non–radio-collared avian predators, only 4 deaths (of barn owls, *Tyto alba*) were attributed to predation (Newton 1998).

In our study, even though few raptors were radio marked, we found many cases of intraguild predation. It involved a variety of species (table 16.9) and was more frequent during the hare decline and early low phases (figure 16.6), especially when predation involved great horned owls. The data in figure 16.6, however, probably underestimate the frequency of intraguild predation after 1991, as diets of the great horned owl were studied more intensively from 1989 to 1991 (see chapter 15) than from 1992 to 1995. We conclude that, although we were unable to describe the exact pattern of intraguild predation during the cycle, it was frequent enough to have played a significant role in the declines in raptor numbers during the hare decline (figure 16.1). Predation by raptors was mostly of young individuals and may have lowered the recruitment of young to raptor populations after the hare decline, perhaps facilitating the recovery of hare numbers in 1995 and 1996. Also, predation in winters with few hares may have severely limited numbers of resident hawk owls, boreal owls, and goshawks (Doyle and Smith 1994, Rohner et al. 1995). Reduced territorial calling by great horned owls and boreal owls during the snowshoe hare decline suggests that these birds are sensitive to intraguild predation risk.

16.4.4   Methodology and Limitations

Our studies of raptors at Kluane were stretched rather thinly over many species, and much more could have been achieved if we had studied each common species intensively. Such an undertaking would, however, have required many more personnel and more resources. Indeed, we were advised by some expert raptor biologists at the beginning of our studies that the task was too difficult to attempt for even the larger and more common species such as horned owls, goshawks, and red-tailed hawks.

A second limitation to our work is that we mostly studied the diets of nestlings, not those of adult birds (but see table 15.6). It is likely that diets of adults differed from those fed to nestlings. A third limitation that we were only able to overcome in the horned owl is that we only studied the territorial segment of the population. It is possible that numbers of nonterritorial birds varied considerably through the cycle in other species.

16.5   Summary

We studied an assemblage of 21 species of raptorial and scavenging birds, 12 of which were residents or frequent summer visitors to the study area. Most larger raptors at Kluane maintained broad diets through the hare cycle, but the horned owl specialized strongly on hares at the hare peak and took a broader range of prey when hares were scarce. Two species, the boreal owl and northern hawk owl, specialized on voles at all times. Northern harriers ate mainly voles and smaller birds, but also took many juvenile hares at the peak. All four larger species for which we had diet data through the cycle (horned owl, goshawk, red-tailed hawk, harrier) showed functional responses to hare densities. One large species, the northern goshawk, and two smaller species, the American kestrel and the northern hawk owl, showed direct numerical responses to snowshoe hare density, and the horned owl showed a delayed response. Three species, great horned owls, northern harriers, and common ravens, produced more young in years of high hare abundance. The reproductive success of several raptors at Kluane at the peak of the hare cycle reached the maximum level reported for the species in any study. In the decline after the hare peak, raptors at Kluane frequently preyed on each other, and this intraguild predation may have contributed to the rapid decline in total raptor numbers within 3 years of the peak of the hare cycle.

Literature Cited

Adamcik, R. S., A. W. Todd, and L. B. Keith. 1978. Demographic and dietary responses of great horned owls during a snowshoe hare fluctuation. Canadian Field-Naturalist **92**:156–166.

Adamcik, R. S., A. W. Todd, and L. B. Keith. 1979. Demographic and dietary responses of red-tailed hawks during a snowshoe hare fluctuation. Canadian Field-Naturalist **93**:16–27.

Birkhead, T. R. 1991. The magpies. The ecology and behaviour of black-billed and yellow-billed magpies. T. & A. D. Poyser, London.

Bonsall. M. H., and M. P. Hassell. 1998. Population dynamics of apparent competition in a host-parasite assemblage. Journal of Animal Ecology **67**:918–929.

Boutin, S., C. J. Krebs, R. Boonstra, M. R. T. Dale, S. J. Hannon, K. Martin, A. R. E. Sinclair, J. N. M. Smith, R. Turkington, M. Blower, A. Byrom, F. I. Doyle, C. Doyle, D. S. Hik, L. Hofer, A. Hubbs, T. Karels, D. L. Murray, V. O. Nams, M. O'Donoghue, C. Rohner, and S. Schweiger. 1995. Population changes of the vertebrate community during a snowshoe hare cycle in Canada's boreal forest. Oikos **74**:69–80.

Bull, E. L., and J. R. Duncan 1993. Great gray owl (*Strix nebulosa*). *in* A. Poole and F. Gill (eds). The birds of North America, no. 62. American Ornithologists' Union, Washington, DC.

Campbell, R. W., N. K. Dawe, I. McTaggart Cowan, J. M. Cooper, G. W. Kaiser, and M. C. E. McNall. 1990. The birds of British Columbia, vol. II. Nonpasserines: diurnal birds of prey through woodpeckers. Royal British Columbia Museum, Victoria.

Craighead, J. J., and F. C. Craighead. 1956. Hawks, owls and Wildlife. Dover, New York.

Crooks, K. R., and M. E. Soulé. 1999. Mesopredator release and avifaunal extinctions in a fragmented system. Nature **400**:563–566.

Delehanty, B. 1995. Effects of food addition on a population of gray jays. MSc thesis. University of British Columbia, Vancouver.

Doyle, F. I. 1995a. Bald eagle *Haliaeetus leucocephalus* and northern goshawk *Accipiter gentilis* nests apparently preyed upon by a wolverine *Gulo gulo* in the southwestern Yukon. Canadian Field-Naturalist **108**:115–116.

Doyle, F. I. 1995b. Bigamy in the red-tailed hawk *Buteo jamaicensis harlani* population of southwestern Yukon. Journal of Raptor Research. **30**:38–40.

Doyle, F. I., and J. N. M. Smith. 1994. Population responses of northern goshawks to the 10 year cycle in numbers of snowshoe hares. Studies in Avian Biology **16**:122–129.

Duncan, J. R., and P. A. Duncan. 1998. Northern hawk owl (*Surnia ulula*). *in* A. Poole and F. Gill (Eds). The birds of North America, no. 356. American Ornithologists' Union, Washington, DC.

Dunning, J. B., Jr. (ed). 1993. CRC handbook of avian body mass. CRC Press, Boca Raton, Florida.

Ehrlich, P. R., D. S. Dobkin, and D. Wheye. 1988. The birder's handbook. A field guide to the natural history of North American birds. Simon and Schuster, New York.

Ethier, T. J. 1999. Breeding ecology and habitat of northern goshawks (*Accipiter gentilis laingi*) on Vancouver Island: a hierarchical approach. MSc thesis, University of Victoria, Victoria, British Columbia.

Grant, V. C., B. B. Steele, and R. L. Bayn, Jr. 1991. Raptor population dynamics in Utah's Uinta Basin: the importance of food resource. Southwestern Naturalist **36**:265–280.

Hayward, G. D., and P. H. Hayward 1993. Boreal owl (*Aegolius funereus*). *In* A. Poole and F. Gill (eds). The birds of North America, no. 63. American Ornithologists' Union, Washington, DC.

Heinrich, B. 1989. Ravens in winter. Simon and Schuster, New York.

Heinrich, B. 1994. Does the early raven get (and show) the meat? Auk **111**:764–769.

Hörnfeldt, B. 1978. Synchronous population fluctuations in voles, small game, owls and tularemia, in northern Sweden. Oecologia **32**:141–152.

Hörnfeldt, B., B. G. Carlsson, and U. Eklund. 1990. Effects of cyclic food supply on breeding performance in Tengmalm's owl *Aegolius funereus*. Canadian Journal of Zoology **68**:522–530.

Hörnfeldt, B., and U. Eklund. 1989. The effects of food on the laying date and clutch-size in Tengmalm's owl *Aegolius funereus*. Ibis **132**:395–406.

Houston, C. S., D. G. Smith, and C. Rohner. 1999. Great horned owl *Bubo virginianus*. *in* A. Poole and F. Gill (eds). The birds of North America, no. 372. American Ornithologists' Union, Washington, DC.

Joy, S. M., R. T. Reynolds, R. L. Knight, and R. W. Hoffman. 1994. Feeding ecology of sharp-shinned hawks nesting in deciduous and coniferous forests in Colorado. Condor **96**:455–467.

Kennedy, A. J., and L. N. Carbyn. 1981. Identification of wolf prey using hair and feather remains with special reference to Western Canadian Parks. Canadian Wildlife Service Report, Ottawa, Ontario.

Kertell, K. 1986. Reproductive biology of northern hawk owls in Denali National Park, Alaska. Journal of Raptor Research **20**:91–100.

Korpimäki, E. 1985. Rapid tracking of microtine populations by their avian predators: possible evidence for stabilizing predator. Oikos **45**:281–284.

Korpimäki, E. 1988. Diet of breeding Tengmalm's owls *Aegolius funereus:* long term changes and year to year variation under cyclic food conditions. Ornis Fennica **65**:21–30.

Korpimäki, E. 1992. Population dynamics of Fennoscandian owls in relation to wintering conditions and between year fluctuations in food supply. Nature Conservation **5**:1–10.

Korpimäki, E., and K. Norrdahl. 1991. Numerical and functional responses of kestrels, short-eared owls, and long-eared owls to vole densities. Ecology **72**:814–826.

Levins, R. 1968. Evolution in changing environments: some theoretical explorations. Princeton University Press, Princeton, New Jersey.

Linsdale, J. M. 1937. The natural history of magpies. Pacific Coast Avifauna no. 25. University of California, Berkeley.

Luttich, S. N., L. B. Keith, and J. D. Stephenson. 1971. Population dynamics of the red-tailed hawk *Buteo jamaicensis* at Rochester, Alberta. Auk **88**:75–87.

Marti, C. D. 1987. Raptor food habits studies. *in* B. G. Pendleton, B. A. Millsap, K. W. Cline and D. M. Bird (eds). Raptor management techniques manual, pages 67–79. National Wildlife Federation, Washington, DC.

MacWhirter, R. B., and K. L. Bildstein. 1996. Northern harrier (*Circus cynaeus*). *In* A. Poole and F. Gill (eds). The birds of North America, no. 210 American Ornithologists' Union, Washington, DC.

McGowan, J. D. 1975. Distribution, density and productivity of goshawks in interior Alaska. Wildlife Research Project Report 10.6R. Alaska Fish and Game Department, Juneau, Alaska.

McIntyre, C. L., and L. G. Adams. 1999. Reproductive characterisitics of migratory golden eagles in Denali National Park, Alaska. Condor **101**:115–123.

McInvaille, W. B. Jr., and L. B. Keith. 1974. Predator-prey relations and breeding biology of the great horned owl and red-tailed hawk in central Alberta. Canadian Field-Naturalist **88**:1–19.

Mindell, D. P. 1983. Harlan's hawk (*Buteo jamaicensis harlani*): a valid subspecies. Auk **100**:161–169.

Moore, T. D., L. E. Spence, C. E. Dugnolle, and W. G. Hepworth. 1974. Identification of the dorsal guard hairs of some mammals of Wyoming. Wyoming Game and Fish Department Bulletin no. 14. Laramie, Wyoming.

Mueller, H. C., D. D. Berger, and G. Allez. 1977. The periodic invasions of goshawks. Auk **94**:652–663.

Newton, I. 1979. Population ecology of raptors. T. & A. D. Poyser, London.

Newton, I. 1986. The sparrowhawk. T. & A. D. Poyser, London.

Newton, I. 1998. Population limitation in birds. Academic Press, San Diego, California.

O'Donoghue, M., L. Hofer, and F. Doyle. 1995. Predator versus predator. Natural History **104**(3):6–9.

Pelech, S. A-M. 1999. Habitat use and nest searching success of red squirrels at a forest edge. M.Sc. thesis, University of British Columbia, Vancouver, British Columbia.

Phelan, F. J. S., and R. J. Robertson. 1978. Predatory responses of a raptor guild to changes in prey density. Canadian Journal of Zoology **56**:2565–2572.

Ratcliff, D. 1997. The raven. A natural history in Britain and Ireland. T. & A. D. Poyser, London.

Restani, M. 1991. Resource partitioning among three *Buteo* species in the Centennial Valley, Montana. Condor **93**:1007–1010.

Rohner, C., and F. I. Doyle. 1992a. Methods of locating great-horned owl nests in the boreal forest. Journal of Raptor Research **26**:33–35.

Rohner, C., and F. I. Doyle. 1992b. Food-stressed great-horned owl kills adult goshawk: exceptional observation or community process? Journal of Raptor Research **26**:261–263.

Rohner, C., J. N. M. Smith, J. Stroman, M. Joyce, F. I. Doyle, and R. Boonstra. 1995. Northern hawk owls in the Nearctic boreal forest: prey selection and population consequences of multiple prey cycles. Condor **97**:208–220.

Sonerud, G. A. 1997. Hawk owls in Fennoscandia: population fluctuations, effects of modern forestry, and recommendations on improving foraging habitats. Journal of Raptor Research **31**:167–174.

Sovell, J. R. 1993. Attempts to determine the influence of parasitism on a snowshoe hare population during the peak and initial decline phase of a snowshoe hare cycle. MSc thesis. University of Alberta, Edmonton, Alberta.

Squires, J. R. S., and R. T. Reynolds. 1997. Northern goshawk *Accipiter gentilis*. *In* A. Poole and F. Gill (eds). The birds of North America, no. 298. American Ornithologists' Union, Washington, DC.

Strickland, D., and H. Ouellet. 1993. Gray jay, *Perisoreus canadensis*. *In* A. Poole and F. Gill (eds). The birds of North America, no. 40. American Ornithologists' Union, Washington, DC.

Widén, P. 1987. Goshawk predation during winter, spring and summer in a boreal forest in central Sweden. Holarctic Ecology **10**:104–109.

PART VI

COMMUNITY AND ECOSYSTEM ORGANIZATION

Testing Hypotheses of Community Organization for the Kluane Ecosystem

A. R. E. SINCLAIR, CHARLES J. KREBS, RUDY BOONSTRA,
STAN BOUTIN, & ROY TURKINGTON

In chapter 3 we outlined the general ideas of community organization through top-down and bottom-up processes. In this chapter we review how each experiment has affected the total biomass or nutrient pool of the trophic levels. We compare these results with the predictions for different models of trophic level interactions discussed in chapter 3. We identify the major components that hold the system together and the direction and strength of their interactions. The need is to identify the components most sensitive to change so that they can be used as indicators for long-term monitoring.

Empirical studies have tested models of community organization that involve variations of top-down (pure predator control) hypotheses, bottom-up (nutrient limitation) hypotheses and hypotheses including reciprocal interactions (control by predators that are dependent on the prey; see Menge 1992, Power 1992). These concepts and some related ones (e.g., donor control) have not always been rigorously defined. Consequently, the same terms can apply to several different models, and so it has been difficult to derive testable and generally acceptable predictions.

Bottom-up hypotheses assume that systems are regulated by nutrient flow from below (White 1978, 1984, Lampert 1985, Strong 1992, Polis and Strong 1997). These hypotheses assume there is a shortage of suitable resources (e.g., areas with accessible nutrients, plant parts with sufficiently high quality, prey that are easy to catch), even if resources appear superficially to be abundant. Bottom-up hypotheses can also be regarded as biological null hypotheses because plants are essential to the levels above, whereas the reverse need not be true. One bottom-up hypothesis predicts that an increase in resources, such as soil nutrients, should lead to a biomass increase in all subsequent levels (Abrams 1993). Another hypothesis predicts that the biomass of the top level increases as well as alternating lower levels, while intermediate levels remain constant (Oksanen et al. 1981, Abrams 1993, Schmitz 1998). Yet others predict that higher trophic levels have neither a regulating effect nor any influence on productivity or overall biomass on the levels below them, although the standing crops of subcomponents that make up suitable resources can be depressed (Hawkins 1992, Hunter and Price 1992, Strong 1992). Technically, bottom-up hypotheses require some form of self-regulation because if there were no self-regulation population densities at each trophic level would either decline toward zero or increase ad infinitum.

There are various top-down hypotheses with different implications for population regulation at all levels. According to the hypothesis of Menge and Sutherland (1976), predation pressure increases monotonically from the top of the food chain downward, which implies that top predators can exploit basal organisms. Caughley and Lawton (1981) argued that strong reciprocal interaction exists between herbivores and plants, whereas predators and herbivores are involved in a bottom-up relationship. Their view has been supported by data from arid and arctic ecosystems (Batzli et al. 1980, Caughley and Gunn 1993).

A converse idea was presented by Hairston et al. (1960), who proposed that a tight, reciprocal interaction occurs between predators and herbivores. Consequently, herbivores only have light impact on plants, which are either self-regulated or involved in a reciprocal interaction with mineral nutrients. These ideas imply that removal of the effective top consumer generates cascading impacts down to plants and nutrients.

An alternative hypothesis, combining reciprocal and bottom-up features, suggests that biomass is regulated from below by nutrient availability but that this effect is strongest at the plant level and becomes weaker at progressively higher levels (Benndorf and Horn 1985, McQueen et al. 1986, 1989, Pace and Funke 1991). Conversely, at the top of the food web, top-down interactions are strong, but these effects weaken with every step down the food chain. Such attenuation could be achieved if the feeding efficiency of predators is reduced by interference, territoriality, or prey refuges (Power 1984, Arditi and Ginzburg 1989, Arditi et al. 1991, Hanski 1991, Leibold 1996).

17.1 Experimental Perturbations of the Boreal Forest

We tested the predictions from the 27 models in chapter 3 by experimentally perturbing the boreal forest food web. In these experiments, each trophic level was perturbed and the subsequent effects on the biomass, productivity, or activity of other levels were measured.

17.1.1 Direct and Indirect Effects

We measured two types of response, biomass change and productivity change. A direct effect occurs in the biomass or productivity at the next trophic level; an indirect effect occurs at one or more levels removed. These indirect effects are a subset of "interaction-chain indirect effects" (Wootton 1993, 1994a,b,c, Billick and Case 1994, Menge 1997) in that they are linear on the food chain. We discuss briefly interactions such as apparent competition (Holt 1977) and indirect mutualism (Dungan 1987) that involve interactions between species within a trophic level and interaction modifications such as effects of predation risk (Hik 1995, Boonstra et al. 1998).

Predictions from the models on the effects of the perturbations are through the direction of change in biomass relative to control areas. For some levels, in particular plants, we also considered changes in productivity as a response to perturbation. The predicted changes in biomass are indicated in table 3.3 as an increase, decrease, or no change. However, where two perturbations are applied simultaneously, both acting in the same direction, an additional prediction can be made on the magnitude of change relative to either single perturbation.

17.1.2 The Experiments

We recapitulate the major experimental perturbations here and highlight the predictions of the trophic level models for each experiment. We refer to the models summarized in table 3.2.

Application of Fertilizer   Fertilizer was applied from the air to two 1-km^2 blocks of forest as described in chapter 4. This manipulation should increase the soil nutrient pool (N). Models that suggest that plants (V) are responsive to the nutrient pool (i.e., those with right or double arrows between N and V) predict increases in biomass of higher trophic levels to varying lengths of the food chain (18 of the models).

Models with top-down effects (left arrow) only (nine models) predict that plant biomass is not limited by nutrients and should not respond to fertilizer inputs. However, productivity of plants and consumption by herbivores could increase. Similarly, productivity of herbivores and consumption by predators could increase.

Food Addition   Commercial rabbit chow was applied ad libitum to two areas (chapter 4). This food is eaten readily by hares and ground squirrels. This treatment had the effect of artificially increasing the food supply for herbivores independently of the natural food and, therefore, increasing the herbivore level independently of the vegetation level. Of the 27 models, 9 assume that herbivores are not limited by food supply and so they predict no response to food addition. The remaining 18 models assume that herbivores are limited by food (bottom-up control indicated by a right arrow). In nine of these models herbivores also have a top-down limiting effect on vegetation (a reciprocal effect indicated by a double arrow).

The Exclusion of Carnivores   Carnivores were excluded from a 1-km^2 area by wire fencing as described in chapter 4. This fence was permeable to hares and squirrels through small holes in the fence. Although some predators occasionally broke into the exclosure, predation by carnivores was considerably reduced within the exclosure. For models predicting top-down effects, herbivores should increase in biomass, plants should decrease, and nutrients should increase.

The Exclusion of Carnivores and the Addition of Food   Carnivores were excluded from a 1-km^2 area by wire fencing, and ad libitum rabbit chow was provided for the herbivores. For top-down models (left arrow) the removal of mammalian predators predicted an increase in herbivores. In addition, for models where herbivores are also responsive to their food supply, we predict an additional increase in herbivore biomass relative to the herbivore increase predicted from the predator exclosure experiment. Hence, in this experiment, there should be increased effects on other trophic levels relative to the single perturbations. The remaining models where herbivores do not respond to food supply predict changes in herbivore biomass similar to that in the predator exclosure experiment.

The Exclusion of Hares   Hares were the dominant component of the herbivore biomass in this system, and they were excluded by fencing a 4-ha area. In the 18 top-down models, exclusion of the top two trophic levels predicted that vegetation biomass should increase. In 12 of these models, top-down effects continued to the nutrient level, predicting a decrease in the soil nutrients.

The Exclusion of Hares and the Addition of Fertilizer   This experiment was similar to the hare exclusion experiment but with the addition of fertilizer. The two perturbations are predicted to have opposite effects on soil nutrients: fertilizer adds nutrients but removal of hares should result in fewer nutrients. Because the outcome could be any value depending on absolute amounts of inputs and outputs, no qualitative prediction for soil nutrients can be made. However, for vegetation biomass one can predict two alternative responses. In models where plants respond to nutrient levels, there should be a greater increase in plant biomass relative to the hare exclosure experiment. In contrast, the other models predict no difference in plant biomass changes between hare exclosure and hare exclosure + fertilizer.

The Removal of Vegetation   Soil nutrient levels were measured in plots where vegetation was removed and compared to an equal number of immediately adjacent control plots with intact vegetation. Soil nutrients should increase in plots where the vegetation has been removed in those models predicting that nutrients respond to plant uptake.

With these seven experiments just described, 21 models out of 27 make unique sets of predictions. There are also three pairs of unique sets (5 and 8, 12 and 19, and 6 and 26; table 3.1; see 3.4.7).

17.2   Methods

Most of the methods for the data discussed here were presented in earlier chapters. We highlight here only a few specific points for methods not described in earlier chapters.

17.2.1   Soil Nitrogen

Fertilizer was applied by air once each year from 1987 to 1994 as described in chapter 4. In 1995, 10 soil samples were collected at sites on experimental treatments and immediately adjacent to them as controls. The treatment grids sampled were fertilizer, food addition, predator exclosure plus food addition, and hare exclosure. In 1996 soils were sampled inside ($n = 100$ samples) and outside the hare exclosure. Also outside this exclosure, vegetation was removed from 156 plots and soil samples were compared with 156 plots with vegetation left intact. Vegetation was killed by using Roundup (Monsanto Corp.) and left in situ. The edges of all plots were cut to a spade-depth to kill roots and so reduce the movement of nutrients into or out from the plot. In 1995, soils were sent to Peace Growers' lab in Fort St. John, British Columbia, and in 1996 to Pacific Soil Analysis in Richmond, British Columbia, for analysis of nitrate nitrogen.

17.2.2   Vegetation

Details of methods are presented in chapters 5–7 as well as in Krebs et al. (1992, 1995) and Turkington et al. (1998). Mammalian herbivores have little influence on biomass of herbaceous vegetation during the summer (John and Turkington 1995). Winter food supply is more likely to be limiting to mammalian herbivores. Winter food plants for snowshoe hares are largely bog birch (*Betula glandulosa*) and gray willow (*Salix glauca*). At certain times when these two shrubs were heavily browsed, hares turned to eating white spruce (*Picea glauca*). We concentrated, however, on measuring the shrubs. Twigs <5 mm diameter were the main food. We measured biomass, growth rates, and browse rates of shrubs as described in chapter 6.

The net effect on plant biomass as defined in this chapter combines the herbivory in one winter and the growth in the subsequent summer. It measures the degree to which an increase in growth rate (g_i) of plant species i in year y compensates for the biomass loss to herbivory (h_i) in the previous winter. The net effect, C_{iy}, is given by:

$$C_{iy} = (1 - h_i)(1 + g_i).$$

If $C_{iy} = 1$, there is complete compensation and no change in biomass, if $C_{iy} < 1$, there is a net loss in biomass, and if $C_{iy} > 1$, there is either an increase in biomass by overcompensation for the herbivory or no herbivory.

17.2.3 Plant Secondary Chemicals

Earlier studies (Sinclair and Smith 1984, Sinclair et al. 1988) have shown that phenolic compounds change over the hare cycle, and of all the secondary compounds, phenols are most sensitive to browsing. These compounds have been identified in other birch species (Reichardt et al. 1984), but they appear to be at low levels in willow species. The product obtained from methanol extraction, as described in chapter 6, can be used as an index of secondary compound content.

17.2.4 Herbivore Biomass

The dominant herbivores in summer are the snowshoe hare and arctic ground squirrel, with red squirrel a minor third species. In winter, although moose are present, they had relatively minor effects, and the snowshoe hare is the dominant herbivore. All smaller mammal species were enumerated by live trapping as described in chapters 8–11.

17.2.5 Predator Activity

For practical reasons we could monitor only the main carnivore species, lynx and coyote, and we could not measure the biomass of these species for each experiment independently. Therefore, we used an index of their total response (numerical × functional) by counting the number of tracks in winter. After each snowfall, the 4-km perimeters of the control, fertilizer, and food addition plots were surveyed. The number of fresh tracks of each species crossing the 4-km perimeter in the previous day was scored and averaged for that winter.

17.3 Direct Effects of Trophic-level Perturbations

Results for each of the experiments are compared with values from the control sites. We discuss in this section only the direct effects. Indirect effects are seen in the level next to the one perturbed, and double indirect effects are seen two trophic levels removed from the perturbation. To measure direct and indirect effects, data from different years were used. Peak herbivore biomass occurred in 1990. Both maximum browsing by herbivores on winter food and peak predator activity were observed in the winter of 1990–1991. The major decline of herbivore numbers took place in winter 1991–1992. Peak growth of winter food plants (growth index (%), see 6.3.1) occurred in 1992, after browsing declined.

17.3.1 Fertilizer Addition

The mean rate of fertilizer application over the years 1987–1994 was 150 ($\pm$30 SE) kg N per hectare per year in the form of ammonium nitrate (Turkington et al. 1998). Figure 17.1 illustrates the soil nitrate nitrogen in mid-June about 3 weeks after fertilizer application. As expected, there were significant increases in soil nitrate in 1995 after fertilizer was added. Measurement of nitrogen in the soil in mid-June 1995 showed 40.3 ppm ($\pm$9.7, 95% CL) nitrate nitrogen compared to 0.97 ppm ($\pm$0.19, CL) in immediately adjacent control samples ($p < .001$; table 17.1).

Figure 17.2 shows the total dry weight biomass of small twigs (<5 mm) of willow and

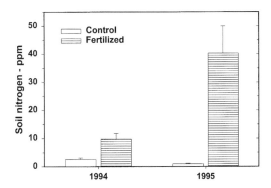

Figure 17.1 Soil nitrate nitrogen (ppm, with 95% CL) on fertilizer and control sites in 1994 and 1995.

birch obtained from our clip plots. Because each grid had a different absolute biomass, we standardized across grids by setting 1990 as 100%. Both control and fertilizer grids showed an increase in biomass of small birch twigs before the hare peak, followed by a decline in 1991 and 1992. The fertilizer grids, however, showed consistently higher values after 1992 than the controls as a consequence of the added soil nitrogen. Thus, in 1992, addition of fertilizer produced biomass values of 149.6 g/m^2 compared to 113.6 g/m^2 on control sites for winter twigs (table 17.2).

Productivity was measured by the growth index (%) in 5-mm twigs (table 17.3; see also figures 6.5 and 6.7). In birch, productivity increased to a peak in 1992, then declined on both fertilizer and control grids. This trend was less evident in willow, but there appeared to be a decline in growth toward the low point of the hare cycle. In both species, however, growth on the fertilizer grids was consistently above that on controls. Productivity of these shrubs in the peak year of 1992 occurred 2 years after peak herbivore biomass. In 1992, birch % growth index was 31.91 (± 1.41 CL) on fertilized grids, compared to 23.88 (± 1.8 CL) on controls. For willow, % growth index was 21.34 (± 1.34 CL) on fertilized grids, compared to 11.98 (± 0.80 CL) in controls.

The percent nitrogen in the growth index of willow twigs on fertilizer grids was consistently higher than that of twigs on control areas (figure 17.3). Thus, the overall availability of high-protein food for hares increased on fertilizer grids. In neither of the treatments was there any trend through the cycle.

In contrast to twig nitrogen, the methanol extract from both willow and birch twigs was similar on fertilizer and control treatments (figure 17.4a,c). Thus, addition of nitrogen did not appear to alter twig defensive chemistry. However, both shrub species appeared to show a response to the hare cycle through an increase of the methanol extract at the hare peak, then a decrease subsequently.

In general, these figures are consistent with models showing direct bottom-up effects (see chapter 3, table 3.1) as a result of increases to the soil nutrient pool.

17.3.2 Addition of Hare Food

On the food-addition sites, hares ate rabbit chow year-round, and arctic ground squirrels ate it in summer. Rabbit chow had the effect of increasing the density of these herbivores. Thus, in spring 1991 at the peak, total herbivore biomass on these sites was 13.34 kg/ha, compared to 2.83 kg/ha on the controls, a fivefold increase. Over the cycle, herbi-

Table 17.1 Nitrate nitrogen (ppm) from soil samples collected at experimental and control sites in 1995 and 1996.

Site	n	Experimental Site	Control Site	p
Fertilizer	10	40.3	0.97	.001
Food addition	10	2.70	3.3	ns
Predator exclosure	10	0.37	0.18	ns
Predator exclosure + food addition	10	0.85	0.33	ns
Hare exclosure	100	8.34	5.2	.0001
Vegetation exclosure	156	6.58	5.2	.0001

vore biomass on food-addition sites was consistently above that on controls (figure 17.5). On average, these values were 6.52 kg/ha and 2.48 kg/ha, respectively (table 17.4).

The impact of hare feeding on winter food plants was measured by the percentage of 5-mm twigs browsed. The effect of fertilizer was to increase hare browsing on willows (table 17.5, figure 6.9). However, birch was heavily browsed on controls, and, if anything, there was slightly less browsing on the fertilizer grids in comparison. Thus, higher hare density on this treatment resulted in higher browsing on willow, but not on birch.

Vegetation growth, but not biomass, was measured on food-addition sites (figure 17.6). After the period of high hare numbers, the % growth index of 5-mm twigs was higher on food-addition sites than on controls for both species. Thus, the higher browsing on willow resulted in a significant ($p < .05$) increase in growth on this food-addition treatment 2 years later.

The net effect of peak browsing and one season's subsequent growth (figure 17.7) showed that neither birch nor willow could compensate for browsing during the hare peak. However, because growth responded 2 years after peak browsing, the net effect in 1992 for birch biomass was 1.22, and for willow biomass it was 1.24, indicating a subsequent increase in vegetation biomass once herbivory declined. The methanol extract of these twigs did not differ between food-addition and control sites (figure 17.4b,d).

The tracks of lynx and coyote crossing the perimeter of each experimental site were counted after each fresh snowfall and averaged as the number per day in each winter (figure 17.8, table 17.6). In winter 1990–1991, there were 3.5 coyote tracks/day on the food-

Table 17.2 Combined gray willow and bog birch biomass of 5-mm twigs (g/m²) on 2-m² quadrats for 1992, the year following maximum browsing, and as a mean for the 10 years, 1987–1996 (1990–1996 for the predator exclosure plus food addition).

Site	Year	Peak Biomass	10-year Mean
Control	1991	133.4	94.5
Control	1992	113.6	94.5
Fertilizer	1992	149.6	133.7
Predator exclosure + food addition	1991	14.3	32.9
Predator exclosure + food addition	1992	6.9	32.9
Hare exclosure	1992	101.9	78.0

To make comparisons across grids, values were standardized to 100 in 1990.

(a) Bog Birch

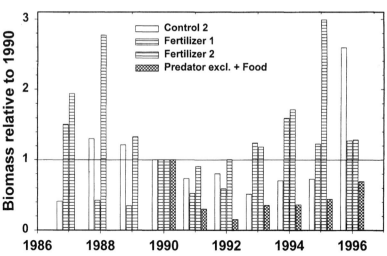

(b) Gray Willow

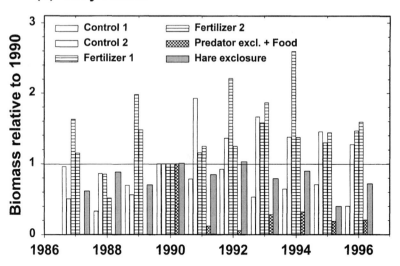

Figure 17.2 Biomass (dry weight) of 5-mm twigs of (a) bog birch and (b) gray willow on all treatments, 1987–1996. Because different areas vary in shrub density, the biomass in 1990 is indexed as 1.0, and all other values for each treatment area are referenced to the 1990 value for that area.

Table 17.3 Percent growth index of 5-mm twigs of gray willow and bog birch in the peak year and as the mean over 1988–1995.

Site	Year	Peak (95% CR)	Mean (95% CR)
Gray Willow			
Control	1991	12.43 (11.70–13.2)	15.37 (15.07–15.67)
Control	1992	11.98 (11.22–12.78)	15.37 (15.07–15.67)
Fertilizer	1992	21.34 (20.00–22.68)	19.95 (19.48–20.42)
Food addition	1992	31.56 (28.22–35.31)	22.72 (21.81–23.63)
Predator exclosure	1992	16.11 (14.82–17.40)	16.77 (16.22–17.32)
Predator exclosure + food addition	1992	26.68 (24.46–29.09)	23.22 (22.42–24.02)
Hare exclosure	1991	11.27 (9.97–12.73)	14.82 (14.26–15.38)
Hare exclosure	1992	14.84 (13.17–16.71)	14.82 (14.26–15.38)
Hare exclosure + fertilizer	1992	16.91 (15.13–18.69)	20.22 (19.75–20.69)
Bog Birch			
Control	1992	23.88 (22.13–25.70)	16.24 (15.85–16.63)
Fertilizer	1992	31.91 (30.50–33.32)	20.56 (20.10–21.02)
Food addition	1992	27.16 (25.23–29.16)	12.08 (11.28–12.88)
Predator exclosure	1992	21.11 (19.41–22.89)	17.54 (17.02–18.06)
Predator exclosure + food addition	1992	36.55 (34.02–39.17)	25.24 (24.44–26.04)
Hare exclosure + fertilizer	1992	19.94 (18.54–21.34)	20.41 (19.69–21.13)

Growth index = new growth/total twig dry mass. 95% CR = 95% confidence range.

addition areas versus 0.50/day on controls, and 2.89 and 0.96 lynx tracks/day for the two areas, respectively.

In summary, increasing herbivore biomass decreased vegetation biomass, and this decrease could not be compensated for by increased growth the following year. However, once herbivore populations had declined 2 years later, growth more than compensated for herbivory. These results are consistent with top-down effects. At the same time, mammalian predator activity increased by a factor of 3 on the food-addition sites, consistent with bottom-up effects.

17.3.3 Predator Exclosure

Our fence excluded the large carnivores such as lynx and coyote. Avian predators were inhibited in their hunting by overhead lines in a small (10 ha) area, but the inhibitory effect was minor for the whole site. Herbivore biomass during the peak years, 1988–1991, in the predator exclosure was higher than that on the controls, indicating a small but consistent effect from the removal of predators (table 17.4). This indicates a top-down effect of mammalian predators.

17.3.4 Predator Exclosure and Food Addition

The predator exclosure and food addition site received the double treatment of removal of carnivores and addition of rabbit chow. Once the treatment was initiated in summer

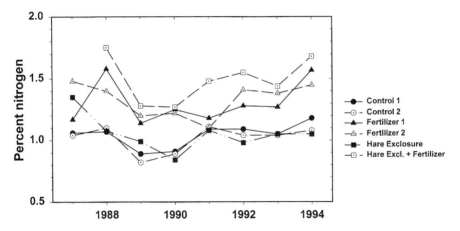

Figure 17.3 Percent nitrogen in gray willow twigs collected at the end of winter each year, 1987–1994.

1988, herbivore biomass increased and remained well above that of the control and predator exclusion sites. However, this treatment maintained similar herbivore densities to those on the food addition up to the peak of the cycle. After the peak, this treatment maintained high densities, while those on the food addition declined (see figure 8.2 and figure 9.10). Thus, food supply was the dominant effect up to the peak, but effects of predator exclusion became detectable after the peak. The mean value for herbivore biomass over the cycle for this joint treatment was 11.30 kg/ha. This density represented a 4.5-fold increase over controls (2.48 kg/ha), a 3.5-fold increase over the predator exclosure alone (3.23 kg/ha), and a 1.7-fold increase over the food addition alone (6.5 kg/ha; table 17.4).

Winter shrub biomass on the predator exclosure + food-addition treatment declined well below that on controls during the period of high hare density. Thus, in 1991 small twig biomass on this treatment was 14.3 kg/ha, one-ninth the biomass on controls (133.4 kg/ha). Mean values over the cycle were 32.9 kg/ha and 94.5 kg/ha respectively, a three-fold difference (table 17.2).

As a consequence of the high herbivore biomass, herbivore browsing was severe in winter 1990–1991. Thus, 88.2% of birch twigs were browsed on the joint treatment, compared to 85.7% on controls, and 62.8% of willow twigs on the treatment, compared to 12.9% on the controls (tables 6.5, 6.6; see also table 17.5). Because browsing on birch was so high on control plots, there was little room for the treatment effects to be higher. However, browsing of willow on the treatment was considerably higher than controls. Consequently, in 1991 the net reduction of biomass in birch twigs was very large but did not differ between treatment and controls (0.15 of the biomass a year earlier on the treatment compared to 0.16 on controls). In contrast, net biomass of willow twigs declined to 0.44 of that a year earlier on the treatment compared to 0.98 on controls. Thus, willow biomass declined to less than half that of controls (figure 17.7, table 17.3). Growth did not compensate for browsing in either species on this treatment.

The joint treatment produced consistently higher growth rates of birch and willow twigs relative to controls. Also, growth of both species on this treatment was higher than

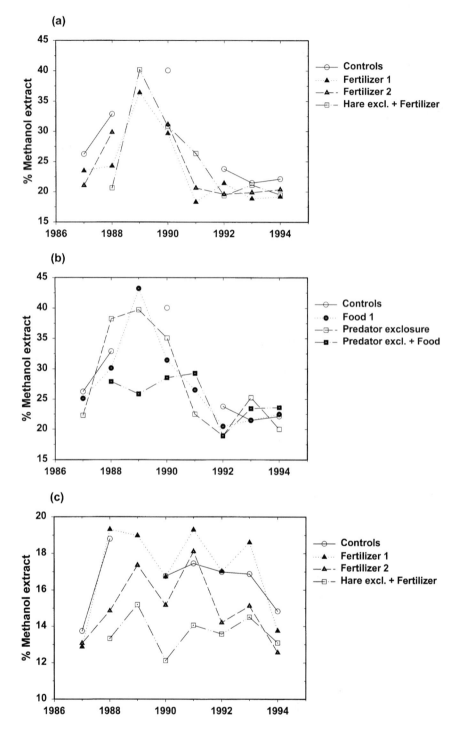

Figure 17.4 Percent methanol extract in twigs collected at the end of each winter. (a, b) Bog birch; (c–e) gray willow. This is a general measure of secondary compounds in the winter twigs.

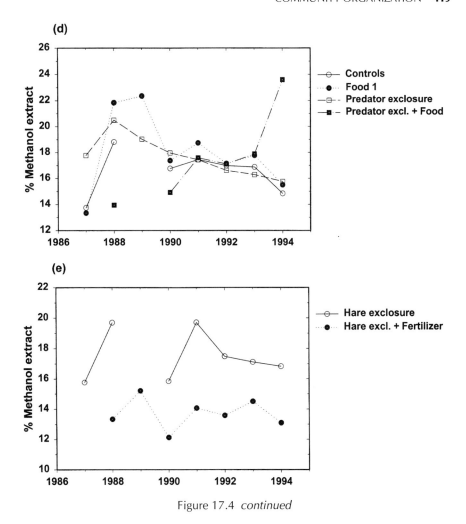

Figure 17.4 *continued*

that on the predator exclosure, and birch growth (but willow less so) was also higher than that on the food addition (figure 6.4, table 17.3).

In summary, this joint treatment can be compared not only to controls but also to the single treatments. The increase in herbivore density relative to food addition alone indicates the top-down effect from removal of predators after the peak in hare numbers had passed. Similarly, the decrease in plant biomass relative to controls indicates the top-down effect of herbivores. The increase in plant productivity was a response to the herbivory, but it was not sufficient to compensate for the loss of plant biomass.

17.3.5   Hare Exclosure

The hare exclosure treatment excluded hares and moose (*Alces alces*), which were the only herbivores in winter. Moose were rare, and the treatment reflected the effects of ex-

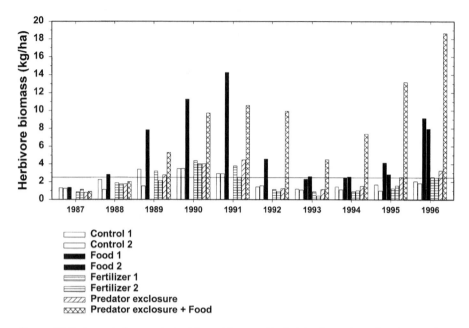

Figure 17.5 Annual changes in herbivore biomass (kg/ha) in the spring of each year from 1987 to 1996 on all control and treatment areas. All herbivore species are included in these totals. The average herbivore biomass for control areas is indicated by the horizontal line.

cluding hares as the dominant herbivore. Willow was the only winter shrub species in the hare exclosure. In 1992, biomass of 5-mm twigs was 101.9 kg/ha in the exclosure, compared to 113.6 kg/ha on the controls. When standardized to 100 in 1990 there was no difference in twig biomass between the exclosure and control in the period 1990–1996 (figure 6.3).

Growth of willow twigs in the hare exclosure did not follow the cycle as it did outside, but rather appeared to decline slightly over the 8 years (figure 17.6b). In 1992 growth in the exclosure (14.84%) did not differ markedly from willow on controls (11.98%), and

Table 17.4 Total herbivore biomass (kg/ha) on the experimental sites in the year of peak biomass and as a mean over 1987–1996.

Site	Year	Peak	10-year Mean
Control	1990	3.34	2.48
	1991	2.83	
Fertilizer	1990	4.22	2.73
Food addition	1991	13.34	6.52
Predator exclosure	1991	4.49	3.23
Predator exclosure + food addition	1991	10.59	11.30

Herbivores include hares, squirrels, small mammals, and grouse.

Table 17.5 The percentage of 5-mm twigs of willow and birch that were completely browsed in the peak winter 1990–1991.

Site	Willow		Birch	
	% Browse	Net Effect	% Browse	Net Effect
Control	12.9	0.98	85.7	0.16
Fertilizer	34.9	0.78	61.7	0.49
Food addition	47.1	0.63	57.3	0.51
Predator exclosure	10.8	1.01	51.5	0.56
Predator exclosure + food addition	62.8	0.44	88.2	0.15
Hare exclosure	0	1.13	—	—
Hare exclosure + fertilizer	0	1.18	0	1.23

The net effect is the net biomass in September 1991 relative to that in September 1990 as a result of browsing followed by growth in summer 1991. Net effect <1 indicates undercompensation, >1 overcompensation of growth.

over the cycle mean growth was similar (14.82% exclosure, 15.37% controls). Nitrogen values of these twigs also did not differ from those of controls (figure 17.3).

In 1991, the year of highest browsing impact, net biomass of willow for the exclosure increased by a factor of 1.13, compared to 0.98 on controls. Thus, the removal of herbivores allowed a small net increase in willow twig biomass. Over the whole cycle, willow biomass showed a net increase in the exclosure by 1.15, compared to 1.09 on controls (figure 17.7d). In general, exclusion of hares showed small but significant impacts of herbivory on productivity but not on biomass of willows. Because birch were absent from this site, we could not measure the equivalent impacts. However, we anticipate that herbivory would be greater on birch.

17.3.6   Hare Exclosure Plus Fertilizer

The growth of willow twigs on the hare exclosure + fertilizer plot was consistently above controls (figure 17.6). Over the cycle, growth remained approximately constant and averaged 20.2% compared to 15.4% on controls. Similarly, for birch, % growth index was usually above that on controls. Over the cycle, birch growth also showed no trend in the exclosure and averaged 20.4% compared to 16.2% on controls.

Both willow and birch had similar high growth rates on the fertilizer grids and on the fertilized hare exclosure (figure 17.6). However, growth on the fertilizer grids increased to a peak, then declined, reflecting the changes in browsing during the hare cycle. Growth in the hare exclosure did not show this pattern, and it merely declined over time. In 1992, after peak browsing, the growth rate (% growth index) on fertilizer plots (21.3% for willow, 31.9% for birch) was significantly higher than that in the fertilized hare exclosure (16.9% for willow, 19.9% for birch). Therefore, browsing by herbivores at peak densities on the fertilizer areas stimulated productivity of the vegetation. Nitrogen values in the twigs of willow in the fertilized hare exclosure were similar to those on the fertilizer plots and consistently higher than those of controls (figure 17.3).

Because there was substantial browsing of twigs on the fertilizer plots, the higher

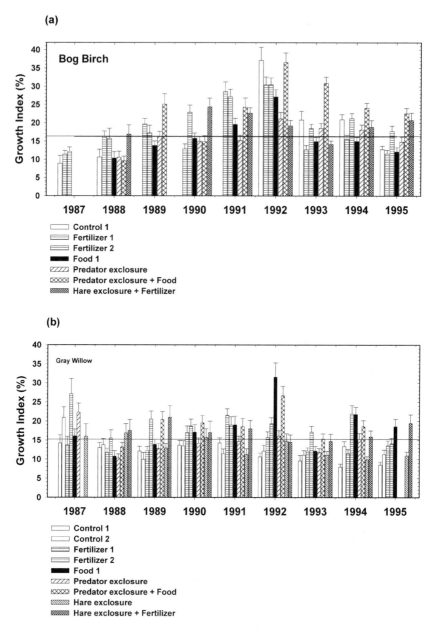

Figure 17.6 Annual changes in the growth rates (±1 SE) of (a) bog birch and (b) gray willow. Growth is indexed as the percentage dry weight of current annual growth to total 5-mm twig biomass. The horizontal line marks the average growth rate for control areas over 1987–1996.

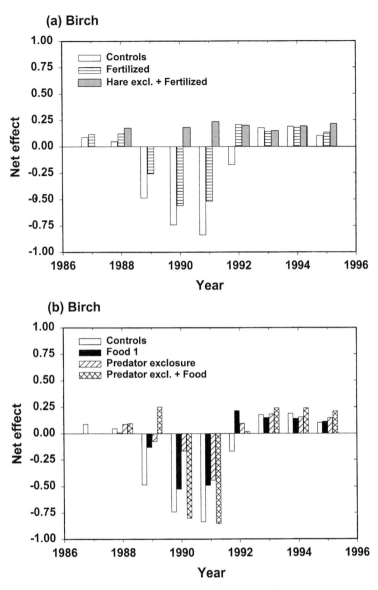

Figure 17.7 The net effect of winter browsing and summer growth on the standing crop of (a, b) bog birch, and (c, d) gray willow. Negative values indicate years in which browsing exceeds growth.

growth rates could not compensate sufficiently for the loss of biomass. Net biomass of willow twigs on fertilized grids decreased by a factor of 0.78, and that of birch decreased by a factor of 0.49 in 1991 (table 17.5). In contrast, on the fertilized hare exclosure, net biomass of 5-mm willow twigs increased by a factor of 1.18 and that of birch increased by 1.23, showing that protection from browsing allowed an increase of vegetation biomass, while exposure to browsing caused a decrease by as much as half.

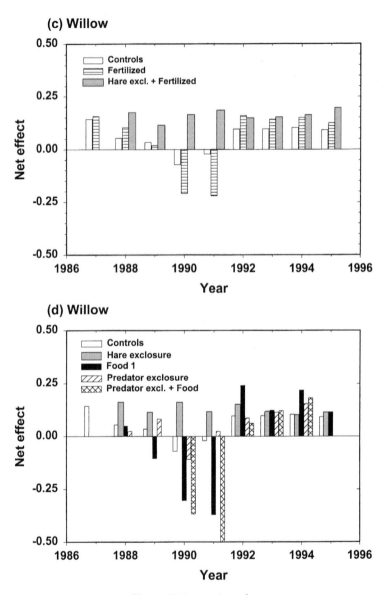

Figure 17.7 *continued*

Comparing the fertilized hare exclosure with the unfertilized hare exclosure, the increase in net biomass of willows did not differ between the two, despite the higher growth rate of the former. Therefore, fertilizer had little bottom-up effect, and hare removal rather than fertilizer was the major effect observed with this experiment. In summary, top-down effects were observed both through the decrease in vegetation biomass and through an increase in subsequent vegetation productivity.

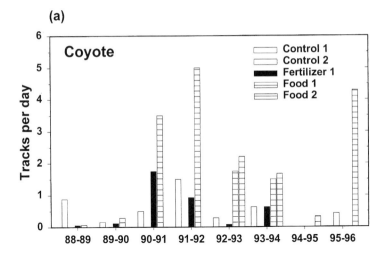

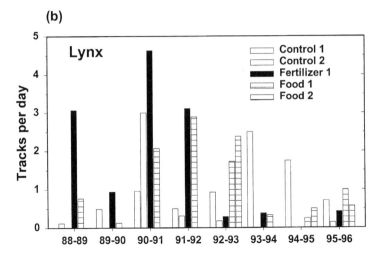

Figure 17.8 Average number of (a) coyote and (b) lynx tracks around experimental and control areas. Transects were 4 km in length and were done only within 24 h after fresh snowfall.

17.3.7   Vegetation Removal

In 1996 soils were sampled from 156 vegetation removal plots and 156 control sites with vegetation present immediately adjacent to them (table 17.1). The nitrate nitrogen value on the 156 removal plots was 6.58 ($\pm$0.41, 95% CL), compared to 5.2 ($\pm$0.34) on the control plots. The removal plots were significantly ($p < .0001$) higher in soil nitrogen compared to those with vegetation, indicating that herbaceous vegetation reduced the nitrogen content of soils.

Table 17.6 Mean snow track counts per day for lynx and coyote over the winter of peak hare numbers (1990–1991) and the mean over the eight winters 1988–1989 to 1995–1996 (95% CL).

	Lynx		*Coyote*	
Site	Peak	8-year Mean	Peak	8-year Mean
Control	0.96	0.76 (0.54)	0.50	0.50 (0.36)
Fertilizer	4.63	1.61 (1.24)	1.75	0.45 (0.44)
Food addition	2.89	1.15 (0.82)	3.50	1.51 (1.08)

17.4   Indirect Effects of Trophic-level Perturbations

17.4.1   Fertilizer Addition

Indirect effects of fertilizer addition would be experienced at the herbivore level. In 1990, the year of peak biomass, herbivore biomass on fertilizer grids was 4.22 kg/ha and that on controls 3.34 kg/ha, a small but significant difference. However, over all 8 years, there was no difference in herbivore biomass (fertilizer 2.73 kg/ha, control 2.48 kg/ha; figure 17.5, table 17.4), indicating that the effect of fertilizer was only apparent at high herbivore levels.

Double indirect effects would occur at the predator level. Predator tracks on fertilizer grids in the peak winter of 1990–1991 were 4.63/day for lynx and 1.75/day for coyote, compared to 0.96/day and 0.50/day, respectively, for controls (table 17.6). Thus, both predator species used fertilizer grids more than control areas in the peak year. Over the 8 years, there was again no significant difference in mean track counts for either species between fertilizer and control grids (figure 17.8). Thus, bottom-up indirect effects were observed at both higher levels of the food chain, but only when herbivore biomass was high.

17.4.2   Food Addition

The indirect effect of increasing herbivore numbers on the food addition grids would be seen at the soil nutrient level. In 1995 soil nitrate nitrogen on the food addition sites (2.7 ppm) was similar to that on controls (3.3 ppm; table 17.1). Therefore, there were no detectable indirect effects of herbivores on soil nutrients.

17.4.3   Predator Exclosure

The indirect effects from excluding predators would be observed at the vegetation level. In winter 1990–1991, browsing impact in the predator exclosure on birch was 51.5% versus 85.7% on controls, and on willow the values were 10.8% in exclosure and 12.9% on controls. Herbivory in the peak year, therefore, did not increase in the predator exclosure (table 17.5), and may even have declined.

Furthermore, willow growth in the subsequent summer of 1991 completely compensated for this browsing in both the predator exclosure and control (net change in biomass was 1.01 and 0.98, respectively). Birch growth, however, did not compensate for the higher browsing rate, and biomass declined in 1991 on both the predator exclosure (net change 0.56) and control (net change 0.16; figure 17.7, table 17.5).

In 1992, after herbivore declines, willows had significantly higher growth rates in the predator exclosure than on controls (16.1% vs. 12.0%, $p < .05$), but birch did not (21.1% vs. 23.9%; table 17.3). Thus, in general we did not detect indirect top-down changes in either willow or birch biomass resulting from the higher herbivore numbers on this predator exclosure treatment.

The double indirect effect would be detected at the soil nutrient level. In 1995 soil nitrate nitrogen was 0.37 ppm on the predator exclosure, which was not significantly different from the 0.18 ppm on controls (table 17.1). Hence, indirect effects resulting from higher herbivore biomass were not detected in soil nutrients.

17.4.4 Food Addition and Predator Exclosure

Indirect effects resulting from the high herbivore biomass would be seen at the soil nutrient level. In the exclosure in 1995, soil nitrate nitrogen was 0.85 ppm, while that on controls was 0.33 ppm. No significant differences were detected (table 17.1).

17.4.5 Hare Exclosure

In 1996 soil nitrate nitrogen averaged 8.34 (± 0.45, 95% CL) ppm, while outside the exclosure nitrogen averaged 5.2 (± 0.36) ppm, a difference significant at $p < .0001$ (table 17.1). A similar difference was detected in 1995 ($p < .02$). Therefore, the indirect effect of herbivores is to increase the uptake of nutrients into the vegetation to compensate for herbivory and so reduce the pool of nutrients in the soil. This result is due to a productivity response at the plant level.

17.5 Discussion

17.5.1 Direct Effects

Table 17.7 summarizes the significant direct effects on biomass derived from each of the experiments. Each of the removal experiments produced an increase in biomass at the level below. The two addition experiments (nutrients on fertilizer grids, food for herbivores on food addition grids) produced an increase in the level above, and for food addition a decrease in the level below. Because we were unable for practical reasons to increase the herbivores directly, we could not test the herbivore (H) → predator (P) direct interaction. However, we can see from the indirect effect of food addition, resulting in a higher herbivore biomass and a consequent higher use by predators on those sites (table 17.8), that there is a positive link from H to P. Therefore, putting these interactions together we see the pattern.

$$N \leftrightarrow V \leftrightarrow H \leftrightarrow P,$$

where N is nutrients, and V is vegetation. This is the pure reciprocal model 27 discussed

Table 17.7 Summary of qualitative direct effects on biomass produced by experimental perturbation at the peak values.

Site	Trophic Level				Predicted by Model 27			
	N	V	H	P	N	V	H	P
Fertilizer	M^+	$+$			M^+	$+$	$+$	$+$
Food addition		M^+-	$+$	$+$	$+$	M^+-	$+$	$+$
Predator removal			$+$	M^-	$+$	$-$	$+$	M^-
Predator removal + food addition		M^+--	$++$	M^-	$++$	M^+--	$++$	M^-
Herbivore removal		$+$	M^-		$-$	$+$	M^-	
Herbivore removal + fertilizer addition	M^+	$++$	M^-		?	$++$	M^-	
Vegetation removal	$+$	M^-			$+$	M^-		

M^+ indicates the experimental addition, M^- the experimental removal. $+$ = increase, $-$ = decrease. N = nutrients, V = vegetation, H = herbivores, P = predators.

in chapter 3, involving two-way interactions at each level. The predictions of this model are provided in table 17.7 for comparison with our results.

17.5.2   Indirect Effects

The above conclusion stems largely from direct effects derived from separate experiments at each trophic level. When we include indirect effects as well, we see other models are equally valid, but no models are entirely consistent with our results. Table 17.8 summarizes the indirect effects on biomass from each experiment. The nutrient addition produced small but positive effects at the peak of the cycle, but over the whole cycle these effects were not evident. In contrast, top-down indirect effects were not detectable despite the strong direct effects. One top-down effect (herbivore removal on soil nutrients [experiment 5]) even went in the direction opposite to that predicted—namely, an increase in the nutrient pool instead of a decrease. This result is most likely due to the fact that although there was less plant biomass, shrubs responded to herbivory by growing more (see below). More growth, therefore, reduced the nutrient pool relative to areas with no herbivory. In contrast, where herbivores were removed, there was less growth and the nutrient pool increased. This result from the nutrient pool illustrates that predictions from simple change of biomass are different from those due to change in productivity.

Table 17.8 presents the models for which the results of each experiment are consistent. Our findings do not match any model completely if we take them strictly at face value because of the anomalous result of herbivore removal on soil nutrients. If we recognize that this nutrient result was due to a productivity response of the vegetation and that there was a top-down effect as seen from the vegetation removal experiment 7 (table 17.7), then models 23, 21, and 27 are all supported.

In general, both top-down and bottom-up indirect effects tend to attenuate rapidly so that there is little effect of perturbations at the far end of the food chain.

Table 17.8 Summary of qualitative indirect effects on biomass.

Site	Observed				Predicted by Model 23				Predicted by Models	Total models
	N	V	H	P	N	V	H	P		
Fertilizer	M+	+	+	+	M+	+	+	+	1–4, 21, 23, 25, 27	8
Food addition	0	M+ –	+	+	0	M+ –	+	+	3, 23	2
Predator removal	0	0	+	M–	0	–	+	M–	4, 15–18, 25	6
Predator removal + food addition	0	M+ – –	++	M–	0	M+ – –	++	M–	20, 23	2
Herbivore removal	+	+	M–		0	+	M–		None	0
Direct Effects (table 17.7)										
Herbivore removal + fertilizer addition	M+	++	M–		M+	++	M–		3, 6, 10–12, 19–23, 26, 27	12
Vegetation removal	+	M–			0	M–			2, 5–9, 11, 13–17, 21, 22, 24–27	18

M+ indicates the experimental addition, M– the experimental removal. + = increase, – = decrease, 0 = no change. N = nutrients, V = vegetation, H = herbivores, P = predators.

17.5.3 Top-down versus Bottom-up

The joint perturbation experiments provide some clues on the relative strengths of top-down and bottom-up effects. In 1991, birch on the fertilized grids (with hares) had a higher growth rate than that on the fertilized hare exclosure by a factor of 1.23, but willow showed no difference in growth rate. This result suggests that the greater hare browsing on birch stimulated growth, but the lesser browsing on willow did not. In contrast, willow on the unfertilized hare exclosure had a much lower growth rate than that on the fertilized exclosure (by a factor of 0.62). This result suggests that in willow, fertilizer produced a stronger stimulation for growth than did hare browsing. Thus, both top-down and bottom-up effects could be detected.

The net effect of browsing and growth on fertilizer grids produced a decline of willow twig biomass by a factor of 0.78, whereas on control areas willow biomass barely declined at all (factor of 0.98). Browsing on birch was much more severe, so that in both fertilizer and control areas net biomass declined to 0.49 and 0.16 of that a year earlier. Thus, fertilizer may have increased growth rate, but it also increased herbivory to an even greater extent so that biomass declined. Therefore, top-down effects outweighed bottom-up effects at the vegetation level in winter. This result is opposite to the summer food situation, where herbivory had a negligible effect on vegetation relative to that of fertilizer addition (John and Turkington 1995).

In the winter of 1991–1992 predators had their most marked effect. Herbivore biomass on the food-addition site declined from spring 1991 to spring 1992 at an instantaneous rate of -1.137, compared to a decline of herbivore biomass of -0.062 on the food addition plus predator exclosure. The greater rate of decline of herbivores on the food addition is, therefore, due to the presence of predators. Furthermore, the rate of decline on control grids in the presence of predators but without the extra hare food (-0.667) was actually less than that on the food addition in the presence of food. Therefore, predation was the dominant process, and food had no detectable effect in mitigating the decline in herbivores. Thus, top-down effects dominated bottom-up effects at the herbivore level.

17.5.4 Productivity and Biomass Responses

Our main index of trophic-level response was through changes in biomass. However, productivity also responded to top-down effects. Did increased growth rate compensate for biomass loss? The hare exclosure, control areas, predator exclosure, food addition, and the food addition and predator exclosure produced a sequence of increasing herbivory in the peak winter of 1990–1991. Subsequent growth of both willow and birch was positively related to browsing intensity in the previous winter (regression for both species combined, % growth $= 13.03 + 0.216$ (% browsed), $n = 9$, $p < .005$; figure 17.9). At least at the vegetation level, winter food plants responded to top-down effects by an increased growth rate.

In no case, however, did this productivity response compensate entirely for herbivory, and in general the compensatory effect was relatively minor. However, at the next level down, soil nutrient content declined under conditions of herbivory (experiment 5) contrary to the predicted increase, a result consistent with the higher productivity of plants that experience herbivory drawing down the nutrient pool.

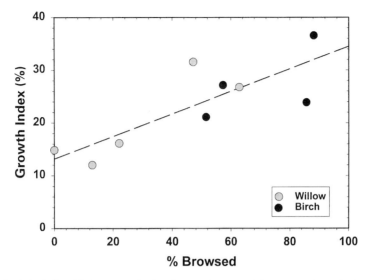

Figure 17.9 Relationship between shrub twig growth rate and the percent browsing in the previous winter for nine different sites in 1991, the year after the snowshoe hare peak.

17.5.5   Other Indirect Effects

Our indices of interaction effects are through changes in biomass over the whole trophic level. Gross biomass values necessarily obscure interactions between species in the same trophic level. In particular, our measures hide the effects of predators through apparent competition or alternative prey, these two processes having opposite results. In apparent competition (Holt 1977), a decrease in primary prey results in fewer predators and an increase in secondary prey. In the alternative prey scenario, a decrease in the main prey causes predators to switch to alternative prey and so cause a decrease in these species as well, though after a time lag.

In our study we have detected some of these indirect interactions. During the hare peak ground squirrel numbers increased (chapter 9), and during the hare decline ground squirrel numbers also declined. Thus, ground squirrels acted as alternative prey to hares. A similar sequence of events occurred with spruce grouse (chapter 11). These birds even reached peak numbers one year ahead of the hare cycle, suggesting that predators were focusing on the increasing hare population and so allowed grouse to reach high numbers.

Another type of indirect effect could arise from the increase in soil nutrients produced by the urine and feces of herbivores at high densities, a result sometimes detected in aquatic studies (Neill 1988). Our soil analyses on the food-addition experiments suggest a minor and nonsignificant increase in soil nutrients. The nitrogen values were low in both treatment and control areas, suggesting that the urine effect is minor relative to other impacts such as plant extraction on soil nutrients.

Exploitation competition is also an indirect effect. At the peak of the hare cycle, hares had little impact on their summer food plants (John and Turkington 1995), and so had little competitive effect on ground squirrels that often ate similar herbaceous species. Hares had a much larger impact on their winter food supplies (chapter 6; Smith et al. 1988). Ground squirrels were absent in winter, moose were very scarce, and red squirrels ate different foods. The only possible competitor, therefore, was willow ptarmigan, a species that eats gray willow buds. The decline in willow ptarmigan during the hare peak could be interpreted as exploitation competition during winter, but it was more likely the result of increased predation (i.e., apparent competition).

17.5.6 The Role of Secondary Chemicals

Secondary compounds in plants that act as antifeedants appear to respond directly to the influence of browsing by hares. The heavier the browsing, the higher the values of crude methanol extract, our index of secondary plant compounds. Thus, regrowth in the following summer acts as a compensatory response to browsing, and elevated secondary compound levels provide an additional deterrent to further browsing. There is experimental evidence that such extracts inhibit both diet selection and digestive abilities (Sinclair et al. 1982, 1988, Rodgers and Sinclair 1997). This effect is most apparent in bog birch and less so in gray willow. Because birch is the preferred species of winter food for hares, chemical defense is perhaps of higher value to this plant. However, where browsing was extremely high, as in the predator exclosure + food-addition treatment, not only was regrowth depressed but also secondary compounds were inhibited. Thus, heavily stressed plants become more prone to browsing.

Fertilizer had the effect of reducing the secondary compounds in both species, although the effect was not great. Our results are consistent with hypotheses proposing that secondary compounds may function to protect nutrients that are hard for plants to obtain (Coley et al. 1985). Where nutrients are provided, there is less stimulus for the plants to produce secondary compounds to defend the nutrients.

17.5.7 The Dominant Pathways in the Vertebrate Community

The Kluane boreal forest community is composed of several species of raptors, carnivores, and mammalian herbivores. In addition, there are herbivorous birds such as grouse and many species of granivorous or insectivorous birds. The herbivores feed on a variety of herbaceous plants in summer and several woody shrubs and trees in winter. There are trophic connections among several components between trophic levels. Despite this set of complex interactions, we find that there are only a few strong interactions that dominate the system. In general, a suite of raptor and carnivore species feed on a single dominant herbivore, snowshoe hare. This species, in turn, inhibits the growth of their main winter food plants—namely, gray willow and bog birch—and they stunt the growth of young white spruce trees. Other herbivores (ground squirrels and grouse species) remain relatively scarce, apparently limited by predators whose numbers are determined by snowshoe hares. These predators treat the less abundant herbivore species as alternative prey. Hares appear to be a keystone species in this ecosystem, maintaining the diversity of predators and herbivores. Removal of hares or a major disruption of the hare cycle may cause a profound change in the whole community.

17.6   Conclusion

We conducted either removal or addition experiments on each trophic level of the boreal forest ecosystem at Kluane, Yukon. We measured the response of other levels in terms of biomass, productivity, and activity. The direct effects of the perturbations on the next trophic level are consistent with the pure reciprocal model. The two experiments that produced simultaneous bottom-up and top-down effects indicated that top-down effects were stronger. In contrast to direct effects, indirect effects, although detectable, were relatively weak at all levels. The experiments suggest, therefore, strong reciprocal direct effects and weak, highly attenuated interaction chain indirect effects at all trophic levels in the Kluane boreal forest ecosystem.

Other types of indirect effects between species at the same trophic level are suggested from results in other chapters. Herbivores such as ground squirrels and willow ptarmigan may be secondary prey for predators living primarily on hares. However, there is little evidence for exploitation competition.

17.7   Summary

We used an experiment in the boreal forest of Canada to test predictions of instantaneous changes to trophic levels and distinguish between competing models of trophic level interactions. Seven different perturbations systematically removed or supplemented trophic levels. The predictions resulting from the perturbations were concerned with the direction of change in biomass at the other levels. The direct effects of each perturbation produced strong top-down and bottom-up changes in biomass. At both the vegetation and herbivore levels, top-down effects were stronger than bottom-up, despite some compensatory growth stimulated by herbivory. The combination of experiments produced results consistent with two-way (reciprocal) interactions at each level (model 27; chapter 3). Indirect effects on one or two levels removed from the perturbation were either very weak or undetectable. Top-down effects were strong when direct but attenuated quickly. Bottom-up effects were less strong but persisted as indirect effects to higher levels. Other types of indirect effects between species at the same trophic level were suggested from results in other chapters. Herbivores such as ground squirrels and willow ptarmigan may be alternative prey for predators living primarily on hares. However, there is little evidence for exploitation competition.

Literature Cited

Abrams, P. A. 1993. Effect of increased productivity on the abundances of trophic levels. American Naturalist **141**:351–371.

Arditi, R., and L. R. Ginsburg. 1989. Coupling in predator-prey dynamics: ratio-dependence. Journal of Theoretical Biology **139**:311–326.

Arditi, R., L. R. Ginsburg, and H. R. Akcakaya. 1991. Variation in plankton densities among lakes: a case for ration-dependent predation models. American Naturalist **138**:1287–1296.

Batzli, G. O., R. G. White, S. F. Maclean, F. A. Pitelka, and B. D. Collier. 1980. The herbivore based trophic system. *in* J. Brown, P. C. Miller, L. L. Tiezen, and F. L. Bunnell (eds). An arctic ecosystem: the coastal tundra at Barrow, Alaska, pages 335–410. Dowden, Hutchinson and Ross, Stroudsburg, Pennsylvania.

Benndorf, J., and W. Horn. 1985. Theoretical considerations on the relative importance of food limitation and predation in structuring zooplankton communities. *in* W. Lampert (ed). Food limitation and the structure of zooplankton communities, pages 383–396. E. Schweizerbart'sche Verlagsbuchhandlung, Stuttgart.

Billick, I., and T. J. Case. 1994. Higher order interactions in ecological communities: what are they and how can they be detected? Ecology **75**:1529–1543.

Boonstra, R., D. Hik, G. R. Singleton, and A. Tinnikov. 1998. The impact of predator-induced stress on the snowshoe hare cycle. Ecological Monographs **68**:371–394.

Caughley, G., and J. H. Lawton. 1981. Plant-herbivore systems. *in* R. M. May (ed). Theoretical ecology, 2nd ed., pages 132–166. Blackwell Scientific, Oxford.

Caughley, G., and A. Gunn. 1993. Dynamics of large herbivores in deserts: kangaroos and caribou. Oikos **67**:47–55.

Coley, P. D., J. P. Bryant, and F. S. Chapin III. 1985. Resource availability and plant antiherbivore defense. Science **230**:895–899.

Dungan, M. L. 1987. Indirect mutualism: complementary effects of grazing and predation in a rocky intertidal community. *in* W. C. Kerfoot and A. Sih (eds). Predation: direct and indirect impacts on aquatic communities, pages 188–200. University Press of New England, Hanover, New Hampshire.

Hairston, N. G., F. E. Smith, and L. B. Slobodkin. 1960. Community structure, population control and competition. American Naturalist **94**:421–425.

Hanski, I. 1991. The functional response of predators: worries about scale. Trends in Ecology and Evolution **6**:141–142.

Hawkins, B. A. 1992. Parasitoid-host food web and donor control. Oikos **65**:159–162.

Hik, D. 1995. Does risk of predation influence population dynamics? Evidence from the cyclic decline of snowshoe hares. Wildlife Research **22**:115–129.

Holt, R. D. 1977. Predation, apparent competition, and the structure of prey communities. Theoretical Population Biology **12**:197–229.

Hunter, M. D., and P. W. Price. 1992. Playing chutes and ladders: bottom-up and top-down forces in natural communities. Ecology **73**:724–732.

John, E., and R. Turkington. 1995. Herbaceous vegetation in the understory of the boreal forest: does nutrient supply or snowshoe hare herbivory regulate species composition and abundance? Journal of Ecology **83**:581–590.

Krebs, C. J., R. Boonstra, S. Boutin, M. Dale, S. Hannon, A. R. E. Sinclair, J. N. M. Smith, and R. Turkington. 1992. What drives the snowshoe hare cycle in Canada's Yukon? *in* D. M. McCullough and R. Barrett (eds). Wildlife 2001: populations, pages 886–896. Elsevier, London.

Krebs, C. J., S. Boutin, R. Boonstra, A. R. E. Sinclair, J. N. M. Smith, M. R. T. Dale, K. Martin, and R. Turkington. 1995. Impact of food and predation on the snowshoe hare cycle. Science **269**:1112–1115.

Lampert, W. (ed). 1985. Food limitation and the structure of zooplankton communities. E. Schweizerbart'sche Verlagsbuchhandlung, Stuttgart.

Leibold, M. A. 1996. A graphical model of keystone predators in food webs: trophic regulation of abundance, incidence, and diversity patterns in communities. American Naturalist **145**:784–812.

McQueen, D. G., M. R. S. Johannes, J. R. Post, T. J. Stewart, and D. R. S. Lean. 1989. Bottom-up and top-down impacts on freshwater pelagic community structure. Ecological Monographs **59**:289–309.

McQueen, D. G., J. R. Post, and E. L. Mills. 1986. Trophic relationships in freshwater pelagic ecosystems. Canadian Journal of Fisheries and Aquatic Sciences **43**:1571–1581.

Menge, B. A. 1992. Community regulation: under what conditions are bottom-up factors important on rocky shores? Ecology **73**:755–765.

Menge, B. A. 1997. Detection of direct versus indirect effects: were experiments long enough? American Naturalist **149**:801–823.

Menge, B. A., and J. P. Sutherland. 1976. Species diversity gradients: synthesis of the roles of predation, competition and temporal heterogeneity. American Naturalist **110**:351–369.

Neill, W. E. 1988. Complex interactions in oligotrophic lake food webs: responses to nutrient enrichment. *in* S. R. Carpenter (ed). Complex interactions in lake communities, pages 31–44. Springer-Verlag, New York.

Oksanen, L., S. D. Fretwell, J. Arruda, and P. Niemala. 1981. Exploitation ecosystems in gradients of primary productivity. American Naturalist **118**:240–261.

Pace, M. L., and E. Funke. 1991. Regulation of planktonic microbial communities by nutrients and herbivores. Ecology **72**:904–914.

Polis, G. A., and D. R. Strong. 1997. Food web complexity and community dynamics. American Naturalist **147**:813–846.

Power, M. E. 1984. Depth distributions of armoured catfish: predator-induced resource avoidance? Ecology **65**:523–528.

Power, M. E. 1992. Top-down and bottom-up forces in food webs: do plants have primacy? Ecology **73**:733–746.

Reichardt, P. B., J. P. Bryant, T. P. Clausen, and G. D. Wieland. 1984. Defense of winter-dormant Alaska paper birch against snowshoe hares. Oecologia **65**:58–69.

Rodgers, A. R., and A. R. E. Sinclair. 1997. Diet choice and nutrition of captive snowshoe hares (*Lepus americanus*): interactions of energy, protein, and plant secondary compounds. Ecoscience **4**:163–169.

Schmitz, O. J. 1998. Direct and indirect effects of predation and predation risk in old-field interaction webs. American Naturalist **151**:327–342.

Sinclair, A. R. E., C. J. Krebs, and J. N. M. Smith. 1982. Diet quality and food limitation in herbivores: the case of the Snowshoe hare. Canadian Journal of Zoology **60**:889–897.

Sinclair, A. R. E., C. J. Krebs, J. N. M. Smith, and S. Boutin. 1988. Population biology of snoeshow hares III. Nutrition, plant secondary compounds and food limitation. Journal of Animal Ecology **57**:787–806.

Sinclair, A. R. E., and J. N. M. Smith. 1984. Do plant secondary compounds determine feeding preferences of snowshoe hares? Oecologia **61**:403–410.

Smith, J. N. M., C. J. Krebs, A. R. E. Sinclair, and R. Boonstra. 1988. Population biology of snowshoe hares II. Interactions with winter food plants. Journal of Animal Ecology **57**:269–286.

Strong, D. R. 1992. Are trophic cascades all wet? Differentiation and donor-control in speciose ecosystems. Ecology **73**:747–754.

Turkington, R., E. John, C. J. Krebs, M. Dale, V. O. Nams, R. Boonstra, S. Boutin, K. Martin, A. R. E. Sinclair, and J. M. N. Smith. 1998. The effects of NPK fertilization for nine years on the vegetation of the boreal forest in northwestern Canada. Journal of Vegetation Science **9**:333–346.

White, T. C. R. 1978. The importance of a relative shortage of food in animal ecology. Oecologia **3**:71–86.

White, T. C. R. 1984. The abundance of invertebrate herbivores in relation to the availability of nitrogen in stressed food plants. Oecologia **63**:90–105.

Wootton, J. T. 1993. Indirect effects and habitat use in an intertidal community: interaction chains and interaction modifications. American Naturalist **141**:71–89.

Wootton, J. T. 1994a. Predicting direct and indirect effects: an integrated approach using experiments and path analysis. Ecology **75**:151–165.

Wootton, J. T. 1994b. The nature and consequences of indirect effects in ecological communities. Annual Review of Ecology and Systematics **25**:443–466.

Wootton, J. T. 1994c. Putting the pieces together: testing the independence of interactions among organisms. Ecology **75**:1544–1551.

18

Vertebrate Community Structure in the Boreal Forest

Modeling the Effects of Trophic Interaction

DAVID CHOQUENOT, CHARLES J. KREBS, A. R. E. SINCLAIR,
RUDY BOONSTRA, & STAN BOUTIN

M odels of biological communities have employed two general approaches: structural analysis of communities based on food webs (e.g., Paine 1980, Pimm 1982, 1991, Cohen 1989) and mechanistic models based on analysis of trophic interactions (e.g., Caughley 1976, 1977, Fretwell 1977, Tilman 1982, see chapter 17). Although both approaches have been used to understand community organization, mechanistic models generally provide more insight into the forces that influence community stability (Tilman 1982, Lawton and McGarvin 1986). Analyses of changes in biomass and productivity associated with experimental perturbation of the boreal forest (described elsewhere in this book) were used to test the nature of trophic-level interactions in this ecosystem described in chapter 17. The effect of trophic-level interactions on community organization were interpreted in terms of the role top-down (predator control) and bottom-up (nutrient limitation) processes played in changing the biomass or productivity of given trophic levels. The emphasis this approach places on biomass and productivity changes subsumes the more specific influence that trophic mechanisms can have on the maintenance of the species complexes within different trophic levels. For example, the observation that predation significantly reduces the biomass of herbivores clearly indicates that top-down processes are an important influence on community dynamics in this ecosystem. What is less clear is the role predation may play in facilitating coexistence among herbivores.

18.1 Trophic Interaction and Species Coexistence

Simple mechanistic models predict that where two or more species compete for limiting resources, the species with the highest rate of resource consumption or with the ability to monopolize resource access will competitively exclude all others (Gause 1934). From this basic premise, most theoretical and empirical studies of multiple species coexisting in communities have focused on identifying the circumstances under which coexistence can occur (Anderson and Kikkawa 1986, Yodzis 1993). A range of mechanisms that facilitate coexistence in communities have been identified (Tilman 1982, Tilman and Pacala 1993). These mechanisms can be divided into those that invoke the effect of limiting agents (predators, pathogens or parasites) or those that invoke limiting resources. Limiting agents facilitate coexistence where the susceptibility of competitors to a limiting agent varies temporally or spatially (Connell 1983), or where the agent constrains the density of competitors below levels where the resources for which they compete become limiting (Schoener 1986, Tilman and Pacala 1993). Limiting resources facilitate coexistence where the rate at which they are supplied to each competing species varies spatially or temporally (Tilman 1982, Tilman and Pacala 1993). The relative influence of limiting agents and limiting resources on community structure and function reflect the relative influence of top-down and bottom-up processes.

A central theme of the research described in this book was that experimental perturbation of factors influencing the abundance of snowshoe hares (a keystone species in the boreal forest) would elucidate the relative influence top-down (predation) and bottom-up (food resources) processes had on the structure and stability of the vertebrate community inhabiting the boreal forest at Kluane (chapters 1, 3). The analyses described in chapter 17 focused on what these perturbations revealed about the relative importance of top-down and bottom-up processes for changes in biomass and productivity at different trophic levels.

In the boreal forest, vertebrates occupying at least two trophic levels (herbivores and carnivores) coexist despite potential competition for food resources. In this chapter we construct simple models of trophic interaction between some of the dominant species in the boreal forest to explore potential mechanisms of coexistence for herbivores and carnivores. In particular, we focus on coexistence of snowshoe hares and ground squirrels as potential competitors for summer vegetation, and of lynx, coyote, and great horned owls as competitors for available prey during the cyclic low in snowshoe hare abundance. Mechanisms of coexistence are interpreted in terms of the influence of the top-down and bottom-up influences described in chapter 17.

18.2   Interaction between Vegetation and Herbivores

Snowshoe hares and ground squirrels use some of the same food resources over summer. However, although this provides some potential for competition, exclusion of one or other species does not occur because (1) the food resources that limit snowshoe hare and ground squirrel populations are different, or (2) predation keeps the density of both species below levels where they compete for food resources. (A third possibility is that social regulation limits the density of both species below levels where they compete for food. However, there is little empirical evidence that either hares or ground squirrels are intrinsically regulated, and this hypothesis is not explored further in this chapter.) These hypotheses are not mutually exclusive. Predation may hold snowshoe hare and ground squirrel densities below certain levels where they are food limited, regardless of whether or not they compete for the same food resource. For example, the analyses described in chapter 17 suggest that herbivores in the boreal forest are influenced by the reciprocal effects of top-down and bottom-up processes, corresponding to the simultaneous or sequential influence of predation and food limitation.

The potential for competition between snowshoe hares and ground squirrels will largely depend on how food availability at different times of the year influences their demographic rates. During the time ground squirrels are active, they rely primarily on summer herbs for nutrition (chapter 9). When preferred herbs are unavailable, ground squirrels will consume grasses and willow, upon which adults can survive but not reproduce (T. Karels personal communication). Hence, in the absence of predation, the availability of summer herbs appears to limit ground squirrel populations. In contrast, snowshoe hares remain active year-round. Snowshoe hares rely on the terminal twigs of shrubs over winter but consume a range of forbs, grasses, and the leaves of shrubs during the summer (chapter 8). Although considerable work has been undertaken on winter feeding by snowshoe hares (chapter 8), summer feeding patterns are less well understood. Both severe browsing effects and mobilization of secondary plant compounds for shrubs over winter have been observed during the cyclic peak in snowshoe hare densities. This has led to the general assumption that any food limitation acting on snowshoe hare populations will be related to shrub availability over winter (Pease et al. 1979, Bryant 1981, Fox and Bryant 1984, Sinclair et al. 1988, Smith et al. 1988). However, in this study, although the biomass of terminal twigs showed a cyclic decline related to snowshoe hare browsing, there was no evidence that the general availability of winter food ever fell to levels that would limit reproduction or survival (chapter 8). Hence, in the absence of predation, it is not known whether shrubs over winter or herbs over summer would eventually limit snowshoe hare populations.

While bottom-up effects may be important for general patterns of variation in herbivore biomass, seasonal patterns in these effects may have important consequences for herbivore species composition. For example, because ground squirrels do not feed over winter, potential competition with snowshoe hares is largely restricted to consumption of herbs over summer. If the availability of shrubs had more influence on predation-free snowshoe hare populations than the availability of herbs, the scope for competition with ground squirrels would be limited. However, if the availability of summer herbs had more influence on predation-free snowshoe hare populations, the potential for competition with ground squirrels would be greater. To evaluate whether snowshoe hares and ground squirrels would persist in the absence of predation, we developed a model of their interaction with summer and winter food resources. If predation is necessary to facilitate coexistence of snowshoe hares and ground squirrels, in its absence we should be unable to maintain both herbivores within the modeled system.

18.3   Interaction between Herbivores and Predators

The three dominant predators in the boreal forest are lynx, coyote and great horned owls (chapters 13, 15). All three predators share snowshoe hares as their most common prey, and cyclic changes in the density and demography of these predators demonstrate that they are fundamentally limited by snowshoe hare abundance during the low-phase of the snowshoe hare cycle. Hence, the role bottom-up processes play in structuring the predator community in the boreal forest is self-evident. Similarly, the clear association between cyclic changes in the abundance of snowshoe hares and the demography of lynx, coyote, and great horned owls suggest that temporal rather than spatial variation in the rate at which snowshoe hares are supplied to these predators facilitates their coexistence. This implies that the increase in prey abundance following the low-phase of the snowshoe hare cycle must occur regularly enough (i.e., every 9–11 years) that lynx, coyote, and great horned owls avoid competitively excluding one another. The cyclic increase in snowshoe hare abundance appears necessary for predator coexistence, despite the diversification of predator diets during the low-phase of the snowshoe hare cycle. For example, although lynx were to some extent able to substitute alternative prey (especially red squirrels) when snowshoe hare availability was low (1992–1995; see chapter 13), the abundance of lynx continued to decline over this period. Hence, although predators could potentially reduce competition during the cyclic low-phase in snowshoe hare density by broadening their diet, they could not sufficiently alleviate the effect that low snowshoe hare availability had on their demography to halt their ongoing decline. It follows that if the low-phase of the snowshoe hare cycle were protracted, competition between these three predators would intensify, and their continuing coexistence would become less certain.

18.4   Modeling

The models described in this chapter are contained in a single Excel spreadsheet on the CD-ROM (frame 71, multiprey accounting model) that accompanies this book (Community.xls). The spreadsheet allows the structure of the model to be explored interactively and the scenarios described in this chapter to be set up and varied. Excel's accounting tools (which allow cell precedents and dependents to be traced) are particularly useful in tracking how constants and variables are related throughout these models.

All models ran for 100 years, with a basic time step of 6 months dividing each year into winter and summer periods. The abundance of all animal species could be switched between densities based on field estimates or densities predicted by the model.

18.4.1 Interaction between Vegetation and Herbivores

Models that linked the dynamics of snowshoe hares and ground squirrels to that of summer vegetation (herbs) and year-round vegetation (shrubs), used the functional and demographic responses of each herbivore to changes in herb and shrub biomass (figure 18.1). Both herbs and shrubs grew logistically, values for maximum instantaneous growth rate (herbs: 0.5, shrubs: 0.15) and biomass at carrying capacity (herbs: 200 kg/ha, shrubs: 1500 kg/ha) being selected to approximate seasonal productivities estimated in chapters 5 and 6 (figure 18.2). Herbs were available and grew only over the 6 months designated as summer. Shrubs also grew over summer but were available year-round.

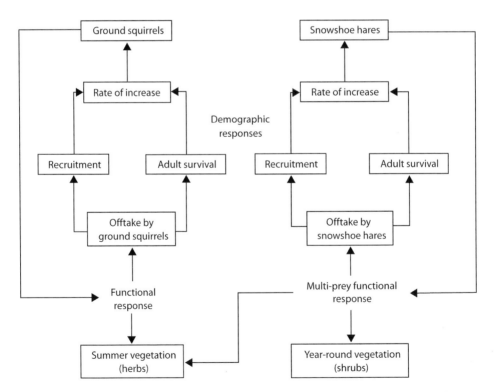

Figure 18.1 Structure of the model linking vegetation to snowshoe hare and ground squirrel population dynamics.

(a)

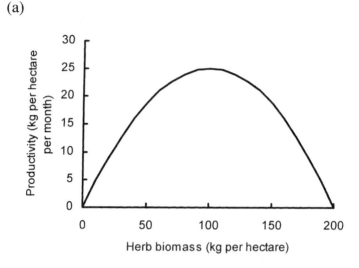

(b)

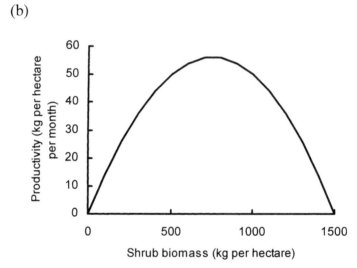

Figure 18.2 Productivity of (a) herbs and (b) shrubs as used in the model linking vegetation to snowshoe hare and ground squirrel population dynamics.

Ground squirrels consumed herbs only, with per capita offtake (I_s) being related to herb biomass (s) by a functional response described using Holling's (1959) disc equation (figure 18.3a):

$$I_s = \frac{(a_s s)}{1 + a_s h_s s)}, \tag{1}$$

where a_s is the rate of effective search of ground squirrels for herbs, and h_s is handling time, which is equal to the time each ground squirrel takes to consume and digest each

(a)

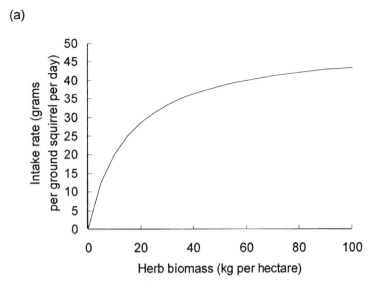

(b)

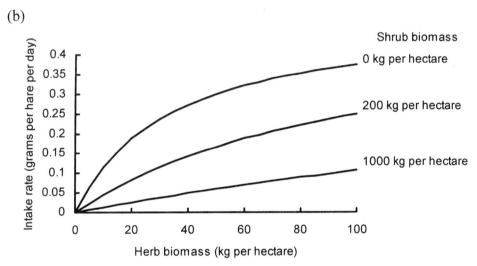

Figure 18.3 Functional responses of (a) ground squirrels and (b) snowshoe hares consuming summer herbs. The range of responses shown for snowshoe hares shows the effect of increasing shrub availability on herb offtake.

kilogram of herbs it finds. Handling time was estimated as the reciprocal of maximum herb intake rate, which was assumed to be 0.3 kg per ground squirrel per day ($h_s = 3.33$). Rate of effective search was derived by using observed ground squirrel densities in the model, then varying a_s until a plausible pattern of variation in herb biomass was achieved. This gave a rate of effective search of $a_s = 0.02$.

Snowshoe hares consumed both herbs and shrubs, with respective offtake related to the instantaneous availability of each by functional responses described using a multiple-prey version of the disc equation (Caughley and Sinclair 1994; figure 18.3b). The functional responses for herbs and shrubs were:

$$I_s = \frac{(a_s h_s s)}{1 + [(a_s h_s s) + (a_w h_w w)]} \tag{2}$$

and

$$I_w = \frac{(a_w h_w w)}{1 + [(a_w h_w w) + (a_s h_s s)]}, \tag{3}$$

where a_s and a_w are the rates of effective search by snowshoe hares for herbs and shrubs, respectively, and h_s and h_w are the handling times, which are equal to the time taken to consume and digest each kilogram of herbs and shrubs found.

Schmitz et al. (1992) estimated maximum shrub intake by snowshoe hares at 0.5 kg per hare per day. It was assumed that maximum intake of herbs (i.e., the maximum intake of herbs if no shrubs were available) would be the same, giving a value of 2 for both h_s and h_w. As with rate of effective search for ground squirrels, a_s and a_w were derived by using observed snowshoe hare densities in the model, then varying each until plausible patterns of variation in herb and shrub biomass were achieved. This gave rates of effective search of $a_s = 0.015$ and $a_w = 0.005$. The relative values for rates of effective search by ground squirrels and snowshoe hares implies that ground squirrels are more efficient foragers for herbs than are snowshoe hares and that snowshoe hares forage more efficiently for herbs than they do for shrubs. Vegetation growth and offtake were accounted monthly for each sequential summer and winter phase (see the worksheets "Herbivore responses" and "Vegetation" of Community.xls on the CD-ROM for details).

We simulated the demography of both snowshoe hares and ground squirrels by linking variation in per capita recruitment (R) and adult survival (S) in the absence of predation to average food intake over each 6-month period. Average rate of food intake over each 6-month period was expressed as a proportion of maximum possible intake to calculate a food intake deficit. For both ground squirrels and snowshoe hares, recruitment and adult survival were related to food intake deficit using an exponential saturation curve that had the form:

$$D = D_{max}(1 - e^{-FDd}), \tag{4}$$

where the demographic rate of interest is D (i.e., recruitment or adult survival), its maximum potential rate is D_{max}, food intake deficit is FD, and d is a measure of demographic efficiency describing the relative effect a decline in food intake has on the demographic rate. The parameter measuring demographic efficiency is a unit-less index that simply controls how rapidly declining food intake rate reduces either recruitment or adult survival. As its value increases, any decrease in food intake (measured here as a change in food intake deficit) will have a less dramatic effect on recruitment or survival. As its value decreases, any decrease in food intake will have a more dramatic effect on recruitment or survival.

Variation in recruitment and adult survival with food intake deficit for snowshoe hares and ground squirrels is shown in figure 18.4a and b, respectively. Maximum recruitment for ground squirrels ($R_{max} = 1.86$) was estimated from maximum litter size and weaning

(a)

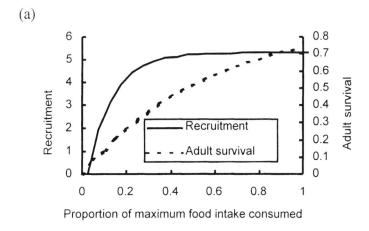

Proportion of maximum food intake consumed

(b)

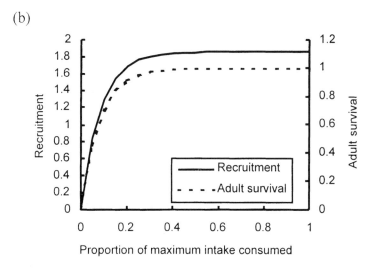

Proportion of maximum intake consumed

Figure 18.4 Demographic responses of (a) snowshoe hares and (b) ground squirrels with de-creasing food intake deficit.

rates on the predator exclosure grid to exclude any indirect effects of predation on repro-duction (chapter 9). Maximum adult survival for ground squirrels over summer was as-sumed to be high (S_{max} = 1), given the low mortality unrelated to predation. Because over-winter survival of ground squirrels is independent of food availability, the variable summer rate was replaced with a constant rate estimated from field data (D = 0.66). We estimated maximum recruitment of snowshoe hares (R_{max} = 5.30) from the product of maximum fecundity and maximum juvenile survival. Maximum adult survival for snow-shoe hares year-round was assumed to be high (S_{max} = 0.9), given low mortality unre-lated to predation (chapter 8). Recruitment and adult survival over each sequential 6-month period of summer and winter were combined in an additive model that predicted instantaneous rates of change (r) in ground squirrel and snowshoe hare abundance (see the worksheet "Herbivore responses" in Community.xls on the CD-ROM for additional details).

18.4.2 Equilibrium Conditions for Snowshoe Hares and Ground Squirrels

To determine whether it was possible for ground squirrels and snowshoe hares to persist indefinitely in the absence of predation (i.e., whether their food resources were sufficiently different to avoid competitive exclusion), we varied relative demographic efficiencies (d values for recruitment and adult survival) for each species independently and together, and changes in their respective densities and biomass of their food resources assessed. For each species, the relative efficiencies for recruitment and adult survival were interchangeable over a wide range of values. Hence, to explore equilibrium conditions for snowshoe hares, adult survival efficiency was set to 0.8 while recruitment efficiency was varied; and for ground squirrels, adult summer survival efficiency was set to 3 while recruitment efficiency was varied. By varying recruitment efficiencies for each species, the models changed the sensitivity of reproduction and juvenile survival (the product of which is recruitment) to fluctuations in food availability, such that declining food intake had a greater or lesser effect on rates of population change.

When snowshoe hares were modeled alone, recruitment efficiencies between 2 and 2.6 produced an equilibrium with food resources, the equilibrium being approached through dampening oscillations at higher values within this range (figure 18.5). At lower recruit-

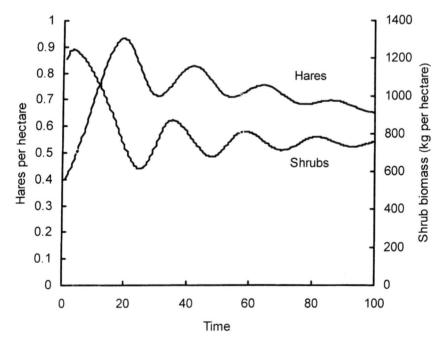

Figure 18.5 Reciprocal variation in snowshoe hares and shrubs when only hares are included in the model. Adult survival efficiency is set to 0.8 and recruitment efficiency to 2.5.

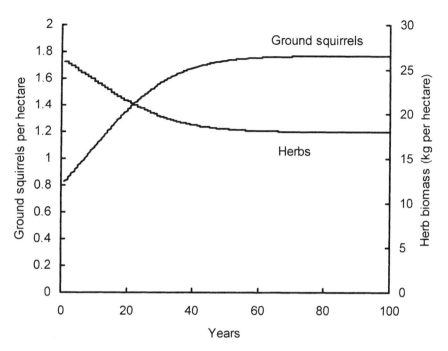

Figure 18.6 Reciprocal variation in ground squirrel and herbs when only ground squirrels are included in the model. Adult survival efficiency is set to 3 and recruitment efficiency to 4.

ment efficiencies, snowshoe hares underwent deterministic extinction, shrub biomass moved to a specified maximum, and herb biomass waxed and waned through the sequence of summers and winters. Between recruitment efficiencies of 2.7 and 3.3, snowshoe hares and their food resources became locked into stable limit cycles, the amplitude of cycles increasing with recruitment efficiency. Above a recruitment efficiency of 3.3, snowshoe hares underwent a sequence of increasing eruptions until their food resources collapsed and they crashed to extinction. The only stable scenarios in which snowshoe hares approached maximum densities seen on control and predator exclosure grids were those producing stable limit cycles of higher amplitude (i.e., those brought about by recruitment efficiencies >3).

When ground squirrels were modeled alone, all recruitment efficiencies >3 produced stable equilibria between squirrels and their food resources. The average density of equilibria increased with recruitment efficiency until efficiency exceeded 6, when ground squirrel density stabilized at an average of about 3/ha (figure 18.6). At recruitment efficiencies <3, ground squirrels underwent deterministic extinction.

When snowshoe hares and ground squirrels were modeled together, coexistence proved possible over a broad range of recruitment efficiencies. However, the range of efficiencies that produced stable outcomes was increased over those when the two species were modeled by themselves. For example, if ground recruitment efficiency for ground squirrels was set to 6.5, the range of efficiencies required for snowshoe hares to achieve some form of equilibrium with their food resources increased from 2–3.3 to 2.2–3.3. This

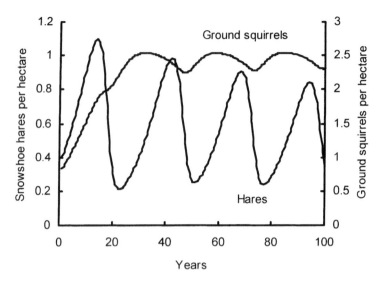

Figure 18.7 Variation in snowshoe hare and ground squirrel density when both are included in the model. Adult survival efficiencies are set to 0.8 and 3 for snowshoe hares and ground squirrels, respectively, and recruitment efficiencies to 3 and 6.6, respectively.

suggests that competition from ground squirrels for summer food resources required snowshoe hare populations to have higher net productivity than necessary for equilibrium in the absence of ground squirrels. Snowshoe hares had a reciprocal competitive effect on ground squirrels, their presence (at a recruitment efficiency of 3) pushing up the range of recruitment efficiencies ground squirrels require for equilibrium from values around 3 to 3.5. Evidence of competition for food resources can be seen in the reciprocal oscillation of snowshoe hares and ground squirrels over a broad range of recruitment efficiencies for both species (figure 18.7).

In the absence of predation, snowshoe hares and ground squirrels appear theoretically capable of coexisting, despite potential competition for summer food resources. Coexistence is consistent with the limiting food resources of each species being sufficiently different to avoid competitive exclusion. This implies that bottom-up processes not only influence herbivore biomass but may also be important for herbivore coexistence.

For example, the degree to which snowshoe hares and ground squirrels compete in these models is determined by the relative efficiency with which each species can use herbs to enhance rates of increase and the effect shrub availability has on the degree to which snowshoe hares will consume herbs as their biomass declines. We set foraging efficiencies in the model so that ground squirrels were better foragers for herbs than snowshoe hares, but so that snowshoe hares preferred herbs to shrubs when both were available. The capacity of snowshoe hares to use shrubs year-round effectively buffers this system from more intense competition between snowshoe hares and ground squirrels for herbs over summer. The effect of this buffering can be seen by reducing the maximum rate of shrub growth, forcing the two herbivores to compete more aggressively for summer food (figure 18.8). This reduces shrub productivity below levels that can sustain viable

(a)

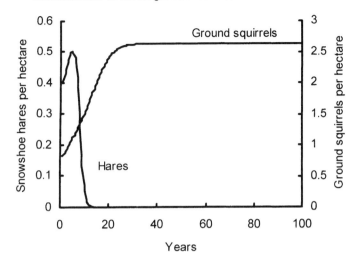

(b)

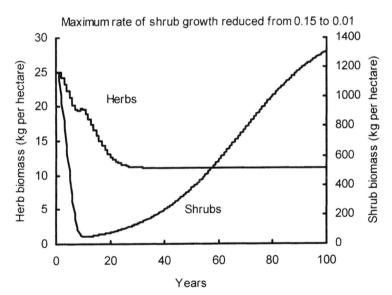

Figure 18.8 The effect of reducing the maximum rate of shrub growth on competition between snowshoe hares and ground squirrels in the absence of predation. Ground squirrels competitively exclude snowshoe hares because their use of herbs is more efficient. Once snowshoe hares are excluded, (a) shrubs recover back toward their maximum biomass and herbs reach equilibrium with ground squirrels, and (b) snowshoe hare and ground squirrel recruitment efficiencies are set to 3 and 6.6, respectively.

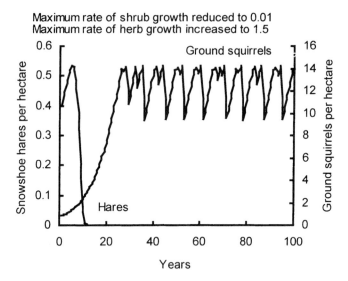

Figure 18.9 The effect of reducing the maximum rate of shrub growth and increasing the maximum rate of herb growth on competition between snowshoe hares and ground squirrels in the absence of predation. Ground squirrels competitively exclude snowshoe hares despite significantly higher productivity at the next trophic level down, vegetation.

snowshoe hare populations, forcing snowshoe hares to compete directly with ground squirrels for available herbs. Because ground squirrels have a more efficient functional response than snowshoe hares, the herb biomass that allows them to generate positive rates of increase is lower than that for snowshoe hares. Hence, herb offtake by ground squirrels reduces herb biomass to levels where snowshoe hares cannot sustain themselves in the modeled system. Even when the productivity of herbs is increased dramatically (maximum herb growth rate tripled to 1.5), competition from ground squirrels excludes hares from the modeled system (figure 18.9).

18.4.3   Adding Predation to Models of Interaction between Herbivores and Vegetation

We added estimated rates of offtake by predators (chapters 8 and 9) to the model of interaction of snowshoe hares and ground squirrels with their food resources. Estimated levels of predation could not be sustained by either snowshoe hares or ground squirrels at demographic efficiencies that produced equilibria between herbivores and their food resources. Increasing the demographic efficiencies of snowshoe hares and ground squirrels allowed them to sustain increasing levels of predation offtake. However, at most, 70% and 40% of estimated predation could be sustained by snowshoe hares over winter and summer, respectively, and 60% of estimated predation could be sustained by ground squirrels over summer. This suggests that either the estimates of predation obtained overstated actual predation rates, or the productivity of the vegetation–herbivore model was too low at some basic level. Productivity of the vegetation–herbivore model is largely determined by vegetation growth rates and maximum herbivore recruitment. Productivity levels for shrubs and maximum recruitment rates for snowshoe hares and ground squirrels were well documented. However, few data were available with which to estimate rates of growth in summer vegetation. Similarly, potential effects of resource-dependent dispersal or habitat use were not included in the model despite both being postulated as important factors linking predation, food availability, and herbivore demography (chapters 8 and 9).

Notwithstanding the limited capacity of the vegetation–herbivore model to sustain observed rates of predation offtake, the general patterns of variation predicted for snowshoe hares and ground squirrels appear plausible. For example, increasing recruitment efficiency for snowshoe hares to a value of 9 and adult survival efficiency to 1.7 produced a cycle in snowshoe hare abundance that mirrors that on control sites (figure 18.10). Similarly, increasing both recruitment and adult survival efficiencies for ground squirrels to a value of 12 allowed them to persist through regular cyclic declines, albeit at a lower density than observed (figure 18.11).

Increasing the demographic efficiency of herbivores reduces the level of food intake they require to generate positive rates of increase. This, in effect, reduces the influence that variation in food availability has on demographic rates. However, in reducing the influence food has on herbivore demography, increased demographic efficiency negates the potential for stable interaction between herbivores and vegetation. For example, if predation offtake is removed from the model that produced regular cycles in snowshoe hare

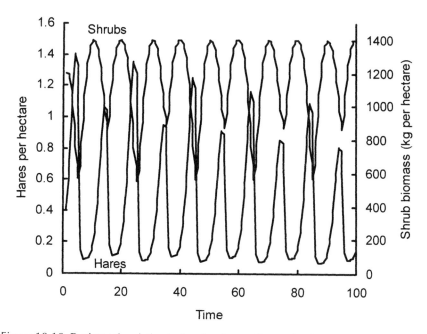

Figure 18.10 Reciprocal variation in the abundance of snowshoe hares and shrubs with 70% of observed predation rates over winter and 40% of predation rates over summer imposed on the model of interaction between herbivores and vegetation. Snowshoe hare recruitment efficiency is set to 9 and adult survival efficiency to 1.7.

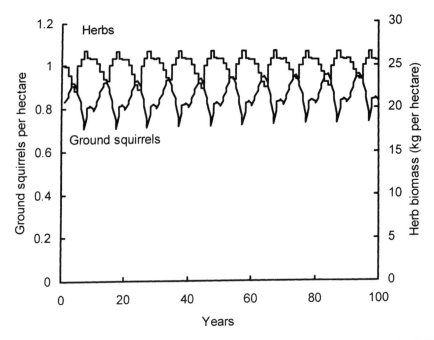

Figure 18.11 Reciprocal variation in the abundance of ground squirrels and herbs with 60% of observed predation rates over summer imposed on ground squirrels. Ground squirrel recruitment efficiency and adult survival efficiency are both set to 12.

abundance (figure 18.10), snowshoe hares erupt to densities where they decimate their food resources before crashing to extinction. This suggests that the demographic efficiencies necessary to sustain the high levels of predation to which snowshoe hares are typically exposed are incompatible with any sort of stable interaction between snowshoe hares and their food resources. This may not be surprising considering that the life history of snowshoe hares has evolved under continuous pressure from predators rather than absolute food shortage and that of the plants they exploit have evolved under conditions where excessive herbivory is apparently rare. Under these conditions, top-down processes would be expected to be more influential on community structure than would bottom-up processes.

18.4.4 The Effect of Herbivore Cycles on Predators

The model that linked observed changes in snowshoe hare and ground squirrel density to the three numerically dominant predators (lynx, coyote, and great horned owls) used the functional and demographic responses of each predator to variation in prey abundance (figure 18.12). We modeled functional responses using the same multiple-prey disc equation used to predict snowshoe hare diet (equations 2, 3; see worksheet "Predator functional responses" of Community.xls on the CD-ROM for additional details and parameter estimates). We used ground squirrels as a surrogate for all non-snowshoe hare prey exploited by these predators. Maximum intake rates and rates of effective search for snowshoe hares

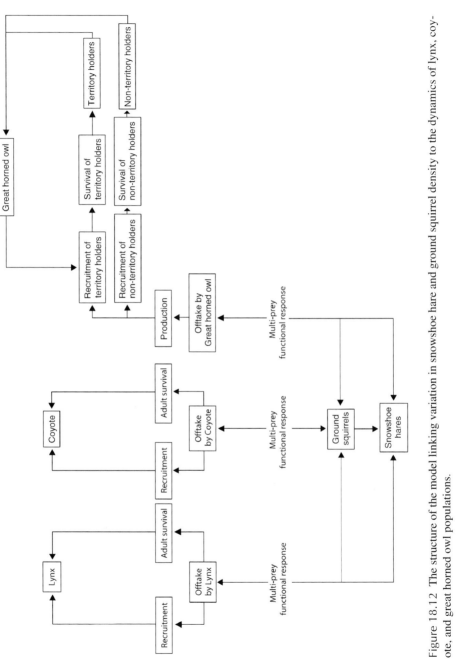

Figure 18.12 The structure of the model linking variation in snowshoe hare and ground squirrel density to the dynamics of lynx, coyote, and great horned owl populations.

were taken from chapter 13 for lynx and coyote and from chapter 15 for great horned owl. We estimated maximum intake rates for ground squirrels by assuming that 15% (by occurrence) of the saturated intake rate of each predator over summer was taken as ground squirrels (see chapters 13 and 15), that 90% of each ground squirrel was consumed, and that ground squirrels averaged 500 g body weight. Rates of effective search for ground squirrels were given values substantially less than those estimated for snowshoe hares to imply that snowshoe hares, when available, were strongly preferred by all predators.

We simulated predator demography by linking variation in per capita recruitment and adult survival to average food intake over each 6-month period, using estimates of food deficit and the same general exponential saturation function used to simulate the dynamics of snowshoe hare and ground squirrel populations. The exponential function used to link recruitment (R) to food intake deficit (FD) was modified to the form:

$$R = R_{max}[1 - e^{-(FD - FD_{min})d}], \qquad (5)$$

where R_{max} is maximum per capita recruitment, d is a measure of demographic efficiency describing the relative effect a decline in food intake has on recruitment, and FD_{min} is the minimum food intake deficit for any reproductive activity to occur. Addition of a minimum food intake deficit allows the apparent threshold in food intake necessary for predator reproduction to be simulated. Recruitment and adult survival over each sequential 6-month period of summer and winter were combined in an additive model predicting instantaneous rate of change (r) for lynx and coyote density. Because the great horned owl population was separated into territory holders and nonterritory holders, a more complex population model was developed. Only territory holders bred, with offspring recruited into either the territory holder or non–territory-holder subpopulations depending on the ratio of prevailing territory holder density to maximum territory holder density (the maximum observed density of territory holders reported in chapter 15). The two subpopulations were subject to different rates of adult survival, with nonterritory holders having higher maximum density but a minimum food deficit below which survival fell to 0.

Predator population models were fitted by varying their component demographic efficiencies until good visual agreement between predicted and observed densities was obtained. In all cases, the demographic efficiencies of population models could be varied to predict predator densities that corresponded closely with estimated densities (figure 18.13). To evaluate the effect that the duration of the low-phase in prey abundance had on the degree of competition between predators, we added four 6-month periods (2 years) to the cycle of observed snowshoe hare and ground squirrel densities, at the point where snowshoe densities reached their minimum observed level (i.e., 0.065 hares/ha in spring 1994). Extending the low-phase of prey abundance resulted in the extinction of lynx and coyote over five protracted cycles and slightly reduced peak densities of non–territory-holding great horned owls (figure 18.14). This outcome is consistent with competition between predators intensifying as the period over which prey are in short supply is extended.

The effect that duration of the low phase in cyclic prey populations has on the potential for coexistence among predators is a specific example of a nonequilibrium outcome for a normal competition model (Tilman 1982). The fact that great horned owls are able to tolerate increases in the duration of the low phase in cyclic prey abundance clearly indicates that they are competitively superior to lynx and coyote. It follows that lynx and coyote are only able to coexist in the normally cycling system because they have a greater

(a)

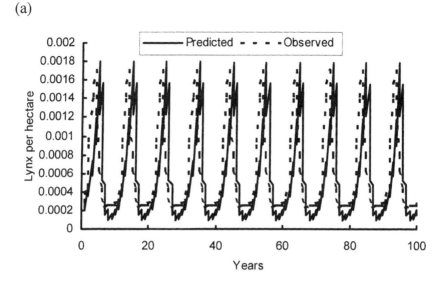

(b)

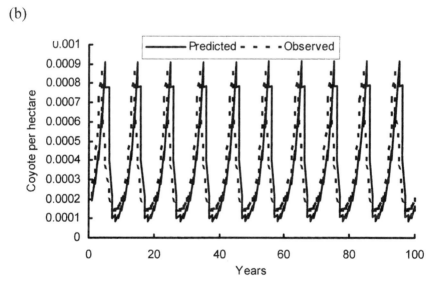

Figure 18.13 Predicted and observed variation in densities of (a) lynx, (b) coyote, and (c) predicted and (d) observed variation in the density of territory holding and non–territory-holding great horned owls.

(c)

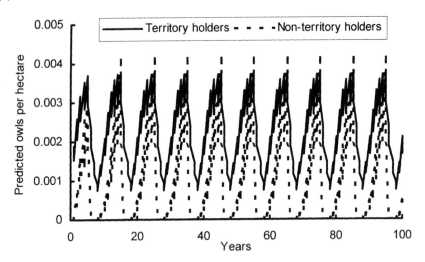

(d)

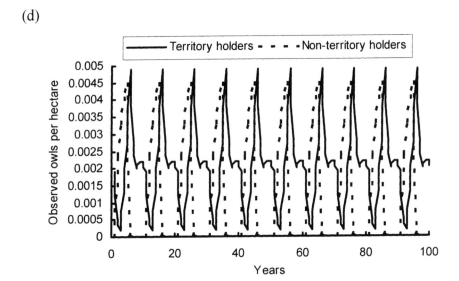

Figure 18.13 *continued*

capacity for increase than do great horned owls when prey are not limiting. The competitive superiority of great horned owls reflects the ability of reproductively active individuals (territory holders) to monopolize resources throughout the prey cycle and their conservative reproductive strategy relative to their mammalian counterparts during the cyclic low in prey availability (chapter 15). Extending the low-phase in cyclic prey abundance to 6 years results in great horned owl extinction over seven protracted cycles, indicating that even these tactics have their limitations.

(a)

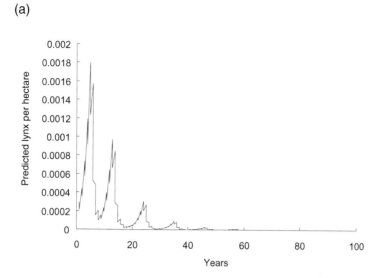

(b)

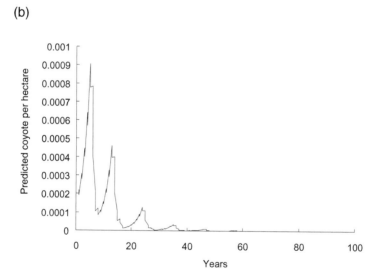

Figure 18.14 The effect on predators of adding 2 years to the low-phase of snowshoe hare and ground squirrel population cycles.

(c)

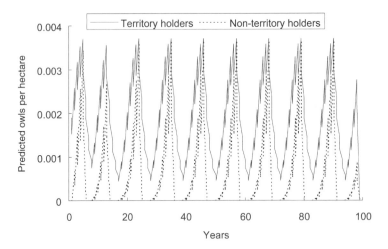

Figure 18.14 *continued*

18.5   Discussion

18.5.1   Unresolved Modeling Issues

The models described in this chapter explored the effect observed rates of predation had on predicted interaction among snowshoe hares, ground squirrels, and their food resources and the effect that observed variation in snowshoe hare and ground squirrel density had on predicted variation in the abundance of their predators. To link the dynamics of predators and prey in a full simulation of interaction among vegetation, herbivores, and predators, observed rates of predation and observed variation in prey abundance would have to be replaced with their predicted counterparts. Despite considering a complete range of demographic efficiencies for both herbivores and carnivores, we were unable to identify parameter values that allowed this replacement while maintaining stable outcomes for predators, prey, and vegetation. This suggests that either parameter values used in other parts of the model are inaccurate or that some important stabilizing mechanisms are missing from the model.

Many parameters used in the model had to be guessed (e.g., rates of effective search for predators consuming ground squirrels), derived indirectly (e.g., demographic efficiencies for all species), or modified from their original values (e.g., predation rates on snowshoe hares and ground squirrels). Hence, although the model is structurally accurate, it is likely that many of the parameter values it uses are not. Similarly, although the model includes the important trophic interactions that influence the structure of the modeled species (i.e., functional and demographic responses), other mechanisms that may enhance community stability by modifying these interactions have been omitted. In particular, the suite of prey species exploited by predators as an alternative to snowshoe hares have been replaced with ground squirrels as a surrogate. Likewise, the tendency for predators to prey upon each other when snowshoe hare and ground squirrel abundance was low has been ignored. Both of these mechanisms would tend to stabilize the interaction between predators and their primary prey by modifying the efficiency of functional and demographic responses at critical phases of the cycle in predator density.

Another potentially important omission in the models described here is the lack of spatial behavior by predators or prey. Large-scale dispersal patterns can to some extent be encompassed by the demographic rates of the species they effect, but smaller scale shifts in habitat preference cannot. These sorts of behaviors may be important influences on the susceptibility of prey to predators and on the efficiency with which predators and prey can use available food resources. Hence, by not explicitly or implicitly representing these effects, the models described may be ignoring important sources of stability in the dynamics of vertebrates in this ecosystem.

It may be feasible to build all or any of this additional complexity into the existing models, but this would require estimates of a whole suite of additional parameters for which few data are available. For example, if predation efficiency is influenced by habitat, the disc equations used in the models described here would need to be modified so that changes in habitat use associated with different phases of the snowshoe hare cycle could be represented. This would be a complex undertaking (particularly for the multispecies equations used for snowshoe hare foraging and for predators) that would be made more difficult when any influence of habitat on predation efficiency varied between prey species. However, regardless of how complex such an undertaking may be, it would be essential if habitat-related constraints on predation efficiency was an important influence on community structure and stability in the boreal forest. This suggests that the view that simple models of interaction between a population and its food resources can be "piled-up" to consider more complex questions about communities may be naïve (Pimm 1991). Where indirect processes such as those related to habitat effects are an important influence on community structure and function, more complex mechanistic models would need to be developed. The data required to parameterize and validate such models will be difficult and expensive to acquire, and the mathematical framework within which to use these parameters will be difficult. For example, the sensitivity of the simple models described here to small shifts in parameter values reflects the relatively coarse time-step we have used (6 months). However, although a continuous-time modeling approach would provide more stability, complex mathematical approaches would be needed to adapt this approach to the seasonal nature of the modeled ecosystem (e.g., La Place transformations; Renshaw 1991). For most real communities, the mathematical complexity required to accommodate the full range of biotic and abiotic processes that influence structure and function rapidly outstrips the capacity of biologists to supply sufficient data. This does not mean that modeling cannot contribute greatly to our understanding of community dynamics, but rather that modeling should focus on the mechanisms that influence the structure of community components, not on the structure of the community as a whole. In essence, this is what most mechanistic models of biological communities have attempted (Tilman 1982).

18.5.2   Mechanisms of Coexistence

Although the analysis of stability for co-occurring snowshoe hares and ground squirrels suggested that their food resources were sufficiently different that they could coexist, the demographic efficiencies used in these models were substantially lower than required to sustain observed levels of predation. When demographic efficiencies were increased to levels where snowshoe hares and ground squirrels could cope with observed levels of predation, their interaction with food resources became unstable. Hence, in the absence of predators, it seems probable that snowshoe hares and ground squirrels would compete fiercely for available food resources and that one species could exclude the other. Under these circumstances, the coexistence of snowshoe hares and ground squirrels would be mediated by predation. Predator-mediated coexistence is considered a widespread phenomenon among phytophagous insects, but it has been less commonly demonstrated among herbivorous mammals (Lawton and MacGarvin 1986). However, recent analysis of predator–prey and vegetation–herbivore interactions in several mammalian communities suggests that predation is important for facilitating the coexistence of at least some mammalian herbivores (Moen et al. 1993, McLaren and Peterson 1994, Sinclair 1995).

The way in which trophic interaction influences competition and coexistence among the three modeled predators is more straightforward. All three predators are extremely reliant on snowshoe hares; none is able to generate positive rates of increase when snowshoe hare abundance is low. The important difference between these predators is the rate at which they decline during the low phase of cyclic snowshoe hare density and the rate at which they increase when snowshoe hare density recovers. The results of the modeling summarized here suggest that great horned owls have the lowest rates of decline over the low phase, but also the lowest rate of increase after snowshoe hare density recovers. This places great horned owls at a competitive advantage over lynx and coyote during the low phase but at a disadvantage over the period of high snowshoe hare availability. The net result of this trade-off is that, although great horned owls will have the same reciprocal cycle in density as lynx or coyote, their cycles will tend to be of lower amplitude. This places them at less risk of localized stochastic effects that may affect their density or that of their principle prey directly. As such, great horned owls may be less reliant than lynx or coyote on dispersal as a means of maintaining their populations throughout the boreal forest.

18.5.3   Top-down and Bottom-up Influences
on Community Structure

The analyses described in chapter 17 did not specifically consider the complexity that nonlinear interactions between trophic levels would introduce to models describing the influence that different trophic levels exert over each other in the boreal forest (i.e., the saturating functional and demographic responses used in the models described here). However, it was argued that these complications would not alter the qualitative conclusions drawn from the simple linear models of interaction used in those analyses. Collectively, the modeling results for herbivores and their interaction with food resources and predators described in this chapter support the reciprocal effects model advocated in chapter 17. Although predation clearly dominates changes in the biomass of herbivores in the boreal forest, food limitation has an important if less dramatic influence on the dynamics of snowshoe hares and ground squirrels.

Literature Cited

Anderson, D. J., and J. Kikkawa. 1986. Development of concepts. *in* J. Kikkawa and D. J. Anderson (eds). Community ecology: pattern and process, pages 3–16. Blackwell Scientific Publications, Oxford.

Bryant, J. P. 1981. Phytochemical deterrance of snowshoe hare browsing by adventitious shoots of four Alaskan trees. Science **213**:889–890.

Caughley, G. 1976. Wildlife management and the dynamics of ungulate populations. *in* T. H. Coaker (ed). Applied biology, vol. 1, pages 183–246. Academic Press, London.

Caughley, G. 1977. Analysis of vertebrate populations. John Wiley and Sons, New York.

Caughley, G., and A. R. E. Sinclair. 1994. Wildlife Ecology and Management. Blackwell Scientific, Boston.

Cohen, J. E. 1989. Food webs and community structure. *in* J. Roughgarden, R. M. May, and S. A. Levin (eds). Perspectives in ecological theory, pages 181–202. Princeton University Press, Princeton, New Jersey.

Fox, J. F., and J. P. Bryant. 1984. Instability of the snowshoe hare and woody plant interaction. Oecologia **63**:128–135.

Fretwell, S. D. 1977. The regulation of plant communities by food chains exploiting them. Perspectives in Biology and Medicine **20**:169–185.

Gause, G. F. 1934. The struggle for existence. Repr. 1971. William and Wilkins, New York.

Holling, C. S. 1959. The components of predation as revealed by a study of small-mammal predation of the European pine sawfly. Canadian Entomologist **91**:293–320.

Lawton, J. H., and M. MacGarvin. 1986. The organization of herbivore communities. *in* J. Kikkawa and D. J. Anderson (eds). Community ecology: pattern and process, pages 163–186. Blackwell Scientific, Oxford.

McLaren, B. E., and R. O. Peterson. 1994. Wolves, moose and tree rings on Isle Royale. Science **266**:1555–1557.

Moen, J., H. Gardfjell, L. Oksanen, L. Ericson, and P. Ekerholm. 1993. Grazing by food-limited microtine rodents on a productive experimental plant community: does the "green desert" exist? Oikos **68**:401–413.

Paine, R. T. 1980. Food webs: linkage, interaction strength and community infrastructure. Journal of Animal Ecology **49**:667–685.

Pease, J. L., R. H. Vowles, and L. B. Keith. 1979. Interaction of snowshoe hares and woody vegetation. Journal of Wildlife Management **43**:43–60.

Pimm, S. L. 1982. Food webs. Chapman and Hall, London.

Pimm, S. L. 1991. The balance of nature? University of Chicago Press, Chicago.

Renshaw, E. 1991. Modelling biological populations in space and time. Cambridge University Press, Cambridge.

Schmitz, O. J., D. S. Hik, and A. R. E. Sinclair. 1992. Plant chemical defense and twig selection by snowshoe hare: an optimal foraging approach. Oikos **65**:295–300.

Schoener, T. W. 1986. Resource partitioning. *in* J. Kikkawa and D. J. Anderson (eds). Community ecology: pattern and process, pages 91–126. Blackwell Scientific, Oxford.

Sinclair, A. R. E. 1995. Population limitation of resident herbivores. *In* A. R. E. Sinclair and P. Arcese (eds). Serengeti: dynamics, management and conservation of an ecosystem, pages 194–219. University of Chicago Press, Chicago.

Sinclair, A. R. E., C. J. Krebs, J. N. M. Smith, and S. Boutin. 1988. Population biology of snowshoe hares. III. Nutrition, plant secondary compounds and food limitation. Journal of Animal Ecology **57**:787–806.

Smith, J. N. M., C. J. Krebs, A. R. E. Sinclair, and R. Boonstra. 1988. Population biology of snowshoe hares. II. Interactions with winter food plants. Journal of Animal Ecology **57**:269–286.

Tilman, D. 1982. Resource competition and community structure. Princeton University Press, Princeton, New Jersey.

Tilman, D., and S. Pacala. 1993. The maintenance of species richness in plant communities. *in* R. E. Ricklefs and D. Schluter (eds). Species diversity in ecological communities: historical and geographical perspectives, pages 13–25. University of Chicago Press, Chicago.

Yodzis, P. 1993. Environment and trophodiversity. *in* R. E. Ricklefs and D. Schluter (eds). Species diversity in ecological communities: historical and geographical perspectives, pages 26–38. University of Chicago Press, Chicago.

19

Trophic Mass Flow Models of the Kluane Boreal Forest Ecosystem

JENNIFER L. RUESINK & KAREN E. HODGES

The Kluane ecosystem project has generated detailed demographic portraits for many species that inhabit the boreal forest. The experimental manipulations have allowed us to test the impacts of food and predation on the population densities and demography of a number of herbivorous species, as well as to compare the strength of bottom-up versus top-down interactions among trophic levels (chapter 17). At Kluane we documented cycles occurring not only in species' abundances, but also in survival, reproduction, and diet composition. A striking general result from the Kluane studies is the huge shift in biomass in the forest throughout the 10-year cycle. Not only does overall herbivore biomass change more than threefold, but also the amount of the herbivore biomass represented by snowshoe hares varies from about 65% at the peak down to about 10% during the low phase (figure 19.1; CD-ROM frame 70). Predator biomass shows a similar overall shift, although no individual predator fluctuates as much in its proportional representation (figure 19.2). The dynamics of this multispecies food web may change substantially over the snowshoe hare cycle as a result of these biomass shifts.

In this chapter, we present a series of food web models to explore the impacts of these changes in biomass on the trophic interactions within the Kluane ecosystem. Specifically, we examine how the connections in this food web, as indicated by biomass flows, change through the cycle, between seasons, and when portions of the food web are manipulated experimentally. We concentrate on the interactions among vertebrates rather than exploring plant–herbivore interactions in any depth, largely because changes in plant biomass are less dramatic than the vertebrate responses due to the large reservoir of biomass in the woody material of trees and shrubs. We therefore restrict our examination of plant–herbivore interactions to analyzing the estimated total biomass of plant material necessary for the offtake by herbivores. For herbivore–predator interactions, we examine biomass flow on a per species basis. The two model outputs of total offtake of plant biomass and species-specific mass balance (i.e., the amount of a species consumed relative to its production) allow us to identify both strong interactions and portions of the food web in need of further study.

The Kluane ecosystem was chosen for an ecosystem study partially for its tractability, in that there are only a handful of species at each trophic level compared with the species diversity in other forested ecosystems. But even with detailed demographic information for many of the Kluane vertebrates, the ecosystem dynamics are not immediately obvious. The approximately 60-fold difference in individual body mass among herbivores (i.e., hares of ~1500 g and microtines of ~25 g) means that even identical demography or abundances of these two groups would have very different implications for their forage plants and their predators. This is because the species eat different plants at different rates, and predators differentially prey on species of varying sizes. Exploring trophic interactions using total biomass instead of abundance does not overcome problems of body size because metabolic rate scales nonlinearly with body size across species (Peters 1983).

To solve these problems, taxonomic uniqueness must be maintained while using a framework with a single and comparable currency. In our food web models, we convert every demographic change into a change in biomass for that species, and predator–prey interactions are similarly modeled as intake of biomass of each herbivore species per unit of predator biomass. In this fashion, demographic changes are all brought into a single

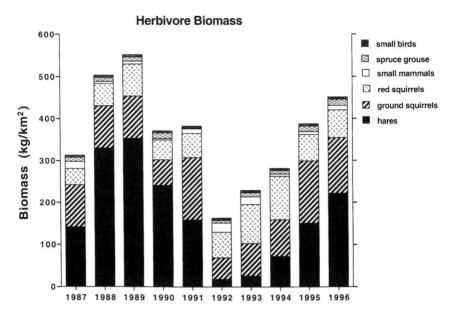

Figure 19.1 Biomass of herbivores in the boreal forest. Snowshoe hares accounted for 11–66% of total herbivore biomass, and ground squirrels 17–39%.

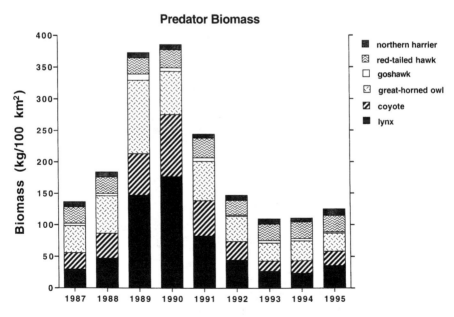

Figure 19.2 Biomass of predators in the boreal forest. Proportional biomass of lynx ranged from 21% to 46%, coyotes 15% to 26%, and great horned owls 18% to 32%.

currency that automatically weights the relative contribution of a species or of a demographic change to the biomass in the entire ecosystem.

19.1   Questions about the Ecosystem Dynamics of the Boreal Forest

As the previous chapters in this book have illustrated, most species at Kluane have extremely variable reproduction and survival, whether related to the hare cycle (e.g., coyotes, lynx, great horned owls, spruce grouse, ground squirrels) or independent of the hare cycle (e.g., red squirrels, microtine rodents). Some species also vary through the cycle in feeding rate and diet composition. This complexity in the ecosystem would allow an infinite series of explorative models. We focused on five main questions that address the large ecosystem questions related to the snowshoe hare cycle and the experimental manipulations.

1. How do trophic interactions change through the four phases of the cycle?
2. What are the impacts of food addition on the trophic interactions?
3. What are the impacts of predator reduction on the trophic interactions?
4. What are the effects of running models from spring to spring versus autumn to autumn?
5. How do the model outputs change when the models exclude one dominant species (snowshoe hares) or many that are rare, small, or noncyclic (most herbivores and raptors)?

The rationale for exploring the cyclic phases is obvious: fluctuations in biomass, relative biomass, and demography could alter the extent to which biomass offtake is balanced by production. We examine the impacts of food addition and predator removal because of the impacts these manipulations had on the demography of several herbivorous species, most notably snowshoe hares and ground squirrels (chapters 8, 9). Food addition primarily affected herbivore production, and its influence should vary by phase of the cycle because the hare peak and low correspond to times when food should be most and least limiting, respectively. Predator exclosure primarily affected depredation of herbivores, and its influence also should vary with cyclic phase, with predation least and most severe during the increase and decline phases, respectively.

We explore seasonality because of the risk that different initial conditions could affect model outputs. Most production (reproduction and growth) occurs in summer, whereas most consumption occurs in winter because winter spans 8 months of the year. Thus, for the species for which we have data from two annual censuses, autumn biomass exceeds spring biomass. The yearly change in biomass also varies by season. For example, snowshoe hare densities (and therefore biomasses) were similar in the spring estimates of 1993 and 1994, but the autumn estimates of 1992, 1993, and 1994 varied considerably from one another.

Finally, given the drama of the snowshoe hare cycle, we explore models of simpler food webs to determine how integral snowshoe hares and other species are to the dynamics observed. We examine models with hares as the only herbivores together with three abundant and cyclic predators. We also explore models in which snowshoe hares are excluded. The results from these reduced-web models should indicate how robust the model outputs of plant offtake and mass balance are to the presence of snowshoe hares, rare and noncyclic predators, and alternative prey species.

19.2   Modeling the Kluane System

To address these five questions, we used a series of four core models that we compared to each other and to the specialized models (table 19.1). The core models use herbivore data from control sites, typically averaged from all control sites that were studied for each species. These core models and most of the specialized models were run using autumn to autumn time periods because the autumn models start immediately after the production of the summer, therefore beginning at the annual maximum in biomass.

To address the impacts of food addition (+ food), we constructed four models, one for each of the four cyclic phases. Similarly, for predator reduction (− predator), we constructed models for each phase. For these treatment models, we varied ground squirrel and hare parameters but kept the other herbivore parameters the same as in the comparable control models. Doing this allowed us to focus on the species that were most impacted demographically by the experimental manipulations. To address seasonality, we constructed a control model that ran from spring to spring instead of from autumn to autumn for each cyclic phase.

Finally, we constructed two types of models of partial food webs using control data. In the first, we excluded all vertebrates except hares, lynx, coyotes, and great horned owls (minimum web models). The cyclic fluctuation of hares dominates the ecosystem dynamics of the boreal forest, and these three predators are all year-round residents that have hares as a dominant part of their diet. We considered including ground squirrels because they were typically the second-largest component of herbivore biomass and they were responsive both to the hare cycle and to the experimental treatments. But whereas hare cycles occur continentally throughout the boreal and sub-boreal forests of North America, ground squirrels are present in boreal forests only in the relatively small region of northern British Columbia, southern Yukon, and a small portion of Alaska. These stripped-down control models therefore are representative of the "typical" cyclic boreal forest species. The second set of partial web models excluded snowshoe hares but retained all other vertebrates as in the control full-web models (no-hare web models).

Because the experimental manipulations were not large enough to affect predator demography, we used the same year-specific values for raptors (goshawks, great horned owls) and mammalian predators (lynx, coyotes) in all models. For snowshoe hares and ground squirrels, both of which showed clear population cycles, we used both treatment- and year-specific variables. For the other herbivorous species and for Harlan's hawks and northern harriers, we used long-term averages because these species were more constant year-to-year (red squirrels, small birds, both raptors), rare (spruce grouse), or had unpredictable dynamics (small rodents).

19.3   Methods for Constructing Mass Balance Models

To construct and analyze mass balance models of the Kluane food web, we used Ecopath, a Windows-based software program developed at the International Center for Living Aquatic Resources Management, based on the original model by Polovina (1984)

Table 19.1 The models constructed for exploring vertebrate trophic interactions at Kluane.

Model	Variable Addressed	Difference from Core Model
Core Models		
Control peak (autumn 1988–89)	Cyclic phase	—
Control decline (autumn 1990–91)	Cyclic phase	—
Control low (autumn 1992–93)	Cyclic phase	—
Control increase (autumn 1994–95)	Cyclic phase	—
Specialized Models		
Spring peak (1989–90)	Season	Beginning season
Spring decline (1991–92)	Season	Beginning season
Spring low (1993–94)	Season	Beginning season
Spring increase (1995–96)	Season	Beginning season
+Food peak	Food	Hare and ground squirrel biomass, production
+Food decline	Food	Hare and ground squirrel biomass, production
+Food low	Food	Hare and ground squirrel biomass, production
+Food increase	Food	Hare and ground squirrel biomass, production
−Predator peak	Predators	Hare biomass, ground squirrel biomass and production, no lynx or coyote
−Predator decline	Predators	Hare biomass, ground squirrel biomass and production, no lynx or coyote
−Predator low	Predators	Hare biomass, ground squirrel biomass and production, no lynx or coyote
−Predator increase	Predators	Hare biomass, ground squirrel biomass and production, no lynx or coyote
Minimum food web peak	Trophic complexity	Only hare, lynx, coyote, great horned owl
Minimum food web decline	Trophic complexity	Only hare, lynx, coyote, great horned owl
Minimum food web low	Trophic complexity	Only hare, lynx, coyote, great horned owl
Minimum food web increase	Trophic complexity	Only hare, lynx, coyote, great horned owl
No-hare web peak	Trophic complexity	No hares
No-hare web decline	Trophic complexity	No hares
No-hare web low	Trophic complexity	No hares
No-hare web increase	Trophic complexity	No hares

All specialized models except the spring models were run from autumn to autumn of the years given for the core models.

(www.ecopath.org; Christensen and Pauly 1993, Pauly and Christensen 1995). Ecopath models require that four of five basic parameters be entered for each taxon. In general, biomass (B), production per biomass (P/B), consumption per biomass (Q/B), and diet composition are inputs, and matrix inversion algorithms in Ecopath determine the attendant ecotrophic efficiency (EE). These parameters are linked by the following equation:

$$B_r(P_r/B_r)(EE_r) = [\Sigma(B_c)(Q_c/B_c)(d_{rc})] + \Delta B_r, \tag{1}$$

where r refers to parameters of the resource (species or guild), c to parameters of each consumer (species or guild), and d_{rc} is the proportion of r in the diet of c. Increases in resource biomass from one time to the next (bioaccumulation, or ΔB_r) can, along with consumption of the resource, be a fate for resource production.

The ecotrophic efficiency of a taxon compares the consumption and accumulation of biomass to the production of biomass. An EE $>$ 1 indicates a problem: more consumption and bioaccumulation is occurring than is biologically possible given the production. High EE values therefore indicate either that the standing biomass is declining or that the model is an inadequate representation of the system. In our models, we initially included biomass increases (as a fate of production) but not biomass declines. For taxa showing biomass declines that also have EE $>$ 1, we can then determine whether the excess consumption matches the empirically observed declines in biomass. An EE $\cong$ 1 indicates that a species' dynamics are largely driven by consumption, whereas low values of EE indicate that production has a greater impact on the dynamics. When EE $<$ 1, there is more production than bioaccumulation and consumption combined; this excess is assumed to become detritus. Low EEs therefore suggest potential for population increase because little of the production is being used in other parts of the food web.

We modeled the plant–herbivore interactions simplistically. We assumed that all herbivores ate from the same stock of plant material. To estimate offtake, we set EE = 1 for plants so that total consumption would equal production; we then set plant production equal to the standing stock. Ecopath solved for standing stock, thus generating our estimate of offtake.

Sources for demographic and trophic information about the Kluane boreal forest ecosystem include published studies and, whenever possible, data presented in previous chapters of this book. The species and interactions we modeled are shown in figure 19.3. General sources for demographic and dietary parameter values can be found in table 19.2. Below, we outline the transformations required to fit the demographic and dietary data into the mass balance ecosystem models.

19.3.1 Parameterization of Mass Balance Models

Biomass   Density estimates were transformed from number per area into biomass per area by multiplying density and individual body mass for each taxon. For all taxa except hares and ground squirrels, these body masses were assumed to be constant throughout the cycle and for all treatments. Our density estimates of the different species come from different time periods throughout the year because density estimates were made at biologically relevant times for each species. We used separate autumn and spring estimates for hares and ground squirrels for each year. Ground squirrel densities were uncertain be-

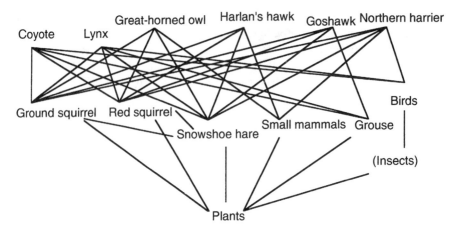

Figure 19.3 Taxa in the Kluane boreal forest food web that were included in the Ecopath models. Lines show dominant trophic relationships (>1% of the diet). Small mammals include *Microtus* and *Clethrionomys* voles. Birds include common small passerines, excluding woodpeckers and corvids. Many birds consume insects, but we did not model this interaction. Instead, birds were modeled as feeding directly on plants.

fore 1991 because of a different trapping regime (T. Karels personal communication), so we used spring 1991 densities for all prior models. Red squirrel numbers showed little variation among years, but autumn densities always exceeded spring densities due to reproduction during the summer, so we calculated average autumn and spring densities separately. For cyclic mammalian and avian predators (lynx, coyote, great horned owl, goshawk), the winter or spring densities were used in models for that spring and for the previous autumn. For example, coyote densities estimated from autumn 1989 to spring 1990 were applied to 1990 spring and 1989 autumn models. Similarly, great horned owl spring estimates were applied to that spring and to the previous autumn. For all other taxa, we used average densities based on all available density estimates from autumn 1988 to spring 1996 inclusive because coefficients of variation were small or because abundances changed erratically.

For models of the food-addition and predator-reduction treatments, we changed the biomasses of only hares and ground squirrels. Other herbivores did not respond to these treatments. For food-addition treatments, predator densities and biomasses were assumed to be the same as for controls because the experimental scale was smaller than the scale of predators' movements. To mimic the predator exclosure treatment, we excluded lynx and coyotes from the models but retained avian predators. For all species, bioaccumulation was calculated as the difference in biomass between years for biomass increases. Biomass losses between years were not incorporated into models initially.

Diet Composition The proportion of various prey items in the diet of each predator varied throughout the cycle. We estimated mammalian diets from scat analysis, with winter

Table 19.2 Data sources for the models.

Species	Demography	Intake rate	Diet
Snowshoe hare (chapter 8)	Krebs et al. 1995 Stefan 1998 Gillis 1998 O'Donoghue 1994 O'Donoghue and Krebs 1992 Hodges et al. 1999 Krebs et al. unpublished data	Pease et al. 1979* Hodges unpublished data	—
Ground squirrel (chapter 9)	Hubbs and Boonstra 1997 Karels et al. 2000 Karels et al. unpublished data	Nagy 1987*	O'Donoghue 1994 Stefan 1998
Red squirrel (chapter 9)	Boutin et al. unpublished data	Nagy 1987*	O'Donoghue 1994 Stefan 1998
Small mammals[a] (chapter 10)	Boonstra et al. unpublished data Krebs and Wingate 1985	Nagy 1987*	—
Spruce grouse (chapter 11)	Martin et al. unpublished data	Nagy 1987*	—
Small birds[b]	Folkard and Smith 1995 Smith unpublished data	Nagy 1987*	—
Lynx (chapter 13)	O'Donoghue et al. 1997 O'Donoghue unpublished data Slough and Mowat 1996*	O'Donoghue et al. 1998	O'Donoghue et al. 1998 O'Donoghue et al. unpublished data
Coyote (chapter 13)	O'Donoghue et al. 1997 O'Donoghue unpublished data	O'Donoghue et al. 1998	O'Donoghue et al. 1998 O'Donoghue et al. unpublished data
Great horned owl (chapter 15)	Rohner 1996 Doyle unpublished data	Nagy 1987*	Rohner 1994, 1996 Doyle unpublished data
Goshawk (chapter 16)	Doyle and Smith 1994 Doyle unpublished data	Nagy 1987*	Doyle unpublished data
Harlan's hawk (chapter 16)	Doyle unpublished data	Nagy 1987*	Doyle unpublished data
Northern harrier (chapter 16)	Doyle unpublished data	Nagy 1987*	Doyle unpublished data

For each species, many of the data are summarized in chapters in this book. A more detailed breakdown of the collated data is available by contacting the chapter authors, as is a complete description of how the demographic data were transformed into the model inputs. Asterisks denote non-Kluane sources. Herbivore diets were not modeled in detail, but the predation of red squirrels and ground squirrels on snowshoe hare leverets was included.

[a]The small mammal category includes the *Microtus* and *Clethrionomys* voles caught at Kluane.

[b]The small bird category includes the 10 most common passerine bird species (excluding corvids and woodpeckers) at Kluane, as described in Folkard and Smith (1995).

diets contributing two-thirds and summer diets one-third to the total diet. Diets of goshawks, Harlan's hawks, and northern harriers were calculated from items found in nests, and we assumed that adults were provisioning young with prey items similar to those they were eating themselves. Great horned owl diet estimates were based on items found in pellets and from prey remains at nests. Most great horned owl data are available for summer diets, which we use as half of the annual diet composition. About 90% of the biomass in the winter diets of great horned owls consists of hares (Rohner 1996), and we assumed the remainder was divided equally between small mammals and red squirrels. The number of individuals of each species was converted to proportion of biomass using a constant body mass for each prey item. Juvenile hares were assigned a biomass of 0.7 kg and leverets a biomass of 0.1 kg because most deaths occur within 5 days of birth (O'Donoghue 1994). We assumed diet composition to be constant across treatments.

Red squirrels and ground squirrels depredated hare leverets <2 weeks old. Although both red squirrels and ground squirrels are primarily herbivorous, we modeled the hares in their diet as well. We calculated kilograms hares consumed per kilogram squirrel annually as

$$C_{Lp} = \frac{L(1 - S_L)(M_{Lp}b_L)}{B_p},$$

(2)

where C_{Lp} is consumption of leverets by squirrels, L is the number of leverets born, S_L is leveret survival rate to 30 days based on data from radiotagged leverets, M_{Lp} is the proportion of leveret mortality due to squirrels, b_L is the mass of a single leveret (0.1 kg), and B_p is total squirrel biomass. We calculated squirrel consumption of leverets separately for ground squirrels and red squirrels. To assign this consumption of leverets to a proportion of each squirrel species' diet, we calculated total consumption rate as described below.

Consumption per Biomass (Q/B) Daily energy needs for most taxa were calculated from Nagy's (1987) allometric relationships between body mass and consumption rate. He calculated separate relationships for birds, mammalian herbivores, and mammalian predators based on data from field studies and representing energy needs of active animals (e.g., animals that are foraging, moving, and thermoregulating).

Energy needs were calculated as kilocalories per kilogram per year. The amount of biomass required to meet these energy needs depends on the energy content of food (which varies with water, fat, and ash) and the consumer's efficiency of use. We assumed an assimilation efficiency of 80% of ingested energy, and energy contents of 1.75 kcal/g fresh weight (FW) for animals and 1 kcal/g for plants. Based on feeding by captive coyotes, metabolizable energy in snowshoe hares was measured directly as 1.4 kcal/g FW (Litvaitis and Mautz 1980), which is equivalent to 80% of 1.75 kcal/g. Feeding rates of migratory avian predators and songbirds were reduced to one-third because these species reside at Kluane for 4 months of the year. Similarly, ground squirrel feeding rates were reduced to one-third because of their winter hibernation.

These allometric estimates of consumption matched independent field data (Pease et al. 1979) and expert opinions on feeding rates of hares and avian predators at Kluane. However, rates at which snowshoe hares were killed by lynx and coyotes varied throughout the cycle, exceeding energy requirements at the peak but not at the low (O'Donoghue et al. 1998). We therefore used direct estimates of annual predation rates on hares for mammalian predators, weighted by the proportion of hares in the biomass of the diet. When we removed species from the models for the minimum food webs and no-hare webs, we altered diet compositions so that the remaining herbivores maintained their relative abundance in predators' diets.

Production per Biomass (P/B) Production per biomass generally represents total annual production over average biomass, but we used initial biomass for taxa that varied in
biomass among years because we wanted to address biomass accumulation and loss. For
taxa in which autumn and spring densities differed in the models, only individuals present
in spring were allowed to contribute to production. For avian and mammalian predators,
production was calculated as recruitment (R) of young into the adult population in units
of individuals per individual (or kilograms per kilogram per year). This definition leaves
out the production of offspring that died before adulthood (therefore $R \leq P/B$). For raptors,

$$R = rfn \tag{3}$$

and for lynx,

$$R = r[(Af_a n_a s_a) + (Yf_y n_y s_y)], \tag{4}$$

where R is recruitment, r is the proportion of females in the current population, f is the
proportion of females breeding, n is litter size or young fledged, and s is survival until ma-
turity. For lynx, yearling and adult reproductive parameters differed, so these were calculated separately, with A and Y representing the proportion of females in each age class
(adult and yearling) and subscripts denoting the age-specific parameters. Birds were assumed to fledge at adult weight, and hence there was no need to incorporate a survival
term. Coyote production was assumed to equal lynx production because we could find no
data on coyote reproduction in boreal forests, and the temporal patterns in hunting group
sizes of lynx and coyotes (an index of reproduction) at Kluane were similar (O'Donoghue
et al. 1997). We calculated small mammal and grouse production (P/B) using similar equations. For small birds we set $P/B = 2$ because fledging rates should at least equal those of
larger avian species.

For other herbivores, information was available on growth and mortality rates of
young, which allowed us to include the biomass of young that did not reach adulthood
into our production estimates. For ground squirrels and red squirrels, production was calculated taking into account the time it would take an emerging young to reach adult size:

$$P = drfn(b\int^t g \cdot e^{mt}) \tag{5}$$

where P is production, d is the spring breeding density, r is the proportion of females in
the spring population, f is the proportion of females breeding, n is litter size, b is individual offspring mass at the starting time, g is the daily growth rate of young, m is the daily
exponential mortality rate, and t is the time at which animals reach the next stage. For red
squirrels and ground squirrels, we used equation 5 to calculate production from emergence
to adult weight. To obtain production per biomass, this production is divided by either
spring biomass or autumn biomass, depending on the season of the model. For ground
squirrels, we used year-specific data for all parameters except size of emerging young
(only 1993 data available). We also used year-specific birth, growth, and survival data for
red squirrels, although interannual differences were not allowed to alter population densities, which were assumed to be invariant.

For hares, mortality rates of leverets differed from those of juveniles, which required
that we distinguish the production accruing from each stage. Specifically:

$$P = dr\sum f_L n_L (b_L + \int_0^{t1} g_L e^{m_L t_1} + \int_0^{t2} g_J e^{m_J t_2} \qquad (6)$$

Variables are the same as in the production equation for squirrels, with L referring to the litter group (i.e., b_L, g_L, and m_L refer to the birth weight, growth rate, and mortality rate per litter group). Juvenile parameters (g_j, m_j) were assumed to be constant for all litter groups. We had juvenile survival data only for 1995–1996, so these values were applied to all years, whereas we used year-specific leveret parameters. The time units t_1 and t_2 represent days to the end of the leveret and juvenile phases, respectively, and were calculated based on the length of time it would take for an individual to grow to the next phase (i.e., for leverets to reach 0.5 kg and for juveniles to reach adult weight), given known initial biomass and growth rate. The growth rate of leverets differed slightly between litter groups 1 and 2, and growth rates of litters 3 and 4 were assumed to be intermediate between them (O'Donoghue and Krebs 1992).

We thus varied production for each taxon except grouse, small birds, and small mammals. For these taxa, interannual variation was small or we had limited or no year-by-year data. We therefore calculated average production for these taxa from the sources given in table 19.2. For hares, we varied production for the + food models by using data from food-addition treatments, but we used control values for the − predator models because we had no data on reproduction for the predator exclosure treatment, and hare-trapping records indicate that reproduction was probably similar to controls (Krebs et al. unpublished data). For ground squirrels, we varied reproductive parameters for both the + food and − predator models by using data from these treatments.

19.4   Results from Mass Balance Models

19.4.1   Consumption of Plant Biomass by Herbivores

Although we did not separate plants by species, a total biomass of between 10,000 and 40,000 kg/km² of consumed vegetation was estimated as necessary to support this food web (figure 19.4). Our estimates of plant consumption tracked hare densities through the cycle. Estimates of necessary plant biomass from autumn models were always higher than from spring models because consumption estimates were based on herbivore biomass present at the start of each model, and herbivore biomass was always higher in autumn than in spring because of summer production.

Estimated plant offtake was lower for reduced web models than for the control models because of lower herbivore biomass. The low phase estimate for the model with hares as the only herbivore was especially affected (figure 19.4a) because hares make up a much smaller proportion of herbivore biomass at this time than they do during the peak of the cycle (~65% of biomass at the peak and ~10% at the low; figure 19.1). In contrast, the low phase no-hare model showed little difference from the control model. Conversely, offtake differed most between no-hare and control models at the peak of the cycle (figure 19.4a), reflecting the high proportion of herbivore biomass composed of snowshoe hares at this time. Both the food addition and the predator exclosure treatments resulted in higher herbivore biomass (chapters 8, 9). In the models, this higher biomass translated into an in-

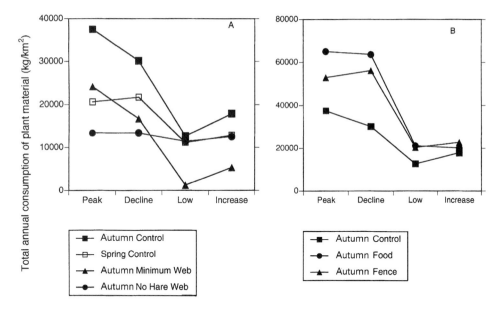

Figure 19.4 Plant biomass consumed by herbivores in Ecopath models of the Kluane boreal forest ecosystem. Feeding rates were calculated from allometric relationships of size versus energy needs for herbivores, assuming an energy content in plants of 1 kcal/g fresh weight. Energy needs of the total herbivore guild appear cyclic, which suggests that hare biomass is critical to total annual consumption. Each model was run for a year, with peak models starting in autumn 1988 and spring 1989; decline, autumn 1990 and spring 1991; low, autumn 1992 and spring 1993; increase, autumn 1994 and spring 1995. (a) Autumn, spring, and reduced web models. The minimum web model includes only plants, hares, and cyclic predators; the no-hare web excludes snowshoe hares while retaining all other modeled species. (b) Control, food addition, and predator reduction autumn models.

creased offtake of plant biomass (figure 19.4b), a result also observed empirically (chapters 5, 6).

The Kluane experiments demonstrated changes in standing biomass of plants, with variation both in annual production (chapters 5, 6) and consumption (chapters 8, 9, 10, 11, 17). Estimates of offtake made via Ecopath ranged from roughly one-quarter to one-half of the standing crop of small (<5-mm diameter) birch and willow twigs (as estimated in chapter 6). However, not all herbivore consumption consists of twigs of these shrubs; boreal forest herbivores additionally eat shrub leaves, the needles, seeds, and twigs of spruce, fruits of various species, and forbs and grasses.

There was thus no obvious food limitation for herbivores. Additionally, our estimates of food offtake may well be high because energy content varies in different plant tissues and species; for instance, the spruce seeds consumed by red squirrels have much higher than our assumed 1 kcal/g FW (Rusch and Reeder 1978). Even so, there may be more subtle food limitations at work, such as the necessity for ground squirrels to obtain particular fatty acids required for hibernation (chapter 9), the reliance of red squirrels on spruce cone crops (chapter 9), and the nutritional versus defensive attributes of species and twig sizes for hares and grouse (Schmitz et al. 1992, Mueller 1993, Rodgers and Sinclair 1997). Modeling these interactions and addressing the relationships of plant growth and plant off-take through the cycle would require more detailed models.

19.4.2 Phase of Cycle Models

Hares Across phases, cyclic species showed dramatic fluctuations in EE, indicating shifts in the extent to which production kept pace with the known fates of biomass (figure 19.5). Ecotrophic efficiencies of hares were >1 during the decline and low phases, indicating insufficient production to account for the known consumption. In contrast, hares during the increase and peak phases had EEs < 1, which indicates that there was potential for even faster population growth than was observed; our models did not fully account for the fates of all the estimated hare production.

If more of a species' biomass is consumed than produced (i.e., EE > 1), the standing stock must decline. When we incorporated the decline in hare biomass as negative bioaccumulation into the autumn decline model, the resulting EE of hares changed from 1.67 to 0.98, indicating that the biomass flow balanced almost exactly when population losses were included and that the high initial EE was indeed associated with a population decline (figure 19.6). The EE was > 1 during the low phase as well. Part of this imbalance arises from a small amount of biomass accumulation; when consumption alone was considered as a fate of hare production, the EE dropped from 1.48 to 1.11. The model results show the importance of predators in causing the decline and in maintaining the low and mirror the experimental results that predation on hares was heavy during these phases (chapters 8,13,15).

The high hare EEs during the low phase indicate that predation has a heavy impact on hares in this phase. This predation effect occurred despite hare survival rates that were greater than during the decline (chapter 8) and the low but stable predator populations during the low phase (chapters 13, 14, 15, 16). The high EEs imply that predation may keep hares rare during the low phase, even though predators are scarce and have partially switched to feeding on other species (chapter 13). Typical models of the cycle have suggested that the low can be extended only if predators remain at high abundance for several years after the hare decline (Keith 1990). In contrast, our model results indicate that this delayed decline in predators is not necessary for predation to be the critical determinant of the duration of the low phase.

For snowshoe hares during the low phase, the high EE reveals a "predator pit" (Pech et al. 1995, Sinclair et al. 1998), in which a species is maintained at low densities by predators because offtake exceeds production. At low densities, hares on control areas simply could not produce quickly enough to swamp the ability of predators to consume the production. During the increase, hare production was slightly higher than during the low phase, partly through the presence of a fourth litter group and partly through an increase in hares weaned per litter (chapter 8; Stefan 1998), and these changes in production may have been just enough to allow hares to escape from the predator pit. We modeled the low phase on controls with increase phase production to see if that could allow hares to escape the predator pit. Doing so changed the EE from 1.11 to 0.73 (in the absence of biomass accumulation), which would allow population increase, thus

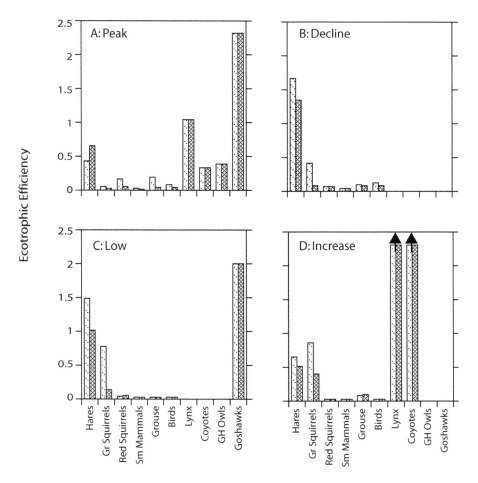

Figure 19.5 Ecotrophic efficiency (EE) of members of the Kluane boreal forest ecosystem. All animals in the model are shown except Harlan's hawks and northern harriers, which were assumed to be invariant year to year and therefore always had EE = 0. Other predators during the hare decline and low show EE = 0 because they had no biomass accumulation. Values of EE > 1 indicate that production cannot account for consumption by predators and for increased biomass. Model years are the same as in figure 19.4. In panel d, the arrows extending above the bars for lynx and coyotes indicate that their EE was infinite (population density increased in the absence of reproduction).

confirming that the combination of low biomass and low production keeps hares in a predator pit.

Based on biomass flows, some hare production had no known fate during the increase and peak phases (autumn EEs of 0.65 and 0.39, respectively). Ecotrophic efficiency was lower at the peak than during the increase phase, despite faster population growth rates during the increase phase. This pattern arises in part because biomass accumulation dur-

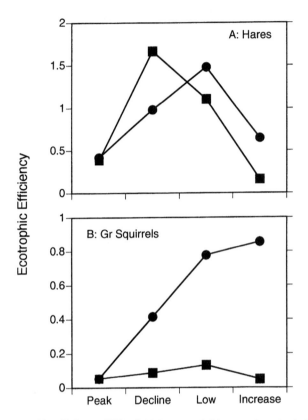

Figure 19.6 Ecotrophic efficiency (EE) of (a) hares and (b) ground squirrels based on Ecopath models that did (circles) or did not (squares) model changes in that species biomass. When circles < squares in terms of EE, biomass has declined. When circles > squares, biomass has increased. No biomass change was modeled for ground squirrels during the hare peak because density was not measured accurately before spring 1991.

ing the increase phase is included as a fate of production, thus increasing EE. When we removed the accumulation of biomass from the autumn increase phase model, hares went from an EE of 0.65 to 0.16 (figure 19.6). This low EE confirms that more production is available for population growth during the increase than at the peak.

The low EE values indicate that hares should have been able to increase even faster during the increase phase and should have been able to continue increasing during the population peak. Empirically, these patterns did not occur, which suggests that the models do not fully capture the system's dynamics. Artificially low EEs could arise if production was overestimated or if consumption was underestimated. Alternatively, low EEs could arise because of mortality sources other than predation (e.g., senescence or starvation), but starvation deaths of hares were rare at Kluane (chapter 8). We doubt that hare production was overestimated because, if anything, our methodology was likely to underestimate leveret survival. We suspect that our estimates of the feeding rates of predators were low for two reasons. First, we modeled predator consumption based on biomass of predators at the inception of the model year; predator production during that year would therefore not be applied to the consumption of hares. Second, we did not incorporate several of the smaller or rarer predator species, such as foxes, wolverine, eagles, boreal owls, and hawk owls. At Kluane, hawk owls appeared to display both functional and numerical responses to the hare cycle, with more birds present and a higher fraction of hares in the diet during the peak in hare densities (chapter 16; Rohner et al. 1995).

Studies in other locations have found functional or numerical responses to hares for some of these other predators as well (Theberge and Wedeles 1989, Dibello et al. 1990, McIntyre and Adams 1999). Our models included numerical responses of four predator species and functional responses of only coyotes and lynx. Clearly, including additional functional and numerical responses in the models would increase our estimates of hares consumed during the peak and could potentially raise EE to a level at which bioaccumulation would not be expected. These patterns raise the intriguing possibility that hares are limited at the peak of their cycle by rare and/or small predators, rather than by the major species that we have included in these food web models. Understanding predator functional responses may be critical to understanding the cycle.

Ground Squirrels and Other Herbivores   Abundances of ground squirrels fluctuated in much the same pattern as did hares; however, this parallelism was not obvious in the 4 years we chose to represent phases of the cycle (figure 19.5). Our models had ground squirrels increasing in all phases except the peak (when we had no information on biomass change and therefore assumed constant biomass). When we eliminated bioaccumulation as a fate of production, ground squirrel EEs ranged from 0.05 to 0.13 (figure 19.6). The highest EE occurred during the low phase and probably reflects the fact that all predators except goshawks shifted their diet composition to include more ground squirrels as hare populations declined. Furthermore, ground squirrels have a known mortality source in addition to predation: approximately 40% suffer overwinter mortality during hibernation in their burrows (chapter 9). Given this mortality rate, an EE > 0.6 would indicate population decline. The observed EEs were still considerably lower than this value, but the impact of predation on ground squirrel populations may depend partly on the level of overwinter mortality.

The remainder of the herbivore guild had low and relatively constant EEs throughout the cycle (figure 19.5). These low EEs strongly indicate that the dynamics of these herbivores are not driven by predator–prey interactions; predators do not use a substantial proportion of their production. Experimental results support the conclusion that there is little top-down control of red squirrels, small mammals, small birds, and grouse (chapters 9, 10, 11, 12). Ecotrophic efficiencies of these herbivores were highest during the peak and decline phases. Comparing EE values of these species at the hare peak to the average of their three EE values for other phases indicates that during the hare peak red squirrels were consumed at 4.1 times their average rate and grouse at 2.8 times their average rate. During the hare decline, small mammals were consumed at 1.8 times their average rate and small birds at 2.8 times their average rate. During the decline, predator diets shifted away from hares and towards alternative prey, raising consumption of these species (chapter 13; O'Donoghue et al. 1998).

Predators   During the decline and low phases of the hare cycle, many predators did not reproduce at all and abundances declined, thus leading to EEs of 0 (figure 19.5). During the increase phase, abundances of lynx and coyotes increased, but no reproduction occurred, giving effectively infinite EE. At the peak of the hare cycle, many predators increased in abundance, and these increases exceeded reproduction in lynx and in goshawks (EE > 1). The excessively high EEs for lynx and coyote during the increase phase suggest that our models for these species need to be improved because recruitment did not match population increases. Two specific things would help: first, our production data for lynx were taken from elsewhere in Yukon (Slough and Mowat 1996), and our production data for coyotes were based on those for lynx because we could find no reproductive data for coyotes in boreal forests. Knowing the values for Kluane would almost certainly help. Second, unlike the herbivorous species at Kluane, lynx and coyotes are highly mobile (O'Donoghue et al. 1997, Poole 1997). The models as written do not incorporate movement into and out of the Shakwak Trench. If significant amounts of immigration occurred during the increase phase, then populations could increase in the absence of reproduction. For example, in 1994–1995, no reproduction was observed locally (based on observation of family groups; O'Donoghue et al. 1997), but biomass estimates rose from 0.24 to 0.35 kg/km^2 for lynx and from 0.20 to 0.23 kg/km^2 for coyotes. Such increases could have come about only through immigration. We explain the high EEs of goshawks during the low and increase phases in a similar fashion. These raptors occurred at low densities, were hard to census, and were also extremely mobile (chapter 16, Doyle and Smith 1994). Therefore the high EE values probably reflect immigration into the Shakwak Trench that we did not incorporate into our estimates of production. Alternatively, we may be seeing the consequences of error in population estimates.

19.4.3   Seasonal Models

Ecotrophic efficiencies of most species showed little difference whether models tracked populations from one autumn to the next or from spring to spring (figure 19.5). In part, this similarity reflects the identical parameters used for many species, because separate spring and autumn estimates were unavailable. Variation between autumn and spring models has two possible sources. First, small differences in ecotrophic efficiency arise due to shifts in diet composition of predators (e.g., predators ate proportionally more grouse in winter 1988–1989, applied to the peak autumn model, than in 1989–1990, applied to the peak spring model). Second, large differences in ecotrophic efficiency arise when biomass accumulation differs between autumn and spring (e.g., for the four years we examined, spring to spring biomass accumulation of hares exceeded autumn to autumn biomass accumulation only during the peak).

19.4.4   Food Addition and Predator Reduction Models

Both the addition of food and the reduction of predators resulted in dramatic differences in the ecotrophic efficiencies of herbivores (figure 19.7). Snowshoe hares had lower EEs on both treatments than on controls in all cyclic phases, with the exception of a higher EE for food addition treatments during the increase phase. The lower EEs of hares under

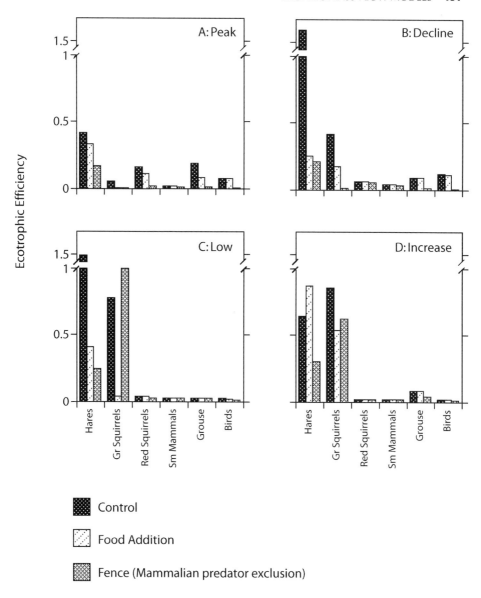

Figure 19.7 Treatment effects on ecotrophic efficiency of herbivores in the Kluane boreal for-est ecosystem. Food models differ from control models in values for biomass and demography of hares and ground squirrels. The −predator models use hare and ground squirrel values from the site where mammalian predators were excluded; lynx and coyotes were not included in the model. Model years are as in figure 19.4.

treatment conditions in part reflect higher hare abundances on these treatments. For instance, at the peak, hare biomass was 343 kg/km^2 on control sites, 609 kg/km^2 on food addition sites, and 495 kg/km^2 on the predator exclosure treatment. Total hare production increases as a function of density because more individuals per area results in more offspring born per unit area, so both the food addition and predator exclosure treatments would have had higher hare production than did control sites. However, when we modeled consumption of this biomass on food treatments, we assumed it to be the same as on control areas. That is, predators were assumed neither to aggregate nor to feed at a higher rate when hare densities were high. Thus, consumption per production is reduced for the food-addition treatment, leading to lower EEs. On the predator exclosure, higher biomass of hares contributed to higher total production and, furthermore, lynx and coyotes removed no production, so EEs of hares dropped even further below those of controls. Theoretically, the low EEs on both treatments could be due to increased production by hares (on a per capita or per biomass basis), but our empirical results do not support this idea (chapter 8), so we suspect the EE differences are indeed due to the underlying biomass differences among treatments.

The low EEs for hares on treatment grids indicate that their dynamics were not predator limited; instead, their population dynamics were more responsive to initial biomass and cyclic changes in production. During the low phase, the extremely high control EE relative to the low EEs of the treatments reflects this difference in their initial biomasses. An exception to the general pattern of lower EE in treatment than control models occurs during the increase phase. The higher EE for the food-addition treatment may reflect the attraction of hares to the food-supplemented sites (chapter 8), thus augmenting abundance beyond the level expected from in situ reproduction. Similarly, population declines on treatment grids occurred despite low calculated values for EE. These declines could be partially due to movement of hares off the grids (chapter 8); because EE does not include movement as a potential fate, this could lead to low values of EE. Either directed movement from high to low density or random dispersal would tend to equalize abundances across treatments and would lower EEs for experimental sites.

Ground squirrels also had higher abundances on the food addition and predator exclosure areas than on control sites. Their EEs were accordingly lower for most treatment models than for control models. The exception (predator exclosure during the low) was a time of particularly high population growth (chapter 9), so this high EE actually reflects high biomass accumulation rather than high predator consumption.

Ecotrophic efficiencies of herbivores such as red squirrels, voles, grouse, and small birds were more affected by exclusion of mammalian predators than by addition of food. Food addition left EEs essentially unchanged, because these species did not respond demographically to the rabbit chow, and the predators were modeled identically to the controls. The slight variations between control and food-addition models arise because the body mass of hares and ground squirrels varied among treatments, and these values were used in calculations of mammalian predator consumption rates and diet composition of all predators, which in turn affects consumption of all herbivores by predators.

In contrast, the exclusion of mammalian predators had substantial effects on EEs of these other herbivores, especially during the cyclic peak. For all of these species, EE dropped sharply from control to − predator conditions when predators were abundant.

This pattern indicates that the predator exclosure treatment reduced predation on red squirrels, grouse, small birds, and small mammals when predator densities were high. The predator exclosure treatment did not have this impact during the cyclic low phase. These patterns are explicable by considering the ways in which alternative prey can affect one another (Pech et al. 1995, Sinclair et al. 1998). In one scenario, high prey densities of one prey species (hares) can have a positive impact on other prey because the predators focus on the abundant prey. A diametrically opposed interaction could also occur, with high densities of one prey species leading to high predator densities, resulting in increasing predation on the alternative prey even if the predators prey preferentially on the dominant prey species.

For grouse, red squirrels, small mammals, and small birds, it seems that the second interaction occurred. At high hare densities, these herbivores had lower EEs on the predator exclosure treatment. This suggests that lynx and coyote predation affected these alternative prey species during the hare peak, whereas the lack of impact of the predator exclosure treatment during the low phase suggests that lynx and coyote impacts were simply too low to affect these alternative prey. This result is somewhat counterintuitive because lynx and coyote diets changed to include a higher proportion of these alternative prey species during the cyclic low (chapter 13). Similarly, EEs for these herbivores in control models were highest when hare (and predator) densities were high, rather than when predator diets included relatively high proportions of alternative prey. These results therefore indicate that the dietary shifts of lynx and coyotes do not overcome their low densities sufficiently to impact the demography of alternative prey during the low phase.

19.4.5 Partial Food Web Models

The minimum food web included plants, hares, and the abundant cyclic predators (lynx, coyotes, great horned owl). The EE of hares changed little between the full and minimum food webs (figure 19.8). At the peak, the EE for hares was 15% lower when the other species were excluded (full: 0.394, minimum: 0.335). During the cyclic low, EE was 14% higher for the minimum food web model (full: 1.481, minimum: 1.685). These opposing effects indicate that during the peak the main effect of removing other species is reduced consumption of hares by predators (due to removing other predators), whereas during the decline and low the main effect of removing species is the lack of alternative prey, which causes the diet composition of the few remaining predators to consist exclusively of hares. Overall, the minimum food web with hares as the only herbivore and the three species of abundant cyclic predators did a remarkably good job of capturing the distinctive dynamics of the full boreal forest food web.

In contrast, trophic mass flows were substantially altered in models that left out hares relative to those including all taxa. In control models, most herbivores had EEs much lower than 0.1, but these values skyrocketed without hares because predators were modeled as consuming other species instead. The values of EE for herbivores neared or exceeded 1 when predators were abundant during the peak and decline phases. Taken together, these reduced web models reiterate the preeminent role that the snowshoe hare cycle plays in structuring the boreal forest food web.

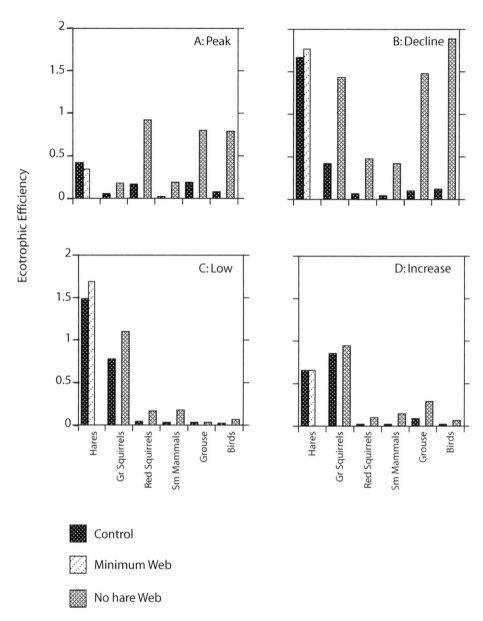

■ Control

▨ Minimum Web

▨ No hare Web

Figure 19.8 Ecotrophic efficiency of herbivores in the Kluane boreal forest ecosystem in control (full food web) models and reduced web models. The minimum web included plants, hares, coyotes, lynx, and great horned owls. The no-hare web differed from the full web only by the exclusion of hares. When some of a species' possible food sources were eliminated from a web, consumption was divided among remaining food sources according to their contribution to the diet. Model years are as in figure 19.4.

19.5   Discussion

19.5.1   Do Predator–Prey Dynamics Structure the Boreal Forest Ecosystem?

These mass balance models allow a way to contrast the impacts of food plants and predators on trophic interactions. As modeled, top-down effects of predators appear to be more important than bottom-up effects of food supply on the cyclic dynamics of hares. Consumption of hare biomass regularly exceeded production, whereas food plant biomass never appeared to be limiting. However, the food supply was not modeled in detail, and poor-quality food or absence of particular food types potentially could be responsible for the observed changes in per capita reproduction by hares.

Ecotrophic efficiency can serve as a proxy indicator of the strength of top-down interactions. When EE > 1 (in the absence of bioaccumulation, which also raises EE), consumption of a species exceeds its production, and biomass should decline. This shift in biomass as a result of predation is one hallmark of a strong interaction (MacArthur 1972, Paine 1992). By this criterion, predators influence population dynamics of hares and perhaps ground squirrels (depending on how much the population is additionally reduced by overwinter hibernation mortality). Predators do not appear to influence population dynamics of other herbivores, despite some shifts in EE of these species through the cycle.

Predator–prey interactions may extend the low phase of the hare cycle. During the low phase, hares are in a predator pit that they escape only when reproduction and starting biomass allow them to swamp the remaining predators. During the decline phase, predators prey on hares at a high rate, and this consumption is sufficient to account for the observed biomass decline. The Ecopath models do not, however, indicate that hares should reach peak biomass when they do, because EE < 1 at the hare peak, indicating that population growth could still occur. Predation at the peak by rare predators or predators that respond functionally to the cycle may account for the switch from increase to decline. We did not model other predators such as fox, wolverine, and eagles, but these species have shown functional responses to the hare cycle (Theberge and Wedeles 1989, Dibello et al. 1990, McIntyre and Adams 1999).

Predator–prey interactions may also cause ground squirrels to track the hare cycle in abundance. The high EEs for ground squirrels when predators are rare suggest that ground squirrels are an important alternative prey for predators during times of hare scarcity (see also chapter 9). Despite predators' heavier use of ground squirrels when hares are scarce, hibernation limits their availability as alternative prey because they are below ground for 8 months of the year. Additionally, relatively high rates of overwinter mortality for ground squirrels (>40%) means that a substantial proportion of each year's production is unavailable to predators. Theoretically, ground squirrel EEs > 0.6 could be associated with a predator-caused decline in biomass; the observed ground squirrel EEs were <0.15 in the core models developed here, which suggests that ground squirrels are probably not strongly predator limited.

Low EEs for most of the other herbivorous species (red squirrels, spruce grouse, small birds, and small mammals) suggest that they were primarily limited by food, weather, or space in the Kluane ecosystem. The predation impact did, however, increase when hare densities were high. The reduced web models show that omitting hares would drastically change the interaction of these species with the boreal forest predators by leading to much higher predation rates. Thus, these other herbivores are food limited only because the predators' attention is elsewhere. The hare–predator interaction dominates the food web structure in the boreal forest at Kluane.

These models explicitly incorporate only the direct effects of predation. Indirect effects may also occur. For example, the snowshoe hare predator pit during the low phase results from the interplay of predation pressure, hare biomass, and hare reproduction. If hares experience sublethal impacts of predation that affect their physiological capacity for reproduction, as has been suggested (Hik 1995, Boonstra et al. 1998), these impacts could prolong the low phase even if hare biomass and predation pressure were at levels conducive for hare population growth. The predator pit dynamics could also occur even with maximum hare reproductive output if predation pressure were high enough (i.e., if predation pressure declined with a lag, as has been suggested for hare cycles in other locations; Keith 1990). Experimental results on the predator exclosure treatment did not, however, lead to an earlier increase. The failure of the predator exclosure to produce an earlier increase probably reflects dispersal of hares from that treatment (chapter 8).

19.5.2 Potential for Other Ecopath Models

The models in this chapter have allowed us to paint a picture of the vertebrate dynamics at Kluane with a very broad brush. At this large scale, there are several more obvious model types to explore, plant–herbivore interactions foremost among them. From the Kluane data, it may be possible to construct a series of models examining plant biomass changes in a species-specific manner. Furthermore, we included no dynamics for detritus, although changes in both plant and animal detritus would be expected through the cycle. Similarly, inclusion of more season–year combinations for the treatments could detail the cyclic EE changes among vertebrates in more depth. These models can summarize cyclic changes in demography (P/B) but do not reveal why they occur.

We suspect that an interesting second generation of models could be derived by devotees of individual species using this modeling system with a much finer-scaled approach. Ecopath does not incorporate any kind of error structure in its inputs. It could be valuable to spend time manipulating variables for a given species to see the sensitivity of that species' EE to changes in any given demographic or dietary parameter. For cases where EE > 1, determining the parameter space that lowers EE below 1 could give insights into cases where empirical estimates may be biased. As an example, most of our raptor dietary data were obtained during summer and may not represent winter accurately; modeling variations in diet would allow us to determine if we could come closer to actual balances in EE. As another alternative focusing on particular species, examining results from a per treatment-grid basis rather than from averages for each treatment could open the door to speculation on site and scale effects.

Overall, the general utility of these biomass flow models lies in their ability to summarize trophic interactions among multiple species in a common framework. The detailed demographic information for each species thus becomes valuable not only in itself, but also in its contribution to our understanding of the linkages among species and between trophic levels. In addition, these models may help to disentangle the types of demographic changes that enhance or reduce bottom-up and top-down interactions by pointing out the sensitivity of EE to particular demographic variables.

19.5.3   Synthesis and Implications

We derive several key points from this modeling exercise. First, the ability of the minimum food web models to capture similar EE values for snowshoe hares indicates the central role of snowshoe hares in the boreal forest ecosystem. It is stunning to be able to start with 12 vertebrate species, remove 5 prey species and 3 predator species, and still derive similar results from the remaining 4 species. The results from the no-hare model offer even more striking evidence that snowshoe hares are centrally important to dynamics in the boreal forest.

Second, hares appear to be in a predator pit during the low phase of the cycle. This pit could be a result of low survival of adults or young hares, low reproductive output, or high predation pressure. There is an increase in reproduction and leveret survival from the low to the increase, which is consistent with the idea that leveret and juvenile dynamics hold the key to escaping from the predator pit. The models cannot distinguish among the various hypotheses for why there is low productivity at this time.

Third, the hare peak as modeled here remains difficult to explain because estimated production exceeded estimated consumption of hares. This conundrum can be resolved if hares die for reasons other than predation or if the predation rate on hares is higher than we modeled. Our results from radio-collared hares (with most hares dying of predation) and the results of studies indicating functional and numerical responses of many predators that we did not model suggest that predation may be higher at the peak than is currently supposed.

Fourth, ground squirrel dynamics appear to be responsive to the predation that accompanies the hare cycle, but with a slight lag behind the hares in that squirrel EEs (not counting bioaccumulation) were highest during the low of the hare cycle.

Fifth, prey species such as small mammals, grouse, red squirrels, and small birds appear to be more affected by mammalian predation when hare densities are high than when they are low. This result is consistent with the idea that high predator density rather than predator diet composition is the main determinant of predator impact on these species. These prey species did not demonstrate responses to the food addition treatment, but it is difficult to distinguish whether their lack of response means that they are not food limited or that rabbit chow was not a suitable food supplement for them.

Sixth, although many of the dynamics of the system were captured by this analysis of production and consumption, some of the model imbalances indicate that animal movements were also important. For predators such as lynx and coyotes, immigration into the Shakwak Trench resulted in increased densities during the early increase phase despite the lack of reproduction. Furthermore, although we did not model rare species, some raptors appear to be present only ephemerally during the cycle (chapter 16). These predator movements indicate that some of the food web dynamics may be integrated across a large spatial scale. At the scale of our experimental manipulations, hares appeared to be attracted to the food-supplemented treatments, resulting in faster population growth than in situ production can explain. Similarly, during the decline phase, there was a high rate of hare dispersal from the predator exclosure treatment.

Finally, the seasonal and treatment models both showed that the starting biomass in the modeled system affected two important dynamics: the amount of plant material required to support the food web and the relative amounts of production and consumption of hares. Some of the differences among treatments in the ecotrophic efficiency of hares may have resulted from differences in herbivore biomass, rather than directly from the annual demographic differences of treatments versus controls. Starting treatments at different times of the cycle, or modeling such an approach, could indicate how much of the treatment impacts are due to this bioaccumulation.

19.6 Summary

Understanding the dynamics of multispecies systems remains one of the major challenges to ecology, particularly when these interactions occur in a nonequilibrium context. In this chapter, we have described the interactions of multiple taxa of the boreal forest food web at Kluane in the single framework of trophic biomass flows. The value in this approach lies in the unification of demographic data into a common currency. However, the framework cannot easily incorporate nontrophic interactions such as movements, nor have we addressed uncertainty and error in parameter estimates.

One central focus of the Kluane project was to identify the causes and consequences of the snowshoe hare cycle. These Ecopath models corroborate the central importance of hares to the boreal forest food web and indicate that the interplay of snowshoe hare production, standing biomass, and consumption is crucial to cyclic dynamics. The transition from increase to decline appears to be predator driven, whereas the extended low phase appears to occur because of a predator pit that is influenced by the low production of hares. Herbivores other than hares were not limited by predation, largely because the predator–hare interaction was so strong. Predators had larger impacts on alternative prey (other than ground squirrels) during the hare peak, when predators were abundant, than during the hare low when predators were scarce but shifted their diets to include more alternative prey species.

Literature Cited

Boonstra, R., D. Hik, G. R. Singleton, and A. Tinnikov. 1998. The impact of predator-induced stress on the snowshoe hare cycle. Ecological Monographs **79**:371–394.

Christensen, V., and D. Pauly (eds). 1993. Trophic models of aquatic ecosystems. ICLARM Conference Proceedings, **26.** International Centre for Living Aquatic Resources Management, Manila, Philippines.

Dibello, F. J., S. M. Arthur, and W. B. Krohn. 1990. Food habits of sympatric coyotes, *Canis latrans,* red foxes, *Vulpes vulpes,* and bobcats, *Lynx rufus,* in Maine. Canadian Field-Naturalist **104**:403–408.

Doyle, F. I., and J. N. M. Smith. 1994. Population responses of northern goshawks to the 10-year cycle in numbers of snowshoe hares. Studies in Avian Biology **16**:122–129.

Folkard, N. F. G., and J. N. M. Smith. 1995. Evidence for bottom-up effects in the boreal forest: Do passerine birds respond to large-scale experimental fertilization? Canadian Journal of Zoology **73**:2231–2237.

Gillis, E. A. 1998. Survival of juvenile snowshoe hares during a cyclic population increase. Canadian Journal of Zoology **76**:1949–1956.

Hik, D. S. 1995. Does risk of predation influence population dynamics? Evidence from the cyclic decline of snowshoe hares. Wildlife Research **22**:115–129.

Hodges, K. E., C. J. Krebs, and A. R. E. Sinclair. 1999. Snowshoe hare demography during a cyclic population low. Journal of Animal Ecology **68**:581–594.

Hubbs, A. H., and R. Boonstra. 1997. Population limitation in Arctic ground squirrels: effects of food and predation. Journal of Animal Ecology **66**:527–541.

Karels, T. J., A. E. Byrom, R. Boonstra, and C. J. Krebs. 2000. The interactive effects of food and predators on reproduction and overwinter survival of arctic ground squirrels. Journal of Animal Ecology **69**:235–247.

Keith, L. B. 1990. Dynamics of snowshoe hare populations. *in* H. H. Genoways (ed). Current Mammalogy, pages 119–195. Plenum, New York.

Krebs, C. J., S. Boutin, R. Boonstra, A. R. E. Sinclair, J. N. M. Smith, M. R. T. Dale, K. Martin, and R. Turkington. 1995. Impact of food and predation on the snowshoe hare cycle. Science **269**:1112–1115.

Krebs, C. J., and I. Wingate. 1985. Population fluctuations in the small mammals of the Kluane Region, Yukon Territory. Canadian Field-Naturalist **99**:51–61.

Litvaitis, J. A., and W. W. Mautz. 1980. Food and energy use by captive coyotes. Journal of Wildlife Management **44**:56–61.

MacArthur, R. H. 1972. Strong, or weak, interactions? Transactions of the Connecticut Academy of Arts and Sciences **44**:177–188.

McIntyre, C. L., and L. G. Adams. 1999. Reproductive characteristics of migratory golden eagles in Denali National Park, Alaska. Condor **101**:115–123.

Mueller, F. P. 1993. Herbivore-plant-soil interactions in the boreal forest: selective winter feeding by spruce grouse. MSc thesis. University of British Columbia, Vancouver.

Nagy, K. A. 1987. Field metabolic rate and food requirement scaling in mammals and birds. Ecological Monographs **57**:111–128.

O'Donoghue, M. 1994. Early survival of juvenile snowshoe hares. Ecology **75**:1582–1592.

O'Donoghue, M., S. Boutin, C. J. Krebs, and E. J. Hofer. 1997. Numerical responses of coyotes and lynx to the snowshoe hare cycle. Oikos **80**:150–162.

O'Donoghue, M., S. Boutin, C. J. Krebs, G. Zuleta, D. L. Murray, and E. J. Hofer. 1998. Functional responses of coyotes and lynx to the snowshoe hare cycle. Ecology **79**:1193–1208.

O'Donoghue, M., and C. J. Krebs. 1992. Effects of supplemental food on snowshoe hare reproduction and juvenile growth at a cyclic population peak. Journal of Animal Ecology **61**:631–641.

Paine, R. T. 1992. Food-web analysis through field measurement of per capita interaction strength. Nature **355**:73–75.

Pauly, D., and V. Christensen. 1995. Primary production required to sustain global fisheries. Nature **374**:255–257.

Pease, J. L., R. H. Vowles, and L. B. Keith. 1979. Interaction of snowshoe hares and woody vegetation. Journal of Wildlife Management **43**:43–60.

Pech, R. P., A. R. E. Sinclair, and A. E. Newsome. 1995. Predation models for primary and secondary prey species. Wildlife Research **22**:55–64.

Peters, R. H. 1983. The ecological implications of body size. Cambridge University Press, Cambridge.

Polovina, J. J. 1984. Model of a coral reef ecosystem. I. The ECOPATH model and its application to French Frigate Shoals. Coral Reefs 3:1–11.

Poole, K. G. 1997. Dispersal patterns of lynx in the Northwest Territories. Journal of Wildlife Management **61**:497–505.

Rodgers, A. R., and A. R. E. Sinclair. 1997. Diet choice and nutrition of captive snowshoe hares (*Lepus americanus*): interactions of energy, protein, and plant secondary compounds. Ecoscience **4**:163–169.

Rohner, C. 1994. The numerical response of great horned owls to the snowshoe hare cycle in the boreal forest. PhD dissertation. University of British Columbia, Vancouver.

Rohner, C. 1996. The numerical response of great horned owls to the snowshoe hare cycle: consequences of non-territorial 'floaters' on demography. Journal of Animal Ecology **65**:359–370.

Rohner, C., J. N. M. Smith, J. Stroman, and M. Joyce. 1995. Northern hawk-owls in the nearctic boreal forest: prey selection and population consequences of multiple prey cycles. Condor **97**:208–220.

Rusch, D. A., and W. G. Reeder. 1978. Population ecology of Alberta red squirrels. Ecology **59**:400–420.

Schmitz, O. J., D. S. Hik, and A. R. E. Sinclair. 1992. Plant chemical defense and twig selection by snowshoe hare: an optimal foraging perspective. Oikos **65**:295–300.

Sinclair, A. R. E., R. P. Pech, C. R. Dickman, D. Hik, P. Mahon, and A. E. Newsome. 1998. Predicting effects of predation on conservation of endangered prey. Conservation Biology **12**:564–575.

Slough, B. G., and G. Mowat. 1996. Lynx population dynamics in an untrapped refugium. Journal of Wildlife Management **60**:946–961.

Stefan, C. I. 1998. Reproduction and pre-weaning juvenile survival in a cyclic population of snowshoe hares. MSc thesis. University of British Columbia, Vancouver.

Theberge, J. B., and C. H. R. Wedeles. 1989. Prey selection and habitat partitioning in sympatric coyote and red fox populations, southwest Yukon. Canadian Journal of Zoology **67**:1285–1290.

20

Conclusions and Future Directions

CHARLES J. KREBS, RUDY BOONSTRA, STAN BOUTIN, & A. R. E. SINCLAIR

The Kluane Region of the boreal forest has been particularly well studied during this 10-year project, and we were fortunate to begin our project with a strong background of ecological research that had been carried out since the 1950s at the Arctic Institute of North America Kluane Research Station. The boreal forest is a large and diverse ecozone, and our studies have been done on one special part of it. This caution must form the background of all our conclusions. As scientists we will generalize to the boreal forests of North America, and we hope that our results will be found to apply in general to this ecozone from Alaska to Newfoundland. But, of course, we do not know if this is correct until further work has been done in many locations to look for generalities and for differences. Ecological research is both blessed and cursed by this dilemma of applicability as it sits midway between the elegant generality of physics and the complex special cases of sociology. We generalize our results as hypotheses for future evaluation.

We categorize our findings under two headings: *primary findings* are major results no one anticipated and are new to this project. *Secondary findings* are important results we or others had speculated about but for which there was no hard evidence when the study began in 1986. The division between these categories is somewhat arbitrary. Together they constitute the take-home message of this book.

20.1   Primary Findings

The food web of the Kluane boreal forest was known only in general when we started this study. We have described and quantified it precisely and have dissected its structure through a set of large-scale experiment that exceed in area and complexity any that have been applied to terrestrial ecosystems. The results summarized in the preceding chapters produced major results at all three trophic levels in this ecosystem.

20.1.1   Predator Trophic Level

Three major results were found at the predator level. First and most critical, we found that the boreal forest community is predominantly a top-down system in which the major changes are driven by the predator trophic level but with reciprocal effects operating at all trophic levels. Predators do not control all the species in the community, but the major influence across many species from hares to spruce grouse is predation, with red squirrels, mice, and voles being the major exception to this rule. Red squirrels are the only significant alternative prey available to predators in winter, and the fact that red squirrels and their predators are largely decoupled means that predators could not survive in the boreal forest in the absence of hare peaks.

Many ecologists have viewed boreal forest dynamics as a lynx–hare interaction, but we found that many predators in the boreal forest operate as a guild with redundancy. If any one of these predators were eliminated, little impact would be observed on ecosystem behavior because of this redundancy. Thus, lynx trapping has no impact on the ecosystem because other predators fill in the gaps. Predator redundancy, however, has its limits. Our predator exclosure results show clearly that avian predators cannot compensate completely for the absence of mammalian predators because snowshoe hare density rose in both areas from which mammalian predators but not avian predators were excluded.

We were surprised to find that, among the mammalian predators, coyotes acted as a specialist predator, much like lynx in this system. The coyote is the archetype of a generalist predator, and its success in North America has been often attributed to this generalist attribute. In the boreal forest, coyotes are constrained to operate as specialists, and they carry on successfully in that role. We found no similar result among the avian predators, which appear to be much more niche diversified.

20.1.2 Herbivore Trophic Level

Five major results were found at the herbivore level. First, snowshoe hares are a keystone species in the boreal forest. Hares account for the dominant biomass in these boreal systems, and we have quantified for the first time the relative biomass of all the main herbivores in this community. Many but not all of the ecological changes in this ecosystem are driven by snowshoe hares. If hares were eliminated, the boreal forest vertebrate community would largely collapse.

Hare dynamics seem to be driven by an interaction between predation and food supplies in a manner that results in most hares dying from predation and very few dying from starvation and malnutrition. We could find no evidence of malnutrition caused by food shortage at Kluane, in contrast to work on hares in Alberta by Keith et al. (1984). Nevertheless, hares impact both summer and winter food plants adversely so that the quality of food available to hares at the peak of the cycle is clearly less than that available in the low phase. The interaction of predation with food supplies may be mediated by predation risk, either through risk-sensitive foraging or through stress effects associated with high density.

The intense stress in snowshoe hares induced by predators was a major finding in this study (chapter 8). No one had applied the stress model to prey–predator interactions before the Boonstra and Singleton (1993) paper. Because we had experimentally manipulated populations, we were able to isolate the effect of predators on the levels of stress and show that it was not a simple food stress. This mechanism has great potential to explain the changes in reproduction that accompany the hare cycle.

We found that red squirrels are a poor alternative prey in the boreal forest. We had expected predators to turn to red squirrels after hares declined, but most predators moved off the study area or starved and thus fell to low numbers before some effectively switched to eating red squirrels. The impact of the hare decline on red squirrel numbers was thus nearly zero. Red squirrel dynamics respond more to cone crops of white spruce than to changes in predator numbers associated with the hare cycle.

Finally, both red squirrels and arctic ground squirrels were major predators of juvenile snowshoe hares during the first 10 days of life, whereas traditional hare predators such as lynx and coyotes were not. This trophic linkage of a "herbivore" eating another herbivore convolutes the food web and was a finding no one had expected. At present we do not know what impact these interactions might have on the hare cycle. Red squirrels and hares occur together virtually everywhere, whereas ground squirrels overlap hares only in a small part of the western boreal forest. Red squirrels may be the only hare predator in the system to show a strong type-3 functional response, and this could contribute to the ability of predators to stop the hare increase.

20.1.3 Plant Trophic Level

Three major results were found at the plant level. First, fertilization of the vegetation produced dramatic increases in plant growth of trees, shrubs, and grasses and forbs. The impact of this increased plant growth ranged from slightly positive for hares to negative for red squirrels and ground squirrels and slightly negative for red-backed voles. We do not know why these population reductions occurred, and our best guesses are that mushroom production was reduced under fertilization (red squirrels), that lupine populations collapsed (ground squirrels), or that perhaps berry production was reduced (voles). It was clear from the different responses of the fertilized plots and the food-addition plots that we could not move the boreal forest plant community very far by adding nutrients, at least in the sense of providing better habitat for the main herbivores in this system.

Protecting 4-ha plots from hare browsing over the 10 years of the study produced only small changes in the vegetation. Although hares remove significant amounts of shrub biomass at the peak of the cycle, the responses of shrubs, herbs, and grasses appear to be so slow that we cannot measure much change in only 10 years. We suspect that 50 years might be a better time frame for these exclosure studies. A larger impact on vegetation, particularly shrubs, seemed to occur when both hares were excluded and fertilizer was added, but this was not quantified. The suggestion is that hares may actually have counteractive effects whereby heavy browsing reduces shrub biomass, but high hare densities stimulate production through a nutrient pulse in the form of urine and feces.

The fire history of the Kluane region showed a time scale of fires over 500 years, with large differences between local areas. No one had expected this dramatic time scale for fire in this part of the boreal region. The time scale of fires in the Kluane regions is long, and perhaps 20–40 times the time scale of the hare cycle. Although the boreal forest in general is thought to be a fire dominated ecosystem, the herbivore impacts of hares and moose occur much more quickly, and herbivory should be added to fire and insect attacks as a major disturbance factor in this part of the boreal landscape.

20.2 Secondary Findings

In this section we summarize findings that are less global than those discussed above, as well as findings that confirm or confront ideas that have been previously presented in the boreal forest literature. Ecologists have developed a skeleton of food web theory for community dynamics (chapter 4), and in part of this section we summarize our findings that were expected from theory and either found or not found in the data from this study. The results summarized in the preceding chapters produced important secondary results at all three trophic levels.

20.2.1 Predator Trophic Level

Four specific results surprised us at the predator level. First, predators kill other predators, especially when hares are declining in abundance. Among both mammalian and avian predators, we were surprised by the amount of intraguild predation. Because this predation can operate to stabilize the predator trophic level, it is critical to document it in relation to cyclic events, and it becomes an important component of any modeling of this community.

We found that the lynx social system was unable to continue through the cyclic decline, in contrast to the great-horned owl social system. Because territoriality in both birds and mammals has a cost as well as a benefit, we can infer that at low hare densities lynx obtain no present or future benefit from defending a fixed territory. Great-horned owls, in contrast, remain on their territories without breeding through the cyclic low phase.

Predators did not respond to hot spots in prey density until the hare decline began in 1991. Predators defend territories that contain far more prey biomass than they can use for most of the hare cycle. Because predators of both mammals and birds can live through more than one hare cycle, territory defense may be oriented to the lean years rather than to the rich years.

Finally, predators surplus-killed through much of the hare cycle. Consequently, estimating the impact of predators on prey population through bioenergetic rules will seriously underestimate how much prey will be taken. Our ability to estimate predator foraging success directly and prey loss rates directly allowed us to determine biases from two directions. Previous studies have seriously underestimated predator offtake potential in boreal forest communities. When the strong functional responses are combined with the modest numerical responses shown by predators at Kluane, the magnitude of changes in killing power are equal to the magnitude of changes in hare density over the cycle.

20.2.2   Herbivore Trophic Level

Five specific findings should be highlighted at the herbivore level. First, the arctic ground squirrel showed a 10-year cycle entrained by predation spinning off from the snowshoe hare cycle. Because ground squirrels hibernate during winter and overwinter losses are constant, their dynamics are driven completely by summer events in this ecosystem. Both food and the social system (female philopatry) are important factors for arctic ground squirrel populations, but their effect was overridden by heavy predation from hare predators such as coyotes and great horned owls. Arctic ground squirrels are primarily mammals of open arctic tundra and rely on clear sightlines to detect predators. The boreal forest is much more marginal for them because of a susceptibility to predation.

Red-backed vole populations continued to show long-term cycles that peaked 2–3 years after the snowshoe hare peak. We postulate that this is driven by fluctuations in berry crops that react to the nutrient cycling that accompanies the snowshoe hare peak density, but this pattern is still a puzzle to us. Vole and mouse abundance in the Kluane ecosystem is low, and the rarity of marten in this system and the rarity of weasels seem to reflect this low average abundance of small mammals.

Red squirrels were the only herbivore to show stable populations throughout the study. This occurred despite extreme variation in spruce cone crops. We suspect that territory size in squirrels, like that of great horned owls, may be determined by periods of low food availability. Consequently, densities are fixed by territoriality. Abundant cone crops lead to increased production of young, but this does not translate into population increase because territories in good habitat are not compressed. Although we did not measure emigration, we suspect that juveniles produced during good years disperse to marginal habitats leading to much wider density fluctuations in these areas. Red squirrels declined in abundance on the fertilized areas relative to control. We were surprised by this decline, and only in the latter part of our study did we connect possible reduction in mushrooms under fertilization to the red squirrel decline. Because we did not quantify mushroom production during this study, this remains a hypothesis in need of testing.

The timing of low-phase cyclic dynamics in snowshoe hares was unchanged by any of the experimental treatments. We had expected the duration of the low phase to be reduced by one or more of the predator exclosure or predator exclosure + food-addition treatments, but the cyclic increase of 1994 began everywhere in our study area and in the rest of the Yukon at exactly the same time.

The reproductive decay in snowshoe hares that begins in the late increase phase and continues through the decline was mimicked in hares taken from the Yukon to Vancouver by Sinclair and raised under ideal conditions in the laboratory (Sinclair unpublished). This reproductive decay has been described by Cary and Keith (1979) and by us in field populations. We have been completely surprised by these findings. Sinclair's results show that the intrinsic reproductive schedule of a female hare seems to be programmed by early life so that it cannot be changed by later experience in favorable environments. This is the first evidence that intrinsic changes may be an essential feature of hare cycles, and it needs to be carefully replicated.

Spruce grouse show a 10-year cycle entrained by the hare cycle in spite of being at low density. Cycles in relatively rare species are not easy to measure, and we think that predators, spinning off from the hare cycle, drive these changes. The picture is probably the same for willow ptarmigan, but our data are less clear for them.

20.2.3 Plant Trophic Level

We highlight four results that were found at the plant level. First, white spruce did not respond to fertilization by changes in the timing of cone crops. In spite of additional growth and at least a slight increase in seed production, the years of high cone crops occurred simultaneously on control and fertilized areas. We presume that cone production is hard wired to climatic parameters (probably temperature and moisture related) and is not a flexible trait in white spruce.

Grasses and most herbs in the boreal forest are controlled primarily bottom-up by nutrients, with little impact top-down by herbivores. The impact of fertilization on herbs and grasses is species specific and long-term. A few herbs are highly selected by snowshoe hares and are strongly affected at the peak of the cycle. It is likely that in 10 years we have seen only transient dynamics of grasses and herbs in response to fertilization of vegetation, and that experiments of a time scale of 50 years or more will be needed to determine the equilibrium conditions for the boreal forest. Snowshoe hares rarely feed on grasses; only *Microtus* spp. feed regularly on them. If herbivory is of only minor importance to grasses and many herbs in this system, an equilibrium set by nutrient levels is possible.

Browsing by hares on willow and birch appears to stimulate shrub production. Heavy overgrazing at the peak of the cycle is followed by a rebound 2–3 years later. The unexpected result was that there was a net increase in shrub biomass over the 10 years of study, rather than an equilibrium between growth and offtake. Some effects through nutrient cycling are implicated, but we do not know the details of how this occurs, and the grazing cycle may be superimposed on a longer-term successional trend.

Secondary compounds in willows and birch seem to respond directly to hare browsing, as predicted by Bryant (1981). Adding fertilizer seemed to reduce the levels of secondary chemicals in shrubs, as predicted by Coley et al. (1985). These changes in the nutritional value of winter shrubs probably contribute to the reduction in hare body weight that occurs in the peak and decline phase of the cycle, but their relative importance is overwhelmed by increasing predation pressure in the Kluane system.

20.3   Opportunities for Further Work

The greatest value of the Kluane boreal forest project was that we could measure all the significant parts of the system in the same place at the same time. Joint studies like this reach levels of insight that are not additive but multiplicative and dependent on the mix and creativity of the people involved. Ours should be a model for further community studies. Nevertheless, we could not study everything in the ecosystem, and in this section we try to identify the main oversights we made in missing opportunities for studying elements that we now think are important pieces of the system dynamics. We recognize in retrospect five major specific opportunities for further work.

Mushrooms vary greatly in abundance from year to year. We began to census them only in 1993. There are years when mushrooms are very abundant in the boreal forest and many more years when mushrooms are difficult to locate. Mushroom crops are thus similar to white spruce cone crops in their irregularity and in their importance for the herbivores in the system. In particular, mushrooms are harvested by red squirrels and are eaten by arctic ground squirrels, voles, and mice. We did not anticipate the potential interference of fertilization with myccorhizae in the boreal forest and whether this might lead to reduced above-ground production of mushrooms for herbivores. These fungi may also be affected by the nutrient pulse associated with hare browsing at the peak of the cycle. This system would well repay study in the boreal forest.

We did not quantify the annual production of berries in our study area, and we now think they are critical for mice and vole population outbreaks. There are a variety of berry-producing shrubs in the Kluane regions from soapberry (*Shepherdia canadensis*) and bearberry (*Arctostaphylos uva-ursi*) to red bearberry (*Arctostaphylos rubra*), crowberry (*Empetrum nigrum*), and cranberry (*Vaccinium vitis-idea*). All these berry producers vary from year to year in productivity for unknown reasons possibly associated with temperature and precipitation levels. Soapberry is an important food of grizzly bears, and we suspect all these berries are important foods for many of the small mammals and birds in the system. We do not know the food habits of these species well enough, nor do we know how their population changes might tie into variations in berry production from year to year. This is an important gap in our understanding.

We were not able to include the larger vertebrates in our study. Moose, grizzly bear, black bear, Dall sheep, caribou, wolves, and wolverines are the main species we have only minimal information on. For the most part, wildlife ecologists study these large species as a separate system with little or no linkages to the snowshoe hare cycle or the smaller scale processes we have detailed in this book. This could well be a mistake because, for example, it does not allow us to cross-reference the predator–prey dynamics of wolves and caribou to the predator–prey dynamics associated with the hare cycle. These may well be two independent systems, but we cannot verify this by keeping studies of these species in two separate worlds. The problems of scale are not easy to overcome because of budget limitations, but we will never overcome these limitations as long as we think we can achieve complete understanding by ignoring half of the vertebrate community.

We concentrated our studies in the Kluane region, which has not had a significant forest fire for more than 70 years and has never had any industrial forestry. This provides good baseline data but does not allow us to answer specific questions about the impact of forest fires on the system, or more important, perhaps, what will happen to the hare cycle when industrial forestry fragments these boreal forests. We bring this out not because we had any possibility of addressing these important questions but because they are two of the three main disturbances that will affect these forests in the future. From 1994 to 1998, just as we were ending our studies, the spruce bark beetle killed approximately 30% of the older white spruce trees in the Kluane region. Insect attacks along with fire and logging must also be understood as a disturbance to the boreal system. Our intuition is that bark beetle kills are much less intrusive on this ecosystem than fire and logging, but this must be tested.

We did not leave in place any long-term experimental areas. Because of the termination of our research funding, we had to take down all our experimental fences after 10 years. The infrastructure required for a longer term of study was not readily available from any of our Canadian funding sources. We regret this. We are reminded of the moose exclosures at Isle Royale National Park in Lake Superior that have been in place for more than 50 years (McInnes et al. 1992), and we believe that similar exclosures in the Kluane area would be equally informative of ecosystem dynamics in 50–100 years. The major problem beyond land use is technical: such exclosures need frequent monitoring because of moose bungling and tree falls against the fence. To add a more interesting treatment of moose and hare exclosure would require more vigilance to keep hares out. We have no current way of funding or managing these kinds of long-term studies.

20.4 Unsolved Problems

There are three aspects of boreal forest dynamics that are as yet poorly understood. First, for the snowshoe hare cycle, which drives so much of this community, we do not understand exactly why there is an interaction between predation and food. There are two hypotheses that tie predation risk to this interaction, one suggesting a food linkage directly by a change in habitat selection (Hik 1995) and a second by Boonstra et al. (1998) suggesting an indirect link through stress caused by predation risk. We have rejected an alternative hypothesis by Bryant et al. (1985) that suggests a direct food-quality impact through secondary compounds and another alternative hypothesis by Keith et al. (1984) that suggests a direct food quantity impact through winter food shortage. We do not know if these alternative hypotheses apply to other regions in the boreal forest.

Our experimental addition of food and removal of predators has created the greatest disruption of the hare cycle of any experiment to date. However, some would maintain that the cycle escaped largely intact. This may be due to the fact that we could not remove all predators and hares liked to venture outside the fence or due to the fact that we missed some essential elements. There is considerable room for researchers with creative minds to overcome the shortcomings of our experiments and to test the robustness of our findings.

Second, we do not comprehend the spatial dynamics of the interactions highlighted in this book. Hare cycles are synchronous across all of western North America, which means that a predator cannot in general improve its chances of survival by moving to a new territory. We do not understand what drives these spatial dynamics. Red squirrels are strongly affected by cone crops, and we do not have a regional or provincial scale of understanding how cone crops vary from year to year. In southern and eastern parts of the boreal forest, plant productivity is higher and arctic ground squirrels are absent. Does this make any difference to how the system operates? Why does this synchronous and highly fluctuating community not go to extinction on occasion? What spatial scale is necessary to prevent extinction of the predators or the prey species in this system? Is conservation in northern Canada similar to conservation farther south? If we protect 12% of the boreal forest, will these community dynamics continue, or will the system collapse if it is reduced (for example) to half its present area? These are all long-term questions and the answers are critical for conservation in the north.

Third, we do not know the sensitivity of this boreal landscape to climate change. There are only two ways to determine this. First, if one knew exactly how climate will change, one could impose an experimental treatment similar in kind. For example, if rainfall is predicted to increase, one could irrigate a section of boreal forest and measure what changes in relation to unmanipulated plots. It is more difficult to do this with temperature and CO_2. Second, one could monitor the plant and animal communities of the boreal forest and measure time trends that might then be associated with measured climatic changes. This is certainly feasible, and the present study can be viewed as a start of a longer term monitoring program. The problem at present is that monitoring in the boreal forest is haphazard, unfunded, and not organized. The key issues of what to monitor, where to monitor, and how to monitor in a standard manner are all undirected. A long-term monitoring program must be coupled with experimental studies to follow up on ideas generated by the correlations that will appear between biological changes and weather trends. The fallacy regarding climate change is that correlation means causation.

20.5 Future Boreal Forest Research

We have completed one 10-year study of the vertebrate community in the southwestern Yukon, and we pause here to reflect on the future research agenda for the boreal forest. First, we recommend that a replicated study be carried out in another part of the boreal forest to test the generality of our conclusions. Ecologists are tempted always to generalize beyond the limits of their immediate data, and indeed they must do this or say nothing about how ecosystems work. But we should be humble enough to recognize that some conclusions are region specific.

Second, we recommend the immediate investigation of how logging and fire may impact the hare cycle, the dominant process structuring the vertebrate community. We need a rigorous before-and-after control–impact design (Green 1979) with suitable replication to answer these questions. The forestry precedent in southern regions is to cut the trees first and then to start to ask questions about impacts on biodiversity. We hope the managers of the Yukon boreal forest areas are more intelligent in their planning.

Third, we suggest a broad program based on adaptive management to ask what level of harvesting is sustainable for game species and furbearers in the boreal region. The government is in the process of promising subsistence hunting privileges to Native Americans in the north without sufficient information on what this means ecologically or the means to determine if it is sustainable. If fur trapping becomes a viable industry, we have no scientific guidelines for how much harvesting is sustainable in a fluctuating system. If snowshoe hares become an important game species, as they are now in Newfoundland, we do not have any guidelines about how much harvesting they can sustain without major impacts on abundance. We suggest a proactive program to determine these limits is more desirable than an a posteriori explanation 50 years from now of why lynx went extinct in the Yukon or Alaska. We emphasize that part of this program must be an attempt to determine the linkages between the large vertebrates and the small vertebrates in the boreal forest. Perhaps we can manage caribou in total ignorance of what snowshoe hares are doing or perhaps not. We assume ignorance is bliss at present, and we may learn the hard way that this is not a good way to manage boreal ecosystems. We should learn from the mistakes in the south and avoid repeating them in the north.

Fourth, in the longer term we need to compare the North American boreal forest with the Siberian boreal forest and the Fennoscandian boreal forest. We have at present little more than anecdotal information about the Siberian boreal forest, and it leads us to believe that it is similar to the Canadian boreal forest so that we would expect similar ecosystem dynamics. The Fennoscandian boreal forest, however, seems completely different. Mountain hares (*Lepus timidus*) are rare in Sweden and Finland (Angerbjörn 1989), and the boreal forest community seems dominated by the 3- to 4-year cycle of voles and lemmings (Hansson and Henttonen 1988, Danell et al. 1998). We do not know why Fennoscandian boreal forest community dynamics differ so completely from North American boreal forest dynamics, and the ecological reasons why one system is dominated by the keystone snowshoe hare while the other is dominated by several species of small rodents are completely unknown. In the global picture this is a key question for ecologists in the twenty-first century.

Literature Cited

Angerbjörn, A. 1989. Mountain hare populations on islands: effects of predation by red fox. Oecologia **81**:335–340.

Boonstra, R., D. Hik, G. R. Singleton, and A. Tinnikov. 1998. The impact of predator-induced stress on the snowshoe hare cycle. Ecological Monographs **68**:371–394.

Boonstra, R., and G. R. Singleton. 1993. Population declines in the snowshoe hare and the role of stress. General and Comparative Endocrinology **91**:126–143.

Bryant, J. P. 1981. Phytochemical deterrence of snowshoe hare browsing by adventitious shoots of four Alaskan trees. Science **213**:1573.

Cary, J. R., and L. B. Keith. 1979. Reproductive change in the 10-year cycle of snowshoe hares. Canadian Journal of Zoology. **57**:375–390.

Coley, P. D., J. P. Bryant, and F. S. Chapin III. 1985. Resource availability and plant antiherbivore defense. Science **230**:895–899.

Danell, K., T. Willebrand, and L. Baskin. 1998. Mammalian herbivores in the boreal forests: their numerical fluctuations and use by man. Conservation Ecology **2**(2):9.

Green, R. H. 1979. Sampling design and statistical methods for environmental biologists. Wiley, New York.

Hansson, L., and H. Henttonen. 1988. Rodent dynamics as community processes. Trends in Ecology and Evoltuion **3**:195–200.

Hik, D. S. 1995. Does risk of predation influence populations? Evidence from the cyclic decline of snowshoe hares. Wildlife Research **22**:115–129.

Keith, L. B., J. R. Cary, O. J. Rongstad, and M. C. Brittingham. 1984. Demography and ecology of a declining snowshoe hare population. Wildlife Monographs **90**:1–43.

McInnes, P. F., R. J. Naiman, J. Pastor, and Y. Cohen. 1992. Effects of moose browsing on vegetation and litter of the boreal forest, Isle Royale, Michigan, USA. Ecology **73**:2059–2075.

Appendix 1

How to Use the Enclosed Kluane CD-ROM

To run the Kluane CD on a PC or Macintosh system: If the program does not start automatically within a few seconds after inserting the CD into the CD drive, double click on the file KLUANE (KLUANE.EXE on the PC) located on the root directory of the CD.

System Requirements

- If running on a PC: Windows 95, 98, or NT on a Pentium Processor
- If running on a Macintosh: Power PC 200 mHz with Mac OS 7.6 or later
- Monitor capable of 800×600 screen resolution, 24-bit color
- Sound card
- 32 MB RAM
- CD reader—minimum speed $8\times$ (preferably $24\times$)

To Run Simulation Models (Optional)

Microsoft Excel 98 is required to run the model COMMUNIT.XLS. A DOS window is required to run the model SIMULATE.EXE (Macintosh users must have a PC emulator program such as Connectix Virtual PC).

Appendix 2

Table of Contents for the Kluane CD-ROM

Opening
Introduction
Personnel and Facilities
Techniques
    Live Trapping
    Radio Telemetry
    Snow Tracking
    Vegetation Sampling
Vegetation
    Classifications
    Fire History
Results
    Fertilizer
    Herbivore Exclosures
    Food Addition
    Predator Exclosures
    Predator Exclosures and Food Addition
    Predators
Trophic Dynamics
Models
Summary
Publication List

Index

Accipiter gentilis. See northern goshawk

Accipiter striatus. See sharp-shinned hawk

Achillea millefolium var. *borealis. See* herbaceous vegetation

Aegolius acadicus. See northern saw-whet owl

Aegolius funereus. See boreal owl

Alaska Highway, 10, 11, 23, 127

Alces alces. See moose

American black bear, 268–69

American kestrel, 390–91, 399

American robin. *See* songbirds

American tree sparrow. *See* songbirds

Anemone parviflora. See herbaceous vegetation

antifeedant, 131, 134–35, 432

Aquila chrysaetos. See golden eagle

arctic ground squirrel
demography, 193–96
diet, 181
effects of weather, 203–4
home range, 181
impact of predation on snowshoe hare, 166
methods, 184–86
natural history, 20, 181
relation to snowshoe hare cycle, 181–82, 193–96, 207, 209

response to fertilizer addition, 196–197, 204–5

response to food addition, 196–202, 205–6

response to predator exclosure, 196–202, 207–8

response to predator exclosure + food addition, 196–202, 208–9

role of fatty acids, 204–5

Arctic Institute of North America, 10

Arctostaphylos uva-ursi. See herbaceous vegetation

arthropods
biomass, 264–65
methods, 263
response to fertilizer addition, 264–65, 269–70

Asio flammeus. See short-eared owl

bald eagle, 385

balsam poplar, 117

beaver, 262, 269

berry crops and small mammals, 230–32, 235

Betula glandulosa. See bog birch

biomass
flow, 30–31
pyramids, 5–6

black-backed woodpecker, 267
black-billed magpie, 379, 395
blackpoll warbler. *See* songbirds
black spruce, 5, 16, 117
bog birch
    biomass, 93, 97–102
    growth index, 102–6
    impact of browsing, 94–95, 97–102,
        106–10, 112–13
    methods, 94, 96, 97–98, 97–98, 106–7
    natural mortality, 106–10
    relation to snowshoe hare cycle, 96, 104–
        5, 106–10
    response to fertilizer addition, 94–96,
        104–5, 107–10
    response to food addition, 95, 104–5
    response to hare exclosure + fertilizer ad-
        dition, 95, 104–105
    response to predator exclosure, 104–5
    response to predator exclosure + food ad-
        dition, 96, 104–5, 110
    role of secondary compounds, 113
    spatial pattern, 93–95
    succession, 110, 112
Bohemian waxwing, 265
Bombycilla garrulus. See Bohemian
    waxwing
Bonasa umbellus. See grouse
boreal chickadee. *See* songbirds
boreal forest, 4–5
    age-class distribution of white spruce in
        study area, 127–28
boreal owl, 393–94, 399–400
bottom-up hypotheses/models, 27–29, 30,
    32–36, 70–71, 72–77, 408–433, 485
browsing
    hare, 106–13, 131
    moose, 106, 112
Bubo virginianus. See great horned owl
Buteo j. harlani. See red-tailed hawk
Buteo lagopus. See rough-legged hawk

Calamagrostis lapponica. See herbaceous
    vegetation.
camphor. *See* antifeedant
Canada lynx. *See* lynx
Canus latrans. See coyote
Canis lupus. See wolf
cascade hypothesis/models, 28
Castor canadensis. See beaver

Catharus ustulatus. See songbirds
chemical defense in plants, 70, 96–97, 113,
    432
chipping sparrow. *See* songbirds
Circus cyaneus. See northern harrier
Clethrionomys spp. See small mammals
coexistence, 438–40, 460
    models of, 447–50
Colaptes auratus. See woodpeckers
common raven, 379, 394–95, 399
community organization, 5–7
community structure, 27
competition, 75–77, 86, 314–16, 439–40,
    448, 460
    apparent, 27, 431
    exploitation, 27, 431–32.
Corvus corax. See common raven
coyote
    coexistance with lynx, 314–16
    demography, 282–87
    diet, 289–93, 304–7
    foraging behavior, 289–97
    functional response, 276–78, 280–82,
        297, 300–307, 310–12
    home range, 288–89
    impact as a predator, 289–93, 313–14
    methods, 278–82
    natural history, 276, 297–300
    numerical response, 276–78, 282–87,
        308–10
    prey switching, 304–7
    relation to snowshoe hare cycle, 276–77,
        282–87, 290–93, 300–301
    response to food addition, 299–300
    social organization, 287–89
cricetids. *See* small mammals

Dall sheep, 23
dark eyed junco. *See* songbirds
Dendroica coronata. See songbirds
density. *See* demography under individual
    species
density-dependence, 32, 371
dilution hypotheses/models, 32–38
dispersal
    coyote, 286
    great horned owl, 348–51
    lynx, 286
    red squirrel, 180
    snowshoe hare, 142, 168–75

ecological zones, 16, 122
Ecopath models. *See* mass balance models
ecotrophic efficiency (EE), 469, 476–84
emigration. *See* demography under individual species
energy flow pyramid, 5, 7
Epilobium angustifolium. See herbaceous vegetation
Erithezon dorsatum. See porcupine

Falcipennis canadensis. See grouse
Falco columbarius. See merlin
Falco peregrinus. See peregrine falcon
Falco rusticolus. See gyrfalcon
Falco spaverius. See American kestrel
feedback, negative, 32
fertilizer addition
    experimental predictions, 38
    methods, 51, 60–61
    summary of results, 412–13, 426
Festuca altaica. See herbaceous vegetation
fire, 117, 123–28, 130
food addition
    experimental predictions, 38
    methods, 51, 58–59
    summary of results, 413–16, 426
food web, 18–21, 27, 432
food web models. *See* mass balance models
forest cover
    classification, 16–18
    historical, 16
    mapping, 123
free fatty acids and ground squirrels, 204–5
Fretwell-Oksanen hypothesis, 28
frost-free period, 12
functional response, 441–44, 452–58
    coyote, 276–78, 280–82, 297, 300–7, 310–12
    great horned owl, 340, 355–59, 369–71
    lynx, 276–78, 280–82, 297, 300–7, 310–12
    northern goshawk, 383–84, 398
    northern harrier, 398
    red squirrel, 209
    red-tailed hawk, 398
fur trapping, 11–12

generalist predator, 276–77, 316–17, 340, 354, 364, 371, 493
glaciation, 10

golden eagle, 396
gray jay, 266, 395
    response to food addition, 267, 270
gray willow
    biomass, 93, 97–102
    growth index, 102–6
    impact of browsing, 94–95, 97–102, 106–13
    methods, 94, 96–98, 106–7
    natural mortality, 106–10
    relation to snowshoe hare cycle, 105–10
    response to fertilizer addition, 94–95, 97, 105–6
    response to food addition, 95–97, 105–6
    response to hare exclosure, 97
    response to hare exclosure + fertilizer addition, 105–6
    response to predator exclosure, 96–97, 105–6
    response to predator exclosure + food addition, 97, 105–6, 110
    role of secondary compounds, 113
    spatial pattern, 93–95
    succession, 110, 112
great gray owl, 396
great horned owl
    demography, 344–51
    diet, 351–55
    foraging behaviour, 351–59
    functional response, 340, 355–59, 369–71
    home range, 360–62
    impact as a predator, 355–59, 369–71
    methods, 341–44
    natural history, 340
    numerical response, 351, 352, 364–66, 371–72
    relation to snowshoe hare cycle, 344–51, 356–59
    response to food addition, 366–68, 372
    social organization, 360–66
grizzly bear, 23, 268–69
grouse
    contribution to herbivore biomass, 253–54
    demography, 243, 246–50
    diet, 241, 242
    impact of predation, 253, 254
    methods, 244–46

grouse (*continued*)
  natural history, 241–42
  predator regulation hypothesis, 243, 253
  relation to snowshoe hare cycle, 243–44, 247–48, 249–50, 253, 255
  response to fertilizer additon, 250–51
  response to food addition, 251
  response to predator exclosure, 252
  response to predator exclosure + food addition, 252–53
  trophic interactions, 254–55
Gulo gulo. See wolverine
gyrfalcon, 396

Haliaeetus leucocephalus. See bald eagle
hare cycles
  food limitation, 142–43
  historical records, 21–22
  hypothetical causes, 142–43
  predation, 142–43
hare exclosure
  experimental predictions, 41
  methods, 52, 61
  summary of results, 419–21, 427
hare exclosure + fertilizer addition
  experimental predictions, 41
  methods, 61
  summary of results, 421–24
Harlan's hawk. *See* red-tailed hawk
herbaceous vegetation
  amino acids response to fertilizer addition, 74–75
  changes in productivity, 78–84
  changes in species composition, 75–77
  and climate change, 87–88
  compensation, 82
  impact of browsing, 76–78, 79–84, 85–86
  response to fertilizer addition, 72–85
  response to herbivore exclosure, 75–84
  response to herbivore exclosure + fertilizer addition, 81–84
herbicide, 41
herbivore dominated hypotheses/models, 37
human impacts, 23, 129

immigration. *See* demography under individual species
intraguild predation, 166, 175, 316, 333, 396–98, 400–401, 494

Junco hyemalis. See songbirds

Keith hypothesis, 143, 209, 277, 498
keystone species, 51, 243, 255, 432, 438, 493, 500
kill rate, 276
  coyote, 280–81, 297, 310, 314
  great horned owl, 355–59, 370
  lynx, 280–81, 297, 310, 314
Kluane region, 5, 10–23

Lagopus lagopus. See ptarmigan
Lagopus leucurus. See ptarmigan
Lagopus mutus. See ptarmigan
LANDSAT, 16, 18, 123
least chipmunk, 219, 390
Lepus americanus. See snowshoe hare
Linnaea borealis. See herbaceous vegetation
Loxia leucoptera, 265
Lupinus arcticus. See herbaceous vegetation
Lutra canadensis. See otter
lynx
  coexistence with coyotes, 314–16
  demography, 282–87
  diet, 289–93, 304–7
  foraging behavior, 289–97
  functional response, 276–78, 280–82, 297, 300–7, 310–12
  fur returns, 21–22
  home range, 288–89
  impact as a predator, 290–93, 313–14
  methods, 278–82
  natural history, 276, 297–300
  numerical response, 276–78, 282–87, 308–10
  prey switching, 304–7
  relation to snowshoe hare cycle, 276–77, 282–87, 290–93, 300–1
  response to food addition, 299–300
  social organization, 287–89

marten, 332
Martes americana. See marten
mass balance models
  biomass parameterization, 469–70
  consumption per biomass (Q/B) parameterization, 472
  core models, 474–80
  diet composition parameterization, 471–72

effects of hare cycle, 476–79
effects of predation, 486
food addition, 480–83
methods, 467–69
partial food web, 483–84
predator reduction, 480–83
production per biomass (P/B) parameterization, 473–74
seasonal, 480
merlin, 396
Mertensia paniculata. See herbaceous vegetation
methanol extraction. *See* chemical defense in plants
Microtus spp. See small mammals
mink, 332
moose, 22, 56, 268–69
mor litter layer, 77
mountain caribou, 22–23
muskrat
abundance index, 268
cycles, 269
methods, 263–64
Mustela erminea. See weasel
Mustela nivalis. See weasel
Mustela vison. See mink
mutualism, indirect, 27

natural disturbances, 117, 128–30. *See also* fire and spruce bark beetle
northern flicker. *See* woodpeckers
northern goshawk, 382–85, 398–400
northern harrier, 388–90, 398–400
northern hawk owl, 391–93, 399–400
northern red-backed vole. *See* small mammals
northern saw-whet owl, 396
numerical response, 276
coyote, 276–78, 282–87, 310
great horned owl, 351, 352, 364–66, 371–72
lynx, 276–78, 282–87, 308–10
nutrients, role of, 204–5, 232–34

Odocoileus virginianus, 269
Ondatra zibethica. See muskrat
osprey, 396
otter, 332

Pandion haliaetus. See osprey
Parus hudsonicus. See songbirds

Passeriformes. *See* songbirds
peregrine falcon, 396
Perisoreus canadensis. See gray jay
permafrost, 10
Peromyscus maniculatus, 216
phenolic compounds. *See* chemical defense in plants
Pica pica. See black-billed magpie
Picea glauca. See white spruce
Picea mariana. See black spruce
Piciformes. *See* woodpeckers
Picoides arcticus, 267
Picoides tridactylus. See woodpeckers
plant defense theory, 96
population limitation, 30, 206–9, 226–34, 361–66, 371
Populus balsamifera. See balsam poplar
Populus tremuloides. See trembling aspen
porcupine, 262
Potentilla fruticosa, 93
predator exclosure
experimental predictions, 41
methods, 51–52, 55–58
summary of results, 416, 426–27
predator exclosure + food addition
experimental predictions, 41
methods, 52
summary of results, 416–19, 427
predator pit, 476–77, 485
prey switching, 180, 207–8, 217, 253, 304–7, 310–11, 340, 354, 369–370, 383–85
ptarmigan
demography, 243, 247
diet, 241
natural history, 241–42
relation to snowshoe hare cycle, 242, 250

rabbit chow. *See* food addition
reciprocal hypotheses/models, 32–34, 38, 408–10, 427–28, 438–460
recruitment. *See* demography under individual species
red-backed vole. *See* small mammals
red fox, 325–27
red squirrel
demography, 186–93
diet, 180–81
home range, 180
impact of predation, 206–7, 314

red squirrel (*continued*)
  impact of predation on snowshoe hare,
    166
  methods, 182–84
  natural history, 180–81
  relation to snowshoe hare cycle, 181–82,
    186–87, 209
  response to fertilizer addition, 190–92,
    204
  response to food addition, 187–90, 191,
    205–6
  response to predator exclosure, 190, 191,
    206–7
  response to predator exclosure + food ad-
    dition, 190–91
  response to spruce cone production, 192–
    93, 205
red-tailed hawk, 385–88, 398–400
Regulus calendula. See songbirds
replication of treatments, 54–55
rough-legged hawk, 396
ruby-crowned kinglet. *See* songbirds

Salix alaxensis, 93
Salix glauca. See gray willow
Salix scouleriana, 93
scavenging, 293–95, 381, 394–95
secondary chemicals. *See* chemical defense
  in plants
seen sheets, 63–64, 244, 246, 263
Shakwak Trench, 10, 15
sharp-shinned hawk, 391
Shepherdia canadensis. See soapberry
short-eared owl, 396
shrew, 219, 220
small mammals, 219–35
  body mass, 223
  demography, 219–26
  diet, 216–17
  methods, 218–19
  natural history, 216
  relation to snowshoe hare cycle, 217,
    219–27, 232–34
  response to fertilizer addition, 220–22,
    225–26, 234
  response to food addition, 220–22
  response to hare exclosure, 220–22
  response to hare exclosure + fertilizer ad-
    dition, 220–22
  response to predator exclosure, 220–22

response to predator exclosure + food ad-
    dition, 220–22
  role of food, 223–24, 228–30
  role of nutrients, 232–34
  role of predation, 224–5, 227–28, 234
  role of weather, 217, 230–32
snow, 15, 55, 113
snowshoe hare. *See also* hare cycles
  cycle, 173
  demography, 142–43, 145–69
  food limitation, 175
  home range, 174
  impact of predation, 162–69, 313
  methods, 143–45
  response to fertilizer addition, 169–73
  response to food addition, 169–73
  response to predator exclosure, 169–73
  response to predator exclosure + food ad-
    dition, 169–73
  trophic interactions, 175
soapberry, 93
soil nutrients, 70, 77, 82, 84–85
Solidago multiradiata. See herbaceous vege-
    tation
songbirds
  demography, 265–67
  impact of predation, 269
  methods, 263
  response to fertilizer addition, 265–67
  response to spruce bark beetle, 268, 270
Sorex spp. See shrew
specialist predator, 276–77, 310, 316–17,
    493
species interactions
  direct effects, 27, 409, 412–426, 427–28,
    486
  indirect effects, 27, 29, 409, 425–27, 428,
    431–32, 486
species redundancy, 26
Spermophilus parryii. See arctic ground
    squirrel
Spizella arborea. See songbirds
Spizella passerina. See songbirds
spruce bark beetle, 18, 120, 129–30, 220,
    264, 267–68
St. Elias Mountains, 5, 10, 12, 15
stochastic events. *See* weather
stress and sublethal effects, 143, 173, 208
Strix nebulosa. See great gray owl
Surnia ulula. See northern hawk owl

survival. *See* demography under individual
    species
Swainson's thrush. *See* songbirds
synchrony in hare, 21–22

Tamiasciurus hudsonicus. See red squirrel
Tamias minimus. See least chipmunk
telemetry
    arctic ground squirrel, 198
    coyote, 285–86, 288–89
    great horned owl, 344, 348–51, 360–64,
        366–68
    lynx, 283–86, 288–89
    snowshoe hare, 159–62
Tengmalm's owl. *See* boreal owl
three-toed woodpecker. *See* woodpeckers
top-down hypotheses/models, 28–29, 30,
        32–37, 71, 77–78, 253, 408–33, 485
trembling aspen, 117
trophic interaction models
    biomass parameterization, 441–42
    consumption parameterization, 441–44
    demography parameterization, 444–46,
        454
    effects of predation, 450–52
    effects of snowshoe hare cycle, 452–58
    herbivore-vegetation interaction, 446–50
trophic-level interactions, 30–38
Turdus migratorius. See songbirds

Ursus americanus. See American black bear

vegetation dominated hypotheses/models,
        37
vegetation removal, 41, 425

vertebrate community, 5–7
Vulpes vulpes. See red fox

weasel, 219, 223, 225, 226, 228, 329–30
weather
    effect on spruce cone production, 203
    effects on ground squirrels, 203–4
    effects on small mammals, 217, 230–
        32
    precipitation, 14–15
    temperature, 12–13
white spruce
    abundance, 118
    cone production, 120–22, 203
    effects on red squirrels, 180–81, 190–93,
        205
    growth, 118, 120, 131–32
    impact of snowshoe hare browsing, 131
    red-backed vole response to spruce seeds,
        223
    response to fertilizer addition, 132–35
    seed production, 120–22, 130–31
    succession, 130–32
white-tailed deer, 269
white-winged crossbill, 265
Wilsonia pusilla. See songbirds
Wilson's warbler. *See* songbirds
wolf, 23, 327–29
wolverine, 331
woodpeckers
    methods, 263
    response to spruce bark beetle, 267–68,
        270

yellow-rumped warbler. *See* songbirds